HISTORICAL RECORD OF THE 14TH (KING'S) HUSSARS

HISTORICAL RECORD

OF THE

14TH (KING'S) HUSSARS

FROM

A.D. 1715 TO A.D. 1900

BY

COLONEL HENRY BLACKBURNE HAMILTON

M.A. CHRIST CHURCH, OXFORD : LATE COMMANDING THE REGIMENT

WITH NUMEROUS ILLUSTRATIONS

Schiffer Military History
Atglen, PA

Library of Congress Catalog Number: 97-67433.

Printed in the United States of America.
ISBN: 0-7643-0351-1

We are interested in hearing from authors with book ideas on related topics.

Published by Schiffer Publishing Ltd.
4880 Lower Valley Road
Atglen, PA 19310
Phone: (610) 593-1777
FAX: (610) 593-2002
E-mail: Schifferbk@aol.com.
Please write for a free catalog.
This book may be purchased from the publisher.
Please include $3.95 postage.
Try your bookstore first.

WITH

HIS MAJESTY'S GRACIOUS PERMISSION

THIS HISTORICAL RECORD OF

THE 14TH (KING'S) HUSSARS

IS MOST RESPECTFULLY DEDICATED TO

OUR SOVEREIGN LORD

KING EDWARD VII.

IN THIS THE FIRST YEAR

OF HIS MAJESTY'S

REIGN

TITLES OF THE REGIMENT

1715. Raised as Dormer's Dragoons.
(Numbered as Fourteenth Dragoons.)

1776. The Fourteenth Light Dragoons.

1798. The Fourteenth, or the Duchess of York's Own Light Dragoons.

1830. The Fourteenth, or the King's Light Dragoons.

1861. The Fourteenth, or King's Hussars.

BADGES AND DISTINCTIONS CONFERRED

Badges

1751. The White Horse of Hanover authorised for the Guidons by Royal Warrant.

1798. Royal authority given to assume the Prussian Eagle as a Regimental Badge.

1832. Authorised to bear the King's Crest on the Appointments, and the Prussian Eagle on the second and third corners of the Regimental Guidon.

Distinctions

1815. 'Peninsula.'

1820. 'Talavera'—'Fuentes d'Onor'—'Salamanca'—'Vittoria'—'Orthes.'

1837. 'Douro.'

1852. 'Punjaub'—'Chillianwallah'—'Goojerat.

1859. 'Persia.'

1863. 'Central India.'

PREFACE

This Historical Record of the 14th (King's) Hussars from its first formation as Dormer's Dragoons up to the present time, is not intended to be a complete and exhaustive history of that distinguished corps, but rather an attempt to supply in a compact form and in chronological order an authentic record of the stations where it has been from time to time quartered; the battles, sieges, and other military operations in which it has been engaged; honorary distinctions which it has gained; changes in its establishment, uniform, and equipment; how it has been recruited at various times; its reviews and inspections; names of its officers; memoirs and services of its colonels and lieutenant-colonels; together with other matters which are likely to be valued not only by the officers and soldiers who have served, who are now serving, and who may hereafter serve in its ranks, but also by their friends and others interested therein. In addition to this there have been inserted coloured sketches of its past and present uniforms and former guidons, as well as some portraits of the colonels and commanding officers from time to time.

The Fourteenth is one of the most illustrious regiments in the British cavalry. Long before it was immortalised by Lever in *Charles O'Malley, the Irish Dragoon*, it had gained for itself a name as a famous fighting regiment with plenty of *esprit de corps*, and was especially noted for the superior style in which its officers and men performed their outpost duties, so that at the close of the Peninsular War it had the reputation of being the finest light cavalry regiment the British army ever possessed. It bears on its banners twelve marks of distinction, beginning with 'Douro,' which is borne by no other cavalry regiment, and ending with 'Central India,' where it fought gloriously and fully sustained its former high reputation. As the regiment is now engaged on active service in the field in South Africa, it may yet achieve further distinctions and add fresh laurels to its present ample store, which is surpassed by only two cavalry regiments and equalled by only two others.

In the process of compiling the historical record of a regiment, one would naturally expect to find a continuous store of information on the chief points required to form the basis of such a work in 'The Regimental Digest of Services,'[1] kept in the orderly-room of every regiment in Her Majesty's service, and duly posted up from time to time. In the Fourteenth, however, these 'Regimental Records' happen to be very meagre: there are some important events omitted, there are several gaps in the continuity of events, and for nearly the first hundred years, that is up to 1808, there is no mention of the places where the regiment was stationed. The explanation is that when crossing the Pyrenees with Wellington's army on the 10th November 1813, the night before they set foot on French territory, the 14th Light Dragoons had their baggage captured by the enemy, and with it disappeared all their precious regimental books and documents.

Under these circumstances it became necessary, when compiling this Record, to obtain the required information elsewhere, and after a very satisfactory search of original War Office documents deposited in the Public Record Offices in London

[1] Commonly called 'Regimental Records.'

and Dublin, the exact stations of the regiment for each year have been ascertained with the exception of those for the years 1721 to 1726 and for 1743 and 1744, and all we know is that the first-named six years were passed by the regiment somewhere in Ireland, and the latter two years in Great Britain. *The Historical Record of the 14th Light Dragoons*, by Richard Cannon, Esqre., published in 1847, has been taken as the foundation of this work up to that date, and all admirers of regimental histories should be very thankful to him for the care and accuracy he brought to bear on his numerous works of that class.

The author has the greatest possible pleasure in acknowledging the obligations he is under to the following kind friends, and at the same time takes this opportunity to return them his grateful thanks for the help and assistance they have given him :—

Captain C. COTTRELL-DORMER of Rousham, Oxon. (late 13th Hussars), for portrait of his ancestor, Lieutenant-General James Dormer, founder of the regiment.

F. A. HAWKER, Esquire of Eaglehurst, Bracknell, Berks, and ADOLPHUS HAWKER, Esquire, late of the War Office, for portrait of General Sir S. Hawker, G.C.H., their late father; and also for the words of the old regimental song of the 14th Light Dragoons used in the Peninsular War when Sir S. Hawker commanded the regiment.

Captain J. H. TREMAYNE, Adjutant, 13th Hussars, for extracts from MS. journal written in the Peninsular War, with reference to the time when the 13th and 14th Light Dragoons were brigaded together and nicknamed the 'Ragged Brigade,' 1813-14.

Major R. M. RICHARDSON, 14th Hussars, for portrait of Colonel J. Townsend, A.D.C., who commanded the regiment

1829-45; and for several portraits of old colonels and lieutenant-colonels from the album of the Officers' Mess at Newbridge, 1897.

Major-General the Honourable HERBERT EATON, late Grenadier Guards, for copy of War Office roll of the officers and men of the 14th Light Dragoons who were among the recipients of the War Medal given in 1848 by Her Majesty Queen Victoria to the survivors of the wars.

Sir ALBERT WILLIAM WOODS, K.C.B., K.C.M.G., Garter King of Arms, and Inspector of Colours to Her Majesty's Army, for the benefit of his kind advice as to the badges of the regiment.

S. M. MILNE, Esquire, of Calverley House, near Leeds, for portraits of Lord Southampton and General Louis Dejean, both colonels of the 14th Dragoons, as well as for much kind advice and valuable information as to the old uniforms of the regiment, a subject on which he is a well-known authority.

Colonel the Right Honourable A. W. B. EARL BROWNLOW, A.D.C., for a portrait of the Earl of Bridgewater, Colonel of the 14th Light Dragoons, his ancestor.

Sir ARTHUR E. HAVELOCK, G.C.M.G., G.C.S.I., G.C.I.E., for portrait of his late father, Lieutenant-Colonel William Havelock, K.H., killed at Ramnuggur when in command of the 14th Light Dragoons in 1848.

Captain H. R. GALL, late 7th Royal Fusiliers, for the perusal of some interesting letters written by his late father, Major-General R. H. Gall, C.B., a distinguished officer of the 14th Light Dragoons, who served through the Punjaub and Central India campaigns.

Captain the Honourable J. BERESFORD, 7th (Q.O.) Hussars, for portrait of Major-General A. Scudamore, C.B., who commanded the 14th Light Dragoons, 1861.

Messrs. INNES and Co., late of Bedford Street, Strand, for permission to use or reproduce maps and plans in *The Sikhs and the Sikh Wars*, by General Sir Charles Gough, V.C., G.C.B., and A. D. Innes, Esq., M.A. (published by Innes and Co., London, 1897).

J. H. SYLVESTER, Esq., F.G.S., late of 2nd Regiment Mayne's Horse, for permission to use or reproduce the map showing the route of the Central India Field Force, published in his *Recollections of the Campaign in Malwa and Central India* (1860).

Lieutenant-Colonel A. J. ENGLISH, late commanding 14th Hussars, for facilities of access to the 'Regimental Digest of Services' at Newbridge in 1897.

Colonel G. H. C. HAMILTON, Major E. D. J. O'BRIEN, and Captain R. G. BROOKSBANK, 14th Hussars, for information as to the campaign in South Africa.

The portrait of General H. E. Doherty, C.B., was kindly given by his nephew, Major D. H. DOHERTY, late 3rd (King's Own) Hussars.

Miss MURRAY kindly supplied the portrait of her late father, General the Hon. Sir Henry Murray, K.C.B., and Mrs. R. O. MILNE that of her late father, General H. Richmond Jones, C.B.

The portrait of Major-General Sir John Burgoyne, Bart., was obtained by the kindness of his great-grandson, Colonel Sir JOHN MONTAGU BURGOYNE, Bart., to whom the original painting by Romney belongs.

Mrs. FRANK VANDELEUR kindly supplied the picture of General Sir J. O. Vandeleur, G.C.B., from which the photogravure was reproduced.

There are also numerous other kind friends and former brother-officers and some now serving in the 14th Hussars who have helped the author by giving him useful hints and information during the compilation of his work, and by replying to his queries, and to one and all of these he takes this opportunity of thanking them sincerely for their kind assistance. He would also add, that of the old officers of the Fourteenth who, from the first, have given him their best encouragement, as well as much useful information, are Viscount Chetwynd, Captain R. P. Apthorp, and the late General C. W. Thompson, Colonel of the 14th (King's) Hussars.

The late Colonel the Honourable G. H. Gough, C.B., when employed at the War Office as Private Secretary to the Commander-in-Chief (Viscount Wolseley), was good enough to supply the author with those interesting anecdotes of the Peninsular War by the late General Sir T. W. Brotherton, G.C.B., and obtained permission from the family for their publication in this book. He also kindly supplied a copy of the Jhansi, Calpee, and Gwalior despatches of Major-General Sir Hugh Rose, K.C.B., from which extracts will be found in the Appendix.

The portrait of Field-Marshal John Campbell, fifth Duke of Argyll, is from an oil painting by Gainsborough, lately in Argyll Lodge, Kensington, which His Grace the late Duke of Argyll kindly gave the author permission to have photographed.

It remains for the author to acknowledge with his grateful thanks the help he has received from the following on various occasions :—

Lieutenant-Colonel C. Greenhill Gardyne, Glenforsa, Isle of Mull, N.B.

Major H. Davidson, Cantray, Croy, N.B.

F. Leach, Esq., War Office, London.

G. Smith, Esq., War Office, London.

Hubert Hall, Esq., F.S.A., Public Record Office, London.
Herbert Wood, Esq., Public Record Office, Dublin.
Mr. H. M. Winder, Royal Hospital, Kilmainham, Dublin.
Mr. Henry Pridgeon (late 14th Hussars), St. Anne Street, Liverpool.

The author had the advantage of conversing with, probably at the time, the oldest survivor of the 14th Light Dragoons, Mr. Sweeney, warder of the Tower of London, who had many long yarns to spin of his old soldiering days which commenced in King William IV.'s reign. He served with the 14th Light Dragoons through the Punjaub and Indian Mutiny campaigns, and died about two years ago, leaving a widow.

All those interested in Light Dragoons in general, and the Fourteenth in particular, should read *Charles O'Malley*, published in 1847 under the authorship of 'Harry Lorrequer,' with illustrations by 'Phiz.' The following is the well-known song which Captain Richard Power of the Irish Dragoons (the 14th) sang at the mess in Dublin when the officers were trying to induce 'Charles O'Malley' to join their corps (he being then a student at Trinity College):—

THE IRISH DRAGOON

(*Air*—'Love is the soul of a gay Irishman.')

Oh! love is the soul of an Irish Dragoon,
In battle, in bivouac, or in saloon—
From the tip of his spur to his bright sabretache.
With his soldierly gait and his bearing so high,
His gay laughing look and his light-speaking eye,
He frowns at his rival, he ogles his wench,
He springs in his saddle and 'chassés' the French—
With his jingling spur and his bright sabretache.

His spirits are high and he little knows care,
Whether sipping his claret or charging a square—
    With his jingling spur and his bright sabretache.
As ready to sing or to skirmish he 's found,
To take off his wine or to take up his ground;
When the bugle may call him, how little he fears
To charge forth in column, and beat the Mounseers—
    With his jingling spur and his bright sabretache.

When the battle is over, he gaily rides back
To cheer every soul in the night bivouac—
    With his jingling spur and his bright sabretache.
Oh! there you may see him in full glory crowned,
As he sits 'mid his friends on the hardly-won ground,
And hear with what feeling the toast he will give,
As he drinks to the land where all Irishmen live—
    With his jingling spur and his bright sabretache.

There are, besides *Charles O'Malley*, two other books more recently published which are entirely about the 14th Light Dragoons and the 14th Hussars, both written by non-commissioned officers who have served in the regiment, which are extremely interesting and well worth reading, viz.—*Scraps from my Sabretache*, by G. C. Stent (W. H. Allen and Co., London, 1882); *A King's Hussar*, by Herbert Compton (being reminiscences of Troop Sergeant-Major Edwin Mole), late 14th Hussars (Cassell and Co., 1896).

SUMMARY OF THE PRINCIPAL AUTHORITIES QUOTED

PRINTED BOOKS

Historical Record of the 14th King's Light Dragoons. By RICHARD CANNON, Esq. London: Parker, Furnival and Parker. 1847.

Narrative of the War in the Peninsula, 1808-1813. By General CHARLES VANE, Marquis of Londonderry, G.C.B., G.C.H., Colonel 10th Hussars. 1828.

History of the Indian Mutiny. By KAYE and MALLESON. 6 vols. London: Longmans, Green, and Co. 1898.

Scraps from my Sabretache. By G. C. STENT, M.R.A.S. London: W. H. Allen and Co. 1882.

Historical Record of the First or Royal Regiment of Dragoons. By General DE AINSLIE, Colonel of the regiment. London: Chapman and Hall. 1887.

Recollections of the Campaign in Malwa and Central India under Major-General Sir Hugh Rose, G.C.B. By J. H. SYLVESTER, Esq., F.G.S., 2nd Regiment Mayne's Horse. Bombay: Smith, Taylor and Co. 1860.

The Sepoy Revolt. By General McLEOD INNES, V.C. London: Innes and Co. 1897.

History of the Highland Clans. By JAMES BROWNE. 3 vols., 8vo. Glasgow. 1843.

The Sikhs and the Sikh Wars. By General Sir CHARLES GOUGH, V.C., G.C.B., and A. D. INNES, Esq., M.A. London: A. D. Innes and Co. 1897.

Historical Journal of the British Campaign on the Continent, with the Retreat through Holland, 1794-95. By Captain L. J. JONES, 10th Regiment. Birmingham. 1797.

Memoirs of the 10th Royal Hussars (Prince of Wales's Own). By Colonel R. S. LIDDELL, late commanding 10th Royal Hussars. Illustrated. London: Longmans. 1891.

Old Memories. By General Sir HUGH GOUGH, V.C., G.C.B. Blackwood. 1897.

Biography of General Sir James Outram, Bart., G.C.B., K.C.S.I. By Major-General Sir F. G. GOLDSMID, K.C.S.I., C.B. 1880.

History of the War in the Peninsula and the South of France. By Lieutenant-General Sir WILLIAM NAPIER, K.C.B. 6 vols. London: Frederick Warne and Co. 'Chandos Classics.'

Narrative of the Second Sikh War, 1848-49. By E. J. THACKWELL, Esq., late A.D.C. to General W. Thackwell, C.B. London: Richard Bentley. 1851. (2nd Edition revised.)

Standards and Colours of the British Army. By S. M. MILNE, Esq. 1893.

Historical Record of Medals and Honorary Distinctions conferred on the British Navy and Army and the Auxiliary Forces from the earliest periods. By Captain T. TANCRED, late Royal Scots Greys. London: Spink and Co. 1891.

Book of Dignities. London: W. H. Allen and Co. 1890.

Dictionary of National Biography. London: Smith, Elder and Co. 1895.

Historical Record of the 13th Light Dragoons. By RICHARD CANNON, Esq. London: J. W. Parker. 1842.

Historical Record of the 16th (Queen's) Light Dragoons (Lancers). By RICHARD CANNON, Esq. London: J. W. Parker. 1842.

PERIODICAL PUBLICATIONS

Army Lists.

Burke's and Foster's *Peerages* and *Baronetages.*

The London Gazette.

Bulletins, Despatches, and *London Gazette.*

Illustrated London News. 1849.

Journal of the Royal United Service Institution, London. 1895.

Dress Regulations for the Officers of the Army. War Office, 1822, 1831, 1846, 1855, 1857, 1864, 1874, 1883, 1894, and 1900.

Times. Army and Navy Gazette. Broad Arrow.

ORIGINAL DOCUMENTS

Marching Orders.
Muster Rolls.
Marching Books.
Army Returns.
MS. Army Lists.
Pay Lists.
General Officers' Letter-Books.
Review Reports and Inspection Returns.
Routes.
Book of Licences (Military).
} Public Record Offices, London and Dublin.

Regimental Digest of Services, 14th Hussars (Orderly-Room, Newbridge, 1897).

War Office Books (Establishment, Moves, Miscellany, Honorary Distinctions, etc.). Pall Mall, London.

CONTENTS

MISCELLANEOUS

LIST OF ILLUSTRATIONS

COLOURED PLATES

BY WILLIAM GRIGGS

PHOTOGRAVURES

By W. L. Colls

ROYAL PATRON

ROYAL PATRONESS

COLONELS

LIEUTENANT-COLONELS COMMANDING

MAPS AND PLANS

BY EDWARD STANFORD

CORRIGENDA ET ADDENDA

At page xiii, line 18, *for* 'nephew' *read* 'cousin.'

At page 55, line 14—
For 'it was defeated by the French' *read* 'it defeated the French.'
For 'its gallant leader killed' *read* 'its gallant leader was killed.'

At page 65, footnote [2], *add* after the word 'succeed': 'Ça Ira,' the name of a French revolutionary air, was adopted as a Regimental March by the 14th Foot after the taking of Famars, in Flanders, in 1793. Both Mr. F. A. Hawker and Mr. Adolphus Hawker are of opinion that the song was used by the 14th Light Dragoons in the Peninsula when their father (Sir S. Hawker) was commanding the regiment, and they consider he became acquainted with the air when serving in Flanders in the 16th Light Dragoons. 'Ça Ira, 1810,' is quoted in *All the Year Round* (New Series), No. 877, p. 60, Sept. 1885.

At page 214, footnote [1], *for* 'Vicars-Maxims' *read* 'Vickers-Maxims.'

ERRATA

At page 346, lines 9 and 18, *for* '8th Hussars' *read* '18th Hussars.'

,, footnote [1], *for* '8th (King's Royal Irish) Hussars' *read* '18th Hussars.'

At page 347, line 28, *for* '8th Hussars' *read* '18th Hussars.'

HISTORICAL RECORD OF

THE 14TH (KING'S) HUSSARS

1715

George I.

George I., 1715.

The regiment now styled 'The Fourteenth (King's) Hussars' was raised by Brigadier-General James Dormer in South Britain, and was originally known as 'Dormer's Dragoons.'

14th, or Dormer's Dragoons.

The first officers were appointed as follows by commissions dated 22nd July 1715, six troops being the establishment of the regiment :—

Captains.	*Lieutenants.*	*Cornets.*
James Dormer (Col.).	James Stevens (Capt.-Lt.).	Edward Stroud.
H. Killigrew (Lt.-Col.).	Henry Lasale.	Thomas Ellis.
Sol Rapin (Major).	Peter Davenport.	Thomas Delahaye.
Henry Pelham.	Jonathan Pirke.	William Hamilton.
William Boyle.	Cuthbert Smith.	Rigley Molyneux.
Beverly Newcomin.	James Fleming.	Andrew Forrester.

Colonel—James Dormer. Lieutenant-Colonel—H. Killigrew.

In November the newly raised men of 'Dormer's Dragoons' were brigaded with 'Pitt's Horse,' under their own Colonel, Brigadier-General Dormer, and advanced with other forces commanded by Major-General Wills towards Preston. An engagement ensued with a body of the Pretender's rebels, and being directed to dismount, Dormer's Dragoons formed as infantry, and did good service in the attack on the avenue leading to Lancaster. They lost three men and sixteen horses killed, and Brigadier-General Dormer and four men were wounded.

Engaged at Preston, November 1715.

Browne, in his *History of the Highland Clans*, relates how that General Wills had been ordered to collect all the forces he could to march on Warmington Bridge, Preston, so as to prevent the insurgents from advancing on Manchester. When the insurgents entered Preston, Wills was at Manchester, and he decided to march at once upon Preston with such troops as he had.

On the 11th November he left Manchester for Wigan with four regiments of dragoons—Honeywood's (11th), Dormer's (14th), Munden's (13th), Wynne's (9th)—Pitt's Horse, Churchill's, Molesworth's, and Preston's regiments of foot. He reached Wigan early in the evening, and met Stanhope's Dragoons and Houghton's Militia there, they having retired from Preston on the 9th inst.

Amongst the Jacobite leaders in Preston were Lord Kenmure, the Earls of Derwentwater, Nithsdale, and Wintour, Lord George Murray, old Borlume (Mackintosh), and Forster. They were taken aback by the news of Wills's advance. Lord Kenmure was apprised by letter, and came to Forster to consult him. It was resolved to send out a party of horse towards Wigan to watch the enemy and to secure the pass into the town by the Ribble Bridge. At daybreak on the 12th November, General Wills left Wigan. His van consisted of Preston's regiment of foot preceded by an advanced guard of fifty musketeers and fifty dragoons on foot. Then followed—

Honeywood's Dragoons, Dormer's Dragoons, Munden's Dragoons, with the baggage in rear escorted by fifty dragoons.

So soon as the approach was known, a select body of 100 well-armed Highlanders under Farquharson of Invercauld was posted at Ribble Bridge, and Forster himself at the head of some horse advanced over the bridge to reconnoitre. Lord George Murray had command of a barrier at the end of a lane leading to a field: this was attacked by Dormer's Dragoons, who acted very bravely though exposed to a murderous fire from the houses. They carried part of the barrier, were repulsed after a time by the defenders, and again pressed forward with a vigorous attack, when, after displaying great courage and resolution (although only newly raised troops), they were a second time beaten back with loss.

The attack on the windmill barrier, which was defended by Colonel Mackintosh, met with a similar fate.

After this, General Wills set fire to the houses, and next day General Carpenter with reinforcements arrived to help the Royalists. This opportune occurrence quite turned the tide of events: the Jacobites began to think of surrender, actually opened negotiations, and on the 14th November, headed by Lord Kenmure and other Scots noblemen, together with Colonel Mackintosh and Mr. Forster, they surrendered at discretion. There were 1468 prisoners, of whom 463 were English. The Jacobites lost seventeen killed, twenty-five wounded; the losses of the Royalists were five times that number.

Preston had been defended by four barriers, but General Wills being short of troops on his first day of assault could only attack two barriers, those commanded by Mackintosh and Murray, and he divided his troops into three bodies for attack:—

The first, under Brigadier Honeywood, who had the foot regiments.

The second, under Brigadier Dormer, who had Dormer's Dragoons (14th), Wynne's Dragoons (9th), and a squadron of Stanhope's Dragoons (afterwards disbanded).

The third, under Brigadier Munden, who had Pitt's Horse (2nd Dragoon Guards), Munden's Dragoons (13th), and a squadron of Stanhope's Dragoons forming in support.

Cannon says: 'The first barrier was carried with great gallantry, the second or inner barricade could not be forced for want of cannon; but after the houses had been set on fire, the troops prevented the escape of the rebels.'[1]

[1] *Historical Record of the 14th Light Dragoons*, p. 3.

Dormer's Dragoons were employed to escort prisoners to Lancaster gaol, and remained at Lancaster for the next six months.

1716

In May the regiment marched into cantonments at Lincoln and neighbouring towns, such as Sleaford, Newark, and Gainsborough.

1717

Regiment proceeds to Ireland.

Dormer's Dragoons marched in May to Bristol and embarked for Ireland, where they remained for the next twenty-five years. Adjutant George Thornborough was reduced with the regiment to two shillings per diem half-pay, being only twenty years of age, as appears in the Army List for 1740.

It is probable the regiment was quartered shortly after arrival in scattered detachments through the west of Ireland, and for part of the year at Foxford (near Ballinrobe), Carrick-drumrush, Boyle, Collooney, and Castlebar. An order, dated 6th June 1717, was issued for it to proceed, on landing in Ireland, to certain quarters, named 'according to the within route,' but the route is missing.

In the Military Licences Book at the Public Record Office, Four Courts, Dublin, there is an entry allowing Major Sol Rapin, of Brigadier James Dormer's Dragoons, 'to be absent from his command for six months, from June 30, in consideration of his wounds received in your service, for the recovery of his health,' dated 5/8/1717. There are also entries as follows:—

Licence, dated 18/9/1717, for Cornet Thos. Ellis (Dormer's Dragoons) to be absent in Great Britain for three months.

Leave from muster, ending 30/9/1717, granted to Captain Beverly Newcomin and Lieutenant Cuthbert Smith of Dormer's Regiment of Dragoons.

Lieutenant-Colonel—Sol Rapin.

Major Sol Rapin was Lieutenant-Colonel in 1717.

To Cornet Rigley Molyneux of Brigadier Dormer's Dragoons, a Licence of 24/12/1717, 'to go into Great Britain about his own affairs, and to be absent out of this Kingdom for the space of three months from this date.'

To Peter Pickering, Chaplain to Brigadier Dormer's Dragoons, a Licence, dated 22/12/1717, 'to continue in Great Britain upon extry occasions, and be absent from his post in your said regiment for the space of three months from this date'; and there is an order, dated and signed 31/12/1717, 'to remove the cheques put upon the several officers and private men of Dormer's Dragoons for three months ending 30/9/1717.'

Major William Boyle became Lieutenant-Colonel *vice* Lieutenant-Colonel Rapin. Lieutenant-Colonel—William Boyle.

1718

Brigadier Dormer's Dragoons received an order, dated 31st May 1718, to march from their present quarters according to the following route:—

1 Troop from Foxford, *via* Ballinrobe, 4th June, to Nenagh barracks, arriving 14th June.
1 Troop from Carrickdrumrush on 11th June to Charleville barracks, arriving 21st June, *via* Elphin.
1 Troop from Boyle, *via* Elphin, 4th June, to Cappoquin barracks, June 16th.
1 Troop from Collooney, *via* Boyle, 6th June, to Tallow Town, 17th June.
1 Troop from Castlebar, *via* Ballinrobe, 5th June, to Mallow barracks, 16th June.
1 Troop from Castlebar, *via* Ballinrobe, 9th June, to Tallow Town, 21st June.

1719

In pursuance of an order, dated 14th March, Dormer's Dragoons moved as follows:—

1 Troop from Tallow to Charleville Town.

1 Troop from Cappoquin to Charleville Town.
1 Troop from Nenagh to Bruff and Kilmallock Towns.
1 Troop at Charleville } already there, remained fast.
2 Troops at Mallow }

In May another move commenced, and the regiment became located as follows:—

1 Troop (the Brigadier's) to Downpatrick (one-half to barracks, one-half to town).
1 Troop (Colonel Boyle's) to Lisburn Town.
1 Troop (Colonel Diggs's) to Carrickfergus (one-half to barracks, one-half to town).
1 Troop (Captain Newcomin's) to Dromore Town.
1 Troop (Captain Stevens's) to Carrickfergus barracks.
1 Troop (Captain Morin's) to Downpatrick.

N.B.—The Lisburn Troop subsequently moved to Killyleagh.

On 9th November an order came for the regiment to proceed to Mullingar: 1½ troops went to barracks, and the remainder (4½ troops) to the town of Mullingar.

It appears in an old book of Military Licences deposited at the Public Record Office, Dublin, that an order, dated 22nd May 1719, was issued to Brigadier Dormer's Regiment of Dragoons, to allow Mr. Henry Pickering, the chaplain, 'to continue in Great Britain about his own affairs, and to be absent out of this Kingdom (Ireland) for six months from 1st January, the time his former Licence expired.'

1720

Colonel—Clement Neville.

The regiment remained on in quarters at Mullingar. On 9th April, Brigadier James Dormer became Colonel of the 6th Regiment of Foot; and Lieutenant-Colonel Clement Neville from the 13th Dragoons became Colonel of the regiment. Major B. Newcomin became Lieutenant-Colonel on 1st December.

Lieutenant-Colonel—B. Newcomin.

1721

For this and the following five years, no record has been found of the stations at which the regiment was quartered in Ireland.

1722

In Ireland.

1723

In Ireland.

1724

In Ireland.

1725

On 25th March, Lieutenant William Hamilton, with Quartermaster Charles Waldegrave and some non-commissioned officers, went to Great Britain for three months for the purpose of buying recruit horses for the regiment.

At this period Newcomin was still the Lieutenant-Colonel, and the captains were Dilkes, Pearce, Bowles, and Smith.

1726

From 1st October to 31st December of this year, Lieutenant William Hamilton, accompanied by 1 sergeant of Lieutenant-Colonel Newcomin's troop, was employed in England in buying recruit horses for the regiment.

1727

GEORGE II.

George II., 1727.

The regiment was quartered, 1 troop (the Colonel's) at Collooney, 5 troops at Sligo.

1728

Colonel Neville's Dragoons marched at the end of May to new quarters, as follows :—

1 Troop (the Colonel's) from Collooney to Nenagh barracks.

1 Troop (Major Smyth's) from Sligo to Loughrea barracks.

1 Troop (Lieutenant-Colonel Newcomin's) from Sligo to Headford barracks.

1 Troop (Captain Pearce's) from Sligo to Portumna barracks.

1 Troop (Captain Dilkes's) from Sligo to Charleville barracks.

1 Troop (Captain Bowles's) from Sligo to Mallow barracks.

The whole regiment was in its new quarters by 12th June.

On 25th June, 4 troops received an order to march early in July as follows :—

1 Troop, Nenagh to Mallow,
1 Troop, Portumna to Loughrea,
1 Troop, Headford to Loughrea,
1 Troop, Charleville to Mallow,

leaving 1 troop in Mallow barracks and 1 troop in Loughrea barracks. The order directed, that 'when the troops above named are reviewed by the Right Honourable the Lord Viscount Shannon, they are to march back to their respective quarters,' etc. The review was conducted by wings, 3 troops being inspected at Mallow and 3 at Loughrea.

1729

On 6th June, Colonel Neville's Dragoons commenced to march as follows :—

The Colonel's troop, Nenagh to Birr, and later to Castlebar.

Lieutenant-Colonel Newcomin's, Headford to Ballyshannon barracks.

Major Smyth's, Loughrea to Ballyshannon barracks.

Captain Pearce's, Portumna to Newport-Pratt barracks.

Captain Dilkes's, Mallow to Headford barracks.

Captain Bowles's, Charleville to Castlebar (half troop to town and half troop to barracks).

Shortly after arrival the regiment got orders to march and be reviewed by General Honeywood as follows, a wing (4 troops) to Loughrea; a squadron (2 troops) to Enniskillen, returning to former quarters after the inspection.

1730

In June the regiment moved as follows :—

Colonel's troop, Castlebar to Longford barracks.

Lieutenant-Colonel's troop, Ballyshannon to Granard barracks.

Major's, from Ballyshannon to Mullingar barracks.

Captain Pearce's, from Newport-Pratt to Tuam.

Captain Dilkes's, from Headford to Phillipstown barracks.

Captain Bowles's, from Castlebar to Navan barracks.

The whole 6 troops were reviewed at Mullingar by Major-General Naper in August, and returned to their respective stations afterwards.

On 30th November the 6 troops marched to Dublin, and were assembled there in barracks by 9th December.

1731

On 17th June, Colonel Neville's Dragoons marched from Dublin *via* Dundalk, etc., to fresh quarters.

The Colonel's and Captain Pearce's troops to Saintfield barracks, near Hillsborough.

Lieutenant-Colonel Newcomin's and Captain Bowles's troops to Carrickfergus barracks.

The Major's and Captain Dilkes's troops to Killough barracks.

Subsequently 2 troops left Carrickfergus and Saintfield (1 from each), and proceeded to Killyleagh and Downpatrick (1 to each).

Lieutenant-Colonel—Henry Desgrangues.

Henry Desgrangues became Lieutenant-Colonel on 12th November.

1732

A route, dated 10th June, directed Neville's Dragoons to proceed from present quarters as follows:—

1 Troop to Lifford.
2 Troops ,, Ballyshannon.
2 ,, ,, Castle Dawson.
1 Troop ,, Collooney.

Three troops were reviewed by General Wynne at Sligo, and 3 troops at Castle Dawson by Major-General Price.

1733

From 4th to 19th June the regiment moved as follows:—

1 Troop from Collooney, and 1 troop from Ballyshannon to Hamilton's Bawn.

1 Troop from Ballyshannon to Dungannon barracks.

The 1 troop at Lifford and 2 troops at Castle Dawson remained there.

The 6 troops were reviewed in June at Dungannon, and on 24th June they marched from their respective stations to Sligo, arriving there 9th July.

1734

The regiment remained in Sligo during this year.

1735

On 7th July the regiment moved from Sligo to Belturbet, Collooney, Roscommon, Cavan, and Ballyshannon.

One squadron at Ballyshannon; and 1 troop at each of the other stations. The march was completed by 12th July.

IRELAND
Donegal B.
Lifford
Castle Dawson
R. Bann
Carrickfergus
Lough Neagh
BELFAST
Donaghadee
Ballyshannon
Lisburn
Saintfield
Dungannon
Enniskillen
Lurgan
Hillsborough
Dromore
Ballyleagh
R. Erne
Armagh
Hamiltons-bawne
Downpatrick
Sligo
Monaghan
Newry
Killough
Collooney
Newtown Hamilton
Foxford
Belturbet
Castleblaney
Newport Pratt
Boyle
Carrickmacross
Castlebar
Cavan
Dundalk
Westport
Elphin
Kingscourt
Claremorris
Granard
Ballinrobe
Roscommon
Longford
Navan
Drogheda
Dunmore
Ballymahon
Duleek
Tuam
Mullingar
R. Shannon
Headford
Athlone
Philipstown
R. Liffey
DUBLIN
Galway
Ballinasloe
Tullamore
Tallagh
Loughrea
Kildare
Newbridge
Portumna
Birr
Mountmellick
Curragh
Castlemartin
Gort
Roscrea
Maryborough
Tulla
Ennis
Nenagh
Donaghmore
Tullow
Ballyragget
Carlow
Ross
Thurles
Kilkenny
Limerick
R. Barrow
Listowel
Tipperary
Cashel
Callan
Bruff
Fethard
Kilmallock
Cahir
Clonmel
Charleville
Dingle
Castle Island
Clogheen
R. Suir
Carrick on Suir
Cullen
Mallow
Doneraile
Waterford
Killorglin
Killarney
Fermoy
Cappoquin
Cahersiveen
Tallow
Kenmare
Cork
Youghall
Ballincollig
Middleton
Carrigaline
R. Blackwater
Bandon
Inishannon
Kinsale
Bantry
Skibbereen
St. George's Channel
Firth of Clyde
IRI
Harrow
Highgate
Paddington
Colnbrook
Kensington
LONDON
Windsor
Hounslow
Mortlake
Lambeth
New Cross
Putney
Twickenham
Richmond
Hampton Court
Wimbledon
Lewisham
Crystal Palace
R. Thames
Bagshot
Sandhurst
Chobham Ridges
Hartford Bri. Flats
Farnborough
Cove Common
ALDERSHOT
Bourley
Farnham
Guildford
Reigate
Scale of Miles.
5 0 5 10
Truro
Flushing
Fal

London

Stanford's Geog¹. Estab¹.

reen & Co.

1736

A change of stations took place in June :—

1 Troop went from Roscommon to Loughrea.

1 Troop from Belturbet to Granard ; 1 troop, Cavan to Gort.

1 Troop, Collooney to Headford ; and 2 troops moved from Ballyshannon—1 to Cavan and 1 to Belturbet.

Major-General Lord Molesworth reviewed 3 troops in June at Cavan ; and 3 troops were reviewed by Lieutenant-General Naper at Loughrea, viz. the Gort, Headford, and Loughrea troops. Lord Molesworth reviewed the Cavan, Granard, and Belturbet troops.

1737

On 20th June, Neville's Dragoons marched as follows :—

1 Troop, Loughrea to Mullingar.
1 ,, Headford ,, Phillipstown.
1 ,, Gort ,, ,,
1 ,, Cavan ,, ,,
1 ,, Belturbet ,, Navan.
1 ,, at Granard remained there.

This change was completed by the 25th of June.

On 27th June, Brigadier Neville went to the 8th Dragoons, and Colonel Archibald Hamilton, from 27th Foot, became Colonel. **Colonel—Archibald Hamilton.**

W. Wright became Lieutenant-Colonel 7th February, and Captain Michael O'Brien Dilkes was appointed Major on 14th February. The regiment was reviewed at Phillipstown, the three detached troops returning to outquarters afterwards. In December the regiment moved as follows :—Captain Bowles's troop from Granard ; the Colonel's troop from Mullingar ; Captain Norris's from Navan ; and Major Smith's, Major Dilkes's, and Lieutenant-Colonel Wright's from Phillipstown, to Dublin, where the whole regiment was then concentrated under its newly appointed commander, Colonel Archibald Hamilton. **Lieutenant-Colonel—W. Wright.**

1738

In May, Hamilton's Dragoons proceeded as follows:—2 troops to Killough, and 1 troop each to Downpatrick, Saintfield, Carrickfergus, and Killyleagh. Subsequently 2 troops moved to Castle Dawson, 2 to Ballyshannon, 1 to Collooney, and 1 to Lifford; and later in the year 2 troops from Ballyshannon went to Hamilton's Bawn, and 1 from Collooney to Dungannon.

1739

The regiment was stationed at Lifford, Castle Dawson, Dungannon, and Hamilton's Bawn.

1740

In May, 1 Troop marched from Lifford to Athlone.
2 Troops „ „ Castle Dawson to Athlone.
1 Troop „ „ Dungannon to Bandon.
2 Troops „ „ Hamilton's Bawn, 1 to Middleton Town (beyond Cappoquin), and 1 (Major Dilkes's) to Castlemartin Town.

In July the 6 troops began to march to Sligo barracks, the last troop arriving by 21st August.

1741

The regiment remained in Sligo.

Captain Richard Bowles became Major, 13th May.

1742

Regiment proceeds to Great Britain.

The regiment came to Great Britain after twenty-five years' service in Ireland, marching in May from Sligo to Belfast and Donaghadee, where the embarkation took place.

On 1st January of this year Colonel Archibald Hamilton was advanced to the rank of Major-General.

1743

There appears to be no record of where the regiment was stationed in England during this and the following year.

1744

At this period the establishment was 6 troops of 59 troopers each, total 354 (but the total strength, with officers and non-commissioned officers, came to 435), and the amount of pay for the whole corps for 366 days amounted to £19,684, 14s.

1745

The regiment was on the move in Scotland and North of England, being employed against the Jacobite rebels.

During this year the young Pretender, Prince Charles Edward Stuart, raised his father's standard in Scotland. The 14th Dragoons were ordered to Stirling; thence they proceeded to Leith, and were stationed there when the rebel army advanced on Edinburgh.

Being ordered to join the troops under Colonel Gardiner, who, with the 13th Dragoons, was retiring before the rebels, the 14th Dragoons marched through Edinburgh on Sunday, 15th September, and effected a junction with Colonel Gardiner at Corstorphine, whence they fell back to Coltbridge, and thence towards Haddington. On the night of the 20th September the regiment, under Lieutenant-Colonel Wright, furnished vedettes and flanking patrols for the king's army at Prestonpans. The army was under the command of Lieutenant-General Sir John Cope, who advanced from Dunbar to confront the insurgents at Edinburgh, where they had established themselves on the 15th. A battle took place on 20th September. Prestonpans, 21st September 1745.
Prestonpans was lost; but Major Bowles, 14th Dragoons, particularly distinguished himself when the outposts were driven in by the Highlanders in the dark morning mist which enveloped the field.

The Battle of Prestonpans or Gladsmuir.

Browne, in his *History of the Highland Clans*, narrates how that at dawn of day Cope's picquets on his left flank heard the tramp of the enemy, and as soon as Cope himself was apprised of the advance of the Highlanders, he gave orders to change his front to the east. In doing this some confusion took place, as some of the advanced guards got in the way of the main body and prevented two squadrons of Gardiner's Dragoons (13th), which had been posted on the right of General Cope's line, from forming properly; so that, owing to want of room, the squadron under the immediate command of Colonel Gardiner drew up behind the one commanded by Lieutenant-Colonel Whitney, and the artillery was now in front of Whitney's squadron. The intervening ground between the two armies was level and good for cavalry, but the Scots came on so rapidly with their attack, that Cope was quite unable to make any use of it in time. There was a thick mist over the cornfields which prevented the two armies from seeing one another. The Camerons attacked Cope's cannon, which was on his right, and routed it. Whitney's squadron attempted to charge, but the Camerons set up a loud shout, rushed past the cannon, and shot at the dragoons. Lieutenant-Colonel Whitney was wounded. The Scottish swordsmen came on, sword in hand, and caused the squadron to fall into disorder, owing to the Highlanders slashing at the noses of the horses, which caused the dragoons to wheel about, and in their retreat they rode over the artillery guard, also in retreat. Colonel Gardiner received orders to charge the Highlanders with his squadron and stop their advance, and he did his best, but his men were shaken, and commenced to retire before the onward rush of the Highlanders as their comrades and the artillery had already done. Colonel Gardiner fought gallantly and was badly wounded. Cope's infantry made no better stand: the men seemed utterly demoralised and unable to stem the furious onset of the Scots, and many threw down their arms and fled. Hamilton's Dragoons (14th) were posted on Cope's left, and behaved no better than the rest of his troops, but followed their bad example in falling back when the sudden and violent onslaught of the Highlanders was made. It is certain, however, that their officers behaved with great courage and bravery, and did their best to rally the men at this critical juncture, and the troopers did halt now and again and faced the enemy, but, according to Cannon,[1] their sudden advance in the dark, their superior numbers and peculiar mode of fighting, struck the Royalists with consternation. Lieutenant-Colonel Wright commanded the 14th Dragoons on this occasion. Several of his officers and some of the private soldiers behaved with great gallantry, and highly distinguished themselves—particularly so Major Bowles, whose horse was killed, and the troopers who rallied round him were all cut down. Major Bowles was surrounded by the enemy, but continued to fight on foot, when, having received eleven wounds, a rebel leader interposed and saved his life. Browne further relates how that even Sir John Cope himself, with the Earls Howe and Loudon, tried to rally 450 of the dragoons on the west side of Preston village at a favourable spot, but this effort proved ineffectual, and in the end the royal troops with their commander were routed. The brave Colonel Gardiner (13th Dragoons) who commanded the cavalry in this battle fell on the field. He was a veteran of Marlborough's army, a truly good soldier, distinguished by personal bravery and many Christian virtues.

Gardiner was first shot in the breast, and was afterwards cut down by a Highlander with a scythe fastened to a pole, and, as he fell, another Highlander gave him a mortal blow on the head.[2]

After this battle the Fourteenth marched to Berwick. They afterwards joined the army assembled under Field-Marshal Wade at Newcastle, and when the rebels came

[1] Cannon's *Historical Record of the 14th Light Dragoons*, p. 6.

[2] Cannon's *Historical Record of the 13th Light Dragoons*, p. 16.

down into Derbyshire, they were employed in covering Yorkshire.

Subsequently, when the young Pretender retreated hurriedly into Scotland, the regiment was sent to Edinburgh, and joined a few regiments assembled there under Lieutenant-General Hawley.[1]

Mr. Peter Vatass was appointed chaplain to the regiment on 24th December.

1746

The rebels besieged Stirling Castle on returning to Scotland, and Lieutenant-General Hawley set the king's troops in motion to oppose the besiegers. On 13th January, Hamilton's Dragoons left Edinburgh and helped, with other forces, to drive a body of the rebels out of Linlithgow. They afterwards encamped near Falkirk.

Battle of Falkirk, 17th January 1746.

It was on Falkirk Moor, on the 17th January, that the King's army met with a repulse in the midst of a heavy storm of wind and rain which nearly blinded them. Browne relates that the Royalists were drawn up as follows :—

Lieutenant-General Hawley, Commander-in-Chief, took the centre.

On the right was Major-General Huske.

On the left was Brigadier Cholmondeley, and Ligonier, now Colonel of the 13th Dragoons, commanded the cavalry, which consisted of three regiments of dragoons, viz. Cobham's (10th), Ligonier's (13th), and Hamilton's (14th).

Colonel Ligonier's own regiment was posted on the extreme left, Cobham's regiment in the centre, and Hamilton's Dragoons on the right. There was no artillery employed. Hawley's force numbered 10,000 men; the enemy had 8000 men, but they had the best ground, and the wind and driving rain were on their backs at the moment the two armies advanced to battle. At the outset, Ligonier's Dragoons were fired at fiercely by Lord George Murray's High-

[1] Cannon's *Historical Record of the 14th Light Dragoons.*

landers, and many were killed and wounded; this caused great panic amongst Ligonier's and Hamilton's Dragoons, who could not maintain their ground. It appears from Cannon[1] that this battle was a surprise: the troops of the Royalists were at dinner in camp when the approach of the rebel army was descried, and hurriedly moved towards some high ground on Falkirk Moor, where the insurgent bands were formed. The action was commenced by a charge of the cavalry which broke the enemy's first line and did some execution; but the second line of insurgents made a furious onslaught which proved successful and repulsed the dragoons. This is the occasion referred to above by Browne, who also narrates that the English losses in the battle were 280, and those of the Scots 200. Lieutenant-Colonel Whitney of Ligonier's regiment (13th Dragoons) was amongst the slain. After this the infantry was brought up into the fight, as stated by Cannon,[2] but a heavy storm of wind and rain beat so violently in the soldiers' faces as nearly to blind them, and the wet prevented their muskets giving fire. Several regiments retired somewhat in disorder, others maintained their ground and repulsed the Highlanders, and after dark the whole Royalist army withdrew to Linlithgow and Edinburgh.

Colonel Ligonier with the dragoons covered this retreat, but being drenched with rain and benumbed with cold, he was seized with an inflammation in the throat, of which he died on the 25th January. He had previously suffered from pleurisy, and was bled and blistered on the 14th January, but no consideration could keep him from his duty, so he left his bed to command the brigade of dragoons at this battle, and led the opening charge of cavalry with great spirit.[3]

The Duke of Cumberland was subsequently sent to command the King's army in Scotland, and made an advance on Stirling, whereupon the rebels hastily retired northwards, pursued by the Royalists. The Fourteenth (Hamilton's

[1] *Historical Record of the 14th Light Dragoons*, pp. 7 and 8. [2] *Ibid.*

[3] Cannon's *Historical Record of the 13th Light Dragoons*, p. 19.

Dragoons) were left behind with orders to patrol along the roads leading westward from Edinburgh, to prevent the rebels obtaining intelligence.

The Duke of Cumberland was victorious, and at Culloden, on 16th April, he completely overpowered the Highlanders and put an end to the rebellion. Subsequently Prince Charles Edward Stuart, with much difficulty, effected his escape into France.[1]

1747

Regiment returns to Ireland.

Hamilton's Dragoons returned this year to Ireland, and remained in that country till 1795, a period of forty-eight years. In 1747 they were stationed at Carrickfergus, Castle Dawson, and Killough, after landing in Dublin in the early part of the year. On 17th August, Mr. Thomas Irving was appointed surgeon to the regiment. He was the first surgeon of the Fourteenth, and remained with the regiment till the 20th October 1774, when he was succeeded by Mr. George Borthwick.

A new set of buff accoutrements was issued this year.

The six troops were thus designated :—

The General's troop.	Captain Clark's troop.
The Lieutenant-Colonel's troop.	Captain Hamilton's troop.
The Major's troop.	Captain Bailie's troop.

Major-General Archibald Hamilton was the Colonel; William Wright, the Lieutenant-Colonel; James Norris, the Major.

Previous to this date, and even to the end of George II.'s reign, our regiments were generally known by the name of their Colonel, who in these early times took a very active part in the affairs of his regiment, and often accompanied it on active service, and more resembled the present Lieutenant-Colonel commanding a regiment than the full Colonel who is purely an honorary officer at the present day.[2]

[1] Cannon's *Historical Record of the 14th Light Dragoons*, p. 8.

[2] Liddell's *Memoirs of the 10th Royal Hussars.*

1748

In June, Hamilton's Dragoons marched from the north of Ireland to Clonmel and Mallow. After being reviewed there, 2 troops remained stationed at Clonmel and 1 at Mallow; the other 3 proceeded—

1 to Cappoquin and Tulla.
1 to Charleville.
1 to Doneraile.

This year a licence for three months was granted for Cornet James Baily of Captain Bailie's troop to go into Ulster with a sergeant, corporal, and drummer of the same troop, to raise recruits for the regiment. Recruiting.

1749

Lieutenant-General Archibald Hamilton, Colonel of the regiment, died on 8th July, and Lieutenant-General James, Lord Tyrawley, from 10th Regiment of Foot, became Colonel. Colonel—James, Lord Tyrawley. Some new swords were issued. The regiment was reviewed by Lieutenant-General Bragg at Phillipstown in June. Three troops remained there, 2 troops marched to Tullamore, and 1 troop to Mullingar barracks. Major Norris became Lieutenant-Colonel. Lieutenant-Colonel—James Norris.

1750

The regiment was quartered at Athlone, Castlebar, and Granard.

1751

The regiment marched to Sligo barracks, and arrived there on the 20th June.

On 1st July, the clothing of the army was regulated by royal warrant of H.M. King George II. as follows:— 14th Dragoons. Clothing scarlet, with lemon-yellow facings.

14TH DRAGOONS.

Coats, scarlet, double-breasted, without lapels; lined with lemon-colour; button-holes worked with narrow white lace; buttons, white metal, set on three and three; a long slash pocket in each skirt; and a white worsted aiguillette on the right shoulder.

Waistcoats and breeches, lemon-colour.

Hats, bound with silver lace, ornamented with white loop and black cockade.

Red forage-cap turned up with lemon colour, and XIV. D. on the flap. Boots of jacked leather, reaching to knee. Cloaks, scarlet, with a lemon-coloured cape; the buttons set on three and three, upon white frogs or loops, with a red and green stripe down the centre.

Horse-furniture of lemon-coloured cloth, the holster-caps and housings having a border of white lace, with a red and green stripe down the centre; XIV. D. embroidered upon a red ground, within a wreath of roses and thistles, on the housing; and upon the holster-caps G.R. with the crown over it, and XIV. D. underneath.

Officers, distinguished by silver lace and embroidery; and a crimson silk sash worn across the left shoulder. Quarter-masters to wear a crimson silk sash round their waists.

Sergeants to have narrow silver lace on the cuffs, pockets and shoulder-straps; silver aiguillettes; and green, red, and white worsted sashes tied round their waists.

Drummers and hautboys,[1] clothed in lemon-coloured coats, lined and faced with scarlet, and ornamented with white lace, having a red and green stripe down the centre; red waistcoats and breeches.

Guidons, the first or King's guidon to be of crimson silk with a silver and red fringe; in the centre, the rose and thistle conjoined, and crown over them, with the motto 'Dieu et mon Droit' underneath; the white horse in a compartment in the

[1] The 'Hautbois' ('oboe' Italian and Spanish) was a French invention, and introduced into our army after the Restoration. It was a large flageolet or clarionet, and gave a sound louder than all instruments except the trumpet. It was used in the dragoons and infantry. In 1759 it was exchanged again for the trumpet.—(*British Army*, by Sir S. Scott, Bart.) Liddell's 10*th Hussars*, p. 4.

first and fourth corners; and XIV. D. in silver characters on a lemon ground in a compartment in the second and third corners. The second and third guidons to be of lemon-coloured silk; in the centre XIV. D. on a red ground within a wreath of roses and thistles on the same stalk; the white horse on a red ground in the first and fourth compartments, and the rose and thistle conjoined on a red ground in the second and third compartments: the third colour to have a figure 3 on a circular red ground under the wreath.

Captain James Bailie was appointed Major, 19th June.

1752

Three troops went to Castlebar, 2 troops to Athlone (of which 1 troop went later to Roscommon), and 1 troop to Granard.

On 27th November, Lieutenant-General Lord Tyrawley went to the 3rd Dragoons, and Colonel Lewis Dejean was appointed Colonel in his place.

Colonel—Lewis Dejean.

1753

In May the 6 troops of Colonel Dejean's regiment were reviewed at Athlone, and marched to new quarters in June as follows:—

3 troops to Gort; 1 troop to Portumna, 1 to Headford, and 1 to Loughrea.

The last troop reached its destination (Portumna) on 5th June; this was Captain Arabin's troop. Major (and Captain) J. Bailie's troop was at Gort.

1754

Thomas Erle was appointed Lieutenant-Colonel, 4th September. He was appointed Cornet in 1707, the year of his birth, and Lieutenant in 1722 at the age of 15 years. He remained at the head of the 14th Dragoons till 1773.

Lieutenant-Colonel—Thomas Erle.

The regiment was quartered at Phillipstown and Tullamore, 3 troops at each place, and was reviewed at Maryborough. Subsequently the whole regiment moved to Dublin.

EXTRACT FROM DAILY PAY OF LAND FORCES, 1754.

Daily Pay, etc.	British Establishment. Dragoons.						Irish Establishment. Dragoons.					
	Pay.			Subsistence.			Pay.			Subsistence.		
	£	s.	d.	£	s.	d.	£	s.	d.	£	s.	d.
Colonel,	1	15	0	1	6	6	1	11	4	1	3	8
Lieut.-Colonel and Captain,	1	4	6	0	18	6	0	19	4	0	14	7
Major,	1	0	6	0	15	6	0	17	4	0	13	1
Captain,	0	15	6	0	11	6	0	12	4	0	9	2
Lieutenant,	0	9	0	0	7	6	0	6	2	0	4	9
Cornet,	0	8	0	0	6	6	0	5	2	0	3	10
Quartermaster,	0	5	6	0	4	6	0	3	0	0	2	9
Chaplain,	0	6	8	0	5	6	0	6	8	0	5	0
Adjutant,	0	5	0	0	4	6		(?)			(?)	
Surgeon,	0	6	0	0	4	6	0	4	0	0	3	0
Kettledrummer,	0	2	3	0	1	9	0	1	6	0	1	0
Sergeant,	0	2	9	0	2	3	0	2	6	0	1	7
Corporal,	0	2	3	0	1	9	0	1	8	0	1	$1\frac{9}{10}$
Hautbois,	0	2	0	0	1	6	0	1	6	0	1	0
Private Man,	0	1	7	0	1	5	0	1	4	0	0	11
Agent, each Troop,	0	1	2				0	3	$6\frac{3}{4}$			
Widows,	0	0	2									

ALLOWANCE PER DIEM OVER THE PAY.

	Dragoons *s.*	*d.*
Colonel, Lieutenant-Colonel, Major and Captain, recruiting each, (When full, by Warrant of 14th August 1747).	2	4
Colonel, for clothing lost by deserters,	15	0
Each troop of 6 troops,	2	6

1755

The regiment remained in Dublin till May. Colonel Dejean was commanding the cavalry in Ireland in the early part of the year; and the Monthly State signed by him in January shows that only two officers of the 14th Dragoons were absent on leave, viz. Lieutenant E. Vernon and Chaplain P. Vatass, and there were 'fit for duty' 130 rank and file, with 143 troop-horses. There were 14 men and 14 horses wanting to complete. Quartermaster Philip Mercier was appointed Adjutant.

In May and June the regiment left Dublin: 2 troops proceeded to Ballyshannon, 1 to Navan, which went later to Thurles Camp, and 3 troops to Belturbet. Of these latter 1 troop went subsequently to Thurles Camp and 1 to Ballyshannon. Lieutenant-Colonel Erle's troop was at Belturbet, one man of which, Private Ralph West, was 'whipt out.'

Two new standards were issued to the regiment.

1756

Some new swords were issued to the regiment.

Captain Marcus Norman became Major, 2nd January.

The regiment moved to Loughrea, and sent a troop to a camp formed near Kilkenny in August.

Colonel Dejean was promoted Major-General.

1757

The regiment was reviewed at Loughrea and proceeded afterwards to Clonmel. Captain Arabin's troop was at Tallow, the Colonel's troop at Portumna, Lieutenant-Colonel Erle's troop at Loughrea.

On 5th April Colonel John Campbell, afterwards Marquis of Lorne, from the 54th Regiment, succeeded Major-General Dejean as Colonel of the regiment. The latter went to the 3rd Irish Horse, now 6th Dragoon Guards (Carabiniers).

Colonel—John Campbell.

Jeremiah Hurley became Adjutant.

1758

The 6 troops, marching by squadrons, arrived at Athlone from Clonmel by 6th June, detaching 1 troop to Granard and 1 to Roscommon.

Captain Robert Brown's troop was at Athlone, also Major Norman's, Lieutenant-Colonel Erle's, and Captain Ford's.

Four quartermasters of the regiment were sent on command during the summer to Granard, Clonmel, Ross, and Roscommon.

1759

Colonel Campbell was promoted Major-General.

The whole regiment was concentrated at Athlone, and marched in June by squadrons to Clonmel. It was encamped near Fethard, and reviewed there with 3 other regiments of cavalry, subsequently proceeding to quarters in Clonmel, Doneraile, and Ross.

Captain J. Ford became Major, 17th September. At this period it appears from the 'Monthly States of Cavalry in Ireland' that there were 10 regiments of cavalry there, 4 horse and 6 dragoons, the totals being as follows:—

Total Horse {Men, 504. Horses, 580. Total Dragoons {Men, 1209. Horses, 1287.

In 'Campbell's Dragoons' in the Monthly States it is reported (November and December 1759) 'Chaplain Vatass ill of a paralytic disorder and unable to attend his duty.' Signed Thomas Erle, Lieutenant-Colonel.

An order dated 5th May was issued giving route for Major-General Campbell's Dragoons to march from Doneraile and Ross to Clonmel, there to be reviewed and to return to same quarters after the review.

Another order was issued on 5th June 1759 for Campbell's Dragoons to march from Athlone (6 troops) to Clonmel, there to encamp. The troops marched by squadrons at a time, the 1st squadron on 15th June, making five days' march and one halt, viz.—

Birr, June 15th; Roscrea, June 17th; Thurles, June 18th; Cashel, June 19th; Clonmel, June 20th.

The 2nd and 3rd squadrons followed on 16th and 18th June.

1760

George III., 1760.

George III.

In June the 2 troops at Doneraile and 1 at Ross joined headquarters at Clonmel, where the review took place, and the 3 detached troops then returned to former stations. In

December, headquarters and 2 troops were moved to Kilkenny for the winter, 2 troops moving to Ross, 1 troop to Carrick-on-Suir, and 1 troop (the one hitherto at Ross) moving to Cullen.

It appears by the Monthly States and Muster Rolls of Cavalry in Ireland, that in January, 11 men were drafted from Campbell's Dragoons (14th) to the Light Cavalry and the Inniskilling Dragoons in Germany, and 1 sergeant was sent 'on command' with them. In February, or earlier, 12 men were drafted towards the augmentation of the 3rd and 4th Regiments of Horse. In April, Lieutenant-Colonel Erle's troop was at Clonmel; Major J. Ford's and Captain J. Mayne's at Doneraile; Captain A. Molesworth's and Captain R. Howard's at Clonmel, and Captain Pepper's at Ross. In May, 'Adjutant Jeremiah Hurley preferred in the Carabiniers,' and in August, Adjutant Hurley 'on command in Germany.' In October, Lieutenant-Colonel Erle's and Captain A. Molesworth's troops were at Kilkenny, Captain T. Pepper's troop at Cullen, and Captain Pearce's at Ross.

In the same month there were 'present and fit for duty' 1 Lieutenant-Colonel, 1 Captain, 2 Cornets, 1 Adjutant, 4 Quartermasters, 6 Sergeants, 4 Trumpeters, 130 rank and file, and 177 horses. There were 17 officers 'on licence' (leave), and 1 employed recruiting.

1761

Major-General Campbell became Lieutenant-General in January, and in April he became Marquis of Lorne on the decease of his uncle, Archibald, third Duke of Argyll. In June the whole regiment came to Kilkenny for review, and then moved to Tullamore 3 troops, and Phillipstown 3 troops.

1762

The regiment was (probably) quartered in Dublin during the early part of this year, and moved later to Sligo. Lieutenant-General the Marquis of Lorne was appointed Commander-in-Chief in Scotland, and Lieutenant-Colonel Erle was promoted Colonel by brevet.

1763

Most of the regiment was at Sligo. Two men of Captain Butler's troop were imprisoned in Sligo Jail. On 26th May, Captain A. Molesworth became Major *vice* J. Ford. In June (probably) the headquarters and 3 troops moved to Castlebar, and 1 troop each to Roscommon, Athlone, and Headford.

1764

In July 3 troops were at Gort, and 3 troops at Loughrea, Headford, and Portumna. The 3 troops at Gort were the Marquis of Lorne's, Lieutenant-Colonel Erle's, and Captain Pepper's.

1765

The regiment was stationed at Tallow, Cappoquin, and Doneraile in July. Major Molesworth's and Captain Pepper's troops were at Cappoquin, Captain Butler's at Tallow, the Marquis of Lorne's at Doneraile, and Captain Howard's at Tallow.

Lieutenant-General the Marquis of Lorne went to the 1st Royal Regiment of Foot, and was succeeded as Colonel of the 14th Dragoons on 11th September by Colonel Charles Fitzroy (afterwards Lord Southampton), who was only twenty-seven years of age at this date.

Colonel—Chas. Fitzroy.

1766

The regiment was at Tullamore and Phillipstown. Captain Howard's and Captain Pepper's troops were at Tullamore, and Captain Grice Blakeney's troop was at Phillipstown. Chaplain Vatass was absent on leave, and Deputy Gethin Crowe performed his duties in his absence. Cornet Arthur Ormsby became Adjutant.

1767

This year the regiment, now called Fitzroy's Dragoons, was stationed in Dublin. The commanding officer, Lieutenant-Colonel Thomas Erle, was now sixty years of age.

There were six troops at this time, and the officers comprised—

1 Colonel.
1 Lieutenant-Colonel.
1 Major.
3 Captains.
1 Captain-Lieutenant.
5 Lieutenants.
6 Cornets, including the Adjutant.

The total strength was 155 officers and men, 119 troop-horses.

There were 6 men and 25 horses wanting to complete the establishment.

There was an officiating chaplain, a surgeon, and the Adjutant (Cornet A. Ormsby) present at the inspection in Dublin, which was made by General O'Brien D. Dilkes on 24th April. The regiment received a new supply of firelocks, bayonets, swords, and six trumpets this year.

New firelocks, bayonets, etc., issued.

In November it moved to Athlone.

The cost of horses for dragoon regiments was raised from 20 to 22 guineas this year.

Price of troop-horses, 1767.

1768

On 23rd May, at Athlone, Lieutenant-General R. Armiger inspected the regiment.

A new set of buff and small leather accoutrements was issued this year, as well as housings and caps.

The uniform, by royal warrant of George III., dated 19th December 1768, still remained scarlet (as prescribed in 1751), with lemon-coloured facings, without lapels, with silver lace, lemon-coloured cloth waistcoats and breeches, silver lace, the button-holes of the waistcoat laced with silver to the bottom; a silver epaulette, buttons not numbered, light boots, silver spurs,

Description of uniform worn.

[1] The officer's coat was made with a slash sleeve and pocket, turned up with lemon-coloured cloth and six button-holes, laced with silver, the cuffs round and cut obliquely.

lemon-coloured serge lining to coat and vest. Trumpeters rode grey horses, men wore black spatter-dashes, silver-laced hat, button and loop; hats were worn well cocked. The General in his review report remarked that the horses were in good condition, but 'not very nimble,' and '39 with long tails.'

The 132 pairs pistols in possession of the regiment were reported as 'bad.'

1769

Major-General the Earl of Drogheda reviewed the regiment at Athlone on 2nd June. Nearly the whole of the men were Irish, out of 136 only two being English. Three of the men had thirty-five years' service, and 7 had twenty-five years' service; 37 men had from eight to ten years, and 51 from five to one year.

The Rev. Elias Handcock officiated as chaplain at Athlone in the absence of Mr. Peter Vatass, who appears by the returns to have been on commanding officer's leave since his appointment to the regiment on 24th December 1745!

Captain-Lieutenant John Mayne died on 25th April.

The inspection report says: 'The horses are nimble and in good condition, 61 with long tails out of 112 troopers on parade.

'One troop is at Nenagh.

'This is a good regiment, and fit for service.

(Signed) 'DROGHEDA, *Major-General.*'

In June the regiment moved to Clonmel.

1770

In February the regiment moved to Birr. It was reviewed there by Major-General Ham Lambert on 30th May. There were 80 horses with long tails.

Detachments at Thurles and Clonmel. The Rev. A. Downey was officiating chaplain. The review report says: 'A very good regiment, and very fit for service.

(Signed) 'HAM LAMBERT, *Major-General.*'

On 30th April, Colonel Thomas Erle was promoted Major-General at the age of sixty-three years, but still remained the Lieutenant-Colonel of the 14th Dragoons.

1771

Captain Thomas Pepper became Major on 14th March *vice* J. Maine.

The regiment was reviewed at Birr on 3rd June by Major-General Peirson.

Detachments were at Nenagh and Donaghmore. Horses reported as being 'nimble' and 'in exceeding fine order and condition.' The Rev. John Croasdale was deputy-chaplain.

After the review, headquarters moved to Tullamore till December, and then went to Dublin. There were 95 long-tailed horses in the regiment. Six new trumpets were issued this year.

1772

In April, 4 troops were at Dublin, 1 at Lurgan, and 1 at Belfast.

The regiment was reviewed at Dublin by Lieutenant-General M. O'Brien Dilkes on the 1st June, and shortly afterwards proceeded to Athlone and outquarters. The Rev. Mr. Hodgkinson was officiating as chaplain for Mr. Vatass.

On 30th October, Colonel Fitzroy, having been promoted to Major-General, was appointed Colonel of the 3rd Dragoons, and Lieutenant-General Daniel Webb from the 8th Foot became Colonel of the 14th Dragoons. Colonel—Daniel Webb.

1773

The regiment remained at Athlone, Navan, and Roscommon. Lieutenant-General Webb died, and was succeeded as Colonel of the 14th Dragoons, on the 11th November, by Colonel George Warde, from the Lieutenant-Colonelcy of the 4th Dragoons. Colonel—George Warde.

Lieutenant-Colonel—John Burgoyne.

Lieutenant-Colonel J. Burgoyne, from the 58th Regiment, became Lieutenant-Colonel, 15th July, *vice* Major-General Thomas Erle.[1] Lieutenant-Colonel Burgoyne had held the rank of Lieutenant-Colonel since 19th December 1762.

Review report of Lord Blayney, 5th June 1773.

The regiment was reviewed at Athlone on 5th June by Lieutenant-General Lord Blayney.

Observations on His Majesty's 14th Regiment of Dragoons.

'Officers made a good appearance: saluted well, well mounted, uniform agreeable to the King's regulations.

Non-commissioned officers steady under arms, but indifferently mounted.

Men clean under arms, hats well cocked, but not in size equal to the King's order.

Arms in very good order.

Complaints, none.

Recruits, good, well certified for; 18 enlisted since last review.

Stock purse account settled to 24th June 1772.

Orderly-books well kept.

Manual exercise performed by word of command, and pretty well.

Trumpeters pretty well mounted and well taught.

Horses in pretty good order. Long-tailed horses, 120 out of a total of 138; 13 purchased since last review.

Clothing well fitted and agreeable to the King's order.

Accounts and horse furniture agreeable to the King's order.

Movements and Manœuvres executed at Inspection.

'Received the General, marched by in squadrons, marched by troops and passed the General, dismount, marched by the General by troops, manual exercise, firings nine rounds, usual

[1] Major-General T. Erle, who was now sixty-six years of age, had held the Lieutenant-Colonelcy of the 14th Dragoons since 1754—nearly nineteen years.

evolutions. Squadrons took ground to the left, wheeled the line to the right, etc. etc.

Part of the regiment is at Navan and Roscommon.

I reviewed this regiment at Athlone on 5th June 1773; found them in pretty good order, the officers well mounted, and the quartermasters but very indifferent. The horses were very active, but the men in general ride too long. None of the hats of the cavalry were agreeable to the King's order in size. (Signed) BLAYNEY, *Lieutenant-General.*'

The regiment moved in June to Birr. The chaplain, Peter Vatass, was still absent on leave, and the Rev. Dean Handcock was the deputy-chaplain.

1774

Reviewed at Birr by Major-General the Earl of Drogheda.

Detachments at Gort and Portumna.

George Borthwick appointed Surgeon *vice* Irving, 20th October.

The regiment moved on the 31st December, 2 troops to Thurles, 2 troops to Tipperary, and 2 troops to Charleville.

1775

This year hostilities between Great Britain and her North American colonies commenced, and it was found that that country was not well adapted for the operations of heavy cavalry; hence the necessity arose for a larger number of light cavalry regiments. During the first part of the year the 14th Dragoons remained at Tipperary, 3 troops and head-quarters, 2 troops at Thurles, 1 troop at Charleville.

In June the regiment was in Queen's County, 3 troops at Maryborough, and 3 at Mountmellick. Inspected by General Cunninghame at Tipperary on 5th June. The buttons on the uniform were numbered regimentally.

1776

14th Light Dragoons.

The 14th Dragoons were now constituted a corps of Light Dragoons in Ireland: the standard height for men and horses was reduced. Cocked hats were replaced by helmets; arms and appointments of a lighter pattern were adopted, and the regiment was designated the '14th Light Dragoons.'

Captain Grice Blakeney was promoted Major, 16th March, *vice* Thomas Pepper.

The regiment remained at Maryborough and Mountmellick for the early part of the year, and on 1st July there were 2 troops and headquarters at Tullamore, and 2 troops at Phillipstown.

Two new standards, and new saddles and bridles were issued this year.

1777

Colonel Warde was promoted Major-General, and Lieutenant-Colonel Burgoyne received the brevet rank of Colonel on the 29th of August.

Regiment remained at Tullamore and Phillipstown.

Reviewed at Tullamore on the 12th of June by Lieutenant-General James Johnston, who reported that the movements, evolutions, firings, and manœuvres, were well performed.

The Lieutenant-Colonel, J. Burgoyne, was thirty-seven years of age; the Major, Grice Blakeney, was thirty-seven; the Adjutant, J. Wilson, was thirty-six. The Reverend Quintin Finlay officiated as Chaplain in the continued absence of Mr. Vatass, on leave.

New pistols were issued this year, also some new swords, carbines, and bayonets.

1778

The regiment remained at Tullamore till August.

Colonel—Robert Sloper.

Major-General Warde went to the first Irish Horse (now 4th Dragoon Guards), and Major-General Robert Sloper was appointed Colonel of the 14th Light Dragoons on the 2nd April.

The regiment was reviewed at a camp near Clonmel by Major-General Joseph Gabbett on 28th September, having moved there from Tullamore in August.

The strength of the regiment was 199 men, 137 troop-horses. Everything had been recently issued new to the Fourteenth on becoming 'Light Dragoons.' They moved to Carlow for the winter in the month of November.

1779

Colonel J. Burgoyne became Colonel Sir J. Burgoyne, Bart.

Regiment went into camp near Carrigaline for August and September.

Reviewed at Carrigaline Camp on 27th September by Major-General Mocher.

The Fourteenth consisted of 6 troops of 34 non-commissioned officers and men, 23 troop-horses, and 3 officers per troop.

Height of troop-horses.

The troop-horses averaged from 14 hands 3 inches to 15 hands 3 inches in height.

In October they moved to Bandon.

1780

Five troops and headquarters remained at Bandon; 1 troop was at Innishannon.

1781

Lieutenant-Colonel—Grice Blakeney.

Major G. Blakeney became Lieutenant-Colonel *vice* Sir J. Burgoyne, Bart., appointed Colonel of the 23rd Light Dragoons, and Andrew Corbet became Major, 19th November.

The regiment remained at Bandon and Innishannon for the whole of this year.

1782

On 30th June the following change of quarters took place: 3 troops to Kilkenny, 3 troops to Carrick-on-Suir; later 1 troop went from Carrick-on-Suir to Ballyragget.

At this period the Muster-Rolls, one for each troop, were most accurately kept, and were rendered quarterly on 15th

April, 15th July, 15th October, and 15th January, for the quarter ending with previous month in each case. All officers present signed the docket, and the commanding officer at each station, and for each troop, signed this certificate: 'I certify that the several men mustered, on furlo', or sick, are effectives.' The names of all officers, sergeants, corporals, trumpeters, hautboys, and privates, were entered on the Rolls. The Muster-Rolls had to be sworn-to in presence of a magistrate, and so stated on reverse, with signatures attached of the Commissary of the Musters and the magistrate.

On 20th November, Major-General R. Sloper was promoted to the rank of Lieutenant-General.

1783

William Richardson was appointed Major *vice* A. Corbet.

In the spring the regiment moved from Kilkenny, etc., and was stationed at Clonmel, 3 troops, Clogheen 2 troops, and Cappoquin 1 troop.

1784

Clothing blue, facings lemon-yellow as before.

On 18th day of May, by an order from the Adjutant-General's office, Dublin, the clothing of Light Dragoons was changed from scarlet to dark blue, as follows:—

Regulations for the clothing of the Light Dragoons.—The clothing of a private light dragoon is to consist of a jacket, shell, under-waistcoat, and leather breeches. The jacket and shell to be of blue cloth; the collars and cuffs of the royal regiments to be red, and those of the other regiments to be of the colour of the facings of the regiment, looped upon the breast, edged with white cord, and to be lined with white, the 11th and 13th regiments excepted, which are to be lined with buff. The under waistcoat to be of flannel with sleeves, and made so as to be buttoned within the waistband of the breeches. The breeches to be of buckskin.

N.B.—The make of the dress and method of placing the cord upon the breast of the jacket to be exactly conformable to the pattern approved by His Majesty.

Officers and Quartermasters.—The dress-uniform of the officers and quartermasters of the Light Dragoons to be made according to the king's regulations of the 19th December 1768, excepting that the coats are to be blue and faced with the same colour as the private men, and that the royal regiments are to be faced with scarlet.

Field uniform of the Officers and Quartermasters.—The jacket and shell to be made up in the same manner as those of the men, excepting that the shell is to have sleeves, and that the looping is to be made of *silver*, the 13th regiment excepted, which is to be of gold.

Sergeants.—The sergeants of the Light Dragoons to be distinguished by gold or silver looping.

Corporals.—The corporals of the Light Dragoons to be distinguished by a gold or silver cord round the collar and cuffs.

Trumpeters.—The trumpeters to have a jacket and shell, the colour and facing of the regiment, with lace instead of looping in front and down the seams.

N.B.—A pattern suit may be seen at the Commander-in-Chief's office at the Royal Hospital, Dublin.

(Signed) F. PIGOT, *Adjutant-General.*

The foregoing orders were sent to the officers commanding the 8th, 12th, 13th, 14th, 17th, and 18th Light Dragoons, to the Earl of Drogheda, to the Major of Brigade for the General Officers, and to the agents, Messrs. Montgomery Wybrants and Cane.

The regiment was reviewed at Clonmel on the 1st June by Major-General Sir Henry Calder, who inspected the regimental standing orders and reported: 'I am informed each troop is furnished with a copy of them.' This is the first mention of any 'standing orders.' At this date there were six troops: one commanded by the Colonel, one by the Lieutenant-Colonel, one by the Major, and three by Captains.

'Standing Orders' existed in the Regiment.

After the review the regiment moved to Loughrea.

1785

Reviewed at Loughrea on 7th June by Major-General Lord Luttrell, who reports: 'The system of discipline and good order originally established by Colonel Sir John Burgoyne is well preserved by Colonel Blakeney and a very active adjutant.—(Signed) LUTTRELL.'

The adjutant was J. M'Gee, appointed 12th September 1780. After the inspection the regiment moved to Athlone.

1786

The regiment remained at Athlone till June, then moved to Kilkenny and out-stations. New sergeants' sashes were issued. Captain Arthur Carter was appointed Major 10th April (*vice* W. Richardson), from 4th Regiment of Horse. Inspected by Major-General Lord Luttrell on 4th June, who reports: 'They have a code of regimental orders, and each officer is furnished with a copy. This regiment has ever been kept in perfect order and fit for service.—(Signed) LUTTRELL.'

Light Dragoons were ordered to have blue cloaks instead of red ones.

1787

The regiment was stationed at Kilkenny.

Reviewed on 5th June by Major-General the Earl of Carhampton. The men wore black spatter-dashes and helmets. The standards in use were issued in 1773.

Lord Carhampton reports: 'This regiment shows a constant and uniform attention of the officers to its appearance and discipline.'

In June the regiment moved to Nenagh and out-quarters.

1788

The regiment moved to the Curragh of Kildare, where Major-General Paterson reviewed it on 4th June. He reports: 'They have a code of regimental orders, and each officer is supplied with a copy.'

The chaplain, Mr. Peter Vatass, who has been on leave since his appointment to the regiment on 26th December 1745, is still shown as 'absent on leave,' he being at this date 66 years of age.

The zealous adjutant, J. M'Gee, being now 50 years of age, is reported as being 'sick at Carlow.'

New pouches and belts were issued.

In June the regiment moved to Carlow and out-stations.

On the 6th June, Lieutenant-General R. Sloper was awarded the dignity of Knight Companion of the Order of the Bath.

1789

At Carlow. Inspected on 6th June by Major-General Prescott, who states: 'A copy of the standing code is in possession of each officer. The regiment is well disciplined, in perfect good order, and fit for service.'

1790

At Carlow. Reviewed by Major-General C. W. Lyon on 26th June. The commanding officer certifies that each officer has a copy of the regimental 'Standing Orders.'

New standard-belts, waist-belts, housings, and caps were issued, also two new standards. One hundred and twenty-six carbines, bayonets, and pairs of pistols were reported in bad order, and new ones were issued. One hundred and thirty-eight swords in possession all reported good.

Lieutenant-Colonel Grice Blakeney was promoted to the rank of Colonel.

In July the regiment moved to Dublin.

1791

The regiment was reviewed in the Phœnix Park, Dublin, on 28th May, by Major-General R. Whyte.

The 138 swords in possession of regiment (issued in 1776) were reported bad, and new ones were supplied.

The complement for the regiment was :—

Swords	138
Firelocks and carbines .	
Bayonets	
Pairs of pistols . . .	
Trumpets . . .	6

This was the time of the French Revolution, and the spirit of republicanism had spread to the French West Indian Islands, where the blacks rose against the planters, committing acts of outrage and spoliation.

1792

The regiment was reviewed in the Phœnix Park, Dublin, on 15th May, by Major-General Richard Whyte. The regimental code was still in use, and each officer had a copy. The chaplain was still on leave, being 70 years of age, and the Reverend Mr. Devereux officiated for him.

Seventeen privates were drafted to Jamaica. In June the regiment moved to Kilkenny, thence in July to Tallow, County Waterford, till November, and thence to Cappoquin in December.

1793

A British army under the Duke of York went to Flanders to act against the French, and British aid was also given to the planters in St. Domingo (Hayti). During this year the regiment seems to have been frequently on the move, and quartered in turn at Tallow, Cappoquin, Clogheen, and Kilkenny, sending other troops on detachment elsewhere.

1794

In January the headquarters of the regiment were at Clogheen. In March a move was made to Bandon and out-stations, and in September to Clonmel.

Major Arthur Carter received the brevet rank of Lieutenant-

Colonel on 1st March, and Colonel Grice Blakeney became Major-General on 3rd October.

Two troops were sent from Ireland to Flanders, and were there attached to the 8th Light Dragoons on arrival. In the muster-roll of these two troops from June to December 1794, they are shown as being in Major-General Vyse's brigade of the army in Flanders. They formed part of the van of the forces under Lieutenant-General the Earl of Moira, which proceeded on the march from Ostend to join the army under His Royal Highness the Duke of York.

The two troops wintered in Holland, taking part in several encounters with the enemy, and after an unusually severe season and a hard campaign, in which several men and six horses were lost, came to Germany early in the following year (1795), and became incorporated with the 8th Regiment of Light Dragoons.

Two troops serve in Flanders and Holland under the Duke of York.

These troops saw a good deal of service. It was on 15th September 1794 that Lieutenant-General Abercromby (under the orders of the Duke of York) marched with the reserve to try and regain the lost position of Boxtell in Brabant, near the river Dommel and the town of Berlicon. He found the enemy so strong that he was obliged to retire, and consequently the Duke of York had to fall back with his army across the Meuse,[1] taking up a position near the old lines of Velpen, three miles in front of Grave. Next day, having crossed the Meuse, he encamped at Wichen, seven miles distant. In the engagement at Boxtell nearly 1500 men were lost, mostly German troops. The squadron of the 14th Light Dragoons lost 2 men who were reported 'missing.' The battle of Boxtell took place on the 14th September 1794: it was undoubtedly a victory of the French over the allied British and Dutch forces under the Duke of York.

Battle of Boxtell, 14th September 1794.

On 6th December 1794 the Duke of York left the army in Holland, and the command of the allied army was taken over by General Walmoden, and of the British by Lieutenant-

[1] Or 'Maas.'

General Abercromby. In the month of December the troops suffered intensely from the inclemency of the weather during the time they were at Grave on the Meuse, near Nimeguen, in Dutch Brabant. The ice was so strong that cavalry and heavy cannon could cross the rivers. The winter clothing for the men came from Arnheim on the river Rhine. In the action of Tuyl, fought on 30th December, the allies were victorious and drove the enemy across the Waal River. There were no casualties among the cavalry, but the 19th, 33rd, 42nd, 78th, and 80th Regiments, as well as the 'Loyal Emigrants,' lost between them 2 officers and 24 men.

Action of Tuyl, 30th December 1794.

1795

Battle of Geldermalsen, 5th January 1795.

On 5th January, at Geldermalsen, where the enemy was defeated with a loss of 200 men, our losses were 4 officers, 64 men, and 11 horses, which included 7 men and 3 horses of the 11th Light Dragoons; but it does not appear that any losses were incurred by the 14th Light Dragoons or 8th Light Dragoons on this occasion, as the cavalry was mostly in reserve.

Battle of Bueren on Lingen, 8th January 1795.

On the 8th January, at a place near Bueren, on the river Lingen, a battle took place. The British and allies were opposed to very superior forces of the enemy, but fought with the greatest gallantry, and lost 3 officers and 18 men killed, 8 officers, 113 men, and 3 horses wounded. No casualties occurred among the cavalry. On this occasion Major-General Lord Cathcart was in command. The 27th and 28th Regiments suffered heavily, and we had to retire across the Lingen to Elst.

Affair near Elst, 10th January 1795.

Another engagement took place here on 10th January, when the British lost several officers and men. Lieutenant-General Abercromby, who was marching on Echlade, suddenly found himself threatened by the French both on his left flank and in rear, the Hanoverians and Austrians being also hard pressed by bodies of the enemy. In this state of affairs

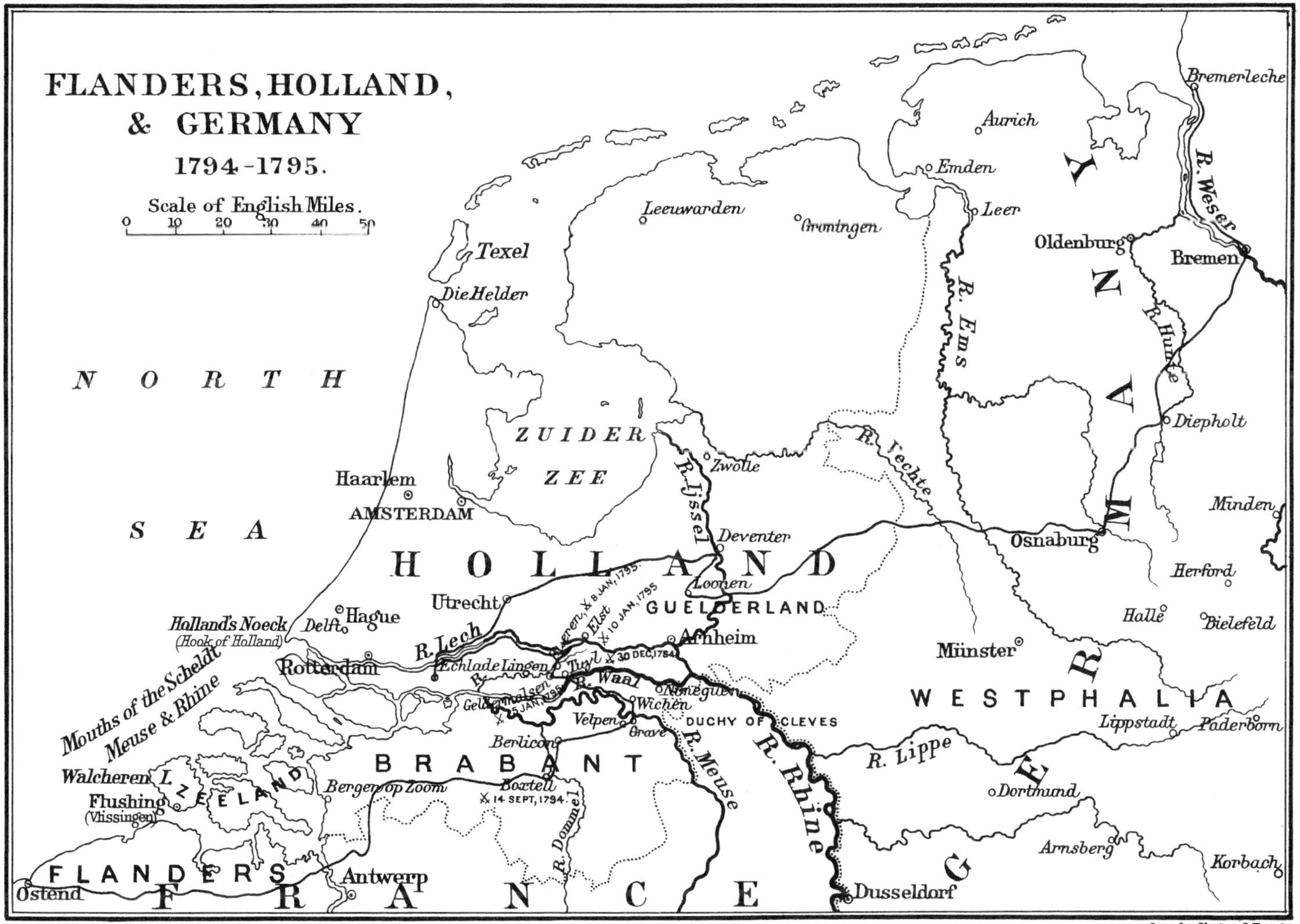

London: Longmans, Green & Co.

Lieutenant-General Abercromby retreated across the Lech, and eventually the whole British army had to retire into Westphalia and Germany, suffering great hardships and privations on their march. They reached Deventer on 14th January, thence proceeding to Loonen and other places in Guelderland, and by the month of March General Abercromby had established the headquarters of the British troops at Osnaburg, and later at Diephelt in Westphalia. Towards the end of March the British marched to Bremen on Weser, and on the 14th April 1795 a large force embarked at Bremerleche, leaving for a time Major-General D. Dundas and Lord Cathcart with a detachment of artillery and the whole of the British cavalry, whilst the remainder sailed for England on the 24th April. In this campaign the superior forces of the enemy completely overpowered the British and their allies.[1]

Retreat into Westphalia and Germany.

The two troops of the 14th Light Dragoons which took part in these operations were by this time incorporated with the 8th Light Dragoons, now the 8th King's Royal Irish Hussars, and so we take leave of them to return to the main body of the regiment which we left in Ireland in County Tipperary.

Two troops of 14th Light Dragoons incorporated with 8th Light Dragoons, April 1795, in Germany.

For some years the spirit of republicanism had run riot in the French West Indian Islands, and numerous acts of outrage and spoliation had been committed by the blacks who had hitherto been slaves, against the properties of their owners. In 1793 the planters of St. Domingo obtained British aid, whilst the revolutionists afterwards received assistance from France. At this period the 7 troops of the Fourteenth at Clonmel were ordered out to St. Domingo. The establishment at this time was 450 rank and file, with 486 troop-horses.

The 7 troops gave up their horses to the 24th Light Dragoons at Clonmel, and in June embarked without horses at Waterford under the superintendence of Major-General John William Egerton, whence they sailed to Bristol, taking with them their appointments and everything ready for service.

[1] The above is taken from Jones's *Historical Journal of the British Campaign on the Continent with the retreat through Holland*, 1794-95 (published 1797).

Regiment proceeds to St. Domingo, June 1795.

They subsequently proceeded to Hayti or St. Domingo, one of the Leeward Islands in the Atlantic, for service there under Brevet Lieutenant-Colonel A. Carter. At Bristol the regiment met the 13th Light Dragoons destined for the same service as themselves, also the 30th Light Dragoons under Colonel Sir John Carden, as well as the 32nd Light Dragoons under Colonel Blake. The two latter regiments were mounted, and had come from Ireland to do duty in England. Those regiments destined for active service, after landing at Bristol were stationed in billets in the adjacent towns, till the vessels which were to convey them to the West Indies were ready to receive them.

On arrival at St. Domingo the regiment was supplied with horses from America, and was soon engaged in active operations against the bands of armed negroes and mulattoes who had enrolled themselves under the banners of the French Republic.

During the years 1795, 1796, and 1797, numerous actions occurred, but against a hundred thousand trained blacks who had been instructed in European discipline, the few British troops on the island were unable to do more than exhibit many examples of discipline and valour.[1]

1796

There was a small party of the regiment stationed at Marlborough in England, and a depôt at Maidstone, and the establishments of men and horses were largely augmented, but the effectives were very far below these numbers, owing to the casualties on service and want of recruits.

The country of Hayti or St. Domingo was close and hilly, and therefore quite unsuited to cavalry operations; the climate, moreover, was unhealthy in the extreme, and after a time yellow fever broke out, and made fearful ravages amongst the British forces. There were seven cavalry regiments serving in

[1] Cannon's *Record*, pp. 16-17.

the expedition, viz. the 13th, 14th, 17th, 18th, 21st, 26th, and 29th Light Dragoons, and these suffered most terribly from the fatal scourge. The deaths were so numerous and rapid that regiments could not bury their own dead, and it is recorded that the 13th Light Dragoons had to obtain help in this duty from the men of the 56th Regiment serving near them. A vivid picture of the state of affairs is given in the early pages of Gleig's *Hussar*, published in 1837.[1]

There is an old parchment muster-roll of the 14th Light Dragoons now preserved in the Public Record Office, Chancery Lane, London, dated May to December 1795, in which Major Arthur Carter and four privates are returned as 'Prisoners of War,' a few men are returned as 'at Halstead,' and others as 'in England.' This muster-roll was sworn to by Major (Brevet Lieutenant-Colonel) Sir J. Dunbar, Bart., on the 26th April 1798, at Marylebone, before a justice of the peace. In the old 'Monthly Returns,' also preserved in the Public Record Office, the 14th Light Dragoons on 1st July 1795 are accounted for as follows:—

'Seven Troops at Marlborough.

'Two Troops on foreign service. (108 Horses).'

and the nine Troops are thus designated:—

The Colonel's (Lieutenant-General R. Sloper's).

The Lieutenant-Colonel's (Major-General Grice Blakeney's).

The Major's (Brevet Lieutenant-Colonel Arthur Carter's on foreign service).

Captain the Honourable James Butler's (on leave).

Captain N. Hutchinson's.

Captain Hamilton Gorge's.

Captain Henry Browne's.

Captain G. H. Montgomery's.

——— vacant (afterwards Captain J. Kearnay's).

Probably this return had reference to the period just before

[1] *The Hussar*, by the author of *The Subaltern*. London: Henry Colburn, 1837.

the seven troops from Ireland embarked in the south of England for the West Indies, and when they were on their way to the coast. It is probable Lieutenant-Colonel Carter with some other officers and part of a troop proceeded early in the year 1795 to St. Domingo, and was followed by the remainder of the regiment in July or later. Colonel Blakeney having been promoted Major-General before the regiment embarked, was practically succeeded in command by Lieutenant-Colonel Carter, the senior Major, and Captain Sir J. Dunbar, Bart., had been promoted Major on augmentation, on 1st September 1795.

On the 3rd May, Lieutenant-General Sir Robert Sloper became General.

1797

Affair of Le Mirebalais, St. Domingo, 2nd June 1797.

Cannon relates that in an enterprise against the post of 'Le Mirebalais,' a town to the north of Port-au-Prince on the west coast of St. Domingo, a detachment of the 14th, 18th, and 21st Light Dragoons, commanded by Brevet Lieutenant-Colonel Arthur Carter of the 14th Light Dragoons, distinguished itself, and was commended in the public despatches.

The bulletins of 1797-98 contain the following relative to the affair at Mirebalais:—

Despatch of Brigadier-General Churchill to Lieutenant-General George Simcox, commanding His Majesty's troops in the Isle of St. Domingo, dated Mirebalais, June 2nd, 1797.

'We were enabled to drive the enemy from a very advantageous position they had taken, which, from their superiority of numbers (about 1200 men), with three pieces of cannon, must in all probability have cost us a number of valuable lives to have carried; but this additional strength[1] gave us an easy victory, for no sooner did they perceive a detachment of infantry and cavalry which I sent to guard the heights and

[1] This refers to the junction of a column under Colonel Dessources at Port Michell on the previous day.

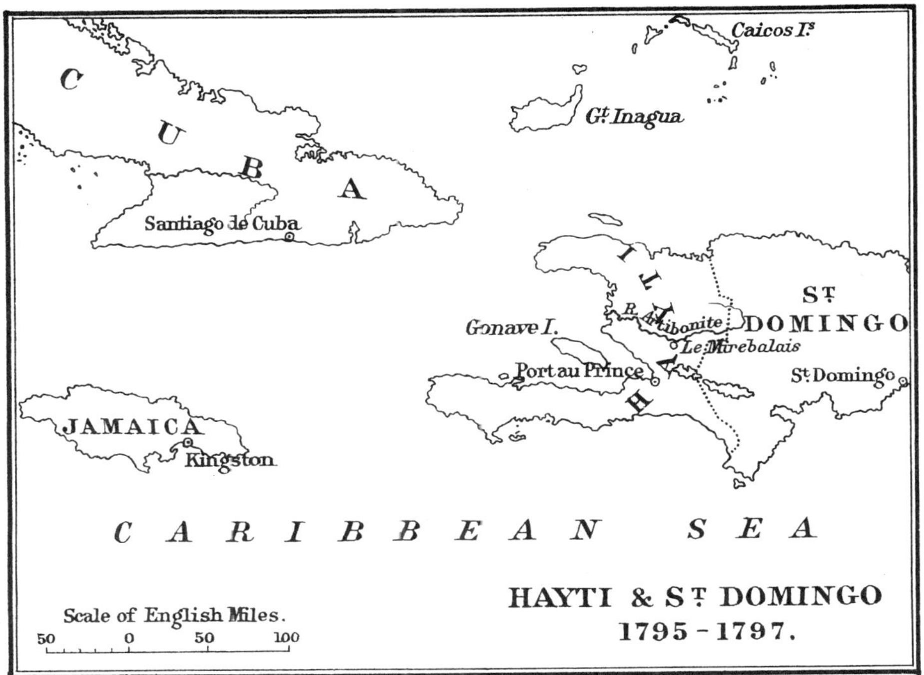

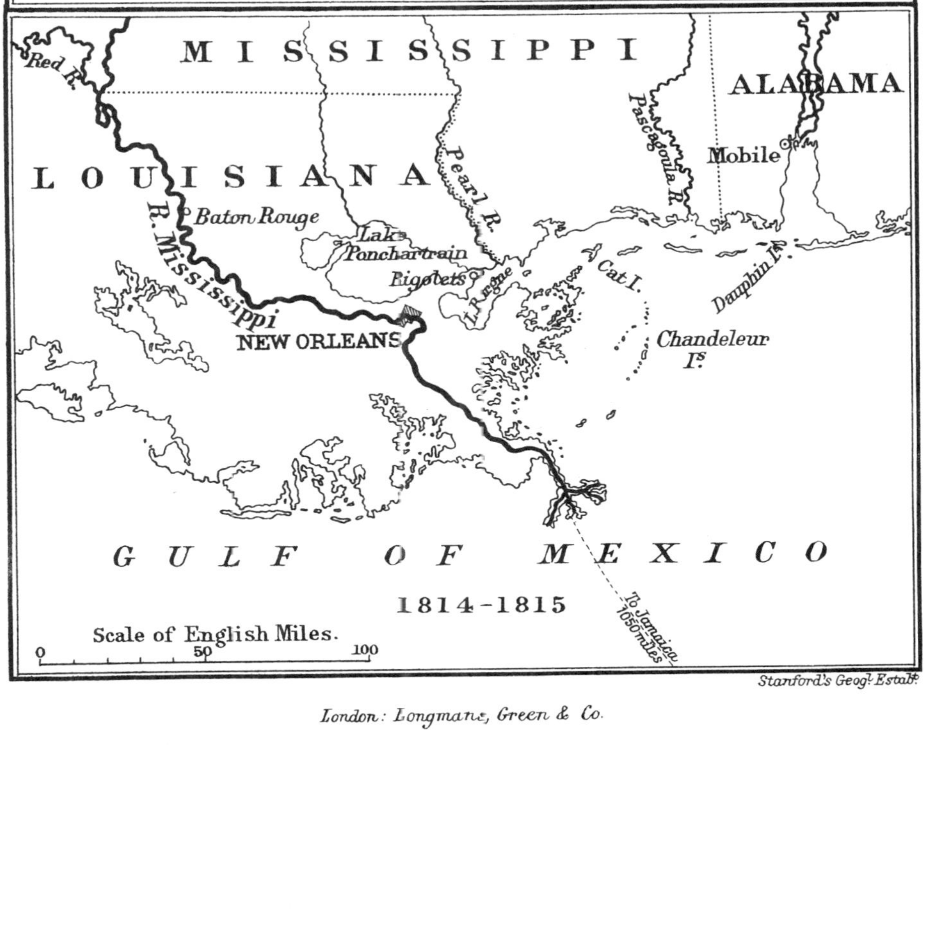

London: Longmans, Green & Co.

turn their right flank, than they immediately fled in the utmost confusion, and with such precipitation, that though Lieutenant-Colonel Carter with the detachments of the 14th, 18th, and 21st Dragoons pursued them with that alacrity and spirit which has ever distinguished him, he could only come up with a very few. He succeeded, however, in driving a great many into the river Artibonite, most of whom perished, and he had the good fortune to take two of the guns with their ammunition, mules, etc. etc. The third gun was most probably lost in the river, the carriage being left behind. We found the fort in the Bourg of Mirebalais as perfect as it had ever been and in no manner destroyed. I enclose a return of the artillery and ammunition found in the fort of Mirebalais, and I am happy to inform your Excellency that the repossession of this important post and district was effected without loss, one sergeant and one private of the dragoons being all our wounded.

'*Return of Ordnance and Ordnance stores taken in the fort of Mirebalais on the 2nd June* 1797.

'2 French 8-pounders badly spiked (since unspiked and rendered serviceable).
2 6-pounders, serviceable.
2 2-pounders, serviceable.
A large proportion of shot for the above ordnance of every description.
The ammunition not ascertained, but stated to be damaged.

(Signed) 'GEO. CHURCHILL, *Brigadier-General.*'

The pestilential climate of the West Indies was highly injurious to the health of Europeans, and the 14th Light Dragoons, like other regiments serving there, were soon reduced to a skeleton. When the order came to return home, those who survived were permitted to volunteer into other corps remaining out longer, and those who remained in the regiment,

Regiment arrives in England, Oct. 1797.

twenty-five in number, were embarked for England, where they arrived in October, and were stationed at Chelmsford.

On the 1st June, General Sir Robert Sloper, K.B., was removed to the Colonelcy of the 4th Dragoons, and Major-General John William Egerton (afterwards Earl of Bridgewater) was appointed to the Colonelcy of the 14th Light Dragoons, from the first Lieutenant-Colonelcy of the 7th Light Dragoons. Major-General Egerton was employed on the staff at Chelmsford when the remnant of the 14th Light Dragoons arrived there from Hayti or St. Domingo, and receiving eight old and worn-out horses with the recruits from the depôt at Maidstone, he had the satisfaction soon after of mounting the regiment afresh, and of seeing it within a few years with upwards of 900 horses in its ranks. There do not appear to be any authentic returns of the actual numbers of the 14th Light Dragoons who embarked for the West Indies, or of those who subsequently returned to England; but in the case of the 13th Light Dragoons there is a very accurate casualty return preserved, and this may be taken as a fair sample of what the casualties of the 14th Light Dragoons probably were in the same campaign, for there is no reason to suppose they fared better.

Colonel—John Wm. Egerton.

'Casualty Return' of 13th Light Dragoons:—

	Men.
Sailed for the West Indies, . .	452
Died, ...	287
Drafted to 20th and 21st Light Dragoons,	113
Returned to England, . . .	52
	452

In the 'Digest of Services' of the 14th Hussars,[1] which are kept in the orderly-room at regimental headquarters, the only allusion to this West Indian expedition is as follows:—'In

[1] The Regimental Records.

June 1795 the regiment was dismounted at Clonmel barracks, and marched to Carrick-on-Suir, from whence it embarked to St. Domingo under the command of Colonel Carter. 1797, the regiment, on arrival in England, went to Chelmsford.'

As the above-quoted records are so very meagre, it may be interesting to relate the following facts taken from Cannon's *Historical Record of the 13th Light Dragoons* (pp. 28-30) with reference to the same expedition, in which that regiment also took part:—

'Two troops of the 13th had embarked about July or August 1795 for Jamaica; then the rest of the regiment gave up their horses at Mallow, in Ireland, and sailed from Cork to Bristol. Whilst waiting for orders at Bristol, the 13th Light Dragoons met the 14th Light Dragoons destined for the same service. The Thirteenth subsequently embarked at Southampton in transports; all the ships then laden with troops proceeded from Southampton to Cove Harbour, Cork, and in February 1796 seven troops of the Thirteenth sailed for Barbadoes, where they arrived early in April. The worst revolt had taken place in the island of St. Domingo, which drove a large body of the planters to obtain aid from the English people and to transfer their allegiance from France to the British Crown. In consequence of this, additional forces were ordered to the West Indies. From Barbadoes the 13th Light Dragoons were moved to St. Domingo; but here the climate was so pestilential, the regiment (which had been supplied with some horses from America) was very soon reduced to a mere skeleton, and lost in six months 20 officers, 7 troop quartermasters, and 233 non-commissioned officers and men, so that the few remaining were removed to Jamaica in December 1796. Here the remnant of the regiment remained until July 1798, and, after transferring some men to the 20th and 21st Light Dragoons, all that remained of the 13th Light Dragoons, 52 souls, embarked for England, and landed at Gravesend in October 1798.'

Mr. Peter Vatass, who had been chaplain of the regiment since 24th December 1745, and was now seventy-five years of age, was removed, and no successor seems to have been appointed. Apparently Mr. Vatass had been *on leave* the whole of his service! The salary of a regimental chaplain in those days was about £120 per annum.

Major-General Egerton, the new Colonel of the Fourteenth, was already well known to them, as he was the Staff Officer at Clonmel in 1795 when they were ordered to give up their horses after being placed under orders for active service in the West Indies; and when in 1797 they arrived at Chelmsford on return from St. Domingo, he was the Major-General in command there. He always took an intense interest in the regiment, and was extremely proud of the high reputation it subsequently gained in the Peninsula, and he remained at its head till his death in 1823. When he was first appointed to the full Colonelcy he took a very active part in the measures adopted for remounting the regiment and recruiting its diminished ranks after service in the West Indies.

Two 'assistant-surgeons' were for the first time appointed this year, viz. Henry Killaly, 1st February 1797; Samuel Newman, 25th April 1797.

1798

The regiment having been recruited and remounted soon mustered 600 sabres: it was divided into 8 troops, and was stationed at Chelmsford. On 26th July 1798 His Majesty King George III. was graciously pleased to approve of the regiment being styled the 14th (or Duchess of York's Own) Regiment of Light Dragoons, in honour of Her Royal Highness the Princess Frederica Charlotte Ulrica Catherina, Princess Royal of Prussia, who had married H.R.H. the Duke of York in 1791. This honour was conferred for services performed by the 14th Light Dragoons in receiving and escorting H.R.H. the Duchess of York on her arrival in England in the same year. The royal authority was given for the regiment to

Styled 14th (Duchess of York's Own) Light Dragoons.

assume the Prussian Eagle as a badge, and the colour of the facings was changed from lemon-yellow to orange, the colour of the livery of the Royal House of Brandenburg.[1] The Fourteenth thus lost for a time the original pale yellow colour, but this was revived in 1861 in the colour of the busby-bags when the regiment became Hussars. In the autumn the regiment moved to Braintree and Halstead, where, on 9th October, Major-General Egerton made an inspection. At this review neither the Lieutenant-Colonel (Major-General Grice Blakeney) nor the senior Major (Colonel Arthur Carter) were present, but the junior Major (Lieutenant-Colonel Sir George Dunbar, Bart.) was in command of the regiment. On the 1st January both the Majors were promoted by brevet, Lieutenant-Colonel Carter becoming Colonel, and Major Sir G. Dunbar, Bart., Lieutenant-Colonel.

Prussian Eagle given as a badge to the regiment. Facings changed from lemon-yellow to orange.

1799

Being now stationed at Canterbury, the regiment was inspected there on the 25th of September by Major-General Garth. Its establishment was 720, but the effectives on that date were only 450 men and 390 troop-horses.

Major and Brevet-Colonel Arthur Carter, who had practically commanded the regiment for several years, went on 5th December on appointment as Lieutenant-Colonel to the 18th Light Dragoons, and Lieutenant-Colonel John Michel from the unattached list, late 30th Light Dragoons, became second Lieutenant-Colonel of the 14th Light Dragoons on augmentation. Major-General G. Blakeney was still continued as the Lieutenant-Colonel of the regiment though a general officer, and Major (Brevet Lieutenant-Colonel) Sir George Dunbar, Bart.,[2] died on the 15th October.

In the army list for this year a Veterinary Surgeon appears

[1] Cannon's *Historical Record of the 14th Light Dragoons*, p. 18. An entry elsewhere gives the date of the royal authority which granted the Prussian Eagle as 7th September 1799. Orange, the colour of the ribbon worn with the Order of the 'Black Eagle,' is still considered the royal colour of Prussia.

[2] Dunbar of Mochrum, Wigtonshire.

for the first time in the list of commissioned officers, his name being Samuel Newman, appointed 9th January.

The first Paymaster, Mr. James Flanegan, was also appointed this year.

There was an order issued on 22nd April authorising queues ten inches in length to be worn by cavalry and infantry, excepting the light companies.

Captain the Hon. James Butler became Major in the regiment.

1800

There were further augmentations in the establishment this year; and there were 3 Lieutenant-Colonels—Grice Blakeney, John Michel, and Samuel Hawker, the latter having been appointed on 12th June.

Establishment.

The establishment was raised to 10 troops of 90 rank and file each. Four new guidons were received.

At Canterbury on 1st November the regiment was inspected by Major-General Wilford. A portion of the regiment had lately been in Swinly Camp under Lieutenant-Colonel J. Michel.

No 'Standing Orders' in the Regiment.

General Wilford mentions that 'no established orders existed in the corps for its direction and guidance.'

The established code of regimental orders which had existed previously in the regiment, and which was so highly spoken of by the Inspecting General officers from 1784 to 1790, appears to have been allowed to drop out about this period, or perhaps somewhat earlier.

'The system of discipline and good order' established and maintained by the carrying out of these orders is referred to by Major-General Lord Luttrell in his inspection reports of 1785 and 1786, and he gives the credit of it to Colonel Sir John Burgoyne, Bart., who commanded the regiment from 1774 to 1781.[1] His successor in the command, Lieutenant-Colonel Grice Blakeney, who remained at the head of the regiment till he became Lieutenant-General in 1802, appears in the earlier

[1] See p. 36.

years of his command to have kept up this good regimental system, and adhered to the 'Code of Regimental Orders,' but the latter was allowed to disappear as time went on.[1] There is no record of the regiment ever having any 'regimental standing orders' again until May 1891, when Colonel H. B. Hamilton at Hounslow introduced the present existing ones, which were then thought to be the first the Fourteenth had ever possessed, as no traces whatever of any previous ones existed, and none apparently had been heard of by any one living at the time.[2] The reference proving the former existence of this old 'regimental code of orders' is very interesting, and was discovered by the author of these pages in an old 'Review-Book' of the eighteenth century preserved in the Public Record Office in Chancery Lane, London, where it was deposited by the War Office.

1801

The regiment remained at Canterbury till March. It was at Newbury in April and May. From June it was at Romford and Hornchurch.

On 12th June, Captain Henry Brown became Major, and on 14th August, James Gambier was appointed to the same rank in the regiment. Mr. Robert Thomson became Veterinary Surgeon this year, and remained in that position in the regiment until 10th November 1814.

1802

In consequence of the Peace of Amiens on 27th March 1802, the establishment of the regiment was reduced by two troops. **Establishment.**

Captain Neil Talbot became Major on 25th June *vice* J. Gambier, and Lieutenant-Colonel S. Hawker was placed on half-pay owing to the reduction of a Lieutenant-Colonelcy in the establishment of the regiment.

[1] The first mention of 'Standing Orders' is in 1784, when Lieutenant-Colonel Blakeney commanded the regiment (see p. 35).

[2] See p. 381.

A squadron was stationed under Captain Talbot at Chelmsford during the early part of this year, consisting of 125 troop-horses, and was inspected there on 31st March by Major-General Milner. Lieutenant-Colonel J. Michel became Brevet-Colonel on 29th April, and then succeeded to the command of the Fourteenth, as Major-General Grice Blakeney was promoted Lieutenant-General on the same date, and removed from the regiment after holding the Lieutenant-Colonelcy since 19th November 1781—upwards of 21 years, but it is probable that he had not exercised the active duties of command since his promotion to Major-General on 3rd October 1794. The second Lieutenant-Colonelcy was not filled up. Captain N. Talbot became Major on 25th June.

Lieutenant-Colonel—John Michel.

1803

Establishment.

War broke out again. Considerable augmentations were again ordered, and on the 10th of March the establishment of the regiment was raised to 664 men and 600 troop-horses.

The headquarters were at Hythe in November, where Major-General Cartwright inspected on 5th December.

Colonel J. Michel was succeeded in the Lieutenant-Colonelcy by Lieutenant-Colonel S. Hawker, who was brought back again into the regiment from half-pay to command.

Lieutenant-Colonel—Samuel Hawker.

The full Colonel of the regiment, Lieutenant-General John William Egerton, became Earl of Bridgewater.

1804

Further augmentations took place this year, and the regiment now consisted of 10 troops of about 90 rank and file each.

The total gross cost of the regiment for one year came to about £37,857.

'Recruiting-parties' sent out.

The troopers required to complete the regiment up to its augmented establishment were obtained partly by subaltern officers being permitted to raise a stipulated number each for promotion to a higher rank, and partly by the aid of 'recruiting-parties' sent out to various centres. The recruits were chiefly obtained from London, Birmingham, Shrewsbury, and Chichester. In December the establishment was definitely fixed as follows:—

10 Troops.	50 Corporals.
54 Sergeants.	950 Privates.
10 Trumpeters.	1064 Troop-horses.

Captain (Brevet-Major) Richard Pigot became Major on 4th August *vice* H. Browne. Major-General Cartwright inspected the regiment on 1st June at Hythe under command of Lieutenant-Colonel Hawker. The various troops of the Fourteenth were much detached in separate stations same as last year, and had no regular place to assemble in for drill.

Major-General Cartwright made a second inspection in November at the same place (Hythe). In December headquarters were at Guildford, and portions of the regiment at adjacent stations.

1805

In June the regiment moved to Hounslow.

On 22nd August, Major N. Talbot became second Lieutenant-Colonel on augmentation of establishment, and Captain Thomas Smith became Major.

There was a detachment at Kensington under Captain P. Keogh, which consisted of 9 sergeants and 34 corporals, selected from the 10 troops of the regiment.

Whilst stationed at Hounslow, Kensington, Hampton Court, Royal Escorts. and other adjacent suburban places, the detached troops of the regiment relieved those of the 9th Light Dragoons, and supplied the travelling escorts and letter parties for His Majesty King George III. and other members of the Royal Family, up to the month of September. There were 640 effective troop-horses by the end of the year.

1806

The regiment left Hounslow in July for the south-western district, and headquarters were at Southampton on August 1st, at Winchester August 14th, and at Dorchester on October 14th, having passed through Basingstoke and Alton *en route*.

Captain F. B. Hervey became Major on 8th May *vice* R. Pigot.

1807

In July the regiment left Dorchester, and, after being employed in the early part of the year on election duty in the county of Sussex, for the Midhurst election at Petworth, Fittleworth, and Pullborough, marched *via* Guildford, Farnham, and Bagshot, to its former quarters at Hounslow, Kensington, and Hampton Court.

New carbines, etc., issued; old arms and bayonets returned, 1807.

A supply of new carbines and pistols was issued from the ordnance stores this year to the regiment, including the two augmentation troops, when the whole of the old pistols, carbines, and bayonets were returned into store.

On 9th July, Captain the Honourable Charles Butler became Major *vice* T. Smith.

In September the headquarters of the regiment under Lieutenant-Colonel S. Hawker were at Blatchington, near Brighton, and detachments were stationed at Eastbourne, Hastings, and Bexhill, so the stay at Hounslow must have been very short.

1808

14th Light Dragoons placed under orders for the Peninsula, 19th October 1808.

On 5th July the regiment left Blatchington, Eastbourne, Bexhill, and Hastings in four divisions, and marched through London to Ipswich, arriving there 25th July, where it remained three months. On 19th October a depôt squadron with heavy baggage was ordered to be left at Ipswich, and the four service squadrons were placed under orders for active service in the Peninsula. The regiment accordingly marched to Tiverton, Taunton, Exeter, and Honiton, one squadron remaining for a time at each place between 16th and 29th November, and on the latter date the four squadrons marched to Flushing, near Falmouth, where they embarked on 5th December on board several transports and sailed for Lisbon. The headquarters were at Romford on 1st November and at Liskeard on 1st December, previous to embarkation.

Regiment embarks for Lisbon, 5th December.

On 23rd December, under command of Colonel Samuel Hawker, the 14th (Duchess of York's Own) Light Dragoons landed at Lisbon ready to take part in the war against the French.

1809

The French Emperor, Napoleon Buonaparte, had already invaded Spain and Portugal, and a British army had proceeded to Portugal to help the Portuguese to expel the invaders. The Portuguese had been successful, and an army under Lieutenant-General Sir John Moore was advancing into Spain, where subsequently, on the 16th January 1809, it was defeated by the French under Marshal Soult, at Corunna, and its gallant leader killed. It was shortly before this catastrophe that the Fourteenth arrived in Portugal to join the British army, of which Major-General Sir Arthur Wellesley assumed command when he arrived at Lisbon on 22nd April 1809.

Colonel Hawker, commanding the regiment, was appointed A.D.C. to His Majesty King George III. with the rank of Colonel in the army on 25th April. The regiment remained quartered about Lisbon till the spring, when it advanced to Bucellas, an outpost of our army, and formed the advance-guard of the British troops on the march to Coimbra in the month of April. In May the Fourteenth were brigaded with the 16th Light Dragoons and 20th Light Dragoons, as well as with the 3rd Light Dragoons of the King's German Legion, under command of Brigadier-General Stapleton Cotton (afterwards Field-Marshal Viscount Combermere), and took part in a review of the army in Portugal which was held at Coimbra[1]

[1] The troops concentrated at Coimbra on 5th May were 25,000 sabres and bayonets, of which 9000 were Portuguese, 3000 Germans, the remainder British. There were also 24 guns. The cavalry division was commanded by Lieutenant-General Payne ; the three infantry divisions by Edward Paget (1st), Sherbrooke (2nd), Hill (3rd). Beresford's corps consisted of the Portuguese and a few British troops —(Napier.)

before Lieutenant-General Sir Arthur Wellesley, K.B. (afterwards Field-Marshal the Duke of Wellington); the other cavalry brigade, 3rd and 4th Dragoon Guards, was under the command of Brigadier-General Fane. The French troops under Marshals Soult and Victor had, in the meantime, invaded Portugal, and Marshal Soult had taken Oporto.[1] The first service undertaken by the British commander was to expel the French from the important city of Oporto. Two squadrons of the Fourteenth under Lieutenant-Colonel Neil Talbot were detached with the Portuguese troops under Marshal Beresford to intercept the French if they should attempt to retreat northwards by Amarante. The remaining 3 squadrons under Colonel Hawker advanced direct on Oporto with the main body of the army, when, being employed with the rest of the cavalry on outpost duty and advance-picquets, they had several combats with the enemy, especially on the 10th and 12th May. It was on the 12th May when Sir Arthur arrived on the banks of the Douro near Oporto, unperceived by the French, who were on the opposite (right) bank. He determined to force a passage across the river, and immediately detached 2 squadrons of the Fourteenth with the German brigade and 2 guns under command of Major-General John Murray 3 miles up the river on the left bank to Barca de Avintas, where they effected a passage in boats. In the meantime Sir Arthur concentrated the main body of his army behind the Serra convent height, where he posted 18 guns in a commanding position on the rock near the convent, and having with difficulty obtained 3 large barges, began to send his troops across. In his careful observation from the high ground the British general had observed the horses and baggage of the enemy amid clouds of dust retreating along the Vallonga road, and no large force seemed near the river, neither were the guards or patrols vigilant along the banks. There was a large building called the seminary, placed admirably for defence, which also caught the

Douro, 12th May 1809.

[1] Cannon's *Historical Record of the 14th Light Dragoons.*

British general's eyes, and large enough to hold two battalions or more.[1] The first troops to cross were only 25 men under an officer, and these seized and occupied the seminary so quietly that the French in Oporto were not roused. By the time the third boat passed, in which was General Paget, leader of the 1st Division, the city was roused to arms, and the seminary was furiously attacked. General Paget, who had mounted the roof, fell severely wounded, and his place was taken by General Hill, whose division, as well as General Sherbrooke's, were crossing the river in all haste. The English guns from the Serra opposite commanded the enclosure of the seminary and swept the ground on one flank with great effect. The struggle, however, was violent, and as the expected help from General Murray did not appear, the position was critical, especially as the numbers of the enemy were so overwhelming. At length Murray was descried coming down the right bank of the river. About this time the French evacuated the lower city; the attack on the seminary became slacker, and large columns of the enemy were passing in haste and confusion along the Vallonga road. Sherbrooke's men had most opportunely come upon the rear of the French at the lower part of the town, and had taken 5 guns. Murray's arrival across the enemy's line of retreat was of the greatest moment, but as he did not open his guns on the crowds of troops passing along the Vallonga road, Brigadier-General Charles Stewart, just at the right time, headed a most effective charge which was executed by the 2 squadrons of the Fourteenth, one in support of the other, gallantly led by Major F. B. Hervey[2] and Major the Honourable Charles Butler, who charged through the enemy's rear-guard as it was pushing through a narrow road to gain an open space beyond, unhorsed the French general Laborde, and

[1] This account of the Douro affair is from Napier's *History of the Peninsular War*, from which numerous extracts have been made and much information gathered and inserted in this Record, bearing on this campaign.

[2] Afterwards Colonel Sir Felton B. Hervey, Bart., C.B., A.D.C.

wounded General Foy, but as no support was at hand from Murray's troops, these brave men had to cut their way back with considerable loss. Major Hervey lost his right arm, and the Fourteenth had 12 rank and file killed, 1 rank and file missing. Captain Peter Hawker, Lieutenant Robert Knipe, Lieutenant Evelyn P. Dormer were wounded, as well as 1 sergeant and 18 rank and file.

Losses of the Fourteenth at the passage of the Douro on 12th May 1809.

Napier says: 'This finished the action. The French continued their retreat; the British remained on the ground they had gained. The latter lost 20 killed, a general and 95 men wounded; the former had 500 men killed and wounded.'

The Fourteenth lost heavily, as we have seen, in this affair, but the gallant conduct of the regiment was highly commended in Sir Arthur Wellesley's public despatch as well as in general orders. The troops had marched 85 miles in 4 days over most difficult country, and during 3 of those days they were constantly fighting. The sudden arrival of Murray's Division was a complete surprise to the enemy, and the gallant charge of the 14th Light Dragoons at the critical moment helped in a great measure to turn the tide of victory completely in favour of the British, and to cause the French to make a hasty retreat.

The following is an extract from the general orders issued by Lieutenant-General Sir Arthur Wellesley, commanding the British forces, dated Oporto, 12th May 1809:—

'The passage of the Douro and subsequent movements on the enemy's flank by Lieutenant-General Sherbrooke with the Brigade of Guards and His Majesty's 29th regiment, and the bravery of the two squadrons of the 14th Light Dragoons under command of Major Hervey, and led by Brigadier-General the Honourable Charles Stewart, obtained the victory which has contributed so much to the honour of the troops on this day.'

It was not till the year 1837 that permission was given

for the regiment to bear the word 'Douro' on its appointments.

The 14th Light Dragoons were now employed with the separate corps under Marshal Beresford in following up the French army under Marshal Soult as far as Ginjo. Here they halted, and afterwards moved to Abrantes on the Tagus, where the British army was concentrated for further operations. On 27th June an advance was made in the direction of Talavera de la Reyna, through Castello Brancho, Placentia, and the valley of the Alberche in Estremadura. On 15th July the British headquarters were at Placentia.

According to Napier, Sir Arthur Wellesley had now about 21,000 men with 30 guns, and Cuesta's Spaniards numbered about 35,000 with 70 guns.

The British had one cavalry division, 6 regiments, 3047 sabres, under Lieutenant-General Payne in three brigades; 3rd and 4th Dragoon Guards under Fane, 14th and 16th Light Dragoons under Cotton, and the cavalry of the King's German Legion under Anson.

Four divisions of infantry as follows :—

1st, Lieutenant-General Sherbroke; 2nd, Major-General Hill; 3rd, Major-General Mackenzie; 4th, Brigadier-General Campbell.

The artillery was commanded by Major-General Howarth.

On the march through Spain the British army suffered great privations from scarcity of provisions; and the incapacity of the Spanish general, as well as the unreliable nature of his troops, caused Sir Arthur much anxiety, and greatly hampered his movements. Towards the end of July two of the British divisions were sent forward, as well as the whole of the cavalry, to support a movement of the Spaniards against Marshal Victor's army, which, according to reports circulated, was falling back on Torrijos and behind the Guadarama river. Lieutenant-Generals Sherbrooke and Payne commanded this force. On the 26th July the Spanish army under Cuesta was at St. Ollalla,

Sherbrooke was at Cassalegas, Sir Arthur Wellesley was at Talavera, and Marshal Victor's army was only a few miles from Cuesta's advanced posts with 50,000 men and 90 guns. The French had also an army of 50,000 men under Mortier at Salamanca. The Spaniards were driven back by Victor, and had to retire on Talavera, supported by Sherbrooke and the cavalry under Payne.

On the 27th July a battle was imminent, and Sir Arthur left Mackenzie's Division with a brigade of cavalry to cover a retrograde movement, whilst he withdrew the allies into a position for order of battle six miles in rear. General Mackenzie was left in a wood lying to the right of the Alberche, which covered his left flank. Between 2 and 3 o'clock this post was attacked by the French, when the 14th Light Dragoons were ordered to advance, and they crossed the river Alberche, sending out a line of skirmishers to cover the retirement of the infantry. The regiment was kept out skirmishing till nightfall, and lost 9 horses killed; one officer, Lieutenant Theophilus T. Ellis, and one private soldier being wounded, and two horses missing.[1] After performing this service the Fourteenth resumed the post assigned them in the allied army, which was in the rear of Brigadier-General Campbell's Division. The position was as follows:—the Spaniards were placed on the right of our line, their right resting on Talavera; on their left came Campbell's Division, in two lines; Sherbrooke's Division came next to Campbell's, in one line only; Mackenzie's Division was destined to be the second line to Sherbrooke's; Hill's Division was to be on the left of the line. The whole line was two miles long, and was an exceedingly well chosen and strong position. The British and Germans mustered about 20,000 men under arms, with 30 guns. The Spaniards had about 34,000 men and 70 guns. The French numbered 80 guns and 50,000 men, and they were hardy veterans, while the allies had only 19,000 genuine soldiers.[2] King Joseph Buonaparte was in command,

Casualties of Fourteenth in front of Talavera, 27th July 1809.

[1] This was the combat of Salinas mentioned by Napier.

[2] Napier.

with Jourdan, Victor, and many other renowned generals under him.

At daybreak on the morning of the 28th July, the British left was attacked by the enemy's artillery and infantry, then the centre, lastly the right of the line became involved in fierce contest. Both sides suffered heavily, the enemy more. At last the French retired in disorder to their original position, still keeping up a powerful fire of artillery.[1]

Battle of Talavera, 28th July 1809; first attack.

Our artillery was no match for the French: it was small of calibre and the guns few in number. When Cuesta was asked for reinforcements he sent two guns, but luckily these were good and well served by the Spaniards. Sir Arthur now sent for his cavalry, which was at a considerable distance; he also obtained some Spanish cavalry, and placed the whole in mass six lines deep, the leading squadrons looking down a valley on his left flank. After this, about 9 o'clock, there was a cessation of hostilities for several hours.[2]

Second attack.

It was not till 12 o'clock that further hostilities took place. Then the French opened the fight with 80 pieces of cannon, followed by the advance of their light troops, who were supported by broad black columns in rear. Campbell's Division was first attacked with fury, but it stood firm; Mackenzie's Brigade and his Spanish battalions gallantly withstood the onslaught, and after repulsing a second attack no less vehement, aided by a flank charge of Spanish cavalry, secured the victory in that quarter. On our left, a brilliant charge of Anson's Brigade of cavalry took place, when the 23rd Light Dragoons under Colonel Seymour and Major F. Ponsonby, with the 1st German Hussars under Colonel Arentschild, charged Villate's troops. In this charge Colonel Arentschild, seeing an impracticable piece of ground ahead of him, pulled up his regiment just in time to avoid a catastrophe; but the 23rd Light Dragoons, continuing their advance, got into difficulties, and still going forward, were at last completely overmatched by the enemy. They then had to

[1] Napier. [2] *Ibid.*

retire, leaving half their numbers killed and wounded—considerably more than 200.[1]

The British centre was very hard pressed by Lapisse's attack, and here it was that Sherbrooke's men bravely drove back their assailants, when, encouraged by success, the Guards and Germans incautiously made a bayonet charge, in which the pursuit was carried too far, until our men were finally repulsed by the French reserves of infantry and dragoons. Confusion ensued, which caused the centre of the line to give way and become hopelessly broken: our right and left flanks, however, remained firm and resolute, Campbell and Hill holding their positions against all attacks. Sir Arthur saw that a critical and perilous moment had come, and was determined to make a final effort to restore the centre. Suddenly the 48th Regiment, led by Colonel Donellan, was seen advancing from the hill right down on the flank of the victorious French columns, pouring in such a steady and destructive fire of musketry as completely checked the enemy's further movements. This enabled the Guards and Germans to rally, whilst our guns played incessantly on their opponents, and to crown all, a brigade of light cavalry under Brigadier-General Cotton, consisting of the 14th and 16th Light Dragoons, was seen rapidly advancing on the enemy's flank. This cavalry attack brought matters to a crisis: the enemy was checked, began to waver, lost all impulse, the battle was won.[2] A general retirement to their former position subsequently took place along the French line. The British were too exhausted by fatigue and weakened by losses to attempt any pursuit, and no assistance could be given by the Spaniards. At 6 o'clock all fighting ceased, and both armies remained on their own positions. Thus it was that the 14th Light Dragoons, 16th Light Dragoons, and the 48th Regiment, by coming forward at the critical time when the Guards Brigade was almost annihilated, contributed in no small measure to

Charge of Fourteenth in Cotton's Brigade.

[1] Napier. [2] *Ibid.*

change the fortunes of the day, and to gain the victory for the British arms.

The Fourteenth lost at Talavera—

Casualties of the Fourteenth at Talavera, 28th July 1809.

Killed—21 horses, 3 men.

Wounded—5 officers { Colonel S. Hawker, slightly. Captain J. Chapman, Captain P. Hawker, Lieutenant W. Wainman, Lieutenant Thomas Smith, } severely.

" 6 men.

" 3 horses.

Missing—13 horses.

Taken prisoner—Lieutenant E. P. Dormer.

Lieutenant-Colonel Talbot and Captain Baker had horses killed under them in this battle.

The total losses of the British at Talavera were—

Losses at Talavera.

Killed { 33 officers, including Generals Mackenzie and Langworth. 800 men.

Wounded { 195 officers, including 3 generals. 3700 men.

Missing { 9 officers, 650 men.

In addition there were about 800 killed and wounded on the 27th, the day before the battle.

The Spaniards lost 1200 men.

The French lost 7389 officers and men.

Colonel Hawker was rewarded with a gold medal, and the Fourteenth subsequently (1820) received authority to bear on the guidons and appointments the word 'Talavera,' in commemoration of their distinguished services in this action.[1]

After the victory of Talavera, the French brought forward such overwhelming armies in different directions under Ney, Soult, Victor, Mortier, and other generals, with the

[1] Cannon.

intention of cutting off and utterly crushing the allies, that Sir Arthur Wellesley decided to act for some time on the defensive, and withdrew his army towards Truxillo and Merida, moving along the left bank of the Tagus, so as to baffle, if he could, the strong combination of armies against him. At first he moved to Deleytoza, crossing the river at Arzobispo, Craufurd's Brigade and six guns being sent to secure the boat-bridge at Almaraz and to cut off the French. About the end of August, owing to various circumstances, principally the scarcity of supplies and the bad conduct of the Spanish army, Sir Arthur fell back into Portugal, and occupied the valley of the Guadiana, his troops being distributed in Badajos, Elvas, Campo Major, etc., but the men suffered considerably from malignant fever in these districts. The Fourteenth were removed to Villa Viçosa, a town in Alemtejo (Portugal), and in December they moved to Santarem, situated on the right bank of the Tagus, in the Estremadura of Portugal. Here they were formed in brigade under Major-General Slade with the 1st Royal Dragoons, a regiment recently come from England.

During this year the rank of Troop-Sergeant-Major was introduced into the cavalry. The establishment of officers in the regiment at this time was—

Establishment.

1 Colonel.
2 Lieutenant-Colonels.
2 Majors.
9 Captains.
21 Lieutenants.
8 Cornets.
1 Paymaster.
1 Adjutant (included in Lieutenants).
1 Surgeon.
1 Assistant Surgeon.
1 Veterinary Surgeon.
1 Quartermaster.

The regimental agents were Messrs. Collyer and Son, London.

Quartermasters became commissioned officers about this year. The first commissioned Quartermaster in the Fourteenth was Mr. Jameson, appointed 4th January 1810.

1810

The 14th Light Dragoons had a regimental song[1] at this period, as follows :—

Old Regimental Song of 14th Light Dragoons, Used in the Peninsular War.

'ÇA IRA!'[2] 1810

Beaten backward in the press
Reeled the Old Fourteenth,
And in triumph shrill arose
The yell of the triumphant foes,
As, where the British Lion flew,
Flaunting 'White and Red and Blue
Full well the fiery Frenchmen knew
The fame of the Fourteenth.

Beaten backward in the press
Reeled the Old Fourteenth,
Cheerily their Colonel spoke
As the red line round him broke,
Laughing, waving with his hand
To the leader of the band,
As again they took their stand,
The men of the Fourteenth.

'Play the Frenchman's March,' he said,
The chief of the Fourteenth;
'Strike it up, strike loud and clear,
As I stand before you here,
We will prove our mettle soon,
Ere yon pale sun rides at noon,
We 'll beat them to their own brave tune,
We men of the Fourteenth.'

On 14th March the Fourteenth left Santarem and returned to the Alemtejo and took the advanced posts of Lieutenant-General Rowland Hill's Corps at Arronches, a town situate at the conflux of the Caya and the Algrette, near the Spanish

[1] This song was kindly presented by Mr. F. A. Hawker, son of the late General Sir S. Hawker, G.C.H. Mr. Hawker also kindly gave the photograph of his father which is reproduced in this Record.

[2] 'Ça ira !' (French) = 'We shall succeed.'

confines. In June the Fourteenth advanced to Almeida in the Beira province, and were attached to the Light Division under Brigadier-General Craufurd, who was behind the Agueda river watching the enemy's movements, when, with the 16th Light Dragoons and 1st Hussars (King's German Legion), they took the outpost duty on this frontier.

Ciudad Rodrigo was at this time being invested by Marshal Ney. Marauding parties of French used to enter the villages of Barquilla and Villa de Puerco; ambuscades were formed to cut them off: the Fourteenth took part in them. On 11th July a portion of the regiment charged a square of French infantry 200 strong: the square withstood the charge and opened a terrific fire. The gallant Lieutenant-Colonel Talbot, Quartermaster M'Cormick, and 11 men were killed close up to the bayonets, and 23 men were wounded. This occurred near Sexmiro, in front of Ciudad Rodrigo, and as Colonel Hawker had gone home wounded after Talavera, Lieutenant-Colonel Talbot was virtually in command of the regiment at the time of his death. Major F. B. Hervey succeeded him as second Lieutenant-Colonel on 2nd August, and he assumed command of the regiment in the absence of Colonel Hawker, immediately after Lieutenant-Colonel Talbot fell, and subsequently when Colonel Hawker was appointed Major-General in 1811, he became the real commanding officer of the regiment, and it was under his able leadership that the 14th Light Dragoons became famous as Light Cavalry, being specially celebrated for the excellent manner in which they performed their outpost duties. Captain J. Chapman was promoted Major on the same date, in succession to Lieutenant-Colonel Hervey.

Sexmiro, 11th July 1810.

Captain Brotherton, late of the 14th Light Dragoons (afterwards General Sir T. W. Brotherton, G.C.B., who died in 1868), has left a graphic account of the death and burial of Lieutenant-Colonel Talbot, amongst other interesting incidents[1] of his experiences in the Peninsula when serving in the Four-

[1] These incidents are in MSS., and were kindly presented by the late Colonel the Hon. G. H. Gough, C.B.

teenth, as follows:—'The most formidable thing for cavalry to deal with is a square of steady infantry—indeed, such a square may be said to bid defiance to cavalry unless the cavalry has the aid of artillery to batter the square before charging it; for the formation, in square, to resist cavalry is a most murderous one when exposed to artillery. At the village of Sexmiro we encountered a square of French infantry. It was lying down, concealed in some high-standing corn, and only rose up when my squadron came within pistol-shot of it, and was beautifully steady. We charged it most gallantly, but they fired a deadly volley into us, and half my men fell killed or wounded. Colonel Talbot, who commanded the regiment, had put himself at the head of the squadron along with me. Poor fellow, he fell pierced by eight balls, literally on the enemy's bayonets. The moment the square had fired into and so sadly crippled us, it moved off to join its support close by, and we were so shattered as not to be able to follow. The French infantry behaved beautifully on this occasion. It was the 61st of the line. Marshal Massena immediately bestowed the Cross of the Legion of Honour on the officer commanding and several of the non-commissioned officers and men. So steady and cool was this little square, that though my horse fell, with the wounded, within two yards of their ranks, not a man moved out to bayonet me, but the square immediately retired in admirable order. We were repulsed, suffered great loss, and left our commanding officer, amongst others, dead on the field. I was sent afterwards to ask for his body, and brought it in. It was taken into a tent in which we messed. We all felt deep grief at his loss, for we all loved him; yet I never shall forget that we ate a hearty meal with our beloved friend's corpse close to us, uncovered. He had a glorious countenance in death. He was a noble-looking fellow, and had died so instantaneously, having had no less than eight balls through him, that his countenance was but little altered. We buried him on the glacis of Fort Conception, and a few days after I saw his body blown into the air, along with the fragments of the fort, when the explosion took place. What became of his remains afterwards we never could ascertain, as several horses and men were killed at the same time by the explosion. I may add that I went out with a flag of truce to fetch his body. When I arrived at the fatal spot where the murderous charge of my squadron had taken place, I saw lying on the ground only three French soldiers, one of whom was dead and the other two much mutilated by our sabres, but this was all the execution we had done in return for our severe loss. I brought poor Talbot's body back, and we buried him (as already related) on the glacis of Fort Conception. A few days afterwards the premeditated explosion of this fort took place, when his body was blown into the air. The blowing up and complete destruction of this important and beautiful little Star Fort which guarded the frontier of Spain, was an operation of extreme delicacy, and of most critical and precarious execution, for Colonel Burgoyne,[1] the talented officer of Engineers, selected for the task, had positive orders not to blow it up till the very last moment (that is, till the advance of the enemy), so that we might make use of it till the last moment, but not leave a vestige of it for the enemy's use. These instructions were carried into effect with extraordinary precision and most thoroughly, though Massena had the meanness and effrontery to say the contrary in his report to Napoleon. I had some little share in the execution of this critical operation. I happened to be on picquet in front of the fort on the night it took place, or rather the morning, at daybreak. As it was a matter of great importance to Massena to preserve the fort, if possible, and prevent its destruction, which he knew was planned, he thought he would best obtain this object by a sudden and rapid advance on our picquets, driving them back at a gallop, and arriving on the glacis of the fort as soon as we did, when, he thought, the officer of Engineers would hesitate to blow it up for fear of destroying our people. The match was always kept ready in the fort for instantaneous explosion.

[1] Afterwards Field-Marshal Sir John Burgoyne, Bart., G.C.B.

Knowing the state of the case, I had only just time to exclaim to an officer close to me (named Wainman), who was beautifully mounted on a thoroughbred horse, to go at speed to Burgoyne and apprise him that we were being driven back most rapidly, and that we had no time to lose. He arrived at the fort only just in time to enable Burgoyne to explode the mine. I found myself on the glacis just at this moment, and lost several horses and men by the explosion, besides the harrowing sight of poor Talbot's body being blown into the air. I had brought the body slung across a troop-horse. He was a delightful fellow, a friend I most deeply regretted, but singular and eccentric, particularly in his dress. He was dressed, the day he was killed, in nankeen pantaloons. Never was anything like the grief for his loss. When we buried him not an eye was dry.'

Napier says that four squadrons of the Fourteenth, under Lieutenant-Colonel Talbot, took part in this charge against the French infantry square.

On 11th July, Ciudad Rodrigo surrendered, the Fourteenth remained in the villages near Fort La Conception until the 21st July, when, as the French were approaching in masses, they fell back to Almeida. Here Brigadier-General Craufurd was bold enough to halt and make a stand against the advancing enemy, which led to the combat of Coa, in which the Light Division suffered heavily, and lost over 300 men.

Outpost affair on the Coa, 24th July 1810.

Early in the morning of the 24th July, after a wet and stormy night on outpost duty, a skirmishing fight took place with the French troops, who were advancing in force, near the passage of the river Coa, when the Light Division was engaged for a considerable time against superior numbers of the vanguard of the French army commanded by Marshal Massena.

On this occasion the Fourteenth were engaged and had the following casualties: killed—1 sergeant; wounded—Lieutenant Blatchford, 1 man and 4 horses.

Brigadier-General Craufurd stated in his despatch—'The retirement of the 14th Light Dragoons from Val-de-la-mula to

Almeida was carried out in the most regular and soldier-like manner, though opposed to a superior force of French cavalry.'

Lord Wellington's headquarters were now at Alverca.

General Brotherton relates the following as to the fight at Coa:—'The combat of the Coa took place on 24th July 1810, and was a very sharp affair; where we were only 6000 strong, against 24,000 brought into action by Massena. There were many gallant and daring deeds done that day, in taking and re-taking the bridge over the river of the same name. In one of these attacks, one of the officers of the 43rd, (Frederick) brother of Sir Richard —— of Barwood Park, was shot through the leg. Happening to be close to him, I jumped off my horse to assist him. He was bleeding profusely, and no surgeon immediately at hand to stop it, I had my canteen slung round me full of strong wine, and put it to his mouth, and made him take a copious draught of it. Just as I had done this the surgeon of the regiment came up, and I told him what I had done, at which he expressed himself displeased, saying that probably I should be the cause of his death; but he (Frederick ——) always said afterwards to everybody that I had saved his life by giving him the wine, as he felt so faint that he felt he was dying. He lived afterwards in excellent health till the 21st June 1854.

'On the same day (24th July 1810) one of the officers under my immediate command, Cornet B——, was hit by a fragment of a shell in the posterior, and as he was rather a soft sort of fellow, I thought, at first, that he made too great a fuss about it, though he turned deadly pale. But he had good reason to complain, for the piece of shell had buried itself deep in his buttock, and caused his death.'

From Almeida to the lines of Torres Vedras the regiment in conjunction with the 16th Light Dragoons and 1st Hussars, King's German Legion, under the command of Lieutenant-General Sir S. Cotton, Bart.,[1] formed the rear-guard of the army.

[1] Afterwards Field-Marshal Viscount Combermere, G.C.B., G.C.H.

On 28th August one squadron, acting with a squadron of the 1st Royal Dragoons at Frexadas, was engaged with a superior force of the enemy and highly distinguished itself.

Frexadas, 28th August 1810.

'The advanced posts of the British army having removed to Frexadas, the French besieged and took Almeida on the 20th August, and on the day following they attacked a squadron of the Royals and one of the 14th Light Dragoons on picquet at Frexadas, under Major Dorville of the Royals. The enemy brought forward a superior force of cavalry supported by infantry, but the two squadrons, undaunted by the greater numbers, charged the French with signal gallantry and drove them from the field with the loss of many men killed and wounded and 8 prisoners. The Royals lost 2 men and 1 horse wounded.'[1]

On 24th September, when the enemy skirmished with our picquets near Mortagao, a squadron of the 14th Light Dragoons under Captain T. W. Brotherton, acting with a squadron of 16th Light Dragoons and a squadron of 1st King's German Hussars, covered the retreat of the Light Division for 4 miles. These 3 squadrons drove back 4 squadrons of French Hussars, and the squadron of the Fourteenth charged the enemy's cavalry, killing 30 men.

Skirmish near Mortagao, 24th September 1810.

On 25th September, Captain the Honourable H. Percy was taken prisoner whilst reconnoitring near the heights of Busaco. On this occasion the regiment, together with the Royals, was employed to cover the retreat of the Light Division to the position of Busaco. Whilst performing this duty against the masses of the French army advancing on Busaco during the 25th and 26th September, the casualties of the 14th Light Dragoons were as follows:—

Casualties of Fourteenth on 25th and 26th September 1810, at heights of Busaco.

Killed—1 horse.

Wounded—1 sergeant, 2 rank and file, 4 horses.

Missing—3 rank and file, 7 horses.

The Fourteenth were now in brigade with the 1st Royal Dragoons under Major-General Slade, and Lieutenant-General Sir Stapleton Cotton commanded the Cavalry Division.

[1] General De Ainslie's *Royal Dragoons*, 1887, pp. 111-12.

Battle of Busaco, 27th September (in reserve).

On the 27th September, at the battle of Busaco, the Fourteenth, together with the 1st Royal Dragoons, were in reserve. Subsequently they were employed in covering the retreat of the army to the strong lines of Torres Vedras. This important battle, after hard fighting, resulted in a victory for the allies. The French lost 800 killed and 1 General, Grain-d'Orge, with a total loss of about 4500, whilst the allies only lost 1300. The position taken up by Lord Wellington was impregnable, and Marshal Massena, after his repulse, marched towards Coimbra, whilst the allies crossed the Mondego near Coimbra, and moved towards Condeixa and Pombal.

General Brotherton relates the following incidents about Busaco:—'At the battle of Busaco, after the charge made on our position by General Simon was repulsed, several of the French soldiers, who had fallen wounded within a few yards of our line, lay gasping in agony and thirst, calling out for water to drink; but such was the galling fire still kept up by the enemy on this point, that it appeared almost certain death for any one to show himself for an instant beyond the shelter which some rocks afforded. I observed, however, a noble young fellow, a Hanoverian belonging to the German Legion, walk coolly and deliberately from behind a rock, and going to the nearest wounded French soldier who was calling out for drink, but lay in a most contorted and painful position (one of his legs, which was broken by a musket-shot, being bent under him), applied his canteen to the poor fellow's mouth, after having, without the least degree of hurry or trepidation (though the fire continued most heavy), settled his head on his knapsack, and otherwise made his position less painful. The fine young fellow did this successively to several other wounded Frenchmen, and then returned to his regiment. When first this young officer stepped out, the enemy, fancying he might be leading an attack, redoubled their fire, but when they perceived what he was doing, the firing immediately ceased, and was succeeded by vociferous cheering at his conduct. A more affecting scene I never beheld in the field, and I only regret that, almost at the same moment, I witnessed a disgusting contrast to it. A staff officer, a German, whose name I shall abstain from mentioning, placed himself in perfect security behind a rock, and with a rifle, with which he piqued himself on being an unerring shot, kept picking off French officers and soldiers by way of amusement! I remonstrated with him on his barbarous conduct, and shamed him out of it, but not before he had hit several poor fellows who were actually employed at the time in burying their dead (it was a working-party sent out for the purpose). The remembrance of such conduct makes my blood curdle in my veins even at this time.

'At this same battle (Busaco) I witnessed an instance of the nervousness and superstition of the bravest soldiers. A battalion of the German Legion (Hanoverians) was sent down to drive the French out of a wood which they occupied in our front. They drove the enemy out most gallantly, but immediately after came running back most wildly and unaccountably till we learnt the cause. It appeared that part of the enemy's troops occupying the wood were part of the German contingent in the French service, and amongst them were some Hanoverians. On finding this out, our Hanoverians fled with the utmost precipitation out of the wood they had so gallantly gained possession of, horrified at the idea of fighting against their countrymen, and perhaps their relatives.'

Nuns and picture conveyed *en croupe.*

Another of the General's stories is well worth relating. He says:—'On the retreat of the army to the famous lines of Torres Vedras, when in command of the rearguards, a whole convent of nuns came running out of their convent, as I passed by it, and implored me to save them from the French. It was impossible for me to stop to protect them, and yet I could not bring myself to leave the poor creatures to the tender mercies of the French soldiers, though they were neither young nor handsome, but old and sallow, from penance and vigils, no doubt; so I resolved on the expedient of placing these poor distracted creatures (22 in number) *en croupe* behind as many dragoons. They had uneasy seats but clasped the dragoons tightly round their waists, and we brought them safe into the lines of Torres Vedras, to their great joy and to the great amusement of all those who saw my convoy—such an one as had never before, I suppose, been escorted in this manner by dragoons. Lord Wellington heard of this adventure, and was much amused by it, and the next time I dined with him, after it took place, he complimented me on my chivalrous affair and laughed heartily about it.

'I was not so fortunate with a cargo of a very different description which I once attempted also to carry off *en croupe*. It was a famous "Murillo" altar-piece at Medina de Ris Secco, in Castile. One of the priests came running out to me as I was patrolling through the streets in search of the enemy, and told me of this famous picture, and advised me, as the French were momentarily expected, to have it taken down, rolled up, and placed carefully between two dragoons, the ends resting on their valises, for it was a large picture. Ill-luck, however, would have it that the French never entered Medina at this period, and knowing how I should get blamed for taking away this picture except to save it from the French, I was obliged, after having carried it a considerable distance, to retrace my steps, having heard of the retreat of the enemy, and replace the picture whence I had taken it. It was afterwards, however, taken away by Marshal Soult, and is, I believe, at this moment part of his dishonestly acquired collection. He was one of the most unscrupulous plunderers in the French army, and this is saying a great deal for him!'

All this time the French legions in overwhelming numbers were still pressing on, and the 14th Light Dragoons formed the rearguard of our army as far as Pombal. On the 1st of October Lord Wellington's outposts were drawn in from the heights of Coimbra, on which occasion 3 troops of the Fourteenth under Major the Honourable Charles Butler constituted the rearguard, and they proceeded through the town in rear of the Light Division, and then acted on the main road leading to Pombal. The remainder of the regiment was acting on the plain with the rest of the cavalry of the army, but had to withdraw before the superior force of the enemy, crossing the Mondego at a ford below the town, and then skirmishing to prevent the passage of that river by the French. Here some sharp fighting took place, and the enemy's cavalry attacked and cut down some of the British in the middle of the river, and altogether 50 or 60 men were lost.

That night (1st October) the British headquarters were at Redinha, having passed through Condeixa, eight miles from Coimbra, and next day they were at Leiria. The retreat now became somewhat hasty and disordered, and plundering commenced, but Lord Wellington took vigorous measures to enforce discipline. At Leiria, three men taken in the act were hanged. On the evening of the 4th October the French drove the English picquets from Pombal, and next morning came so suddenly upon Leiria as to create general confusion. There were daily encounters going on between our rearguard and the enemy's advanced bodies, and the Fourteenth had frequent opportunities of proving their valour, which they certainly did not fail to take advantage of. At Rio Mandevilla, together with the 1st Royal Dragoons, the 16th Light Dragoons, the 1st German Hussars, and Captain Bull's troop of artillery, they repulsed a very superior force, on which occasion the 1st French Hussars were nearly annihilated. The English lost 3 officers and 50 men, and the French many more: it is believed the enemy had 36 squadrons opposed to 10. The casualties of the Fourteenth at Rio Mandevilla were—6 men killed; 8 men, 12 horses wounded, and also on 4th, 5th, 8th, and 9th October they lost 1 man wounded, and 12 men and 2 horses missing.

Rio Mandevilla, October 1810.

Casualties at Rio Mandevilla, etc.

On the 10th, Lord Wellington occupied the fortified lines of Torres Vedras, where the Fourteenth took charge of the outposts on the line from the Sobral road in front of Torres Vedras so long as the French army under Marshal Massena, Prince of Essling, remained opposed to us in the vicinity of these stupendous lines of defensive works. Here the two armies

Lines of Torres Vedras occupied, 10th October 1810.

watched one another for a considerable time, but at length, after frequent reconnoitring, the French commanders declined to attack, and during the night of the 14th November, Massena retired and established his army upon the heights of Santarem, where he remained till the night of the 5th March 1811.

The morning of the 15th November was foggy, and the retreat of the enemy was not discovered for several hours after daybreak. The Fourteenth were ordered forward along the Cartaxo road, and their advanced patrols took a number of stragglers prisoners. The British headquarters were soon re-established at Cartaxo, where the regiment was intrusted with the outposts extending from the causeway and bridge over the river.

Sobral, November 1810.

Captain Brotherton relates the following incident with reference to the affair at Sobral:—'On the last day of our retreat into the famous lines of Torres Vedras before Massena's army, we had a very sharp affair at a place called Sobral, so much so that we were hotly engaged and literally intermixed with the enemy, particularly the 71st Foot, the Rifles, and ourselves, the 14th Light Dragoons. The enemy were in very superior force, and we were giving way very fast. At this moment Colonel (then Captain) Perceval fell close to me, pierced by two balls, one through his leg and another through his arm. He was on the point of falling into the hands of the enemy. In those days I was particularly active, and as we were running away, I could use my legs as well as anybody, so I dismounted, and put poor Perceval on my horse, and joined in the scramble on foot, till I came up to a mounted dragoon of my own regiment. As, in action, the presence and exertions of an officer are more valuable than those of a private, it is not only justifiable, but it is incumbent on an officer, sooner than leave the field, to dismount a private, and take his horse. This I accordingly did without having time to ask any questions. When the fight was over, the late General Sir Denis Packe, a warm-hearted but very passionate man, whose orderly I had unknowingly dismounted, came up to me and lectured me most severely and harshly for having done so; and certainly, had I known the man was orderly to a General Officer, I would not have dismounted him, as it is essential for a General to have his orderly in action. I had presence of mind enough to make no reply, but bear the reprimand in silence. The General went away, but his aide-de-camp, Captain Synge (now Colonel Synge), having informed him of the circumstances under which I had taken his orderly's horse, he returned and made me many apologies for having reprimanded me, and praised what I had done. It was a heartfelt satisfaction to have saved poor Perceval, who, to the last day of his life, was grateful for it. He died at Brussels in 1838, and his wife has told me since that almost with his last breath he exclaimed, "*Generous Brotherton; he once saved my life!*" This was very gratifying to me.'

On 20th October a Trumpet-Major was authorised to be borne on the establishment of cavalry regiments with the pay of a sergeant. Cavalry regiments of ten troops to have one trumpet-major and nine trumpeters. **Establishment altered.**

The following is one of General Brotherton's anecdotes of what occurred when the Fourteenth were at Torres Vedras:— **At Torres Vedras.**

'When we were in the famous lines of Torres Vedras, I had gone to dinner, to some friends of the Guards, on a mule, and returning to my regimental bivouack at night, I became apprehensive of going into an enemy's picquet by mistake. I came to a turning which I thought I knew well, and tried to turn my mule to the *left*, which I thought was the right road, but he insisted on going to the *right*. We had a great battle, but all I could do was in vain, and he carried me his own way, to the *right*, and I got safe to my camp. I had the curiosity next morning to go to the spot where the mule and myself had differed in opinion, when I found that he was not only right, but that I was so wrong, that, if I had had my own way, I should have gone right into the enemy's camp.' **The sagacity of a mule.**

1811

In the Army List for this year the regiment appears with—

1 Colonel.	1 Paymaster.
2 Lieutenant-Colonels.	1 Veterinary Surgeon.
2 Majors.	1 Adjutant.
10 Captains.	1 Surgeon.
22 Lieutenants.	2 Assistant-Surgeons.
7 Cornets.	1 Quartermaster.

Agents—Messrs. Collyer, London.

On the 15th February, Viscount Wellington's headquarters were at Cartaxo. The Fourteenth were still in brigade with the 1st Royal Dragoons, and their Brigadier was Major-General Slade.

On 6th March, at daybreak, Viscount Wellington discovered that the French had retreated and left their camp at Santarem. Marshal Massena was really forced to take this step, his army having become so wasted by sickness and privation. Our troops accordingly advanced in pursuit, the Fourteeth still being employed on outpost duty, and forming part of the advance-guard of the army.

On 8th March, Captain Babington's squadron, supported by the remainder of the regiment under Lieutenant-Colonel Hervey, made a brilliant and most successful charge against four squadrons of the 11th and 26th French Dragoons at Venta de Serra, capturing 14 men and 14 horses, and losing only 2 men and 2 horses.

Venta de Serra, 8th March 1811.

The regiment was now engaged in the different skirmishes and actions which were fought by our troops against the rear of the retreating enemy, the principal of which were :—

Pombal, 10th March.
Redinha, 12th March.
Casal Nova, 14th March.
Foz D'Aronce, 15th March.
Battle of Sabugal, 3rd April.
Miranda de Colvo.
Tay D'Aortos.

In these various engagements in which the Fourteenth took part, they escaped without any casualties.

The French army were heavy losers in the fight at Sabugal, where they were unskilfully handled by Reynier, and lost 1500 men to the allies' 200. Wellington said, 'This was one of the most glorious actions British troops were ever engaged in.' The whole affair did not last an hour. It took place on the banks of the Coa: Reynier had to attack the British up hill. The 43rd and 52nd Regiments of Light Infantry particularly distinguished themselves under Brigadier-General Beckwith, and captured a howitzer. Brigadier-General Colville's Brigade of the 3rd Division by a resolute fire on the French left decided the victory, and our cavalry pursued the flying enemy in their retreat to Rendo and Alfayates. The larger portion of the French army had reached Ciudad Rodrigo about the 4th April, and from thence Marshal Massena, Prince of Essling, continued his retrograde movement to Salamanca, which he occupied. Sabugal, 3rd April 1811.

Wellington was now on the confines of Portugal, and invested Almeida. The Light Division occupied Gallegos and Espeja, whilst the rest of his army was disposed in villages on both sides of the Coa, and the headquarters were settled at Villa Formosa on the frontiers of Spain and Portugal. The Fourteenth furnished the outposts on the left bank of the Agueda, at Villa del Egua in the Spanish province of Leon, and intelligence was brought in from the Spanish town of Ledesma to the effect that the French army had been reinforced and reorganised, and that it was advancing. A squadron of the Fourteenth was hastily despatched under command of Captain Brotherton to Santa Esperita, but this was soon driven back behind the Agueda by the advancing columns of the enemy.

General Brotherton relates the following amusing anecdote with reference to the experiences of his patrol on this occasion: *Sans culotte.*
—'I had been sent (in May 1811) patrolling to a distance from the army, in search of the enemy, when we were behind the

Agueda, and had not found him after a most harassing and fatiguing day, in most sultry weather, and could not get back to the army that night, but put up at San Felices, where I considered we were out of reach of the enemy. I put up at the priest's house, placing the men and horses, twelve in number, under sheds, in a large back yard, and felt so fatigued myself that I was tempted by the sight of a nice clean bed, and after a hearty supper which the priest gave me, to turn into bed, where I soon fell fast asleep. In the middle of the night I was woke by the priest coming to my bedside, and telling me that some French cavalry were passing through the town. I jumped up, and went to the window, and, by the light of the moon, which shone brightly, sure enough, I saw French cavalry very composedly walking through the streets, and just commencing to billet off, knocking at the different doors, and at the same moment came a loud rap at the priest's door. I had not even time to put on my breeches, but scampered off with only my cloak and my sword, and got down just in time to jump on my horse, and get my party out of the back gate, and galloped off in an opposite direction to that which the enemy had come from. I was not followed, and the enemy changing his intention marched through the place without halting in it, so I returned to fetch my breeches, etc., and to thank the honest and hospitable priest who, though frightened at first, laughed heartily at my *sans culotte* adventure, which was matter of mirth throughout the army.'

The object of this advance of the Prince of Essling was to relieve Almeida, which the allies had besieged. On 25th April he reached Ciudad Rodrigo.

On the 3rd May, whilst we were retiring in the face of very superior numbers of the enemy, Lieutenant John Townsend[1] of the 14th Light Dragoons was in charge of the picquets, and he had to bring them in gradually under a heavy cannonade towards Fuentes d'Onor. The main body of the Fourteenth was engaged the same day behind Gallegos, and a squadron

Affairs at Gallegos and Poço Velho, 3rd May 1811.

[1] Afterwards Colonel J. Townsend, A.D.C., commanding 14th Light Dragoons.

commanded by Captain Brotherton had a sharp skirmish near Poço Velho. Wellington had concentrated the main body of his army behind the Duas Casas river, and the French had to cross the Azava river, which was swollen and difficult to ford: this delayed them a few days, and the British advanced posts fell back on Fuentes d'Onor, where the main body occupied a tableland between the Turones and Duas Casas, their left being at Fort Conception, their centre opposite the village of Alameda, and their right behind Fuentes d'Onor.

Battle of Fuentes d'Onor, 5th May 1811.

On the 4th May, Wellington extended his right to Nave d'Aver, which, excluding the circuit of blockade round Almeida, made his line of battle 7 miles in length, but this gave him a safer line of retreat. Our cavalry was very weak at this battle, and the enemy was particularly strong in that arm: ours did not exceed a thousand sabres. The French attack commenced two hours after daybreak on the 5th May, by Montbrun turning the right of Wellington's Seventh Division, and then charging the British cavalry, which had moved up in support.

The attack made on our position in the rear of the village is thus alluded to by Napier:—'The French with one shock drove in all the cavalry out-guards, and cutting off Captain Ramsay's battery, came sweeping in upon the reserves of horse and upon the Seventh Division. But their leading squadrons approaching in a disorderly manner, were partially checked by the British, and, at the same time, a great commotion was observed in their main body. Men and horses there closed with confusion and tumult towards one point, a thick dust arose, and loud cries, and the sparkling of blades, and the flashing of pistols, indicated some extraordinary occurrence. Suddenly the multitude became violently agitated, an English shout pealed high and clear, the mass was rent asunder, and Norman Ramsay burst forth at the head of his battery, his horses, breathing fire, stretched like greyhounds along the plain, the guns bounding behind them as things of no weight, and the mounted gunners followed in full career. Captain Brotherton of the 14th Dragoons seeing this, rode forth with a squadron and overturned the head of the pursuing troops, and General Stewart joining in the charge, took the French General Lamotte, fighting hand to hand.'

After this the British cavalry had to retire behind the Light Division, which was thrown into squares. The Seventh Division fell back from Nave d'Aver, taking up a fresh position across the Turones river by Freneda, and during this retrograde movement the right flank was covered by the 14th Light Dragoons and the Royal Dragoons, who retired in good order by alternate squadrons under a heavy cannonade. One squadron of the Fourteenth charged some French artillery with great gallantry, but was repulsed, and it was here that Captain Knipe, commanding the squadron, fell mortally wounded, and was succeeded by Lieutenant (afterwards Colonel) John Townsend, who took command of the squadron. Lieutenant-Colonel Hervey at the head of the regiment had his horse killed under him, and received a severe contusion.

Casualties at Fuentes d'Onor.

The following casualties were incurred in this battle by the Fourteenth :—

Killed:
- Captain Knipe.
- 4 men.
- 6 horses.

Wounded:
- Lieutenant-Colonel Hervey.
- Captain Milles.
- Lieutenant Townsend.
- Lieutenant Gwynne.
- Lieutenant Badcock.
- Lieutenant Ellis.
- 13 sergeants.
- 28 men.
- 23 horses.

Missing:
- 3 men.
- 4 horses.

The battle of Fuentes d'Onor was a hardly contested

one. The French were superior in cavalry, having 5000 to our 1200, and having 40,000 infantry, 36 pieces of artillery, and a battery of the Imperial Guard to our 32,000 infantry and 42 guns. What took place on our extreme right has already been described. At the same time, nearer the centre, and in front of the village of Fuentes d'Onor, a fierce battle also raged, and two companies of the 79th Regiment were taken, Colonel Cameron being mortally wounded. It was here the French general, Drouet, made his furious attack on our lines, when he captured the lower part of the village notwithstanding the gallant stand made by the 71st, 79th, and 88th Regiments, who, though overmatched at first by sheer numbers and the fierce vigour of the attack, never quite relinquished the whole village, but rallied and then made a charge in which large numbers of the enemy fell. Here the fighting went on till evening, when the French at last retired some distance from the stream, and the British remained holding on to the crags and chapel.

On the extreme left, near Fort Conception, the allies maintained their position; and when at length the fighting, which had been desperate, came to an end, both armies remained as it were in observation the one of the other.

The total losses of the allies were 1500 men and officers, of whom 300 were taken prisoners. The enemy's loss was estimated at 5000, but this was over the mark. By the 10th May, Massena had retired beyond the Agueda, having been foiled in the attempt to relieve Almeida, and shortly afterwards Marmont assumed command of the French army operating towards Portugal. Both sides claimed a victory at Fuentes d'Onor, and Napier says, 'more errors than skill were observable on both sides' in this battle.

Lieutenant-Colonel Hervey received a gold medal, and the Royal authority was granted in 1820 for the Fourteenth to bear on its guidons and appointments the words 'Fuentes d'Onor,' as a special mark of His Majesty's approbation of the conduct of the regiment on this occasion.

General Brotherton relates several interesting episodes about Fuentes d'Onor:—

General Brotherton's anecdotes about the battle of Fuentes d'Onor.

1. 'At Fuentes d'Onor the Adjutant-General of the army (Lord L——) was near me, particularly in one remarkable instance in which he joined in a charge I made to protect and rescue Captain Ramsay's guns of the Horse Artillery, as mentioned in Napier's *History*. At another period of the battle he ordered me to go to the assistance of Don Julian Sanchez, whose guerillas were getting roughly handled by some French cavalry. Of course I immediately obeyed, though it seemed to me an injudicious order, for on this memorable day our great inferiority in cavalry (the enemy having fully 4000 in the field of their very best, a large proportion of it of the Imperial Guard, and commanded by Montbrun, one of their best officers) rendered it advisable to keep the little we had constantly together, and detaching any of it to a distance a dangerous step. However, as I before said, I instantly obeyed, and started at a brisk trot; for, in action, the least hesitation or slowness in executing an order is inexcusable in an inferior officer. I had not proceeded one hundred yards, when Lord Wellington, who was just arriving on this part of the field, rode up to me and asked me where I was going. I told him the orders I had received from Lord L—— (then General S.). He made no further observation than "Go back!"

Death of Captain Knipe.

2. 'At Fuentes d'Onor we had a very fine fellow, Captain Knipe, killed through his gallant *obstinacy*, if I may so call it. We had the night before been discussing the best mode for cavalry to attack batteries in the open field. He maintained, contrary to us all, that they ought to be charged in front, instead of the usual way in gaining their flanks, and thereby avoiding their fire. Poor fellow, the experiment next day, in support of his argument, was fatal to him. He had the opportunity of charging one of the enemy's batteries, which he did by attacking it immediately in front, and got through the discharge of round-shot with little loss; but the enemy having most rapidly reloaded with grape, let fly at his party,

at a close and murderous distance, almost entirely destroying it; he himself receiving a grape-shot, passing through his body. The shot went through his lungs. I was with the poor fellow the next morning, as long as he survived. He could speak distinctly, and was most composed and resigned, and even argued the point over again. His chief anxiety, however, was to be permitted to write a line to his mother, and he expired in the very act of attempting it. We buried him in the same grave with another gallant soldier who fell that day, Colonel Cameron of the 79th Highlanders.

3. 'Captain Badcock, commanding a squadron of the 14th Light Dragoons, was sitting on his horse at the head of his squadron, when he took for Spaniards running away (a very usual occurrence) some cavalry rapidly approaching him in line, and remained perfectly steady intending to charge those who appeared to be following the supposed Spaniards, the moment the latter had passed him. He was, however, not very agreeably surprised by being undeceived by a cut across the face from the French officer (for the supposed Spaniards were French). Badcock, however, who was an excellent officer, contrived, notwithstanding his surprise, to drive the enemy back in gallant style, with the loss, however, of two of his teeth; but he never thought of his wound till he had completed his duty, and then even never left the field for one moment. Captain Badcock's adventure.

4. 'Colonel Hervey, at the battle of Fuentes d'Onor, escaped losing his right leg by having put a thick book (Quenedo's *Works*, which he had taken from a private house the day before) into his sabretache. An eight-pound shot entering the sabretache, went through the horse, and just appeared on the other side of his body, without coming through the skin, and it was evident that the thick book prevented it from going through and taking off Colonel Hervey's leg. Poor fellow, he had already lost his right arm; and his leg, from the blow, immediately swelled to an immense size, but though the horse fell down dead, and in the fall again hurt him, he would not leave Lieutenant-Colonel Hervey's escape.

the field, but had himself placed under a tree, where he remained during the remainder of the battle.'

Equine sagacity.

5. It was either Fuentes d'Onor or the affair of Espeja,[1] near Ciudad Rodrigo, about which General Brotherton says :— 'I had my charger shot under me, and got on a troop-horse which was also shot under me, through the head, by the pistol of a French officer, so closely that my own face was singed. The animal fell, and a sergeant behind me dismounted and gave me his horse, and I thought no more of the animal that was shot through the head, supposing that he never rose again ; but on rejoining the main body of the regiment I found that the poor animal had arisen by an effort, gone back to where the regiment was formed, placed himself in the ranks in his own squadron, and then fell down dead ! This fact, almost incredible, can be vouched for by any officer or private belonging to the 14th Light Dragoons at the time.

The brave Don Julian.

6. 'I commenced the battle of Fuentes d'Onor by running away with 2 squadrons, for about 2 miles, pursued by a *brigade* of French cavalry. I had been sent the night before to the village of Nave d'Aver, which was occupied by that humbug, Don Julian Sanchez, with his corps of infantry and cavalry. It was a strong post, on an eminence, surrounded by stone wall enclosures, similar to those in Ireland, and no cavalry alone ought to have carried it. I arrived there late at night, and could not see what arrangements Don Julian had made for defence ; but he assured me all was secure, and that he meant to defend himself most obstinately, before he retired. Just at daybreak in the morning, however, having requested him to show me where his picquets were posted, he pointed out to me what he said was one of them, but I observed to him that it appeared to me in the dusk of the morning too large to be one of his picquets, but he persisted. However, the sun rising rapidly, as it does in these countries, dispelled the fog and the illusion the same moment, for what Don Julian pointed out to me as his picquet, proved to be a whole regi-

[1] Or Espejo (Cannon).

ment of French cavalry dismounted. They mounted immediately and advanced. I still thought the Spaniards would make a stand, as cavalry alone never ought to have carried the village; but the *brave* Don Julian, as the Spaniards called him, took himself off immediately with his whole force to the mountains, and left me with my 2 squadrons to shift for myself. The consequence was that I was pursued by the whole French cavalry towards the position at Fuentes d'Onor, where the army was drawn up, and the advance-guard of which was at Poço Velho, which, as I approached, I saw occupied by red-coats, and began to breathe and feel secure. As I approached I found our infantry posted with great regularity and steadiness, but as they did not commence firing on the French cavalry that were closely pursuing me, I rode up to the first officer I could approach, and asked him why he did not fire and stop the progress of the enemy. He replied with astonishment, "Are those the French?" I told him I knew it to my cost, having sustained considerable loss from them during my retreat. He immediately commenced firing on them, and most effectually checked them, bringing down numbers of men and horses. I found this was the 85th regiment, only just come up to the army, and never having seen the enemy before. There was, however, no want of steadiness and bravery when once they were told it was the enemy. This gave me the liberty of retiring leisurely to the position where the army was drawn up, and the battle then commenced in earnest. At this battle the numerical superiority of the enemy, in cavalry, was *four to one*, and of the best description, a considerable proportion being cavalry of the Guard; and some of the most distinguished of the French cavalry generals were commanding it—Montbrun, Fournier, etc. This was an eventful and critical battle.'

We have thus seen how the attempt to relieve Almeida had signally failed; but on the night of the 10th May the garrison blew up the fortifications, destroyed the guns, and breaking through the picquets in one column, skilfully led by

General Brennier, the governor of the fortress, made good their escape and joined the main body of the French army which then withdrew to Salamanca.

Lord Wellington now proceeded to Estremadura, leaving a large portion of his army on the Portuguese frontier, near Ciudad Rodrigo. The Fourteenth and Royal Dragoons, brigaded under Major-General Slade, took the outposts on the Agueda, covering the front between Villa del Egua, Gallegos, and Espeja.

Subsequently Marshal Marmont, Duc de Raguse, advanced once more with a numerous army and drove back the British posts from Ciudad Rodrigo, having introduced a convoy into the town. On the morning of the 6th June two French columns appeared, when the Light Division was directed to retire from Gallegos upon Nave d'Aver, and subsequently upon Alfayates. A squadron [1] of the Fourteenth acted with the Royal Dragoons under Lieutenant-Colonel Clifton to cover the retrograde movement. This retreat is described by General de Ainslie as follows, in his *Historical Record of the Royal Dragoons* :—'The Royal Dragoons, under Lieutenant-Colonel Clifton, together with a troop of the 14th Light Dragoons, assembled at 3 A.M. at Gallegos for the purpose of covering the retreat. At 7 A.M. the enemy, numbering 2000 cavalry, 6000 infantry, and 10 guns showed themselves, and this overwhelming force was met by the British cavalry in a most resolute and able manner. The celebrated French cavalry general, Montbrun, in vain endeavoured to outflank the Royals and 14th Light Dragoons. His squadrons were twice attacked and defeated, and the retreat of the Light Division was effected with little loss. Lieutenant-General Sir Brent Spencer, commanding the forces, in the absence of Viscount Wellington in Estremadura, thus reports to his lordship on these events :—"It is with the greatest pleasure I have to mention the very admirable conduct of 'the Royals' under the command of Lieutenant-Colonel

[1] Cannon, p. 34, says it was a squadron, but General de Ainslie in his *Historical Record of the Royal Dragoons* (p. 120) says it was only a troop of the Fourteenth.

Clifton, and one troop of the 14th Light Dragoons, which, being all that were employed in covering the front from Villa del Egua to Espeja, were assembled at Gallegos, and retreated from thence agreeably to my directions, and notwithstanding all the efforts of General Montbrun, who commanded the French cavalry, to outflank the British, pressing them at the same time in front with eight pieces of cannon. Their retreat to Nave d'Aver merits the highest commendation."'

Blockade of Ciudad Rodrigo, September 1811.

Marshal Marmont afterwards went to Spanish Estremadura with his large army, and the British general withdrew the troops he had in front of Badajos, and brought them to the vicinity of Ciudad Rodrigo, from which they had previously had to retire into Portugal. In the blockade of that city the Fourteenth took part as usual in the outposts, and when the French army advanced to relieve the blockade the regiment was stationed at Espeja, on the lower Azava, with advanced posts at Carpio and Marialva. The French marshal managed to get supplies into Ciudad Rodrigo, and brought up such enormous forces that eventually the blockade was raised and the British had to fall back. On the 23rd September, Marmont, advancing from Tamames, encamped behind the hills north-east of Ciudad Rodrigo. On the 24th the convoy entered the town, and 2 columns of French crossed the hills. On the same day Lord Wellington brought up a division of the allies to the position of Guinaldo, the rest of his army being disposed at various adjacent places such as Elbodon, Pastores, on the Vadillo (a river which falls into the Agueda, 3 miles above Rodrigo), Nave d'Aver, Espeja, and the lower Agueda.

Affair at Carpio, 25th September 1811.

On the 25th September, contrary to the British general's expectation, the French advanced from the Vadillo. Soon after daybreak 14 squadrons of the Imperial Guards drove the outposts from Carpio across the Azava; the Lancers of Berg crossed the river in pursuit, but they were charged and driven back by a squadron of the Fourteenth and 2 squadrons of the 16th Light Dragoons, upon which Carpio was again occupied by the British. The same day another strong body of the enemy under Montbrun attacked the British at Elbodon, from which they had to retire, the Fourteenth also falling back from Carpio, and a succession of retrograde movements of the whole of the allies now took place, until on the morning of the 28th they took up a strong position in front of the Coa, their right resting on the Sierra de Mesas, their centre covered by the village of Soita, their left at Rendo, upon the river. There had been frequent fighting during these three days without heavy losses on either side, and there were many brilliant instances of bravery and heroism in the actions at Carpio, Elbodon, and Aldea Ponte. In consequence of the unreturned fire at Elbodon, delivered by the infantry squares and the Portuguese artillery on the masses of French cavalry which charged the British so frequently and so vehemently, under Montbrun, the losses of the enemy were far greater than those of the allies. The Fourteenth had Lieutenant Hall, 2 private soldiers, and 5 horses wounded in the action at Carpio and the subsequent movements; and the conduct of the commanding officer, Lieutenant-Colonel Felton Hervey, was commended in public despatches, the excellent behaviour of Captain Brotherton being also mentioned.[1]

Casualties at Carpio, etc.

Affair at Espeja, September 1811.

Marmont's army being very short of provisions, he took it back to the valley of the Tagus on the same day that the allies took up their position behind Soita. It was either during these movements or a day or two earlier at Espeja that the memorable affair between the Fourteenth and the enemy's Lancers took place. The 14th and 16th Light Dragoons with the German Hussars were in brigade together under command of Count Arentschild of the German Hussars. The enemy's Lancers ('The Polish Lancers'), a crack corps in the French cavalry, were observed drawn up on some rising ground, when Arentschild rode up to Lieutenant-Colonel Hervey and said, 'Sir, you will charge them.' Two squadrons of the Fourteenth immediately advanced to the attack, the Lancers awaiting them on their own ground with lances 'advanced,'

[1] Cannon's *Historical Record of the 14th Light Dragoons*, p. 35.

thinking the Light Dragoons would never get inside them. However, the Fourteenth charged, broke through their ranks, and sabred more than 60 of them. It is said the Fourteenth were offered lances as a compliment on their return after the war, but refused them, giving as a reason what occurred on this memorable occasion. It was always thought Count Arentschild wanted to spare his 'pet' German Hussars, and so sent the Fourteenth against the Lancers.

French chivalry.

Napier mentions a fine chivalrous act which took place at one of the many cavalry encounters which occurred during these three days' fighting with the enemy at Carpio, Elbodon, and Aldea Ponte:—'A French officer, while striking at Felton Hervey of the 14th Light Dragoons, perceived he had only one arm, and with a rapid change brought down his sword into a salute and passed on.'

Establishment. Reductions in troop-horses, rank and file.

In December of this year (1811) a schoolmaster-sergeant was for the first time appointed to the regiment. There were some reductions made in the establishment, viz., the corporals were reduced from 50 to 40, the privates from 950 to 760, and the troop-horses from 964 to 864.

Lieutenant-Colonel—Felton B. Hervey.

On 4th June, Colonel S. Hawker had been promoted Major-General on the staff at home, and Lieutenant-Colonel Hervey succeeded to the command of the Fourteenth, which he had virtually held since the death of Lieutenant-Colonel Neil Talbot a year previously. It was not till the year 1817 that Lieutenant-Colonel Hervey's name appeared in the Army List as the only Lieutenant-Colonel of the regiment, for in the Army Lists up to and including the year 1816, Major-General Hawker's name continued at the head of the regiment as Lieutenant-Colonel, though holding the rank of Major-General.

Establishment reduced to 6 troops on service.

During the winter the strength of the regiment on foreign service was reduced from 8 to 6 troops. Accordingly, on 5th December 2 troops embarked at Lisbon for England, where they arrived on the 8th January 1812, and having disembarked at Portsmouth, joined the depôt at Radipole Barracks, Weymouth.

Napier says that after the combats about Guinaldo, the

allied army was extensively cantoned on both sides of the Coa.

Ciudad Rodrigo was distantly observed by the British, and so closely by Julian Sanchez, that he actually captured the governor, who had come out with too weak an escort, and also took a large number of oxen. In consequence of this the French army under Thiebault, coming from Salamanca and Tamames, advanced when the Agueda was flooded, revictualled Ciudad Rodrigo, leaving a new governor there, and returned on 2nd November before the waters had subsided; so that Wellington was unable to oppose him, the only bridge available being at Ciudad Rodrigo, in possession of the enemy.

Later in November the French made another advance under Dorsenne. The British and allies then crossed the Agueda near Zamara, whereupon the French retired, harassed in rear by the guerillas under Carlos d'España and Julian Sanchez.

State of the Army in the Peninsula.

After this, owing to want of supplies from the country between the Coa and Agueda, and the failure of the transport promised by the Portuguese, Wellington was forced to spread out his cavalry even as far as the Mondego and valley of the Douro, or they would have been starved. At this time the British army was not in good plight. The last reinforcements received by Wellington consisted of infantry that had served in the recent Walcheren expedition, who were so enfeebled that exposure to night air or hardship at once threw them into hospital by hundreds, whilst the recently arrived cavalry regiments, being inexperienced and not acclimatised, were found, both men and horses, so unfit for duty that they had to be sent to the rear. Added to this, the pay of the army was three months in arrear, the supplies were very scanty, half and quarter rations were often served; often there was no bread for three days consecutively, and the men's clothing was so patched that scarcely a regiment could be known by its uniform. Chopped straw, the only forage, was very scarce; the regimental animals were dying of hunger; corn was rarely distributed save to the generals and staff, and even the horses of the artillery and the old cavalry suffered. The cantonments about the Coa and Agueda were unhealthy from the rains; 20,000 men were in hospital, and only 54,000 men of both nations, including garrisons and posts of communication, were under arms. But the change of position worked wonders: the new cantonments gave abundance of supplies and dry weather, for in Beira the first rains usually subside in December, and the sickness stopped in consequence. At this critical time the army was lucky in having such a capable military secretary as Lord Fitzroy Somerset.[1] It was now that Wellington decided to besiege with vigour and then to storm Ciudad Rodrigo on the first opportunity. He had 35,000 men available to do it with.

1812

Hitherto Ciudad Rodrigo had not been regularly invested, but on the 8th January the redoubt of Francisco was stormed and taken by selected companies of the Light Division led by Colonel Colborne, 52nd Light Infantry, after which a regular siege and investment took place, and by the 19th two breaches became practicable. The assault, after desperate fighting and heavy losses, was successful. Lieutenant Gurwood,[2] one of the first to enter, received the governor's sword at the castle, though himself severely wounded in the head during the fight in the streets. Three hundred French fell, 1500 were made prisoners, 150 pieces of artillery were captured, and the allies lost 1200 men and 90 officers in the siege. Generals Craufurd and Mackinnon were killed. It was a desperate affair: no less than 60 officers and 650 men were killed or hurt at the breaches.[3] After this gallant affair Lord Wellington was created Duke of Ciudad Rodrigo by the Spaniards, Earl of Wellington by the English, and Marquis of Torres Vedras by the Portuguese. The siege had lasted only twelve

Ciudad Rodrigo besieged and captured, January 1812.

[1] Afterwards General Lord Raglan.—(Napier, Book xvi. ch. 2.)

[2] Lieutenant Gurwood, 52nd Regiment led the forlorn hope.—(Napier.)

[3] Napier, from whose *History* this account of the siege is taken.

days, during which time the weather had been intensely cold with severe frosts, but from this time up to the end of February violent and continuous rain fell in the Peninsula. Ciudad Rodrigo was captured on the 19th January, and the garrison marched out as prisoners of war.

On the 30th January, Captain Charles Massey Baker (afterwards Lieutenant-Colonel of the Fourteenth), and on 26th March, Captain T. W. Brotherton (afterwards General Sir T. W. Brotherton, G.C.B.), became Majors in succession to the Hon. C. Butler and J. Chapman. The cavalry had not taken any active part in the capture of Ciudad Rodrigo, but they were at hand in its vicinity to carry out reconnoitring duties as required, and to be in readiness for any unforeseen emergency.

After the place fell into our hands the British army was kept on the Coa for some time, and Lord Wellington's headquarters remained there till 5th March, by which time the main body of his army was well on the way to the Alemtejo and the vicinity of Badajos, Marmont's army being at Salamanca.

Siege of Badajos, 17th March 1812.

The Fourteenth proceeded to Estremadura, and was stationed near Badajos when the siege of that fortress commenced. The heavy rains which fell at the equinox considerably interfered with military operations, and in consequence the commencement of the siege was delayed till the 17th March.

The British headquarters were at Elvas by the 11th, and by the 15th pontoons were laid over the Guadiana river, and the investment of Badajos was completed soon after. There were several French armies hovering about: Soult was before the Isla, but Drouet's division, 5000 strong, was at Villa Franca, while Daricau, with a like force, was near Medellin; in consequence of which Lieutenant-General Sir Thomas Graham[1] was despatched with 3 divisions of infantry and 2 brigades of cavalry to march upon Llerena, by Valverde and Santa Marta, and Lieutenant-General Sir Rowland Hill was sent upon Almendralejos, moving thither from Albuquerque by Merida. The Fourteenth were with the covering army under Sir Thomas Graham, and when the French

[1] Afterwards General Lord Lynedoch, G.C.B., G.C.M.G.

army under Marshal Soult advanced, the British fell back upon Albuhera. The regiment was employed in covering this retrograde movement, and it had an encounter, whilst skirmishing, with the enemy's advance-guard near Villa Franca, which is mentioned below. Badajos was captured by storm on the night of the 5th April, when the French relieving army fell back. Capture of Badajos, 5th April 1812.

The Fourteenth were present at the siege of Badajos up to the 1st April, on which date they were relieved by the 11th Light Dragoons, and proceeded with the covering army under Sir Thomas Graham. Whilst in front of Badajos, on 20th March the garrison made a sortie, and two men of the Fourteenth were wounded.

After the fall of Badajos the Earl of Wellington proceeded to the north, but Lieutenant-General Sir Rowland Hill was left in command of the army in Estremadura. The British and Portuguese losses at Badajos were very heavy: it was a desperately contested assault. Five thousand men and officers fell in the siege, of whom, including 700 Portuguese, 3500 fell in the assault, 60 officers and more than 700 men being slain on the spot.

It was a few days after the capture of Badajos that the Fourteenth were engaged in an enterprise against several regiments of French cavalry. The regiments had moved on the night of the 10th April from Villa Franca upon Usagre, and afterwards along the road to Llerena. On the 11th the Light Brigade skirmished with the French, until the Heavy Brigade turned their flank. The enemy was then charged, overthrown, pursued, and many prisoners taken. On the night of the 12th April a party of the Fourteenth, under Lieutenant Edward Pellew, took a picquet of 22 French dragoons prisoners. The regiment had upwards of 20 men and several horses wounded in these affairs, and the conduct of Lieutenant-Colonel Hervey was commended in the despatch of the commander of the cavalry, Lieutenant-General Sir Stapleton Cotton, Bart. The affair at Usagre was very skilfully managed by Sir Stapleton Cotton. Napier says the advance-guard of the Affair at Usagre, near Villa Franca, 11th April 1812.

Light Brigade commenced the action; the French fell back before Le Marchant's Heavy Brigade could intercept them, but as the heights skirting the Llerena road prevented them from seeing Le Marchant, they again drew up in order of battle behind the junction of the Benvenida road. The numbers on each side were about 1900 sabres, and Cotton, seizing an accidental advantage of ground, kept the enemy's attention engaged with Ponsonby's (light brigade) squadrons, while Le Marchant, secretly passing at the back of the heights, sent the 5th Dragoon Guards against their flank, and the next moment Ponsonby charged their front. They gave way, and being pursued, lost several officers and 128 men prisoners, and many were killed in the field. The loss of the British was 56 men and officers, 45 being of the 5th Dragoon Guards. The French retreated on Drouet's infantry, then at Llerena, but all now fell back behind the Guadalquivir. The French cavalry general was Peyrezmont, belonging to Soult's army. From Estremadura the Fourteenth marched towards the Agueda, and after being some time in Portugal and on the frontiers of Spain, formed the advance-guard of Sir Thomas Graham's column in the march towards Salamanca, near which city it skirmished with a body of the enemy on the 16th June, losing 1 sergeant and 1 trumpeter killed on that occasion, 4 privates and 5 horses of the regiment being wounded. The French under Marshal Marmont retired beyond the Duero,[1] and the allies followed up to the banks of that river, where the 14th Light Dragoons were formed in brigade with the 1st German Hussars, and took the outposts at Tordesillas. The general advance of Wellington's army from the Agueda towards the Tormes commenced about the middle of June, when the rains ceased. His army numbered 24,000 men. He marched in 4 columns, and by the 17th June his army, now concentrated, occupied the mountain of San Christoval, 5 miles in advance of the city of Salamanca. In the middle of July, on the 15th and 16th, Marshal Marmont with his large army

Wellington crosses the Agueda and advances towards Salamanca in June 1812.

[1] Douro in Portugal, Duero in Spain.

PENINSULA AND

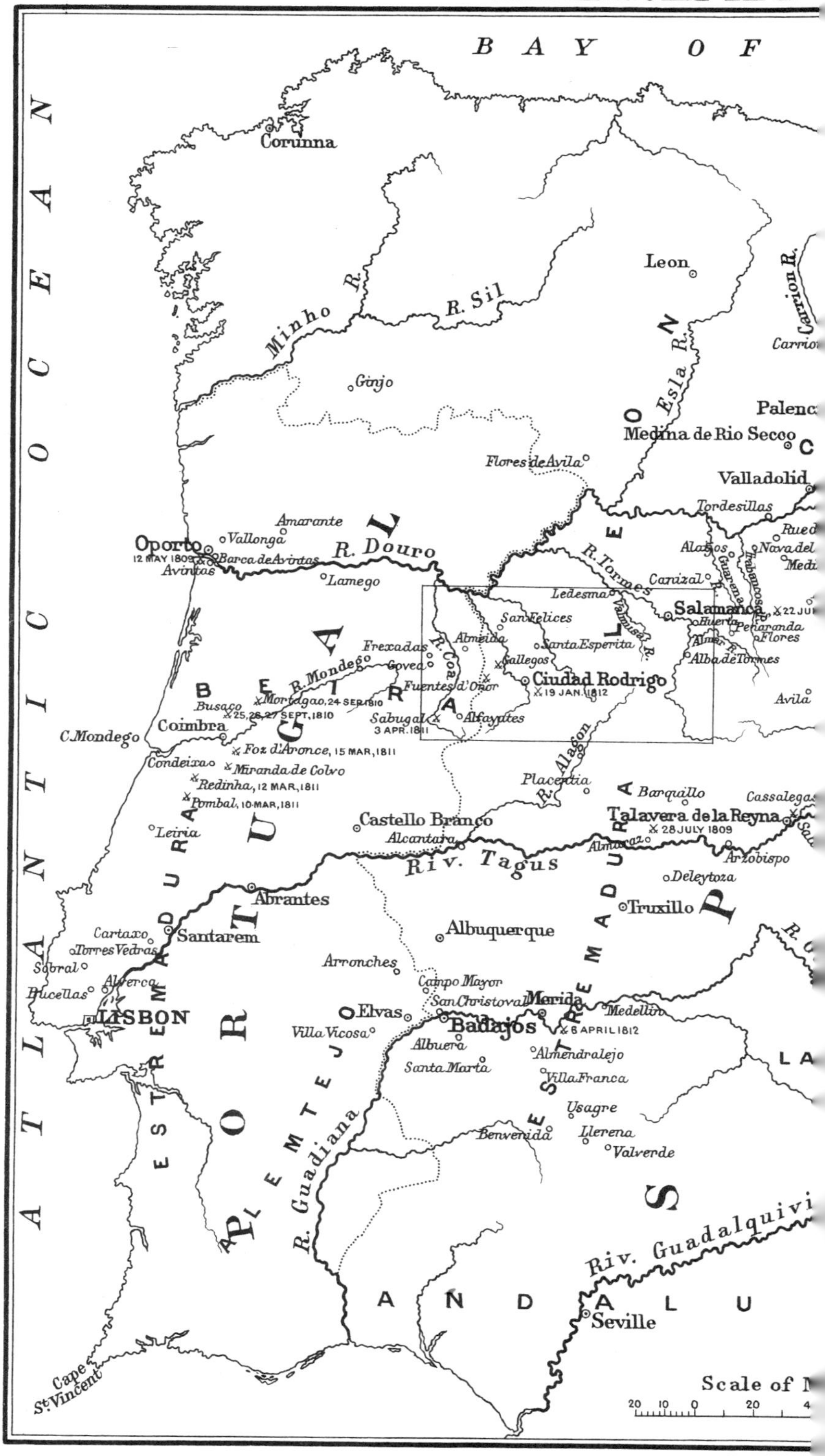

Lond

OF FRANCE 1808-1814.

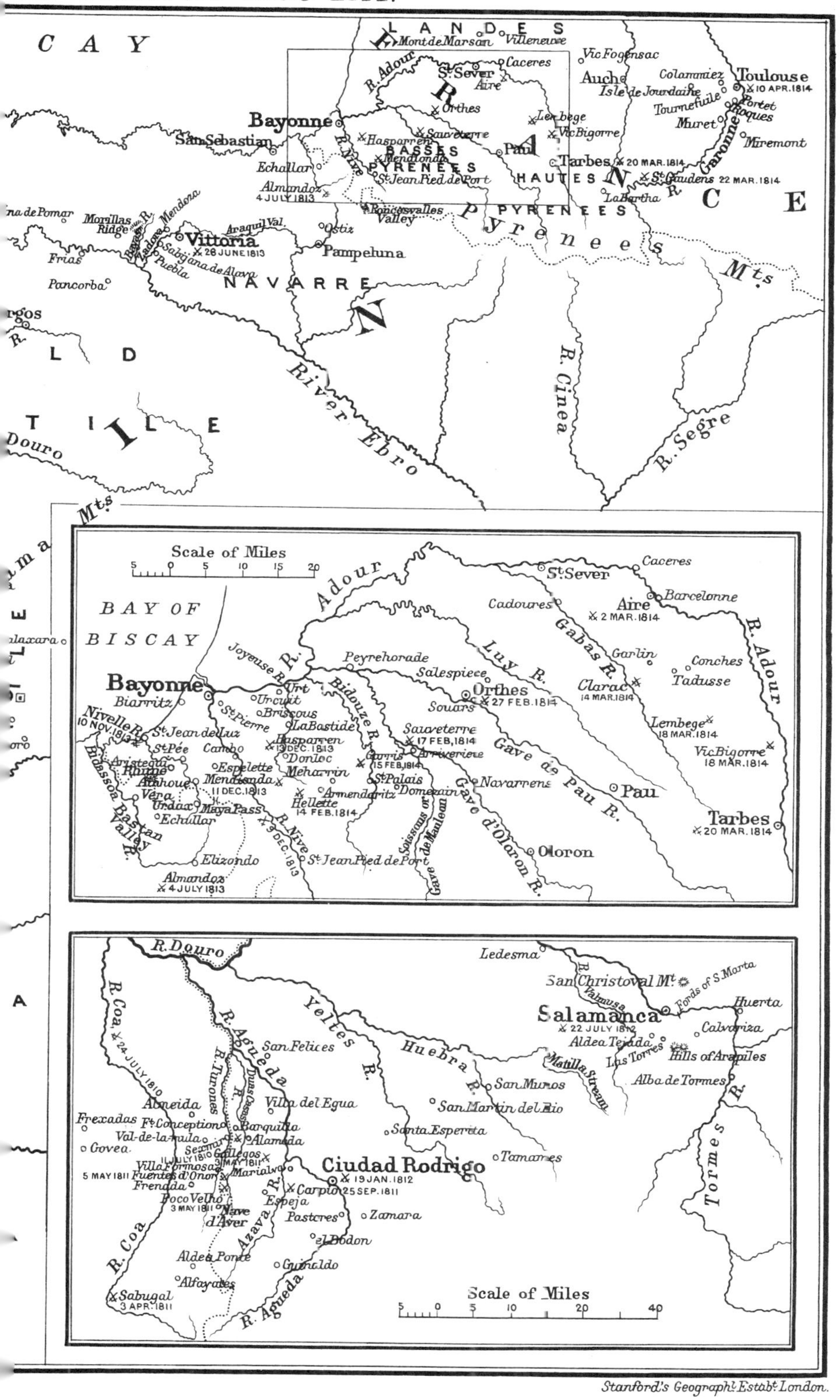

Stanford's Geograph! Estab! London.

reen & Co.

commenced offensive operations against the allies, and crossed the river Duero at several points, so that Lord Wellington found it advisable to take up a position at Canizal, on the Guarena stream, where he united his centre and left, leaving Sir S. Cotton with the right wing (composed of the Fourth and Light Divisons and Anson's cavalry), on the Trabancos.

The French occupied Nava del Rey on the 17th, and on the same day the Fourteenth and the 1st German Hussars, who had acted as rearguard and covered the retreat from Rueda behind the Guarena, moved to Alaijos so as to cover the retrograde movement of the right wing also, and Anson's cavalry from Castrejon. On the 18th some sharp skirmishing took place, and the troops at Castrejon fell back behind the Guarena. The Fourteenth had to retire from the plain near Alaijos under a heavy fire, and moved to Castrillos.

When the French army came up to the opposite bank of the Guarena, General Clausel sent a brigade of cavalry under General Carier across, supported by a column of infantry, with the intention of attacking the British left. On this occasion Major-General Victor Baron Alten, commanding the brigade, led the 14th Light Dragoons and the 1st Hussars of the King's German Legion against the French cavalry, and some sharp encounters took place. Subsequently they charged the enemy's infantry most successfully. General Carier was taken prisoner, and the enemy was driven back. During this engagement the 27th and 40th Regiments, supported by a Portuguese brigade, broke the enemy's infantry by an impetuous bayonet charge after the Fourteenth and German Hussars had repulsed the cavalry, and it was now that our cavalry charged the broken infantry and sabred a number of them, pursuing and making some prisoners. The Fourteenth lost 18 men and 20 horses killed; 34 men and 18 horses wounded, as well as the following officers:— Casualties at Castrillos, 18th July 1812.

Captain Brotherton.
Lieutenant John Gwynne.
Lieutenant Francis Fowke.

During the next three following days, the 19th, 20th, and 21st July, the regiment was actively employed in the operations which ensued between the opposing armies, and they had several skirmishes with the enemy.

Battle of Salamanca, 22nd July 1812.

On the 22nd July, at the battle of Salamanca, the Fourteenth began at daybreak to skirmish with the French outposts, in company with the 1st German Hussars, acting as the advance-guard of the 3rd Division of Wellington's army, and they afterwards took their place in the line of battle. They were subsequently engaged successfully with the 3rd Division in its attack on the French left, when Major-General Victor Baron Alten was wounded.[1] The French commander, Marshal Marmont, endeavoured by several changes of position to turn our right, so as to gain the road leading to Ciudad Rodrigo. These movements occupied many hours, and it was near three o'clock in the afternoon when a report came to Wellington that the French left was actually pointing towards that road, and that it was rapidly moving away from the centre of the French army. He instantly repaired to the high ground and watched the movement intently for some time, and when at length he was quite satisfied that their left wing was entirely separated from their centre, he lost no time in taking advantage of such a flagrant fault in military tactics, and, ordering his divisions forward, commenced the battle in real earnest. The allies' position had to be suddenly reversed from what it had previously been. In the first line as now constituted were the 4th Division, with the 5th on their right, Bradford's Portuguese on the right of the 5th Division, and Le Marchant's heavy cavalry on their right. In the second line were ranged the 6th and 7th Divisions, flanked on the right by Anson's light cavalry. This second line was now prolonged by the Spanish troops in the direction of the 3rd Division, which, with D'Urban's Portuguese cavalry, having

[1] The Fourteenth were in Alten's Brigade with the 1st German Hussars. After being shot in the thigh the Brigadier was able to rejoin his brigade at Madrid six weeks later, and commanded it in the retreat to Salamanca and Portugal. See p. 113.

passed the Tormes river by the fords of Santa Marta, was posted near Aldea Tejada, and so placed as to command the main road leading to Ciudad Rodrigo. The allies' position having thus been reversed to what it originally was, their left now rested on the English Hermanito, their right on Aldea Tejada; the rear had become the front, and the interval between the 3rd and 4th Divisions was quickly filled by a simple counter-march, with Bradford's Portuguese infantry, the Spaniards and the British cavalry, all massed about the village of Las Torres.[1] Marmont's arrangements occupied several hours, he all the time masking his real intentions from the British commander, who had almost ceased to watch him until the false movement already alluded to was detected.

The 3rd Division was now reinforced by Arentschild's German Hussars, which, with D'Urban's horsemen, closed the extreme right at Aldea Tejada. A reserve, composed of the Light Division, Pack's Portuguese, Bock's and Alten's cavalry, remained in heavy masses on the highest ground behind all. The 3rd Division, with its cavalry and 12 guns, was ordered to advance and cross the enemy's line of march; the remainder of the first line, with the main body of the cavalry, was directed to advance whenever the attack of the 3rd Division was developed, while Pack's brigade was to assail the French Hermanito,[2] so soon as the left of the British line should pass it.

Marmont now used all his endeavours to hasten up his troops who were still behind, and at the same time to delay the progress of his left wing, and he was still hopeful of success until he observed Pakenham with the 3rd Division coming forward against his left; and it was when he was hurrying thither that he received a severe wound from an exploding shell which rendered him completely *hors de combat*. This contretemps had a serious effect upon the French troops, owing to the confusion which ensued in the giving of orders.

[1] Napier.

[2] There were two rugged hills on the field of battle called the Hermanitos or Arapiles. One was christened the 'English' and the other 'Marmont's' or the 'French Hermanito.'

Bonnet assumed command, but he too was soon after wounded, when Clausel succeeded him in the supreme command. It was about five o'clock when the 3rd Division fell upon the French left, commanded by Maucune and Thomières (the latter was killed); and two of our batteries of artillery, most skilfully posted on high ground, took them in flank. The French fought with great bravery, but the 3rd Division drove their opponents back in confusion upon the supporting columns. It was just at this juncture the French cavalry assailed the flank of the 3rd Division, and were most gallantly charged by D'Urban's and Arentschild's horsemen, on which occasion the Oporto regiment under Watson attacked a square of infantry unsuccessfully and retired, leaving Watson wounded on the ground.[1] The Fourteenth were present with the 3rd Division in this important attack, which led to the complete discomfiture of the enemy's left; and two squadrons under Lieutenant-Colonel Hervey reinforced D'Urban's Portuguese Brigade, and thus took an active part in the successful turning movement which was effected. In the result, this repulse of the French left flank contributed more than anything else to our glorious victory at Salamanca, whereby the enemy was defeated with great loss and driven from the field. Whilst this movement had been going on, the 4th and 5th Divisions and Bradford's Brigade were hotly engaged, but steadily gaining ground; Le Marchant's heavy cavalry, Anson's light cavalry, and Bull's troop of artillery were advancing at a trot upon the left of Pakenham's Division, whilst, as already related, on Pakenham's right D'Urban's cavalry had turned the disordered masses of the enemy's left. Although not more than half an hour had elapsed since the actual battle commenced, the French were already losing ground, their left being in complete confusion. They began to fire at random; and when the British cavalry charged forward, coming between the interval of the 3rd and 5th Divisions, and then forming line, Le Marchant's heavy horsemen and Anson's Light Brigade

[1] Napier.

were seen to break forth at full speed, and next moment 1200 French infantry were trampled down by the charging squadrons, who rode onwards, sabring hundreds of them with their long straight swords. In this onslaught the cavalry lost heavily, Le Marchant and many other officers fell; Cotton and his staff were still at their head and galloped on, though opposed to a terrible fire; and Lord Edward Somerset, who persistently continued the charge at the head of one squadron, captured five guns. The heavy cavalry met with the greatest opposition and suffered in proportion. Anson's Light Brigade had suffered little in the charge, so they still pressed on, joined by D'Urban's horsemen, and, united with the 3rd and 5th Divisions and the guns, engaged the enemy with vigour. Meanwhile a fierce battle raged in the centre also. There Clausel made a surprising effort. Our 4th Division had driven back Bonnet's troops, who got mixed with the disordered masses of Maucune's and Clausel's Divisions, now retreating before Pakenham and the cavalry. The French Hermanito was assailed, but unsuccessfully, by Pack's Portuguese about the time of Le Marchant's charge. Clausel, when he assumed the command of the army, brought up Ferey's Division from Calvariza, and concentrated it in the centre behind Bonnet's troops, who were still strongly fighting; he also brought towards the same point the light cavalry, Boyer's dragoons, and the two divisions so long expected from the forest. By these dispositions he presented a mass for the broken left wing to rally upon, and he caused Sarrut's, Brennier's, and Ferey's unbroken Divisions, supported by the whole of the cavalry, to cover the line of retreat to Alba de Tormes, while another division was in mass close behind Marmont's Hermanito, and Foy remained still intact on the right.

Pack had failed to take the Hermanito with his Portuguese: he was driven back by the French reserves hidden from view behind the rocks on his front and left flank; this was a very critical moment in the battle. The men of the 4th Division were also driven back from the southern ridge, and were menaced by Clausel's troops in rear and on their left, when the gallant 40th Regiment,[1] wheeling about, with a rough charge cleared the rear, and thus saved that quarter. The front of the 4th Division was, however, driven back, as well as the front line of the 5th, which Boyer's dragoons had menaced. Cole had fallen, also Leith, both severely wounded; Beresford brought up some Portuguese troops, but he fell desperately wounded; and Boyer's dragoons now charged with success, in consequence of Anson's cavalry having been checked by a heavy fire of artillery. This was the real crisis of the battle. Wellington now brought up the 6th Division from the second line, which made a well-sustained and vehement charge, losing heavily; Hulse's Brigade on the left lost hundreds, the 61st and 11th Regiments being exposed to a withering fire. The Fifty-third were disordered by an impetuous charge of Boyer's dragoons and lost many men, but bravely held their ground. The fighting still continued, and the changing current of battle now turned for the British. Clausel, however, skilfully protected a line of retreat by the roads leading to the fords of Huerta and Encina, and the road leading to Alba de Tormes. Thither, accordingly, the French army, driven by the 3rd, 5th, and 6th Divisions, fell back in disorder. Wellington assailed Foy's Division, which was covering the retreat, with his Light Division and some cavalry, supported by the 1st Division and two brigades of the 4th Division, whilst as a reserve he sent the Spaniards and the 7th Division.

The French retired by alternate wings in good order before these troops, firing upon the Light Division from every rise of ground, and this retrograde march continued for two miles, but luckily, owing to the twilight which baffled the French aim, our casualties were not large. Just before it became dark, however, the 6th Division was involved in a fight with Maucune's Division, and attacked a strong position, when they lost heavily from the fire of the French guns; but eventually the enemy got completely under cover of the forest

[1] Now the South Lancashire Regiment, 1st Battalion.

and were lost to view, when the battle came to an end, and the French made good their retreat to Alba.

In this battle, Captain Brotherton of the 14th Light Dragoons, who, when fighting on the 18th July at the Guarena amongst the foremost, as he was always wont to do,[1] had a sword thrust through his side; yet he was again on horseback on the 22nd, and being denied leave to remain in that condition with his own regiment, secretely joined Pack's Portuguese in an undress, and was again hurt in the unfortunate charge at the Hermanito

The casualties of the Fourteenth were not great at Salamanca—they had 4 men killed, and 6 men and 7 horses wounded.

After the battle Clausel passed the Tormes by the narrow bridge of Alba and the fords below it, and at daylight was in full retreat upon Penaranda. Wellington, having brought up his German dragoons and Anson's cavalry, also crossed the river with his left wing at daylight, and came upon the rear of the French army near the Almar, a small river at the foot of a height near the village of La Serna. The cavalry pursued and attacked some squares of infantry, whom they took by surprise, and gained a success, but lost a hundred of their number killed by the fire of the French muskets.

The Light Division also came up and pursued to Nava de Setroval, near which place such large bodies of the French cavalry covered the rear of the army, that the allied cavalry, who were reduced in numbers and fatigued with fighting and marching, did not make any further attack, and the French made good their retreat in the direction of Flores de Avila and Arevalo. The Fourteenth were in the pursuit on 23rd July, when two squadrons of the regiment had a sharp engagement and took several prisoners near Penaranda. Penaranda, 23rd July 1812.

The total losses of the allies at Salamanca and in the operations before and after the battle amounted to— Casualties at Salamanca,

[1] Napier, vol. iv. p. 275.

1 Field-Marshal; 4 Generals; and nearly 6000 officers and men killed and wounded.[1]

The losses of the opposing French army during the same period, 18th to 30th July, were:—

1 Marshal; 7 Generals; and 12,500 men and officers killed, wounded, and taken.

The French also lost two eagles, several standards, twelve guns, and eight carriages.[2] It was on the 18th July that Marmont's army crossed the Duero in its advance, and on the 30th it recrossed that river in retreat, and finally had to retreat to Burgos, when the allies took possession of Valladolid.

For its gallant bearing at Salamanca the Fourteenth received the royal authority in 1820 to bear the word 'Salamanca' on its guidons and appointments, and Lieutenant-Colonel Sir F. B. Hervey, Bart., received a gold medal in recognition of his services, and as a mark of royal favour and approbation.

Gallant exploit of Corporal Hanley's patrol, 26th July 1812.

It was on the 26th July that a patrol of 3 dragoons of the Fourteenth, and 4 men of the German Hussars, under Corporal William Hanley of the 14th Light Dragoons, detached to Blasco Sancho, captured a party of the enemy, consisting of 2 officers, 1 sergeant, 1 corporal, and 27 mounted dragoons, with 1 private servant and 2 mules with baggage, when they had been sent forward to observe the movements of the enemy. Corporal Hanley and his gallant patrol were most highly complimented by the Commander of the Forces, who himself, through Lieutenant-Colonel Sir F. B. Hervey, Bart., at Madrid, presented a pecuniary reward for the men of the patrol, and Corporal Hanley afterwards received a special

[1] The British lost 1 General killed—Le Marchant; 5 Generals wounded—Beresford, Cole, Leith, Cotton, and Alten.

The French lost, killed, 3 Generals of Brigade—Thomières, Farey, and Desgraviers; wounded, Marmont and Bonnet, severely, and Clausel, slightly.

The French Marshal Marmont had with him 44,000 men. He was joined on 8th July by Bonnet from the Asturias, and later by King Joseph from Madrid.

[2] The account of the battle of Salamanca is mostly taken from Napier.

medal for his gallantry. The French horses were given to the Fourteenth and the German Hussars to complete deficiencies. The men captured by Hanley's patrol had really been left there by King Joseph of Spain, who had quitted Madrid on 21st July, and was at Blasco Sancho on the 24th, and thence made a forced march to Espinar.

The following extract from Tancred's *Historical Record of Medals* has reference to Corporal Hanley's medal:—

Hanley's medal for Blasco Sancho.

'Corporal William Hanley, 14th Light Dragoons, was presented by the officers of his regiment with a silver medal at a full dress parade, as an honourable testimony and to commemorate a brave action. The medal bears on

Obverse (within a wreath)—

"Fortitudine Blasco Sancho, 26th July 1812. Peninsula."

Reverse (within a wreath)—

"William Hanley, Corporal, 14th Light Dragoons."'

(This medal was in the late Stewart Mackenzie's collection, but is now in that of Major-General the Hon. Herbert Eaton, late Grenadier Guards.)

General Brotherton relates some interesting episodes concerning events which happened at Salamanca, both before and after the battle, as follows:—

Fording the Duero.

'The river called Douro in Portugal, is called Duero in Spain. When, in the summer of 1812, Lord Wellington and Marshal Marmont were manœuvring against each other on the banks of this river, towards Tordesillas, the river being much swollen by rains, it became essential, but very difficult to ascertain the fords, in order for one party or the other to know where to cross, either to attack or defend the passage. As the picquets of the respective armies were placed close to the banks, it was a ticklish thing to attempt the trial of any ford. I was determined, however, to attempt it, and had recourse to a little ruse in order to accomplish it. I rode out one evening late, with my orderly dragoon behind me, to the border of the river on the left bank, on which side the British army was then posted, and immediately opposite to a French picquet.

I crossed the river and found out the ford. The French picquet looked on most composedly, thinking, no doubt, that myself and my orderly were deserters, and when we reached the opposite bank the French officer came up to me, and asked what we were and what we wanted. He advanced about fifty yards from his picquet, which was drawn out. I saw that it was necessary to get out of the scrape by stratagem, and I gave him some vague reply, upon which he seized hold of my bridle, perceiving that he had been humbugged by allowing me to cross the ford. I immediately drew my pistol, cocked it, and might have shot him dead, but did not like to commit such an act in cold blood. He staggered back and ran to his picquet, which he ordered to fire. I had to dash into the river again, and in the hurry mistook the direction of the ford, and had to swim back across, during which time, the river being broad, myself and orderly had to undergo a very sharp fire of musketry. The trumpeter was shot through the body, and kept howling aloud till we reached the opposite shore. Thus I succeeded, though at no little risk, to ascertain what it would have cost the lives of many men to have ascertained by force, and I believe I got some credit for this act.

Single combat.

'The first time we entered Salamanca I was commanding the advanced guard, and we gave a "View halloo" when we came within sight of the town which so astonished the French skirmishers with whom we were engaged that they took it for the "Hurrah" of a charge, and went back rapidly. I had an encounter, in single combat, this day with a very young French officer, between the two lines of skirmishers, French and English, who stood still, by mutual consent, to witness it. The French officer showed great cunning and skill, seeing the superiority of my horse, for he remained stationary to receive me, and allowed me to ride round and round him, whilst he remained on the defensive. He made several cuts at the head of my horse, and succeeded in cutting one of my reins and the forefinger of my bridle-hand, which was, however, saved by the thick glove I wore, though the finger was cut very deeply to the joint. As my antagonist was making the last cut at me, I had the opportunity of making a thrust at his body which staggered him, and he made off. I thought I had but slightly wounded him, but I found, on inquiry the next day, when sent on a flag of truce, that the thrust had proved mortal, having entered the pit of his stomach. I felt deeply on this occasion and was much annoyed, as I had admired the chivalrous and noble bearing of this young officer. He was a mere youth, who, I suppose, thought it necessary to make this display as a first essay, as French officers usually do on their first appearance in the field, and indeed, I believe it is expected of them by their comrades. I shall never forget his good-humoured, fine countenance during the whole time we were engaged in this single combat, talking cheerfully and politely to me, as if we were exchanging civilities instead of sabre-cuts. There was a singular coincidence this day. We, the 14*th Light Dragoons*, wore an *orange* facing, and the French regiment to which we were opposed proved to be the 14*th French Chasseurs*, and also wore an *orange* facing. The cut I received on the forefinger of my bridle-hand proved a great grievance for some time, as it prevented me from playing the violin for weeks—a great deprivation, as I always played in bivouac at night.

'Early in the morning we found ourselves (that is General Alten's brigade, composed of the 14th Light Dragoons and 1st Hanoverian Hussars) in presence of a very superior force of the enemy's cavalry, with whom we commenced skirmishing, and who drove us back across the Guarena stream, a small river with steep banks. When we had crossed this stream with the whole brigade, we formed and waited till the enemy had crossed also, and then attacked him in "succession of squadrons from the right." The two first squadrons that charged failed to make an impression on the enemy, and were repulsed. In leading the third squadron to the charge (which was mine), I was run through the body, from the right side to the navel, about six inches. When the point of the sword

18th July 1812, and Salamanca.

came out, and as I staggered and fell, my antagonist, instead of withdrawing his sword from my body altogether, drew it up a little and then made another thrust, which went into the cavity of my chest. I was then led off the field faint and sick, and I well remember one of my best old soldiers offering his assistance. He was wounded also, but said "it was nothing, only a little stab in the stomach."

'Such, however, is the mortal nature of wounds with the point (the regiment we had charged was a heavy cavalry one with straight swords), that the poor fellow, as he was leading me off the field, suddenly staggered, vomited blood, and fell down dead. I must mention that I received my wound in the act of uplifting my arm and making a cut at the head of my antagonist, on his near side. He wore a brass helmet, and the blade of my sabre broke in two on it, which left me quite at his mercy. I forgot to mention that, in the early part of this eventful day to me, the enemy cannonaded us when we were formed in line, and the mare I rode, a most valuable one, a pure Arabian of the highest caste, and known to the whole army for her great beauty, had her thigh shattered by a shell which fell close to me and burst. I immediately dismounted one of the troop-sergeant-majors and took his horse, sending him to the rear. She was at first considered so desperately wounded that I was advised, and was on the point of shooting her, but she afterwards miraculously recovered, and I was taken prisoner on her on the 13th December 1813, when her head was cut open.

Battle of Salamanca.

'I was bled twice, profusely, during the night, as the effects of inflammation were apprehended, these sort of wounds never bleeding much of themselves. I was, of course, much weakened, but determined not to lose the glorious battle of Salamanca, which took place on the 22nd instant. I got on my horse, having slept in the town of Salamanca on the previous night, and I joined my regiment (in the field), which I found on the point of being engaged. I remained with it only a short time, as Colonel Hervey threatened to put me under arrest if I did not quit the field immediately, conceiving I was not in a fit state to remain. I left the regiment. It was, however, impossible to quit such a field at such a moment, and I repaired to the Arapiles hill, of which we then had possession, the enemy occupying the other immediately in front of it.

'General Packe's brigade being ordered to attack the latter, and perceiving one of the Portuguese regiments giving way, I could not resist the temptation of attempting to rally them, and rode down to the valley for that purpose, but my horse was shot under me, and in the very weak state I was in, I felt very unequal to further exertion. Still, it was impossible to leave such a field at such a moment, and I remained to the last, having joined in the very last attack made by the 6th Division on the rocky heights to which the French had retired before they entirely gave way, and retreated. This was late in the evening, and quite dark. I returned to Salamanca that night, where I slept, but started early next morning, looking like a ghost, and overtook the army pursuing the enemy. I was, however, obliged to lie down and rest every quarter of an hour. When I reached my regiment Colonel Hervey again remonstrated against my remaining with it, but my wound beginning to suppurate and do well, the surgeon said there was no risk, and I remained. We marched to Madrid, driving the enemy before us, and entered this romantic place amidst such enthusiasm on the part of the inhabitants, that it was more like a tale in the Arabian Nights than reality.'

Young Generals at Salamanca.

At the battle of Salamanca only two generals in the field were more than forty—the Duke himself and another.

Corporal Hanley's patrol.

'During the war in the Spanish Peninsula, in 1812, a patrol of four men of the 14th Light Dragoons (now the 14th Hussars), and four men of the 1st German Hussars, under a man whose name was Hanley, and whose rank I will tell you by and by, entered a village which some French soldiers they were watching had left shortly before. The patrol rode through the village, and on arriving at the further end saw three French dragoons returning from foraging, and making for a house standing by itself in the plain. Galloping after them, the patrol took them prisoners and then rode towards the house. From each end of the house ran a rather high wall forming a courtyard, with a stable in rear, but the only entrance was through the door of the house, and a narrow passage. The door was closed, but was opened by firing into the lock. Inside were French dragoons feeding their horses and preparing them for night.

'Hanley made his men fire quickly down the passage so as to make the French believe his party was numerous. At this moment the French officer in command of the post suddenly fired at Hanley, through a window on the ground floor, but when Hanley was about to return the compliment, the Frenchman surrendered. He was made to give up his sword and pistol, and as the only way out of the room was into the passage, he would have been sure to be killed had he tried to escape. One of the Germans of the patrol, who could talk French, was then sent with the French officer into the court to tell the French that they had better surrender, because the English cavalry brigade was coming, and unless they surrendered at once the thatched roof would be set on fire, and all in the place burnt to death. After a few minutes, the French officer and the German hussar returned, saying the men agreed to surrender. The prisoners were ordered to leave their sabres in the courtyard, and come out one by one through the narrow passage, leading their horses. The passage was only broad enough for one man at a time. As each came out his carbine was taken from him, the stock broken and the pieces thrown away. One by one they came out, twenty-seven in all. Imagine their disgust when they found only nine English and German soldiers outside! They were made to mount their horses and cross their stirrups, then they marched off in fours, three of the escort on one side, four on the other, and Hanley and one trooper bringing up the rear, the French officer riding alongside Hanley, who held his reins for him. This party had just left the house, when up came from some French troops on their march to this place a French lieutenant-colonel. He had seen the party from a distance, and thought they were English prisoners. Slapping Hanley on the shoulder as he passed him, he called out, "Good-day, Englishman," but before he knew where he was Hanley had whipped the Frenchman's sword out of its scabbard and made him prisoner. Then came up the colonel's orderly with a couple of mules with the colonel's baggage, and they all had to join the party, which after a march of some dozen miles rejoined the brigade.'

The march of the army to Madrid now took place, the 14th Light Dragoons being in brigade with the 1st German Hussars (Hanoverians). The Marquis of Wellington himself entered Madrid on the 12th August, but the Fourteenth passed Segovia in Old Castile and bivouacked near Escurial, which is 26 miles north-west of the Spanish capital, where there is the magnificent palace built by Philip II. and used as a monastery. The headquarters of the regiment were established at Getafe, and the men were employed in outpost duty. Lieutenant Cust commanded a post of observation at Consingia, in La Mancha, and Lieutenant Ward was with a post of communication between that place and Madrid. Advance on Madrid, August 1812.

When the Marquis of Wellington left Madrid for the siege of Burgos, the regiment remained on for some time in the vicinity of the capital; but about the 24th of October, when, owing to a concentration of the French armies under King Joseph, Soult, and other marshals, the siege of Burgos was raised and a retiring movement forced upon the allies, the 14th Light Dragoons with the 1st German Hussars assembled at Guadalaxara, fell back on Madrid, and formed part of the rearguard of Lieutenant-General Sir Rowland Hill's Corps from thence to Alba de Tormes. For several days the Fourteenth were constantly engaged in manœuvring and skirmishing to retard the advance of the enemy. The French moved upon Arevalo, Fontiveros, and thence to Alba de Tormes. Wellington coming from Burgos reached San Christoval, near Salamanca, on the 6th November. Retirement from Madrid, October 1812.

At this time the king and his marshals had an army of

90,000 combatants on the Tormes, mostly veteran troops, of whom 12,000 nearly were cavalry, and they had 120 guns. Wellington had 68,000 combatants, and 70 guns. On the 14th November the French crossed the Tormes and took post at Mozarbes. The next day Wellington began his retirement towards Portugal on Ciudad Rodrigo, finding the French were too strong for him. He moved away in three columns across the Junguen,[1] and then, covering his left flank with his cavalry and guns, defiled in order of battle before the enemy at little more than cannon-shot. Owing to a thick fog and heavy rain which was all in his favour, having possession of the high roads, while the enemy had only the fields and by-ways, he was enabled to bring his whole army in one mass quite round the French left, and gained the Valmusa river. This dangerous movement was rendered necessary owing to the time that had been lost by the allies whilst they waited on the Arapiles at Salamanca, and thus allowed the French under Soult to cross the Tormes and turn their position.

Retreat from Salamanca, 15th November 1812.

On the 16th the allies retired by the three roads which led across the Matilla stream through Tamames, San Munos, and Martin del Rio, the Light Division and cavalry closing the rear. The 14th Light Dragoons, as before, still took part in the picquets and other duties of the rearguard, all through this retirement of the allies from Salamanca to Ciudad Rodrigo; and until the army went into cantonments behind the Agueda, they continued to perform them. The surrounding country was one huge forest filled with vast herds of swine. For two days the number of our stragglers was enormous, for the hungry soldiers, who were very short of rations and supplies, broke away by hundreds from their colours to shoot the wild pig, and the forests resounded with the roll of musketry as if the enemy were attacking. Lord Wellington took stringent measures to enforce discipline, and had two men hanged, but even so the soldiers did not desist, and in consequence there were no less than 2000 stragglers of Wellington's army taken by the enemy.

[1] A rivulet running into the Tormes opposite Salamanca.

On the 16th November, when near Matilla, the French Lancers pressed our rearguard hotly, but were checked by the light companies of the 28th Regiment, and were afterwards gallantly charged and driven back by the 14th Light Dragoons under Lieutenant-Colonel Sir F. B. Hervey, Bart., who again distinguished himself and was nearly made a prisoner, the enemy being in very superior numbers. The Fourteenth lost 1 corporal and 2 horses killed; 1 man wounded and 1 taken prisoner.

Matilla, 16th November 1812.

Casualties at Matilla.

On the 17th, a large body of the French cavalry[1] surprised the picquets in front of the Light Division. The division was immediately formed in columns: a squadron of the Fourteenth and one of the German Hussars came hastily up from the rear, Julian Sanchez's cavalry, in small parties, formed on the right flank, and all precautions were observed to secure the retreat. This checked the enemy from making an attack, but his squadrons rode up near the flanks of our retreating infantry, and a good deal of baggage was taken and several men killed and wounded, nor did the enemy desist till finally driven off by our artillery. The French succeeded in taking General Paget prisoner as he was riding in the midst of his own men, concerning which Napier remarks that it might have been Wellington who was captured, for he also was continually riding between the columns and without an escort.

Soon after this the main body passed the Huebra, where the Light Division was assailed again by Soult's troops, but effected its passage with small loss. On the 18th, after a long and tiring march, Wellington reached Tamames, and next day Rodrigo and the neighbouring villages were occupied. This retreat beginning at Burgos had cost the allies a large number of men in killed, wounded, stragglers, and missing, besides prisoners captured by the enemy, as well as a large quantity of looted baggage. Victor Alten's Brigade of Cavalry, in which were the Fourteenth and the 1st and 2nd German Hussars, was attached to the Light Division and remained behind the

[1] Seventeen squadrons.

Agueda, Captain Badcock being detached with a reconnoitring party from the regiment to the Sierra de Francia and river Alagon.[1] The remainder of the British cavalry occupied the valley of the Mondego.[2]

On the 14th May, Mr. Charles M'Carthy became Quartermaster of the regiment *vice* Jameson. Quartermaster M'Carthy was afterwards, on the 11th March 1813, promoted to be Cornet, and subsequently, in May 1814, he became Lieutenant and Adjutant.

Retreat from Madrid.

The following incidents, mentioned by General Brotherton, refer to the occasion of the retreat from Madrid, when the 14th Light Dragoons formed the rearguard of General Hill's force:—'On our retreat from Madrid, the first day we reached a large village called Valdemoro, famous for good wine and extensive wine-vaults. *Of course* our men on arriving broke into these vaults and got drunk. The French, who were closely following us, followed their example, and found numbers of our drunken soldiers in the vaults, a large number stretched insensibly drunk. However, instead of fighting each other, they fraternised and embraced. *I* was sent back to this village by Lord Hill to try and get our men out of it. The scene I beheld in these vaults beggars all description, and it was, moreover, a service of danger to go amongst this motley and drunken crew of both nations, and I wonder I escaped. After almost fruitless endeavours I succeeded in bringing away but very few of these drunken brutes.

Pig Shooting.

'A remarkable circumstance, which gave rise to much displeasure from Lord Wellington, though a trivial one, if not a *ludicrous* one, happened on the first night of our retreat from Salamanca. The army was left *totally* without rations of any sort, and almost starving, owing to neglect in the commissariat department. The forest in which we were bivouacked abounded in large herds of pigs, amounting to many thousands—tempting

[1] The Alagon joins the Tagus above Alcantara.

[2] A large portion of the narrative of the Peninsular War in this Record has been taken from Napier's *History* almost *verbatim*, and many details are also taken from Cannon's *Historical Record of the 14th Light Dragoons.*

objects to a starving army. Many of these droves passed along the front of our army as if saying, "Come, kill me." No wonder that volley after volley were let fly at them, laying thousands prostrate. This, of course, when so close to the enemy as we were, our vedettes almost touching each other, was a dreadful irregularity. Lord Wellington roused out of his sleep and rode immediately to the front, thinking the enemy were attacking. His indignation on finding the cause of alarm was excessive, and the consequence was that he, next day, issued a most severe censure. The pork, I well recollect, was most delicious, hunger being the sauce, besides which these pigs feed on nothing but acorns and chestnuts, which abound in these forests. The scene in front of the line was a most extraordinary one. As the night had been very dark when these droves of pigs rushed past the front, the men fired their volleys at random, and many in front, particularly the cavalry, suffered. I myself saw two heavy dragoons and one horse lying dead. I shall never forget the singularity of the scene at dawn of day, close to the bivouac of the 14th Light Dragoons, and near where Lord Wellington himself had bivouacked, surrounded as it was by dead pigs strewed on the ground, dead dragoons, dead horses, etc. etc.!

Flag of Truce. Devotion. Desertion.

'I went out one morning, very early, during the disastrous retreat from Salamanca to Ciudad Rodrigo, *en parlementaire*, to inquire at the French outposts as to the fate of a gallant fellow who was orderly dragoon to Colonel Hervey, who, the night before, going his rounds with this orderly, went by mistake (the night being very dark and the enemy's picquets and ours very close together) up to a French picquet, and finding his mistake galloped off; but his gallant orderly, in order to save him from his pursuers—he, Colonel Hervey, having only one arm—sacrificed himself and kept sabring with the enemy to give time to Colonel Hervey to escape, which he did, though with difficulty. The orderly did not rejoin him, and we were anxious next morning to ascertain his fate, and know whether he was dead or alive. Accordingly, I proceeded early in the morning, accompanied by a trumpeter, towards the enemy's outposts. I found the enemy advancing, with skirmishers extended. I turned round to the trumpeter and told him to sound something to show I was a flag of truce. He thoughtlessly sounded the "charge," upon which (it being a forest and a foggy morning) the enemy imagined we were advancing in force and precipitately retired. I kept following them to endeavour to undeceive them, and at last they halted; and when I came up to the officer he seemed rather ashamed and vexed at what had happened. I shall never forget the *figure* he was. It was a very wet morning, and we had bivouacked without any shelter. He wore nankeen *trousers*. The French officers are not very particular about dress. To the sorrow of the whole regiment I found the gallant fellow (Sergeant Puss, a picked man) had perished in his noble struggle to save his colonel, having been cut to pieces. In reference to the above, the trumpeter with me was a German, and consequently, not attached to us by any feeling. Whilst I was talking to the French officer he galloped off towards the French lines, evidently with the intention of deserting, for the great temptation to do so was that deserters were allowed to sell the horses on which they deserted.

'The French officer, a chivalrous fellow, was indignant at the infamous conduct of the trumpeter, and immediately sent one of his men in pursuit of him, to bring him back and deliver him up to me for punishment; but as I knew he must have been hanged if I took him back to camp, I declined to receive him. The German and other foreign soldiers were not to be trusted always, as they changed sides as suited their convenience, at the risk of being hanged, either by the French or ourselves.

Equine Sagacity.

'When the 14th Light Dragoons were cantoned in Portugal in 1812 at Fundão, a large proportion of the troop-horses were turned out to grass more than five miles from the town. They had previously been groomed and fed every day at particular hours in the great square. The day after they were turned out they all came galloping in at the accustomed hour of feeding, and placed themselves in the square as if they had been led there!'

1813

After passing the winter in cantonments among the Portuguese peasantry, the Fourteenth once more crossed the confines of that kingdom and formed part of the centre column of the allied army in the general advance which commenced in May. Wellington had with him 70,000 men and 90 pieces of artillery. The French armies scattered through Spain numbered probably not more than 160,000 men altogether, but of these there were probably not more than 110,000 in the united armies opposed to Wellington. During the advance the allies were at first divided into three parts, and the Fourteenth entered Spain in the advance-guard of the centre column, arriving at Salamanca on 26th May, when the bridge and streets were found barricaded,[1] and a force under General Villates was formed on the heights above the ford of Santa Marta, consisting of a division of French infantry, 3 squadrons of cavalry, and some artillery. A British brigade passed the river at the ford, and the Fourteenth, with the 1st German Hussars, in brigade under Major-General Baron Victor Alten, removed the barricades and pushed through the town. The enemy fell back, but was overtaken, and lost about 200 men killed and wounded, as well as 200 prisoners. The line of the Tormes was thus gained, and that of the Duero soon afterwards, so that the allied army now firmly advanced and speedily passed the Carrion and the Pisuerga. During these movements the Fourteenth formed, as usual, part of the advance-guard of the army, and was engaged on 12th June near Burgos, where the Light Division, Grant's Hussar Brigade, and Ponsonby's Brigade of Dragoons turned the French right, while the rest of the troops attacked the whole range of heights extending from Hormillas to Estepar, held by the French under Reille,

Engagement near Burgos, 12th June 1813.

[1] Cannon's *Record*, p. 41.

who barred the way to Burgos. Reille began to fall back for the bridge of Baniel on the Arlanzon,[1] but during this movement Gardiner's Horse Artillery raked his columns, and Captain Milles of the Fourteenth, at the head of a squadron, charged and took several prisoners and a gun. The Fourteenth lost 1 man and 1 horse killed, 1 man and 5 horses wounded. After this the French destroyed Burgos castle, and fell back with tumult and confusion behind the Ebro, the British following up towards the sources of that river, and marching through a wild and beautiful district completely turned the enemy's position, cutting him entirely off from the sea-coast. On the 15th June the Fourteenth, being in advance, crossed the Ebro at the bridge of Frias,[2] and a patrol fell in with a body of the enemy near Pancorba.

Casualties.

Passage of the Ebro, 15th June 1813.

The French under King Joseph and Marshal Jourdain were concentrated in front of Vittoria. Wellington had only 60,000 Anglo-Portuguese sabres and bayonets with him, the 6th Division, 6500 strong, being left at Medina de Pomar. The Spanish auxiliaries were above 20,000. The approximate number of the French was about 60,000, but in the number and size of their guns they had the advantage. The enemy had many thousand carriages and impediments of all kinds heaped about Vittoria, which blocked all the roads and created confusion among the artillery parks.[3] On the 20th June the Marquis of Wellington examined the position taken up by the French army, and that day the Fourteenth skirmished with the enemy near the village of Huarte. The troops under Lieutenant-General Sir Rowland Hill, amongst whom were the Fourteenth, in the battle of Vittoria on 21st June, were told off to attack the enemy's left. They consisted of 20,000 men, composed of Morillo's Spaniards, Silveira's Portuguese, and the 2nd British Division, with cavalry and guns. The scene of the battlefield was a very rugged country on the banks of the Zadora and Bayas rivers, and there were no less than seven bridges within the area of operations.

Battle of Vittoria, 21st June 1813.

[1] Napier, vol. v. p. 108. [2] Cannon's *Record*. [3] Napier.

At daybreak on the 21st June, the weather being rainy with a thick vapour, the troops moved from their camps on the Bayas, and the centre of the army advancing in columns passed the ridges in front, and slowly approached the Zadora. The left column pointed to Mendoza, the right column skirted the Morillas ridge, on the other side of which Hill's Corps was marching, and the latter General seized the village of Puebla about 10 o'clock. Hill having crossed the mountains, where he was vigorously opposed by the enemy under Gazan and Villate, won the village of Subijana de Alava on the other side, and eventually outflanked the enemy's left late in the afternoon. At first the Fourteenth supported the attacks of the infantry and artillery, and later in the day were detached to help in the successful turning movements on the left flank of the French. Graham led the attack on the French right, while Wellington himself, at a time when he observed the French central positions somewhat denuded of troops, brought up Picton's 3rd Division against the weakened positions with crushing effect, and after a long and fiercely contested battle, which lasted till past 6 o'clock in the evening, the French army was completely routed and driven in confusion off the field.[1] Napier says: 'Never was a victory more complete. The trophies were innumerable. The French carried off but two pieces of artillery from the battle (and one of these was afterwards taken at Pampeluna). Jourdain's baton of command, a stand of colours, 143 brass pieces, all the parks and depôts from Madrid, Valladolid, and Burgos, carriages, ammunition, treasure of enormous value—everything fell into the hands of the victors.' The loss of men was about 6000 on the French side and 5000 on the side of the allies. The plunder and spoil was immense, chiefly carried off by the followers and non-combatants. In the evening the Fourteenth, serving in Victor Alten's Brigade,[2] were employed in the pursuit of the wreck of

[1] Napier.

[2] At Vittoria, Victor Alten's Brigade consisted of the 14th Light Dragoons, as well as the 1st and 2nd regiments of Hussars of the King's German Legion.

Pursuit towards Pampeluna.

the French army along the Pampeluna road, following the direction taken by King Joseph, who had continued his retreat up the Borundia and Avaquil valleys all night, and they passed the whole of the enemy's baggage, which had been abandoned in the flight. On the 24th they came up with the French rearguard at a pass two leagues from Pampeluna, when the leading squadron of the regiment under Major Brotherton charged and captured a tumbril. In this pursuit the Fourteenth were supported by Colonel Ross's Light Artillery troops. The main body of the French army now retreated into France by the valley of Roncesvalles, pursued by the British.

Major Brotherton captures a tumbril.

This battle of Vittoria was the crowning victory of the war in the Peninsula. The actual scene of the battle was on ground quite unsuitable for cavalry movements, but the Fourteenth and other cavalry during several days afterwards took a very active part in the pursuit of the enemy right into the Pyrenees. The rain for two days following the battle was most vehement and incessant, and as the Pampeluna road was swampy and much blocked, and the fugitive enemy set fire to the villages behind them, the pursuit was necessarily very difficult. On the 28th June, Lieutenant Ward, with a patrol of 3 men of the Fourteenth, got as far as the village of Ostiz, where he found 25 French foot-soldiers regularly armed and formed up at the village, but these surrendered themselves prisoners of war. Another patrol of the Fourteenth, consisting of 6 men under Lieutenant Clavering, penetrated still further into the mountains, and encountered on the 1st July a body of infantry of the French rearguard, on the road leading from Roncesvalles to St. Jean-Pied-de-Port, in the Basses Pyrenees. This patrol dispersed the enemy and took 18 of them prisoners.

Pursuit across the Pyrenees.

'The Emperor.'

It was probably in this pursuit, commenced on the evening of the victory at Vittoria, along the Pampeluna road, that the Fourteenth earned a title to that elegant and historical piece of silver plate known as 'The Emperor,' which has so long adorned the officers' mess. It was the property of His Majesty Joseph Buonaparte, King of Spain, and the royal arms are still discernible upon it. 'The Emperor,' in the piping times of peace, is seldom seen except when filled to the brim with the choicest brands of champagne, and in this condition it has passed through successive generations of the regiment, and done duty as a loving-cup in the officers' mess on many festive occasions.

'Vittoria' borne on guidons and appointments of regiment (1820).

For its gallant bearing at Vittoria the Fourteenth were subsequently (in 1820) rewarded with the royal authority to bear the word 'Vittoria' on their regimental guidons and appointments, and an additional honorary distinction was conferred on the commanding officer, Lieutenant-Colonel Sir F. B. Hervey, Bart.

The following anecdotes are related by General Brotherton regarding the advance to Burgos and the battle of Vittoria:—

'Charge extraordinary.'

'On the advance of the army to Burgos we had been engaged the whole day with the enemy, and my charger was quite worn out with fatigue towards night. A few days before, I had received from England a young horse, quite unbroke and wild. The enemy having unexpectedly advanced again late in the evening, after we had retired to our camp, thinking the work quite over for that day, on the "alert" being given (my charger being too much knocked up) I jumped on the young horse, who had only a snaffle-bridle on, not thinking there would be much to do. However, we had a stiff struggle with the enemy to drive him back, in which, charging at the head of my squadron, this said young horse ran away with me, and was taking me right into the French squadron opposed to us, when at the distance of thirty or forty yards, I threw myself off, to escape being carried into the ranks of the enemy. The few men who were following me dragged me away, as I clung to their stirrups. In the meantime my horse proceeded straight forward, and literally dashed through the centre of the French squadron, where I need not say he was secured, and proved a most acceptable acquisition, being a remarkably fine, valuable young horse. Those who were looking on on this occasion thought me most rash, not knowing I was run away with, and Lord Wellington himself was much amused when he learned the real state of the case.

Scene after the battle of Vittoria.

'Riding over the field of battle, I stopped to give a wounded French officer a drink of water, which he was piteously calling out for. He was mortally wounded, and the blood gushing out from his wound. On my speaking a few words of consolation and comfort to him he could not help, even in his agonies, expressing his indignation at King Joseph and Marshal Jourdain, the general commanding. He said their army was more *un bordel ambulant* than a fighting army.

'And indeed it was a strange scene. The plunder of years collected, hordes of the Spanish women who had been seduced by the French officers and men, and lastly, King Joseph's own seraglio of courtly ladies, cooped up in carriages of which we had taken possession, forming a procession nearly a mile long; the carriages ransacked by our soldiery, and every valuable taken out of them, but a bottle of brandy seized by them more eagerly than precious jewels.

Plate plundered at Vittoria.

'I had collected a considerable quantity of plate, plunder at the battle of Vittoria, which I purchased at the auctions of it that took place in the different corps. When I was taken prisoner all this was sold along with my other effects, as it is the custom in the British service.

'I never could get back any part of this plate, though I had applied to those who had purchased it and offered a high value for it, wishing to preserve it as a memorial of the battle. Amongst others, a Captain —— of my own regiment had purchased a coffee-pot, and I offered him the choice of London for a similar article, if he would return it to me, but he would not. I call this *churlish*, particularly in a brother officer.

'Talking of plate, I will just mention an anecdote showing that there is "nothing new under the sun," and that *fashion* in plate, as well as in everything else, constantly revolves and returns to old things again. I purchased of an old Spaniard, whose ancestors had resided in South America, some silver spoons made in that country more than *a hundred years ago*, of exactly the fashion and make of the present day.

Plunder at Vittoria.

'The result of the battle of Vittoria was a total rout of the

French army, and the capture of all their baggage, artillery, etc. etc. The rich plunder was to an enormous extent, but the troops (that is the "combatants"), however, profited comparatively but little, as they were too busily employed in following the enemy, and could not stop to plunder, so that all valuables fell to the share of the "non-combatants" and civilians attached to the army. Amongst the former may be classed the farriers of the cavalry, who are never in the ranks, and who frequently stop behind to plunder the wounded and the dead under pretence of shoeing horses. On this occasion our farriers of the Fourteenth took advantage of this trick to stop behind and plunder the very carriages which the regiment had passed and not touched: such was the severity of discipline in such cases, that not a single man dismounted to touch anything. Not so the farriers who lingered behind. They all carried what are called "churns," viz. large leather cases placed where the dragoon's holsters for his pistols are placed, and these they crammed with the abundant spoil of jewellery, etc., which they got by handfuls out of the carriages; and moreover, they had overtaken some mules belonging to the Pay Department of the French army, but abandoned by those who led them, carrying large boxes of dollars and other monies. When we arrived at the end of our pursuit of the routed French army, which terminated at Pampeluna, Colonel Hervey, who was aware of the immense booty in possession of the farriers, who, as "non-combatants," were far less entitled to it than their comrades in the ranks, who had fought hard for it, and who had first come up to it, but were restrained from plundering by their exemplary discipline, [Colonel Hervey] in order to pounce upon the plunder of these farriers, who, he knew, had not had time to dispose of it, suddenly ordered the "Rouse and Assembly" to be sounded in the camp, and the whole regiment being assembled, farriers and all (these little suspecting the object of the sudden parade), he ordered a square to be formed, brought all the farriers to the middle of it, dismounted them, had their churns taken off, and the plunder disgorged from them, to the joy and glee of the whole corps, who were justly entitled to it. I shall never forget the scene! There was perhaps more valuable plunder at Vittoria than at any other battle of modern times.

'The French army was dragging after it, as it was evacuating Spain, the accumulation of the plunder and robbery of *years*, during the whole time of their occupation of the country. The civilians of the army, viz. the commissariat, storekeepers, and paymasters, made enormously. One commissary got hold of a chest of gold doubloons, said to amount to £10,000. He thought he would secure it by throwing it (for the time), unperceived, as he imagined, in a deep river, the Zadora, which ran through the field of battle, but he had been observed by some Spanish muleteers, who reported the circumstance. The box was fished up, and the contents credited to the legitimate captors, viz. the army at large.

Promise to resist amputation of a friend's limbs.

'An intimate friend in the infantry had often made me promise that, if ever he was so severely wounded as to require amputation of a limb, and he happened to be in too weak a state to resist the operation, I would not allow it to be performed, as he said he would rather die than undergo the operation. He was shot through both legs, on the advance to Vittoria, and immediately sent for me, and earnestly reminded me of my promise, as the surgeons insisted on amputating in spite of his remonstrances, and he felt too weak to offer resistance. It was an awkward position, I having solemnly promised to interfere, but the surgeons assured me that he must die if the operations were delayed. The poor fellow, whose name was Burgess, looked at me pitifully and reproachfully, as he saw that I felt disinclined to fulfil my promise. Whilst this painful scene was going on a sudden bleeding came on, and he died in my arms.'

The valley of the Bastan.

Soon after the battle of Vittoria, King Joseph had sent a strong body of French troops into Spain again, to hold the valley of Bastan in the Pyrenees,[1] which was a very fertile spot and also full of strong military positions. Lieutenant-General Sir Rowland Hill's Division, consisting of 2 brigades of British

[1] Val de Bastan, in Navarre.

and 2 brigades of Portuguese troops, was at once marched there to endeavour to drive out the French. The Fourteenth were attached to this division, and during the four days from 4th to 7th July vigorous movements took place, in which the troops under Lieutenant-General Sir Rowland Hill succeeded in driving the enemy from all the positions occupied without incurring much loss themselves. The Fourteenth took an active part in these operations, and on 4th July, Major Brotherton, with a squadron of the regiment, had an encounter with the French near Almandoz.

Affair at Almandoz, 4th July 1813.

By this time the allies had established themselves in mountain positions along the edge of the Pyrenees, and occupied the whole line of the Spanish frontier from Roncesvalles to the mouth of the river Bidassoa. They had also invested the towns of Pampeluna and St. Sebastian. King Joseph's reign had practically ceased after the battle of Vittoria, the day of humiliation for France, and the downfall of Napoleon was drawing near. The Fourteenth took the outpost duties in front of Maya, a pass of the Pyrenees, and furnished posts of correspondence during several weeks. After a while the French army was reinforced and reorganised, and Marshal Soult took the supreme command, and advanced to oppose the allies. Important actions were fought on the 25th and 26th July near the Maya Pass, and the allies were forced to retire. On the 26th the Fourteenth were employed in carrying off the wounded from the field during the action, and were publicly thanked by Lieutenant-General Sir Rowland Hill for the efficient manner in which this duty was performed.

Actions in the Maya Pass.

On 30th July, when the post at Arestegui was attacked and an action took place, Major Brotherton's squadron of the Fourteenth was engaged, and Captain Milles's squadron was employed in carrying off the wounded from the field. In August the Fourteenth, forming the van of Lieutenant-General Sir Rowland Hill's Division, were engaged with the enemy in the valley of Bastan, in Navarre. Fighting took place on the 1st and 2nd of August, when the French were

At the Pass of Maya, 5th August 1813.

repulsed. On 5th August the regiment again took the outpost duty in front of Maya.

The siege of San Sebastian, the blockade of Pampeluna, and other sieges and military operations on the confines of Spain and in the Pyrenees occupied a large portion of the allied army for the next two months. On the 8th October, Wellington's army crossed the river Bidassoa and entered French territory. During the same month headquarters were fixed in Vera, and the army was organised in three grand divisions.[1] The right was commanded by Lieutenant-General Sir R. Hill, and extended from Roncesvalles to the Bastan; the centre column, occupying Maya, the Echallar, Rhune, and Bayonettes mountains, was under Lieutenant-General Beresford;[2] and the left, extending from the Mandale mountain to the sea, was under General Sir John Hope.[3]

Passage of river Bidassoa, 8th October 1813.

By the end of October the troops in the mountain passes above Roncesvalles were knee-deep in snow, provisions were scanty, and the privations of the allies were greater even than those of the French army under Marshal Soult on the opposite side of the Pyrenees. Finer weather set in on 6th and 7th November, and Wellington determined to advance into France, but on the 8th heavy rain caused a postponement of the forward movement. On the 10th November the troops descended the Pyrenees, traversing the mountain passes by moonlight, where they halted when they reached the line of picquets, and at daybreak crossed into France. The Fourteenth were attached to Lieutenant-General Sir Rowland Hill's Division, and one squadron of the regiment formed Marshal Beresford's advance-guard. The enemy's position on the Nivelle was forced, and on the following day the regiment was united at Espelette. The regimental baggage, which was attached to the 2nd Division, fell into the hands of the enemy in the rear of the pass of Maya, and on that occasion 1 troop sergeant-major and 2 privates of the Fourteenth were killed.[4]

The Fourteenth enter France, 10th November 1813.

Passage of the Nivelle forced, 10th November 1813.

[1] Napier. [2] Afterwards General Viscount Beresford, G.C.B., G.C.H.

[3] Afterwards General Lord Niddry, and later Earl of Hopetoun, G.C.B.

[4] Cannon's *Record*, p. 46.

The Fourteenth lost their regimental documents and papers in charge of the paymaster and adjutant, as well as the officers' personal baggage. The escort made a very gallant defence under a staff officer who was in charge on this occasion. The loss of these official documents belonging to the Fourteenth was a very serious matter for the regiment, and probably many interesting and valuable records which could not be replaced disappeared on this occasion. General Brotherton relates as follows concerning the incident:—

Regimental baggage and documents lost in Pyrenees.

'Never form a hasty opinion of any man's courage in the field! In a tried corps, such as the 14th Light Dragoons was towards the close of the Peninsular War, we knew each other pretty well, and our comparative value, individually, in the field, both officers and men; but still prejudices were entertained against certain individuals who, if not absolutely shy in action, were considered not very forward, to say the least of it. One of the most remarkable instances of the sort that occurred, in which the judgment proved erroneous and unjust, was that of a troop-sergeant-major, a *nervous* man only, of no great reputation in the field, and who on that account was usually selected to assist in the command of the baggage, which generally followed at some distance in the rear of the regiment. On one occasion, however, the last on which the poor fellow saw an enemy, the baggage was surprised, and all taken. It was in the Pyrenees, on the first day of our descent into France, 10th November 1813, a brilliant day for the British arms. Our baggage went astray, and got separated from us by one of those deep, narrow precipices so frequent in these mountains, and we had the mortification of seeing our baggage assailed immediately under our own eyes, and almost within carbine range, by a battalion of French "voltigeurs," who took the whole of it, but not till after a *desperate resistance* on the part of the baggage-guard under the command of Sergeant-Major B——, whom we distinctly saw most bravely distinguishing himself to save the baggage, till he and most of the guard fell. We witnessed this tardy but brilliant justification of his character with mixed emotions of admiration, but deep regret at having so

Bravery of Troop Sergeant-Major B——, 14th Light Dragoons.

long done him an injustice. It was a lesson that ought never to be forgotten, and should teach all soldiers to be charitable, and not hasty in their judgments upon their comrades in the field.'

Passage of the Nive at Cambo, 9th December 1813.

From the 15th of November to the 9th of December the Fourteenth furnished the outposts on the river Nive,[1] and formed the advance-guard to Sir Rowland Hill's Corps at the passage of that river, which was effected at the fords of Cambo, where the stream being deep, 2 men and 2 horses belonging to the regiment were unfortunately drowned.

Mendionda, 11th December 1813.

On 11th December a squadron of the Fourteenth under command of Major Brotherton was engaged with a body of the enemy near Mendionda, and captured a convoy of corn, wine, and salt, making 4 men as well as 4 horses of the escort prisoners.

Hasparren, 13th December 1813.

On 13th December, at Hasparren in the Basses Pyrénées, the Fourteenth as well as part of the 13th Light Dragoons were engaged with the French light cavalry commanded by General Pièrre Soult (brother of the Marshal), and on this occasion Major Brotherton, 1 sergeant, and 1 private were wounded, and Lieutenant the Honourable H. Southwell's horse was killed. Major Brotherton, Lieutenant the Honourable H. Southwell, and 1 man were taken prisoners by the enemy. The Fourteenth drove the enemy out of Hasparren. Referring to the affair at Hasparren and the forcing of the Nive on the 9th December, the following incidents are related by General Brotherton:—'The most guarded and considerate judgment should ever be passed on the deportment of soldiers in the field, young or old, as the following anecdote will show:—

Moral and physical courage. Nive, 9th December 1813.

'General Sir —— was known in the army as a gallant soldier, and though somewhat slow, and not brilliant in action, yet he had always done his duty. On the 9th December 1813, the day on which we forced the passage of the river Nive, in the south of France, at the bridge of Cambo, Sir ——'s brigade being in line, but lying down as the enemy were throwing

[1] The 13th and 14th Light Dragoons were engaged in the operations by which the passage of the Nive was effected on the 9th December 1813.—Cannon's *Historical Record of the 13th Light Dragoons*, p. 59.

shot and shells at it, the General, also sitting down in front, surrounded by his staff, and taking the opportunity of eating a mouthful, and drinking a glass of wine before commencing the attack—he beckoned me to partake of his grog, which I willingly did, having eaten nothing since noon on the preceding day. I squatted myself down with the party. The enemy's fire was doing but little mischief; still, it was necessary to keep out of sight as much as possible, in order not to attract more fire. However, just as we were filling our mouths, and helping ourselves to a glass of wine, a shell came over us, and fell within four or five yards of our group, but buried itself so deeply in the ground that it was easy, with the least presence of mind, to see that its explosion would be harmless, or nearly so. The General had his glass of wine up to his mouth at the very moment the shell fell; he instantly threw away his glass of wine on the ground, and himself *flat on his belly*, which is the usual mode resorted to in order to escape the effect of the explosion of a shell, which, when it bursts, cannot hurt you when you are in this position. *We*, the group round the General, could not help roaring with laughter at the risible operation the General had gone through from his nervousness, and indeed the whole *Brigade*, lying down in line, roared with laughter also at their General's ridiculous predicament, sprawling on the ground. A poor young fellow, his aide-de-camp, young —— of the 3rd Guards, who joined heartily in the joke, was shortly afterwards killed, on another occasion.

'I relate this anecdote, not in disparagement of, or with any disrespect towards General ——, and his reputation can well bear that such an anecdote be told of him, but only to show the occasional *nervousness* which the best seasoned soldier (and General —— had been a great deal under fire during a long military career) will sometimes involuntarily betray under fire.

'Had this happened to a young untried soldier, he would have felt ashamed of himself perhaps, though without reason, for his *moral* courage to do his duty might be equal to any man's, notwithstanding his physical nerves failed him in this way.

'One of the bravest men of his day, L'Amiral Coligni, in Henri Quatre's time, used to say, "*On n'est pas brave également tous les jours,*" and no doubt this is perfectly true. One's nerves are not equally strong every day. Let us therefore put the most favourable and charitable construction on the deportment of men in battle, till we find out they are really good for nothing.

Hasparren, 13th December 1813.

'The night before the battle (for the 13th December 1813 *was* a battle, and a sanguinary one too), as the Duke of Wellington had promised me a step of rank on the first occasion it could be granted—and I was then only a Major—Sir Hussey Vivian, who was the General commanding our brigade, being a great friend of mine, and anxious to afford me an opportunity of distinguishing myself, arranged that I should lead the charge (a sort of forlorn hope) with a certain number of picked men, to drive the enemy out of the village of Hasparren, which they occupied in force. Accordingly, early in the morning of the 13th, I descended towards the village with my party, immediately supported by a half squadron, and found the enemy, the 13th Chasseurs, and Chamboran (2nd) Hussars, posted behind a narrow bridge, at the entrance of the village. I immediately ordered the trumpeter to sound the "charge," so that those behind us who were to support should advance at the same time, and putting myself at the head of my men, rode at the enemy; but as the bridge was a very narrow one, only myself, Lieutenant Southwell (a distinguished officer, whom I had chosen to accompany me), and my orderly, could pass over at a time, which we accordingly did. The enemy received us with a volley from their pistols and carbines, when we were close upon them. Southwell's horse fell dead, and he fell under him. However, myself and orderly closed with the enemy. The orderly had his bridle-hand nearly chopped off, and was run through the body, and I was then left alone amidst the enemy. I was belaboured with cuts and thrusts from all sides, defending myself as long as I could against such odds. However, after receiving eleven thrusts, *three* of which *only* wounded me (as I wore a buffalo leather cuirass which I had made at Madrid, after having been run through the body at Salamanca), I was wounded through the neck, in the right hip, or to speak more plainly, in the bottom, on the right side, and another stab in the thigh, which would have proved the worst of all, had it not been for a bunch of letters which I had that morning received from England, and which I had put into one of the pockets which were then worn with pantaloons. The sword penetrated the letters, and went a quarter of an inch into the thigh, close to what is called, I believe, the femoral artery, which, had it touched, probably it would have proved fatal; but the blow which rendered it impossible to make further resistance was a sabre-cut, aimed at my head, which fell on the peak of my helmet with such force that it bent it on my nose, which it flattened and nearly broke, and completely stunned me. As I said, this blow disabled me from further resistance, and, indeed, no signs of any assistance appearing, rendered it useless to resist any longer. Surrounded as I was by fellows cutting and thrusting at me in all directions, and so occupied was I in parrying, that I had not time for assaulting in my turn. It was my intention to surrender, but a little circumstance caused me to be much more roughly treated than I otherwise should probably have been. I had, previously to advancing to the "charge," twisted my silk sword-knot round and round my wrist, by way of securing my sword the more effectually; and when stunned by the cut on my helmet, which I have just before mentioned, and summoned on all sides by vociferations to surrender (*rendez vous*), my sword was seized, but as it was so tightly fastened to my wrist, this was taken for an intention not to surrender it; and a fellow cocked his pistol, and put it to my head to blow out my brains, when I had just sufficiently recovered to articulate *Je me rends!* I was then secured, and *tied* on my horse, being too faint to sit on it otherwise, and galloped off to the rear (like Mazeppa), for by this time Sir

Hussey Vivian and his whole brigade were advancing to rescue me. The mare I rode (the same as was wounded near Salamanca) got her head cut open on this occasion, but behaved most gallantly. She reared and literally *pawed*, when on her hind legs, at my antagonists. I will add a little anecdote that happened when the French surgeon was dressing the wound in my neck, as it is to his credit, and shows that chivalrous spirit with which war was then carried on :—

'I always wore eight or ten gold pieces, of about £3 each, in a piece of black silk, round my neck, in case of need, if taken prisoner. I had wore this so long, that the silk was a good deal worn out, and showed the gold. On taking off my stock to look at the wound, the surgeon, perceiving this, immediately covered my neck again, and whispered to me (for the French Hussars who had brought me to the rear were looking on) that I had better conceal the money. I desired him, however, to take it and give it to the men, whose lawful prize it was.

'I tried to recover my mare, and offered any sum to get her and send her back, but in vain, for she was evidently of the purest Arabian breed : she was sent to one of the Imperial studs. My exchange for a French officer of equal rank was arranged, the very day after I was taken, and he was sent over to his army; but through the bad faith of Marshal Soult, I was detained under the pretence, at first, that wounds disabled me from being sent, but, as I found out afterwards, through an idea that I had seen so much of the country, through which our army was likely to advance, as to be enabled to give information disadvantageous to the enemy.

'When taken prisoner on the 13th December 1813, and taken to the rear of the French army, and smarting and lame with my wounds, on one of the day's halts I stopped at Tarbes. This is a town famous for cutlery, and I had a *billet de logement* (the troops in France being always quartered in private houses) at a rich cutler's who, not expecting, I suppose, the further advance of the British army, and the occupation of Tarbes by

British troops, was extremely brutal and unkind to me, which, in my helpless state, I felt most keenly, but was soon revenged on the brute, for, one morning early, that I was sent off further to the rear, on the report that the British army was advancing, "mine host" came to me in a most sneaking humble way, very different from the insolent one he had hitherto behaved to me, and asked me for a letter that might be some sort of protection to his person and property on the entrance of the British troops into Tarbes. I thought it an excellent opportunity of "paying him off," as the saying is, and, as he did not understand a word of English (and very few people in this part of France do), I wrote a "protecting pass" for this scoundrel as follows:—

'"The bearer of this is one of the greatest brutes that ever breathed. He treated me like a dog during the time I was lying wounded in his house, and I strongly recommend him to the special notice of all British soldiers, who, no doubt, will treat him as he deserves, for his infamous treatment of their wounded comrade."

'By a strange fatality it so happened that it was a party of my own regiment, the 14th Light Dragoons, that formed the advance-guard of the British army and first entered Tarbes, and to whom the cutler presented my certificate. I need not say that he was "paid off." They nearly killed him!

'When I was taken prisoner on 13th December 1813, I was a Major in the 14th Light Dragoons, and was promised my rank of Lieutenant-Colonel on the first battle. I was given a sort of "forlorn hope" to give me an opportunity of distinguishing myself. I was severely wounded, and if I did not succeed it was not my fault. Being a prisoner, however, I was left out of the Brevet that was given for the battle.

'This was a hard case, though it is the custom of the service. I never recovered the ground I lost by this disappointment.

'The charger I rode during most of the Peninsular war was the one I rode when taken prisoner, as related above. She was bought by my father (who was a great judge of horses) at the sale of the King's stud, at three years old. She was of the

Life of 'Fatima.'

purest Arabian blood, and perfect symmetry, fifteen hands high, dark brown, a perfect picture, most graceful in all her movements, but very conceited. As she walked along she looked to the right and to the left, as if to see who was admiring her. She was the admiration of the whole army. She was so sagacious that marvellous stories were told of her. She always wore a silken net to protect her from the flies that maddened her when she hadn't it on. She was wounded several times. At Salamanca a shell shattered her stifle or thigh, and I was nearly advised to shoot her as incurable, but the stud groom of Lord Charles Manners effected a perfect cure after a long time, only leaving an immense scar and dent. She was twice wounded by sabre-cuts on the head. The last time was in the *mêlée* on December 13, 1813, when I was taken prisoner, when she actually reared and pawed my antagonist, as if to defend me. She had her head cut open in a dreadful way. Still, I offered any sum to purchase her, but she became the property of the French Government, and was sent to a breeding stud, where her pure blood was most valuable. I was often tempted, by large offers, to sell her. Sir Charles Stewart, since Lord Londonderry, had offered me 300 guineas for her. She was, though of excellent temper, difficult to ride, from her fiery disposition. In bivouac, when lying down beside me, she would lift up her head to see if I was sleeping, and if she saw I was she would immediately lie down again, for fear of disturbing me. She was particularly fond of raw beef-steaks, and it was difficult to keep the men's rations from her, even if suspended on trees as they usually were, by way of safety. Her name was "Fatima."'

After the victory of St. Pierre, near Bayonne, on 13th December, Wellington was firmly established on the Adour. On the 14th December the Fourteenth took the outpost duty in front of Urt, near the river Adour, and on 21st December parties from both the 13th and 14th Light Dragoons shared the outpost work at Urcuit and La Bastide, both the latter places being somewhat to the south of Urt. The two regiments (13th and 14th Light Dragoons), known as the 'Ragged Brigade,' were now under Brigadier-General Vivian (afterwards Sir Hussey Vivian), and he was succeeded in the beginning of 1814 by Major-General Fane.[1] The army under Lord Wellington now took a short rest in quarters during very severe and inclement weather before resuming work again in the middle of February.

The 'Ragged Brigade' and its Brigadiers, 1813-14.

After the battle of Vittoria and the pursuit to Pampeluna the Fourteenth were brigaded with the 13th Light Dragoons, when they were nicknamed the 'Ragged Brigade,' first under the command of Major-General Long, then in November 1813 under Brigadier-General Grant, and subsequently under Brigadier-General Richard Hussey Vivian (afterwards Major-General Sir Hussey Vivian, K.C.B.), during the advance into France, and in 1814 under Major-General Sir Henry Fane, K.C.B., who commanded the brigade at the battles of Orthes, Aire, and Toulouse. The origin of the sobriquet of 'Ragged Brigade' was owing to the ragged state to which those two famous fighting regiments were reduced as a result of their long-continued service in the field, and their frequent marches, bivouacs, and exposure in all weathers.[2]

It is a curious coincidence that both these regiments, who became such firm friends when on service in the Peninsula and afterwards in peace time, should have had an early history very similar one to the other. Both regiments went through those extraordinarily lengthened periods of consecutive service in Ireland during the eighteenth century, of twenty-five years at one time, and of forty-eight years at another; both regiments were engaged in the Jacobite rebellions of 1715 and 1745; in the latter year both regiments were brigaded together at Corstorphine, Coltbridge, and Haddington, under

[1] During the Peninsular War the Fourteenth served in brigades at various times under the following officers as brigadiers, viz.:—1809, Cotton; 1810-11, Slade; 1811, Arentschild; 1812-13, Victor Alten; 1813, Long, Grant, and Vivian; 1814, Fane and Doherty.

[2] Cannon's *Historical Record of the 13th Light Dragoons*, p. 63.

Colonel Gardiner of the Thirteenth, which regiment was then called 'Gardiner's Dragoons,' and the Fourteenth were called 'Hamilton's Dragoons,' Colonel Archibald Hamilton being their Colonel. At Prestonpans the two regiments fought together, on which occasion their experiences were very similar.

In 1795 they met once more at Bristol, both being without horses, when preparing to go on service in the West Indies, where both regiments were subsequently employed in that disastrous campaign—which reduced them to skeletons by yellow fever, and from which they returned home in very reduced numbers when the operations were over. The next recorded meeting of the two regiments was in the Peninsular War, when they were christened the 'Ragged Brigade,' in 1813-14, and their next two meetings were in 1841 and 1897, as related further on in these pages.

The 13th Hussars have in their possession a manuscript journal written by an officer in that regiment at the time of the Peninsular War, and the following extracts, taken from that journal, were kindly given to the writer of these records by an officer of the 13th Hussars,[1] as they contain so many allusions to the doings of the Fourteenth in 1813-14, when brigaded with the Thirteenth.

Extracts from Manuscript Journal in possession of 13th Hussars (1897)

'On 9th November 1813, at Elizondo, in the Bastan Valley, Navarre, the 13th Light Dragoons were in brigade with the 14th Light Dragoons, Colonel Grant commanding the brigade in place of Colonel Long, who had been appointed to a command in England, and the two regiments marched together through the

[1] Captain J. H. Tremayne (Adjt.), 13th Hussars.

Pyrenees by the Maya Pass, thence to Urdache,[1] near the river Bidassoa, which divides France from Spain, and having crossed that river, entered French territory early on the morning of the 10th November. Here, at 3 A.M., they bivouacked after a tedious night-march across the mountains of nearly six leagues. After a few hours' rest the brigade mounted, and advanced with the infantry, one squadron of the 14th Light Dragoons forming an advance-guard for Marshal Beresford's Corps, but the brigade really belonged to Lieutenant-General Sir Rowland Hill's Division. In a short time the advanced posts of the enemy became visible; these were all driven in, and our army came in view of the very strongly entrenched position held by the French under Marshal Soult, on the Nivelle. This was defended by numberless redoubts bristling with cannon, and was considered impregnable—so much so that the French had erected huts in which they imagined they might securely pass the winter and bid defiance to the British advance. As the nature of the country was not suited for cavalry operations, the brigade could only witness the gallant and determined conduct of the infantry, by whom, though exposed to cannon-shot, position after position, hill after hill, forts and redoubts, were successively attacked and carried in the face of Soult's entire force, which, covering the hills on every side, lined the entrenchments, and manned the forts. The attacks were generally made at the point of the bayonet, and ultimately the enemy broke and fled in all directions, pursued by our infantry, and supported by Grant's cavalry brigade, which could only move along the roads, and even there the enemy had placed felled trees and other impediments which prevented our men from acting with effect. Towards evening the brigade retired and bivouacked in a wood, after marching about four leagues. During the operations of this day the Spaniards had been posted on the extreme right of the British line, where they were attacked and repulsed by

The 'Ragged Brigade' enters France, 10th November 1813.

[1] Probably Urdax on the Nivelle, opposite to the heights of Ainhoa (Basses Pyrénées).

the enemy, in consequence of which a very large quantity of baggage belonging to the British, including that attached to the 2nd Division, which on the previous night had come safely through the Maya Pass, fell into the enemy's hands when approaching Espelette (Basses Pyrénées), and on this occasion the 14th Light Dragoons unfortunately lost their baggage, as well as a Troop-Sergeant-Major and two privates who were killed.

'On 11th November the brigade marched onwards. Putting up at the town of Arrina,[1] they moved forward again in the evening, and bivouacked near Espelette, after proceeding about three and a half leagues. On 12th and 13th November they shifted their ground and bivouacked in front of Espelette, at an advanced post two leagues to the front.

'On 13th a sharp cannonade and musketry firing took place between the British, aided by the Portuguese troops, and the enemy, on the banks of the Nive, near Cambo. After dark our infantry fell back, and a frightful storm of rain, which baffles all description, continued the whole night. From 15th November to 9th December the brigade was employed in furnishing the outposts on the river Nive.

Battle of the Nive, 9th December 1813.

'On 9th December the brigade turned out about 2 A.M., got into the positions ordered on the banks of the Nive, with the 14th Light Dragoons forming the advance-guard to the forces under Lieutenant-General Sir Rowland Hill, and by daylight a general attack commenced. The river, which was deep and strongly defended by the enemy, was forded at various points in front of our line, and the enemy was forced back on all sides. Our brigade forded the Nive, forming up on the opposite banks: the Fourteenth lost two men and two horses drowned in the deep stream of the river near Cambo. The main body of the infantry crossed the Nive by a bridge, formed on the opposite bank, advanced, and drove the French from every position occupied. Owing to the enclosed and hilly nature of the country the cavalry could not be brought to act with effect, so

[1] Perhaps Ainhoûe is intended.

our brigade merely moved on in support of the infantry. Our brigade continued to take the outpost duties, and Brigadier-General Hussey Vivian (afterwards General Lord Vivian, G.C.B., G.C.H.) had succeeded Brigadier Grant in command. On 21st December the brigade was at Urcuit and La Bastide, somewhat to the south of Urt.

1814

'In January 1814 our brigade was under the command of Major-General Fane, and a movement was made up to the high ground two leagues in front of Briscous, but the outposts there were attacked and driven in, and the ground retaken, on the 3rd January. The weather continued exceptionally severe at this season, particularly so on the 4th and 5th January, when the brigade was literally bivouacked in mud; the horses, as they stood, were covered with mud almost up to their hocks and knees. Forage had to be procured from near the enemy, and force used to obtain it. Food also was very scarce, the usual supply being a little biscuit with spirits and water; the rain was soaking, and no baggage had come up when the brigade bivouacked on the night of the 5th January, on which day His Grace the Duke of Wellington passed by and saw for himself the state of affairs. On the 6th we moved our ground in the morning and took up a better position, and that evening Lieutenant-General the Honourable Sir Lowry Cole's and Lieutenant-General Sir Thomas Picton's Divisions came up, having been much delayed by the flooded state of the country and the incessant rains.

'On 7th January the 13th Light Dragoons proceeded to Briscous and the Fourteenth to Urt, about four miles further to the north. Both regiments now reposed for a short time in these quarters, where the weather continued very severe, forage for the horses very scarce and exceedingly difficult to obtain. The usual food procurable for our horses was furze, cut and pounded and made into a sort of paste.

'In February military operations were recommenced, the 14th Light Dragoons taking the van in the advance against the enemy's left flank, which led to the subsequent actions fought at Hellette, Garris, and Sauveterre.

'On 11th February the brigade moved its quarters, the Thirteenth to St. Pée and thence to Urcuray; the Fourteenth going a different route, but rejoining the 13th Light Dragoons on 13th February at a point on the St. Palais road.

Garris, 15th February 1814.

'On the 15th February the brigade continued its march in rear of the infantry columns, this district of the Basses Pyrénées proving quite impracticable for cavalry movements; but the infantry drove the enemy from every position occupied, and this notwithstanding the clever tactics displayed by Soult, who disputed the ground inch by inch. At length, towards evening, it was ascertained that the French had halted, and were in position on three mountains to our front in considerable force, so our army halted and hasty dispositions were made for attack. Before the brigades had reached their several points of impact, darkness had begun to set in, but, notwithstanding this, and also that the mountain sides and their summits were crowded with the enemy's troops, nothing could resist our attack; the French were driven at every point, whilst the cheering "huzzas!" which soon rose on all sides, reverberating from mountain to mountain, proclaimed the British victory. The contest was not over till long after dark, and the enemy fought with desperation. These heights being over the town of Garris, the action goes by that name; it is near St. Palais (Basses Pyrénées).

'On 16th February the brigade marched and crossed the ground where the attack had taken place the night previously; the ground was literally filled with dead. The Thirteenth were ordered to put up at St. Palais, but this town was found to be so crowded with Spaniards that it was preferred to bivouac in a wood close by, and the Fourteenth remained at Garris.

'On 17th February the Thirteenth were ordered to protect

the artillery and infantry and continued in front; the country was clear and open and the enemy was soon descried in view, strong in numbers, particularly in cavalry. The brigade with horse artillery was hurried to the front, and shortly afterwards the guns unlimbered and came into action; the enemy replied, and a smart cannonade ensued, to which our brigade was for some time much exposed. The infantry eventually came up and formed to attack a village in front, on the opposite side of the river (Gave d'Oleron), and as the bridge had been destroyed by the French, His Majesty's 92nd Regiment, supported by other British troops, dashed into the water and got safely over to the opposite bank, where they re-formed and rushed forward to the attack. The enemy was now driven out of the village and pursued towards the town of Sauveterre. The infantry, with great gallantry, drove the enemy before them until he took shelter under his own guns, when our line of outposts was established, and after dark the brigade bivouacked.

Sauveterre, 17th February 1814.

'On 18th February we took up a position near the village, and on 19th marched to Nabas.[1] On 20th, owing to the Spaniards having fallen back, our outposts were also thrown back.

'On 21st February a reconnaissance in force was led by Lieutenant-General Sir Rowland Hill in the direction of Navarrenx. On 24th, after crossing the Gave d'Oleron at Villenave, Captain Townsend, 14th Light Dragoons, with a detachment of his regiment, encountered a body of French cavalry and took an officer prisoner, and brought him into the brigade at our bivouac that evening, when both regiments were together. The outpost duties to the right flank were taken by the Thirteenth, those to the left flank by the Fourteenth. A severe frost set in, and forage-parties were sent out after dark, but scanty supplies were obtained.

'On next day (25th February) we marched to the heights above the town of Orthes, by which flows the Gave de Pau.

[1] On the Gave de Mauleon.

On 26th the brigade was ready on its alarm-post, momentarily expecting to be called into action.

Battle of Orthes, 27th February 1814.

'On 27th February the battle of Orthes was fought. Our brigade was with Lieutenant-General Sir Rowland Hill's Corps. The 14th Light Dragoons operated against the enemy's left, after passing the stream above Orthes and advancing towards the great road leading to St. Sever, but subsequently the two regiments received orders to fall back as the French had brought artillery fire to bear from the opposite heights, and their shell and shot came amongst the brigade and through the ranks, wounding several men and horses. The fighting continued till dark, by which time the enemy had been forced at every point, and our victory was as complete as at Vittoria. The losses of the French are computed at from 8000 to 10,000 men in killed and wounded. Our army rested on the field, where the brigade bivouacked.

'On 28th February, at daybreak, the brigade mounted, and with the horse artillery marched in pursuit through the town of Cadoures, thence six leagues onwards, taking a number of prisoners from the enemy who were retreating in disorder.

'On 1st March the brigade reached the banks of the Adour, and the 14th Light Dragoons, forming the advance-guard of Lieutenant-General Sir R. Hill's Corps, forded the river, followed by the rest of the troops; but when they had advanced a league orders for our brigade to countermarch were given, and we were told to recross the river and put up wherever we could find cover, and to send out two squadrons on outpost duty under command of the field-officer of the day. After a march of five leagues, during which the rain had fallen incessantly, the brigade found a resting-place just before dark. Before daybreak on 2nd March we assembled in brigade, and received orders to advance in front of the infantry towards Aire in Landes. The usual advance-guard of one squadron furnished by the 14th Light Dragoons was thrown out, and it was soon reported that the greater part of the enemy were not *in* the town of Aire, but were posted in force at a point on this side of it. Lieutenant-General Sir R. Hill at once reconnoitred their position, and as the country was not suitable for cavalry operations, he immediately ordered up the infantry and made the necessary dispositions for attack. The action was completely successful, the enemy was driven from every position occupied, and ultimately from the town of Aire. A violent and incessant rain fell with but little interruption all day and night. On the 3rd March the rain still continued to fall, accompanied by sleet and snow, but we advanced in the face of the elements, though by slow degrees, as there was also an enemy in great force, close at hand. We reached the Commune Lagos on 3rd March. On 7th March we were at Garlin: here the enemy, advancing from Conches, endeavoured, but without effect, to turn our flank. On the 8th March we were visited on our outposts by Lieutenant-General Sir R. Hill, who ordered up three companies of the 57th Regiment to support us. On the 10th we advanced from Garlin to Tadusse, which post was very far advanced and contiguous to the enemy's lines. On the 12th we reached the great high-road leading from Conches to Lembege where the enemy's cavalry, 800 strong, were discovered by our advanced squadron, and our brigade was retired again to Tadusse. On 13th March the enemy advanced, in considerable force of all arms, along the Lembege-Conches road. Several encounters took place between the advanced parties of our brigade and those of the enemy. The Spanish infantry came up, but they were powerless against the superior forces of the French, and we had to fall back almost to Garlin on 14th March.

Action at Aire, 2nd March 1814.

'At last the British infantry arrived, and the 28th Regiment quickly drove off the enemy's sharpshooters and their supporting battalion. More British regiments came up and the enemy retreated, and our troops took possession of the heights which commanded Conches. On this day (14th March) the 14th Light Dragoons were engaged with the enemy in two affairs on the Pau road, in which they behaved with their usual bravery; in one of them Captain Babington was wounded and

Engaged near Pau, 14th March 1814.

taken prisoner, and a few days before this Captain Townsend had been taken prisoner near Pau.

'On 15th March the two regiments of the "Ragged Brigade" (13th and 14th Light Dragoons), under command of Major-General Fane, turned out and proceeded to the alarm-post on the Pau road, by daylight. Here they remained exposed to severe storms of rain and snow till between 3 and 4 o'clock P.M., when they returned to quarters. On the 16th March the two regiments again assembled at daylight at the alarm-post on the Pau road, and remained again till evening exposed to the most inclement weather, when the Thirteenth returned to their quarters at Garlin, and the Fourteenth went on outpost duty at Tourniquet and Clarac, where they were subsequently relieved by a squadron of the Thirteenth.

'On 19th March (after being separated a few days) the two regiments came up with one another again near Vic Bigorre, where Lieutenant-General Sir Thomas Picton with his 3rd Division had forced the enemy from their quarters; and here the brigade once more bivouacked together in a neighbouring wood without baggage or tents, using their cloaks for bedding, after a long march of six leagues and a half. During the night the Heavy Brigade, consisting of 3rd Dragoon Guards and Royal Dragoons, joined the Light Brigade, and the four regiments formed a division of cavalry under Major-General Fane. The command of the Light Brigade was given to Colonel Doherty, 13th Light Dragoons, and Lieutenant-Colonel Arthur Clifton of the Royals got command of the Heavy Brigade. At 6 A.M., 20th March, the Cavalry Division proceeded to Tarbes, flanking the infantry on their right. The cavalry forded the Adour river and was formed in contiguous columns of half squadrons on the opposite bank; the infantry crossed by a bridge.

Affair at Tarbes, 20th March 1814.

The enemy now appeared in considerable force, strongly posted on the heights to our front; our infantry attacked. The country was unsuitable for cavalry operations. The enemy was forced and driven from every position by the determined conduct of the British infantry. Our cavalry could only look on and admire their gallant conduct, and when the fight was over bivouacked close by for the night. In the evening the outpost duty was taken by the 14th Light Dragoons.

'On 21st March the Light Brigade put up at the village of La Bartha. On 22nd March the affair of St. Gaudens took place. St. Gaudens, 22nd March 1814. The day was wet, rain poured incessantly. The Fourteenth were in advance, and towards evening put up at a village to the left of the line of route, the Royals in a village to the right, the 3rd Dragoon Guards being halted more to the rear, and the Thirteenth more to the front, near to St. Gaudens, where they distinguished themselves in an engagement with the enemy. On 23rd March a halt was made, and the men endeavoured to clean up after their recent hard work. On 24th March the Division proceeded. On 25th the Heavy Brigade, as well as the 14th Light Dragoons, were put up in several adjacent villages, but the Thirteenth occupied one on a different flank.

'On 26th March, in pouring rain, the Division assembled on the great Toulouse road; the heavies and the 14th Light Dragoons passed the night at the town of Muret, and the Thirteenth went on further, driving the enemy before them along the Toulouse road, through Roques and Portet. On 27th March the Division was halted in villages adjacent to Villeneuve.

'On 28th, outposts were established in front of Portet. On 31st the 5th Dragoon Guards came up, and a portion of the Cavalry Division moved on to Mieremont. On 1st April, Villeneuve was occupied. On 4th April the Division was approaching Toulouse, and a chain of outposts was established. On 8th April the 14th Light Dragoons, being in advance, arrived in front of Toulouse, and the other regiments of the Cavalry Division followed. Forage became very scanty and very indifferent in quality. On 10th April, at the battle Battle of Toulouse, 10th April 1814. of Toulouse, the 14th Light Dragoons acted with the troops under command of Lieutenant-General Sir Rowland Hill, and took part in the battle which ended in the repulse of the French army, when it was driven off its ground. On the 12th April it was discovered that the enemy had completely evacuated the city of Toulouse in the night. The Duke of Wellington entered Toulouse in triumph. Our Division under Major-General Fane took part in the triumph, and entered the city. Everywhere the British were received by the populace with most joyful acclamations, the white cockade was worn by all, and the white flag was displayed, the air resounded with cries of "Vivent les Anglais! Vive le Roi!" After marching through the town the Cavalry Division crossed a canal, when some of the troops came in contact with portions of the enemy, but the latter retired. On 13th April news arrived that Napoleon Buonaparte had abdicated, the Bourbon dynasty was restored, and the war was terminated.

'On 18th April it was announced in general orders that hostilities were over.

End of the war.

'On 22nd our brigade, 13th and 14th Light Dragoons, marched for Aire, and passing through Toulouse halted, the Thirteenth at Tournefuile, the Fourteenth at Columniez.[1] On the 23rd the brigade marched to Isle de Jourdaine, on 24th to Guinat. On 25th April a halt was made; 26th April the brigade marched, the Thirteenth putting up at or near Ordain, and the Fourteenth at Auch. On 27th April the brigade marched to Vic Fogensac, on the 28th to Vigora, and on the 29th to Aire, where orders were received to continue the march to Mont-de-Marsan,[2] where both regiments arrived on 30th April 1814.'

(*Here the Extracts from the 13th Hussars' Journal end.*)

At Mont-de-Marsan the brigade was broken up. On 13th May the 14th Light Dragoons marched for Bordeaux, and the two regiments, which had served together so long as companions in arms and become such firm friends, were once again separated, and the historic 'Ragged Brigade' became a thing of the past.

Wellington's advance in the middle of February had been much retarded by the late arrival of new clothing for his troops, and as there was no means of transport, he had to

[1] Or Colummiez, in Haute-Garonne.

[2] Mont-de-Marsan is in Landes, 65 miles south of Bordeaux.

send regiments in succession to the stores to fetch it. His first operations were merely to turn the rivers beyond the Nive, at their sources, with Lieutenant-General Sir Rowland Hill's Corps, in which the Fourteenth were acting as advance-guard, being in brigade with the 13th Light Dragoons, now under Major-General Fane's command.

Hill's Corps, consisting of 20,000 combatants and 16 guns, was operating against the left flank of Marshal Soult's army, and took post on the 12th and 13th February about Urcuray and Hasparren. On the 14th they marched in two columns, one by Bouloc,[1] towards the Joyeuse, the other by the great road of St. Jean Pied de Port towards Hellette. Harispe, the French general at the latter town, retired skirmishing towards St. Palais, thus leaving open the great road to St. Jean Pied de Port, while the Joyeuse was passed by the other column.

On the 15th, Hill marched through Meharin upon Garris, but as the road was bad for artillery, the guns went to the right by Armendaritz. Harispe's rearguard was overtaken and driven back fighting.

The affair at Garris ensued: the French general was in position in advance of the Bidouze river on Garris mountain near St. Palais. The fighting began quite late in the day; the 39th and 28th Regiments greatly distinguished themselves, supported by the Spaniards and Portuguese, and the position was carried by the allies, who lost 160 men, but the French losses were 500 men, of whom 200 were taken prisoners.

Garris 15th February 1814.

On the 16th February, Hill crossed the Bidouze, the cavalry and artillery by the bridge at St. Palais, the infantry by the fords.

On the 17th, Hill passed through Domenzain towards the Soissons, meaning to attack the French at Arriveriete, who, under General Paris, were defending the Soissons above its confluence with the Gave d'Oleron.[2] The French outposts were driven across the Gave, and the British, 92nd Regiment, took the bridge of Arriveriete, where the allies halted. The French retired; and on the 18th the allies seized the great

[1] Or Bonloc.

[2] Or Gave d'Oloron.

road running from Sauveterre to Navarrens[1] up the left bank of the Gave d'Oleron. Harispe was reinforced by Generals Paris and Villate, and took up a strong position at Sauveterre, occupying a bridge-head on the left bank, and supported by a brigade of cavalry. Hill now sent a force up the Soissons to guard the fords, and with Fane's cavalry as well as British and Portuguese infantry spread out between that river and the Gave d'Oleron, he occupied the villages along the road to Navarrens, and opened a cannonade against the Sauveterre bridge-head. After this Marshal Soult sent Pierre Soult with his cavalry brigade and some infantry to operate between Oleron and Pau, fearing that the allies designed to march on the latter place, whilst he decided to hold the Gave d'Oleron and Gave de Pau so long as he could, and then to fall back on Orthes.

By the 23rd February, Wellington had six divisions of infantry and two brigades of cavalry concentrated beyond the Gave de Mauleon on the Gave d'Oleron between Sauveterre and Navarrens. Meanwhile there was a large body of the French army at Orthes and Sauveterre feeling towards Navarrens, and on 24th the allies began to move across the Gave d'Oleron.

On the 25th a large portion of Wellington's army was massed in front of Orthes, including five regiments of cavalry, amongst which were the Fourteenth, whilst another large column with Lord Edward Somerset's brigade of cavalry was at Berenx, five miles lower down the Gave de Pau, and two divisions, as well as Vivian's cavalry, were in front of Peyrehorade; there were also five British regiments at St. Palais, or in that direction.

On the 26th, Marshal Beresford with two divisions and Vivian's Hussar Brigade crossed the Gave de Pau near Peyrehorade, the 18th Hussars having secured the passage of a ford.

Battle of Orthes, 27th February 1814.

On the 27th at daybreak, two more divisions crossed near

[1] Or Navarrenx, on the Gave d'Oleron, in the Basses-Pyrénées, 12 miles south of Orthes.

Berenx by a pontoon-bridge thrown in the night; other portions of the army afterwards crossed, and the French position on the north bank being vigorously assaulted, a fierce battle raged for a considerable time. Hill with 12,000 men, cavalry and infantry, had remained before the bridge of Orthes, waiting for orders to force the passage of the Gave when a favourable opportunity occurred. At last his opportunity came: he was unable to force the bridge, but forded the river above at Souars, drove back the opposing enemy there, and seized the heights above, cut off the French from the road to Pau, and thus turned the town of Orthes. He then menaced Soult's line of retreat by Salespice on the road to St. Sever, and fell successfully upon his left flank. The Fourteenth shared in these successful operations of the troops under Lieutenant-General Sir Rowland Hill. In the result the French were completely overpowered and driven from the field. They were pursued by our cavalry and the troops under Lieutenant-General Sir Rowland Hill for a considerable distance, but ultimately escaped across the Adour to St. Sever, Caçeres, and Barcelonne. In this battle Soult lost 4000 men killed, wounded, and taken prisoners, whilst many thousands of conscripts threw away their arms. The pursuit would have been more vigorous had not Wellington himself been wounded above the thigh by a musket-ball which interfered with his riding. The allies lost 2300, of whom 50 with 3 officers were taken prisoners. Besides Wellington, there were also wounded Lord March, afterwards Duke of Richmond, as well as Major-Generals Walker and Ross. Next day, Wellington continued the pursuit in force. On 1st March, Hill seized the magazines at Aire on the Adour, which river was crossed by the Fourteenth in pursuit on the same day, and on the following day the regiment was engaged in the combat of Aire, when the French were completely repulsed and driven by Sir Rowland Hill's troops from a strong position though fighting with courage and vigour.[1] From this point the enemy retreated unpursued along both banks of the Adour.

Casualties at Orthes.

Combat of Aire, 2nd March 1814.

[1] The above account is taken almost *verbatim* from Napier.

Sergeant Vernor and Privates Craig and Rose distinguished.

It was here that Sergeant Vernor, Private Craig, and Private Rose of the Fourteenth, specially distinguished themselves. As a reward for the gallant conduct of the regiment at the battle of Orthes, the royal authority was granted in 1820 to bear on its guidons and appointments the word 'Orthes,' and the commanding officer, Lieutenant-Colonel Sir F. B. Hervey, Bart., received another honorary distinction.

On the night of the 7th of March, Marshal Soult sent a body of French troops to Pau with the intention of arresting some nobles, favourable to the house of Bourbon, who were assembled there to welcome the arrival of the Duc d'Angoulême. Major-General Fane had arrived at Pau with his brigade of cavalry and a strong force of infantry, thus checkmating this movement of Marshal Soult's. The 14th Light Dragoons, with two guns attached, furnished a strong patrol to Pau on 7th, and on the following day (8th March) fell in with the enemy. On this occasion Captain Townsend and four privates were taken prisoners.[1]

Captain Townsend and 4 men taken prisoners at Pau, 8th March 1814.

In the actions which took place at Aire, 2nd March; Clarac, 14th March; Vic de Bigorre, 19th March; Tarbes, 20th March; and also in some other smaller affairs, the 14th Light Dragoons were engaged with the enemy, and formed the advance-guard of Lieutenant-General Sir Rowland Hill's Corps.

On the 13th March, as well as at Clarac on the 14th March, attacks were made by the enemy; and General Pierre Soult with three regiments of French cavalry moved to Clarac, on the Pau road, to cut off communication with that town and to threaten the right flank of the allied army. The picquet of the 14th Light Dragoons at Clarac repulsed the enemy on the morning of the 14th March, when Captain Babington was unfortunately taken prisoner. On the same evening Captain Badcock, with his picquet, was attacked by the whole of the 5th Regiment of French Chasseurs; he was reinforced by a squadron under command of Captain Milles, and they kept their ground until another squadron under Captain Anderson

Clarac, 14th March 1814. Captain Babington taken prisoner.

[1] Cannon's *Record*, pp. 47 and 48.

arrived, when the French were repulsed with considerable loss. Captain Milles was subsequently rewarded with the brevet rank of Major for his services on this occasion.

Castel Paget, 16th March 1814

On the 16th March the Fourteenth greatly distinguished themselves in an affair at Castel Paget, where they repulsed an attack of the enemy.

Affair on the Lembege road, 18th March 1814.

On 18th March the regiment was actively employed in reconnoitring in different directions: the leading squadron, under Captain Anderson, was engaged with the enemy on the Lembege road, the French right was turned by the valley of the Adour, and their outposts were driven back upon Lembege. In this affair Lieutenant William Lyons was killed.[1]

Vic de Bigorre, 19th March 1814.

Captain Milles's squadron was attached to the division under the command of Lieutenant-General the Honourable Sir W. Stewart, and was engaged near Vic de Bigorre on the 19th March.

There is a handsome memorial placed in the porch of the English church at Biarritz to the officers and men who lost their lives in the south-west of France during the Peninsular War. It was erected in 1882, and was visited by Her Majesty Queen Victoria in 1889. The inscription says that 'Lieutenant William Lyons, 14th (King's) Light Dragoons, and one soldier of the same regiment were killed at Vic Bigorre on 19th March 1814.' This is not quite in accordance with Cannon's *Record*, which says Lieutenant Lyons fell on the Lembege road, on the day previously. It is also stated on the Biarritz memorial that one man of the 14th Light Dragoons was killed in action on the 13th March. This may possibly refer to the action at Castel Paget of the 16th March, recorded by Cannon, or to some other affair on 13th March, not recorded.

Battle of Toulouse, 10th April.

The British army had invested Toulouse on 23rd March, but it was not till 10th April 1814 that the action which decided the surrender of the city was fought. The 14th Light Dragoons were present in this battle, serving amongst the

[1] Cannon's *Record*, p. 49.

troops under Lieutenant-General Sir Rowland Hill in Fane's brigade of cavalry, and they took part in the operations by which the French army was driven from its ground.

'On the morning of the 14th April, Hill's Division entered Toulouse at 8 A.M., when Fane's Brigade of Cavalry filed over the great bridge and marched straight through the city, amidst great apparent enthusiasm displayed for the Bourbons and the British army.'[1]

Hostilities soon afterwards terminated, and the Emperor Napoleon had already abdicated in favour of the restoration of the Bourbon dynasty.

Thus ended, with glory to the British arms, a war in which the Fourteenth had taken a noble part, and gained seven honorary distinctions, which have since been inscribed on their guidons and appointments, viz.—

'Douro'—'Talavera'—'Fuentes d'Onor'—'Salamanca'—'Vittoria'—'Orthes'—'Peninsula.'

After leaving Toulouse the Fourteenth marched in brigade with the 13th Light Dragoons towards Mont-de-Marsan, where they arrived on 30th April, and remained for a short time in quarters preparatory to their return to England.

The following copy of the War Office roll of the officers and men of the 14th Light Dragoons, who were among the recipients of the war medal and clasps for the Peninsular campaign, given in the year 1848 by Her Majesty Queen Victoria to the survivors of the wars, was kindly supplied for this book by Major-General the Honourable Herbert Eaton, late Grenadier Guards.

In the list of the officers the rank given below their names is that they held at the time the medals were given in 1848; the other rank (placed immediately after their names) is the rank they held when serving in the Peninsula.

[1] De Ainslie's *Royal Dragoons*, p. 144.

ROLL OF OFFICERS AND MEN OF THE 14TH LIGHT DRAGOONS who were Recipients of the Medal given in 1848 by Her Majesty Queen Victoria to the Survivors of the Wars.

OFFICERS OF 14TH REGIMENT LIGHT DRAGOONS

RANK AND NAME.	Corunna.	Talavera.	Busaco.	Fuentes d'Onor.	Badajos.	Salamanca.	Vittoria.	Pyrenees.	Nivelle.	Nive.	Orthes.	Toulouse.
Jones, William, Captain, . . . Captain H.P. 52nd.			I		I		I		I			
Rofe, Samuel, Paymaster, . . .							I	I		I	I	
Southwell, Hon. A. F., Lieutenant, . Lieut.-Colonel H.P. unattached.							I	I		I	I	
Sheil, Theodore, Lt., ...				I								
Thursby, John Harvey, Lt., . .					I		I	I		I		I
Thomson, Robt., Vet. Surgeon, . .		I										
Thompson, S. Perronet, Lieutenant, Lieut.-Colonel unattached.									I	I	I	I
Widner, Christopher, Asst. Surgeon, Staff Surgeon unattached.		I	I	I		I	I					
Wainman, William, Captain, . . Late Brevet-Major.		I	I	I		I	I	I			I	I
Wandesford, Hon. C. H. Butler, Major, Late Hon. C. H. Butler.		I	I									
Anderson, F., Captain, . . . Late Captain.		I				I	I			I	I	I
Babington, John, Captain, . . Late Lieut.-Colonel.		I	I						I	I	I	
Cust, Hon. Sir Edward, Lieutenant, . Lieut.-Colonel unattached.				I	I	I	I	I	I	I		
Clifton, Thomas, Cornet and Lieut., .			I	I								
Douglas, Archibald, Lieutenant, . Late Lieutenant.					I		I	I	I	I	I	I
Davies, Daniel Owen, Asst. Surgeon, Surgeon H.P. 18th Foot.[1]					I I			I		I	I	I
Foster, Augustus, Lieutenant, . . Late Captain.		I	I			I	I	I				
Gwynne, S., Lieutenant, . . .		I		I		I						
Hawker, Peter, Captain, . . . Lieut.-Colonel.		I										
Humphreys, Charles E., Lieutenant, Lieutenant H.P.						I	I	I	I	I	I	I

1 Barrossa, in 18th Foot.

NON-COMMISSIONED OFFICERS AND MEN OF 14TH LIGHT DRAGOONS

RANK AND NAME.	TROOP.	Corunna.	Talavera.	Busaco.	Fuentes d'Onor.	Badajos.	Salamanca.	Vittoria.	Pyrenees.	Nivelle.	Nive.	Orthes.	Toulouse.
Cordal, John, ...		I[1]					I	I					I
Clark, John, ...							I						I
Giles, John, ...			I			I		I	I	I	I	I	I
Nicholas, Thomas, . .			I	I			I	I					I
Wheeler, Benjamin, . .						I		I	I				I
Brown, Francis, . . .					I		I	I				I	
Benson, Robert, . . .							I	I					I
Bitner, John, . . .								I	I	I	I	I	I
Chambers, Thomas, . .			I	I	I	I							
Colley, Robert, . . .						I	I	I				I	I
Dry, William, . . .	Capt. Badcock, .								I		I	I	I
Darling, John, . . .	Capt. Milne, . .								I			I	
Dakin or Deakin, John, .	Capt. Badcock, .								I				I
Elvey, James, . . .			I		I		I	I				I	
Elliott, Elias, . . .	Capt. Townsend, .											I	I
Day, Joseph, . . .	Capt. Mills, . .												I
Fairbairn, And., . . .	Capt. Badcock, .											I	I
Frake, James, . . .							I	I				I	I
Fisher, George, . . .	Capt. Maerman, .						I	I	I	I			I
Forbes, Matthew, . . .												I	I
Hughes, John, Troop Sergt.-Major, ...			I	I	I								
Hallett, Joseph, . . .			I	I	I	I	I	I	I			I	I
Hainso, John, . . .	F troop, . .		I	I		I	I	I					I
Jackson, Will., . . .	Hon. Capt. Capel, .		I				I						
Kinghorn, L. Clark, . .												I	I
Mannering, Henry, . .							I	I					I
Manton, John, . . .									I			I	I
M'Kay, James, . . .	Capt. Knipe, . .		I	I	I			I	I			I	I
Moore, Will., ...	G troop, . .		I			I		I	I		I	I	I
Reeves, Joseph, . . .			I	I	I	I	I	I	I			I	I
Randell, John, . . .			I		I		I	I	I				I
Noris, Thomas, . . .	2nd troop, . .						I	I					I
Silvester, James, . . .	Capt. Brotherton,		I				I	I				I	I
Stone, John, ...			I			I	I	I					
Tilley, George, . . .				I	I	I	I		I			I	I
Wells, Jas., Trumpeter, .	D troop, . .						I	I					I
Clifford, Thos., Sergt., . .	Lt.-Col. Hervey, .						I	I		I	I		I
Giles, Joseph, . . .	" " .							I	I		I	I	I
Harrington, Joseph, . .	Capt. Babington, .		I				I	I					
Morris, Thos., Sergt., . .				I	I	I	I	I	I			I	I
Smith, John, Sergt., . .	Capt. Neville, .		I		I	I	I	I	I				I
Shepherd, Sam., . . .	Capt. Chapman, .		I	I	I	I	I		I	I	I	I	I
Smith, Will., ...	Capt. Neville, .		I	I	I	I	I	I	I	I	I	I	I
Sumner, Geo., Sergt., . .	Capt. Sterling, .		I	I	I	I	I	I	I	I	I	I	I

[1] Corunna, in royal wagon-train.

NON-COMMISSIONED OFFICERS AND MEN OF 14TH LIGHT DRAGOONS—*Contd.*

RANK AND NAME.	TROOP.	Corunna.	Talavera.	Busaco.	Fuentes d'Onor.	Badajos.	Salamanca.	Vittoria.	Pyrenees.	Nivelle.	Nive.	Orthes.	Toulouse.
Shippey, Thos., Sergt., . .	Capt. Sterling .		I	I	I	I	I	I	I	I	I	I	I
Sactidge, John, . . .					I		I	I				I	I
Smith, Joshua, . . .	Capt. Deymont, .		I	I	I	I	I	I	I				I
Story, Joseph, . . .	Capt. Sterling, .		I		I	I	I	I	I			I	
Sims, Isaac, ...	Capt. Townsend, .			I			I	I					
Tuck, John, ...	Capt. Neville, .						I	I					
Topper, Will., . . .	Capt. Baker, . .		I			I		I					I
Verner, Alex., Sergt-Major, .	Capt. Badcock, .		I		I	I	I	I	I				I
Trower, Will., . . .	Capt. Kepple, .		I		I		I	I					I
Worvall, John, . . .	Capt. Baker, . .				I					I	I	I	I
Waddell, Sam., . . .	Capt. Badcock, .		I	I	I	I	I	I	I	I	I	I	I
Wooland, Ed., . . .	Capt. Townsend, .		I									I	
Ward, Mathew, . . .	" "		I	I		I		I	I				I
Wood, Arthur, . . .	Capt. Hawker, .		I	I	I	I	I	I	I	I	I	I	I
Walton, Isaac, . . .	" "		I			I			I				
Westbrook, James, . .	Capt Mills, . .		I	I		I		I	I			I	
Walter, Robert, . . .	Capt Smith, . .				I			I					
Wilson, John, Trumpeter, .	Capt. Mills, . .			I	I								
Young, Jas., ...	Capt. Harvey, .			I	I			I	I				I
Yates, Jas., ...	Capt. Townsend, .											I	I
Oliver, Peter, Sergt., . .	Capt. Chapman, .		I	I	I			I				I	I
Poole, Will., ...	Capt. Southwell, .									I	I	I	I
Pridgeon, Thomas, . .	Capt. Mills, . .		I			I	I	I					I
Peach, Thomas, . . .	Capt. Brotherton, .				I								
Payne, John, ...	Capt. Mills, . .		I					I					
Palmer, George, . . .	Capt. Brotherton, .		I	I		I		I	I				I
Pain, W. John, Sergt.-Maj., .	Capt. Badcock, .			I	I	I	I	I	I	I	I	I	I
Powell, Thomas, . . .	Capt. Anderson, .			I	I	I	I		I	I	I	I	I
Reade, Thos., . . .	Capt. Mills, . .		I		I		I	I					I
Ratcliffe, Jas., Sergt., . .			I	I	I	I	I	I	I	I	I	I	I
Read, Will., ...	Capt. Nevile, .		I		I								
Rick, Wm., ...	Capt. Knight, .		I		I		I	I	I				I
Reis, John, ...	Capt. Mills, . .		I					I					
Rivers, Richard, . . .	Capt. Anderson, .							I					I
Rosier, James, . . .	Capt. Hawkins, .		I	I	I	I	I	I	I	I	I	I	I
Randle, Thos., . . .	Lord Manners, .		I					I					
Richard, John, . . .	Capt. Townsend, .		I				I	I	I				
Rycroft, Wm., . . .			I		I			I				I	
Smith, Thos., ...	Capt. Neville, .		I					I				I	
Stone, Thos., ...	Capt. Capell, . .			I		I	I	I	I				I
Summers, John, . . .	Capt. Hawkins, .		I	I	I		I	I				I	I
Surman, Thos., Tr. Ser.-Maj., .	Capt. Badcock, .											I	I
Kinch, George, Sergt., . .	Capt. Hawker, .		I	I	I	I	I	I	I	I	I	I	I
Lomas, James, . . .	Capt. Knightley, .		I		I	I	I	I	I				I
Lawrence, Edw., . . .	Capt. Townsend, .			I		I	I		I				I
Lomas, Charles, . . .	Capt. Hawker, .		I	I	I		I	I	I	I	I	I	I
Lane, Richard, . . .	Capt. Brotherton, .			I	I	I	I	I	I				

NON-COMMISSIONED OFFICERS AND MEN OF 14TH LIGHT DRAGOONS—*Contd.*

RANK AND NAME.	TROOP.	Corunna.	Talavera.	Busaco.	Fuentes d'Onor.	Badajos.	Salamanca.	Vittoria.	Pyrenees.	Nivelle.	Nive.	Orthes.	Toulouse.
Measey, James, . . .	Capt. Townsend, .		I				I	I					I
Morris, James, . . .	Capt. Knightley, .		I	I		I	I	I	I			I	I
Middlet on, Roland, .	Capt. Neville, .		I	I			I	I	I			I	I
Massey, Joseph, . .	Capt. Mills, . .		I				I	I				I	I
Montague, John, L.-Segt., .	Capt. Brotherton, .		I	I	I	I	I	I	I	I	I	I	I
Marshall, John, . . .	Capt. Milles, . .		I	I		I	I	I	I				I
Milles, Henry, . . .	Capt. Badcock, .							I	I				I
Mackall, Dan., Corp., . .	Capt. Babington, .		I	I		I	I	I	I				I
Mouldy, Wm., . . .	Capt. Mills, . .		I	I	I		I	I				I	I
Mantle, Wm., . . .	Captain Mahoney, .		I		I								
Manning, Chas., . . .	Capt. Chapman, .		I	I	I		I	I	I	I	I	I	I
Marner, Jas., ...	Capt. Knightley, .		I				I	I					
Newman, Wm., Tr. Sergt.-Maj., ...	Capt. Chapman, .		I	I	I	I	I	I	I	I	I	I	I
Dudley, Thos., Tr. Sergt.-Maj.	Capt. Mills, . .		I	I	I	I	I	I	I	I	I	I	I
Duell, Thos., ...	Capt. Brotherton, .		I										
Darbyshire, Jas., . . .	Capt. Babington, .		I	I									
Davidson, Jas., . . .	,, ,, .		I		I		I	I					I
Davis, Jam., ...	Capt. Mills, . .				I	I		I	I			I	I
Edgers, Edw., . . .	Capt. Dawson, .		I		I	I		I					I
Forrester, David, . . .	Capt. Hawker, .				I		I	I	I	I		I	I
Flint, James, ...	Capt. Babington, .							I					
Fulham, John, . . .	Capt. Knight, .		I				I	I	I				I
Gibson, Joseph, . . .	Capt. Babington, .		I				I	I	I			I	I
Gunn, Will., ...	Capt. Wainman, .		I	I	I	I	I	I	I	I	I	I	I
Groom, Thurlow, . . .	Capt. Badcock, .					I	I	I				I	I
Harris, John, ...	Capt. Baker, . .		I	I	I	I	I	I	I	I	I	I	I
Humphreys, Will., . . .	Capt. Townsend, .		I				I	I				I	
Hanly, W., Tr. Sergt.-Maj., .	Capt. Chapman, .		I	I	I	I	I	I	I	I	I	I	I
Hicks, Thom., . . .	A and H troop, .						I	I				I	
Heatly, Henry, Trump.-Maj.,	Capt. Baker, . .		I	I		I	I	I	I	I	I	I	I
Hayelock, Abraham, . .	Capt. Wainman, .		I			I	I	I					
Harrison, George, . . .	Capt. Knight, .				I		I	I	I			I	I
Hussey, Thom., . . .	Capt. Anderson, .		I	I		I	I	I					I
Higgs, Joseph, . . .			I	I	I	I	I	I	I	I	I	I	I
Ivalts, Henry, Sergt., . .	Capt. Chapman, .		I	I	I	I	I	I	I	I	I	I	I
Jackson, Ab., ...	Capt. Badcock, .		I	I	I	I	I	I				I	I
Jeffery, Rob., ...	Capt. Brotherton, .					I	I	I	I			I	I
Jevons, Chas., . . .	Capt. Babington, .		I	I	I	I	I	I	I	I	I	I	I
Hendley, Chas., . . .	Capt. Knight, .		I					I	I			I	I
Allen, George, . . .	Capt. Mills, . .									I	I	I	I
Allen, John, Sergt., . .	Capt. Chapman, .		I	I	I	I		I	I	I		I	I
Anstead, Jas., . . .	Capt. Neville, .		I	I			I	I	I			I	I
Arcote, Dan., ...	Capt. Wainman, .						I	I	I				I
Bayliss, Thom., . . .	Capt. Badcock, .											I	I
Bevan, Will., ...	Capt. Townsend, .							I	I				
Branch, Will., . . .	Capt. Knight, .												I

NON-COMMISSIONED OFFICERS AND MEN OF 14TH LIGHT DRAGOONS.—*Contd.*

RANK AND NAME.	TROOP.	Corunna.	Talavera.	Busaco.	Fuentes d'Onor.	Badajos.	Salamanca.	Vittoria.	Pyrenees.	Nivelle.	Nive.	Orthes.	Toulouse.
Barrett, John, . . .	Capt. Townsend, .		I	I	I	I	I	I				I	
Beddy, James, . . .	Capt. Chapman, .					I		I					
Benson, Rob., . . .	Capt. Mills, . .						I	I					I
Browes, John, Sergt., . .	Capt. Brotherton, .		I					I					I
Bradshaw, *alias* Bumidge, John, Sergt., . . .	Capt. Baker, . .		I	I	I		I	I				I	I
Brazier, Edw., . . .	Capt. Harvey, .						I						I
Barnes, John,. . . .	Capt. Baker, . .					I	I	I	I				I
Butler, Joseph, . . .	Capt. Anderson, .		I		I	I	I		I			I	I
Birch, John, Tr. Sergt.-Maj.,.	Capt. Baker, . .						I	I	I		I	I	I
Burke, David, . . .	Capt. Major, . .		I			I		I	I				I
Bill, John, ...	Capt. Milles, . .		I		I		I	I	I			I	I
Barnes, George, . . .	Capt. Babington, .						I	I					
Curtis, Rich., ...	Capt. Baker, . .		I		I		I	I					I
Clarke, Walter, Sergt., . .	Capt. Knightley, .		I			I		I	I			I	
Chase, Henry, . . .	Capt. Mills, . .		I	I	I		I						
Casey, Rich., Farrier, . .	Capt. Hawker, .			I	I	I	I	I	I	I	I	I	I
Conway, Will., . . .	Capt. Anderson, .								I				I
Connor, James, . . .	Capt. Knipe, . .		I		I								
Craig, Moses, . . .			I		I	I	I	I				I	

Cannon in his *Record of the 14th Light Dragoons* (p. 50) says :—'The 14th (Duchess of York's Own Regiment of Light Dragoons) had acquired a high reputation for the excellent *esprit-de-corps* which pervaded the ranks, and especially for the superior style in which the officers and soldiers had, during several years, performed the duties of picquets, patrols, vedettes, and other services which devolve upon a corps employed in the outpost duty.' Cannon's encomium of the Fourteenth.

The Fourteenth, under Captain Milles's command, as Lieutenant-Colonel Hervey had obtained leave of absence, remained in quarters at Mont-de-Marsan a short time, and marched on 14th May to Bordeaux, where they were reviewed by Major-General Lord Dalhousie, who highly complimented them for their appearance, discipline, and fitness for further service. Regiment reviewed at Bordeaux, May 1814.

The Fourteenth had been selected, owing to their great reputation acquired in the Peninsular War, from among the other cavalry corps for service in the United States of America; but

Regiment embarked at Calais for England, 15th July 1814.

this order was subsequently countermanded for the present, and on the 10th June they marched from Bordeaux *en route* to Calais, where they embarked on the 15th July and sailed on the 16th for England.

Two troops had remained in England in 1808, and 2 troops had embarked at Lisbon in December 1811, arriving at Portsmouth on the 8th January 1812, and joined the depôt. Six troops landed at Dover on the 17th July and marched to the neighbourhood of London.

On the 21st July the Commander-in-Chief, H.R.H. the Duke of York, reviewed 3 squadrons of the regiment on Hounslow Heath, and was pleased to compliment Colonel Sir F. B. Hervey, Bart., on the appearance and efficiency of the several troops.

At the inspection, Cannon relates how the Duke of York said to Colonel Hervey, 'They appear as if they had never been on service.' After the review the 3 service squadrons joined the depôt at Weymouth, and were quartered there in Radipole Barracks.

During the five and a half years the regiment had been on foreign service, the non-effectives, including men invalided and sent home, and including horses cast and sold were: 654 men, 1564 troop horses.

MEN:—Embarked with regiment	770	
Sent from depôt to Peninsula	445	
Total	1215	
Non-effectives	654	1215
Returned with regiment	561	
HORSES:—Embarked with regiment	720	
Remounts	664	
Received from other corps	381	
Taken from the enemy	63	
Spanish	14	
Total	1842	
Non-effectives	1564	1842
Returned to England	278	

The Fourteenth had been on service with the army of the

Recruiting of the regiment, from 1795 to 1814.

Peninsula under the command of the Duke of Wellington from 23rd December 1808 to 17th July 1814. Since leaving Ireland in 1795 the recruits had been obtained principally by aid of parties detached from the regiment to the counties of Worcester, Warwick, Salop, and Berks, with some volunteers received from the Fencible Cavalry in 1800, and from the Royal Wagon Train in 1810. It should be mentioned here that the veterinary surgeon of the Fourteenth, who had held that position since 1801, was Mr. Robert Thompson, and he was succeeded by Veterinary Surgeon Alexander Black on 10th November 1814. Before closing the records of the Fourteenth in the Peninsular War, there are some further interesting anecdotes, written by General Sir T. W. Brotherton, having reference to his experiences with the regiment during this eventful period, which may fittingly find a place here.

General Brotherton's anecdotes.

Anecdotes related by General Brotherton, relative to events which occurred in the Peninsula when serving in the 14th Light Dragoons, 1808 to 1814.

Retreat. 'Corunna Races.'

1. 'No army becomes more disorganised and unmanageable, on a retreat, than a British army! We had several deplorable examples of this in the Peninsula. Sir John Moore's retreat was the first. I did not *belong* to this army, but witnessed its disastrous retreat, being then attached to the Spanish General Marquis de la Romana, when he separated from Sir John Moore the 22nd December 1808, at which time "Corunna Races," as this retreat is appropriately called, commenced. It would be endless to enumerate all that came under my notice on this occasion, besides the hardships that I personally experienced in the rapid and incessant retreat that Romana's army had to make, before the French corps detached to pursue it. I lost my health through it, and I think was only saved from death by Romana's excessive kindness to me, in literally sharing with me all his comforts. I lived with him, and when I could not get a separate comfortable quarter, he made me sleep in his own room! He was a delightful fellow, and the only Spanish general that the Duke speaks highly of,

not so much for ability as for integrity and devotedness. He was the man who brought the Spanish army from the north of Germany by stratagem, where they had been most cunningly sent by Napoleon for fear of defection.

Risk of hanging.

2. 'The soldier who pretends that he never felt fear, is a humbug not to be believed. It is his duty to conceal his feelings as much as possible, however. But there are situations in war so trying to the nerves that the stoutest must feel appalled. I never felt so nervous as I did when I ran the risk of being *hanged* as well as run through the body! I was with the Spanish General, the Marquis de la Romana, in Galicia, just after his separation from Sir John Moore, and, although our daily operations consisted only in running away whenever the enemy appeared, the Marquis had thought proper to get manifestoes printed in French, German, Italian, and Spanish, to be distributed amongst the French army to seduce them to desert. The Marshal, Soult, who was following us, highly indignant at this proceeding, which had already caused much desertion from his army, proclaimed that he would *hang* the first person that was taken circulating these papers. One day that we were, as usual, running away from the enemy, I lingered behind with my orderly dragoon, the better to observe the enemy's numbers and intentions, trusting to my good English horse to get away. Accordingly, I remained dismounted till the very last moment, and allowed a few French dragoons to come into the very field where I was, well knowing that, by jumping my horse over one of the stone walls (with which that part of the country is enclosed), I could always bid defiance to them. But they chased me, and having my pockets full of the proclamation above alluded to, I was desirous of getting rid of them, in case of accidents, and, passing by a ditch, I took them out of my pocket, and tried to throw them into the ditch. It happened, however, to be a very high wind, and they were scattered wide and afar, and picked up by the French. I confess that a thrill of dread ran through my veins, which no danger in action could have produced, perhaps, for to be taken

prisoner, and *hanged* into the bargain, was a contingency of war which I had never contemplated. I escaped, however, with ease, owing to my being so well mounted, but never again carried any such papers about me.

3. 'When Lord Wellington's headquarters were at Govea, in the north of Portugal, near Castello Branco, and the enemy close thereto, in superior numbers, menacing to enter Portugal again, the 14th Light Dragoons and 1st German Hussars were in front watching the enemy, who, making a sudden movement in advance (in the middle of the night, in order to conceal it from us), but which movement, if not counteracted by a counter movement on the part of our army, would have compromised its safety, it became urgent that Lord Wellington should be as quickly as possible apprised of it. I was selected to proceed, as quickly as possible, to his headquarters, eight leagues off. I started on one of my own horses to perform this duty, and naturally chose my *best* horse, a thoroughbred one of great value, which my father had just sent me out from England, having bought him, a colt, at the sale of the King's stud, and broke him in himself, which rendered him of additional value to me. I was obliged to urge him, to perform my important mission, to such a pitch that he dropped under me, when I had reached only half-way to my destination, and I had to get a troop-horse from a cavalry regiment on the road, to conclude my journey, which nearly killed him also. I reached Lord Wellington's headquarters early in the morning, still dark. They were situated in an old convent; not a soul stirring in the place except the sentry pacing before the gate, who was not a little surprised at my dismounting and knocking loudly at the gate to obtain admittance, which I did with great difficulty and a long delay, and then I had to go on my way in the dark to one of the bedrooms, where I found one of the aides-de-camp (the late General Fremantle),[1] and, upon telling him my

Ride to Govea.

[1] Afterwards Major-General Fremantle, C.B., who brought home the Duke's despatches from Vittoria and Orthes, as well as the French colours and Marshal Jourdain's baton captured at Vittoria. Captain Fremantle became Adjutant of the Coldstream Guards after the battle of Talavera, and was aide-de-camp to the Duke

business (not at first being aware of its extreme urgency), he hesitated as to going to wake Lord Wellington, who, he said, had retired to bed very tired, and in very ill-humour. I said that if I were obliged myself to open the door of his bedroom and wake him, to deliver my message, I *must* do so, if he would point it out to me. Fremantle persisting in his refusal to wake Lord Wellington, I actually, myself, went and knocked at the door, and on being admitted, and having delivered my message, Lord Wellington told me to go immediately to Sir George Murray's room (the Quartermaster-General) and bring him instantly with me to his room. All was immediately bustle and stir, orderlies and aides-de-camp despatched in every direction, and a general movement of the army ordered. It was a most critical moment, and if Lord Wellington had not been apprised of the enemy's movement, to counteract it, our army would have been turned at Castello Branco, the key of Portugal on that side.

'The only compensation I received for my valuable horse was £35, the mere regulation compensation. The present Lord Londonderry had offered me 300 guineas for him. I would add that what accounts for Fremantle's scruples about waking Lord Wellington, is the notorious fact that his aides-de-camp and those about him were all *afraid* of him.

A Monk's revenge on the French.

4. 'However justifiable it might seem that the Spanish and Portuguese *peasantry* should retaliate, revengefully, on the French soldiery for their excessive cruelty and plundering, and though one could not blame them for it or check it altogether, yet, I confess, that to witness their *clergy* joining in these murderous retaliations was revolting, owing to their sacred character. I happened, one day when detached in Portugal to watch the movements of a French column that was winding its way through a most precipitous and rocky part of country, to observe a monk in his ecclesiastical costume (a jolly fellow, resembling Sir Walter Scott's "Friar Tuck," both in character

of Wellington, 1812-16. The latter took him to the Congress of Vienna as he could talk German. He was much liked by the Duke, and probably feared him less than the others. General Sir Arthur J. Lyon Fremantle, G.C.M.G., C.B., is his son.

and dress), ensconced *securely* behind a rock, elevated above the road along which the French were marching, whence he could, and did, deliberately take deadly aim at individual French officers and soldiers. He knocked half a dozen over in my presence, and seemed vastly to enjoy the sport, and uttered a ferocious exclamation of joy at each victim he laid low; whilst he, as I before said, was in perfect safety behind an *inaccessible* rock, for the French had not time to dislodge him. The peasantry around gave a wild and ferocious cheer at each deadly shot! However, making every allowance for *due* revenge, I could not help upbraiding this rascally monk, and I did so on the ground that he was committing a cowardly act, when in perfect safety himself, thus to butcher human creatures, for, after all, nothing can justify, even in war, taking the life of an enemy except in defence of your own. This sanctified character was, however, not only callous to my remonstrances, but even *insolent*, which I could not resent owing to his sacred character, and the blind veneration of the people in this country, even for the *crimes* and immorality of their priests.

French Atrocities.

5. 'On Massena's retreat, one day, on following one of their last columns closely, we espied something stuck at the end of one of the men's bayonets, which we at first took to be a loaf of bread, carried as French soldiers usually carry their ration bread, but what was our horror on approaching nearer to find it was a small infant! Incredible atrocity, but too true! The first opportunity we had of communicating verbally with French officers we spoke to them of this revolting fact. They did not deny it, but said it was the deed of an Italian and not a French soldier! What an excuse for such an act, as if every soldier in an army was not equally responsible for such barbarities, or, at least, for not preventing them.

Un mauvais cuisinier.

6. 'When we were following the enemy towards the Pyrenees we, one day about noon, came up to a bivouac which he had just left. I happened to be in the advance, and on approaching the bivouac, which I expected to find quite abandoned, I heard some loud cries in it, without at first

perceiving whence they immediately proceeded. On nearer examination I beheld a man strapped to the stem of a large chestnut-tree, who was roaring most lustily, expecting, I suppose, we should kill him. On approaching nearer I found it to be a negro, and there was a placard, in large letters, over his head with these words, "C'est un mauvais cuisinier." I asked the fellow to explain, and he said he was cook to a French officers' mess of the 53ème de Ligne, "que les messieurs étaient ce matin de très mauvaise humeur non pas parceque j'ai mal accomodé le dîner mais parceque messieurs les anglais ne leur avaient pas donné le temps de le manger et ils m'ont ainsi traité si brutalement!" We naturally thought that no real good cook would have been thus treated, as, in these hard times, such an individual was a precious personage, and we hardly thought it worth trying him. We did, however, and found him an excellent one, but, on the first favourable opportunity, he decamped to the enemy again.

Three hundred and eighty miles' ride with despatches.

7. 'The longest ride I ever took, without stopping, was from St. Jean de Luz to Madrid, 95 Spanish leagues (380 English miles). I carried despatches from the Duke of Wellington to the British Ambassador at Madrid, Sir Henry Wellesley, and there were relays of horses ready all the way for me. The most inconvenient and irksome part of the business was that I rode in uniform, and with that most cumbersome of all head-dresses, the bearskin helmet of the British Light Cavalry, at that time. This dress to me was particularly uncomfortable to ride such a distance in, in very hot weather (month of May), as I had been prisoner of war and laid up with wounds since 13th of December, and was quite unaccustomed to such paraphernalia.

Extraordinary wound.

8. 'On a foraging party, in France, at the end of the year 1813, after we had got our forage and were reposing on it, some French peasantry were firing at us, but at such a considerable distance as hardly worth noticing, and here and there a spent ball alighted on us. We were much fatigued, and lying on the bundles of hay which we had foraged in the surrounding country. The orderly dragoon, who was holding my horse

and himself lying down, suddenly started up and cried out, "I am wounded!" As he happened to be one of those who are always well known to their comrades, after a few campaigns, as *rather nervous*, no one attended to him, or believed him, but he kept calling out that he was very much hurt, holding his hand over his right eye. Knowing the man myself as not worth much, I thought he was making much ado about nothing, and on making him take his hand off his eye, with much difficulty, and perceiving no blood, I felt confirmed in the idea that he was more frightened than hurt. I sent him, however, to the rear, and his wound proved to be not only a most extraordinary one but a fatal one. A spent ball had entered the corner of the eye, turned the ball of the eye in its socket without producing a drop of blood, and was taken out at the roof of his mouth. He died of inflammation in forty-eight hours.

Confidence in the Duke.

9. 'Nothing could equal the confidence the army felt in Lord Wellington. It was to such a pitch that if it was occasionally separated from him it felt uneasy, for though we had many brave and good generals amongst us, they were, by comparison with him, so immeasurably inferior, in our estimation, that we used to apply to him the famous turf story about the racer "Eclipse," whose owner won an immense stake, at Salisbury races, by taking immense odds that he would not place every horse in a sweepstake, which he did, and won, by placing "Eclipse" first and all the rest of the horses *distanced.* So it was Lord Wellington might be placed, distancing all generals, either of our own army or of others to which he was opposed. Civilians, in those days, used frequently to put the question to us—who we thought ought to replace Lord Wellington if by any chance he were killed or wounded? We were at a loss to answer this question, as we knew of none equal to succeed to such a man, and we were inclined to reply to it by the well-known exclamation of the French soldiers, on the day upon which Turenne was killed by a cannon-shot, and doubt and discussions arose amongst the troops, in the heat of action, as to the man most proper to succeed him in the command. He, like our chief, had long ridden a favourite horse in all his

battles, as Wellington did his favourite "Copenhagen," and, as the army had no very high opinion of the officer entitled, by seniority, to succeed Turenne, they exclaimed, "Donnez-nous son cheval; il nous mènera à la victoire!" Turenne was *beloved* by his soldiers. Wellington was *feared*, but esteemed most highly. He had no winning ways with him, such as are recorded of Turenne and Napoleon, and, above all, he never made speeches to them as Napoleon did, nor did he ever utter those few words attributed to him at the crisis of the battle of Waterloo, "Up, Guards, and at them!" And as to apostrophising the Pyramids, as Buonaparte did on the field of battle in Egypt, with the famous speech of which the French are so proud, viz., "Soldats du haut de ces Pyramides vingt Siècles vous regardent!"—why, our soldiers, instead of being moved by such trash, would have called out "Fudge!" They want no such "blarney" to make them fight, and the less you say to them at such moments, the better. Nothing more than the caution "Steady" is seldom, if ever, wanted in action with British soldiers. Wellington knew this.

The Duke's severe discipline.

10. 'The severity and peremptoriness of the Duke of Wellington's discipline, on first entering France, was such, that I one day saw a Spanish soldier in the hands of the Provost-Marshal (the executioner of military justice), leading him to execution, *i.e.* to be hanged on the nearest tree. He seemed to be a fine fellow, with little fear of death at such an awful moment, but only indignant that he should suffer death merely for having taken a fowl out of a farmyard, when he, as he said, had had his father and his mother bayonetted by French soldiers, his sister violated, and his home burnt by them, merely for having joined in the patriotic defence of his country. He was a fine fellow, and his deportment and demeanour on the verge of death was noble. It touched me much, and I would have given the world to have saved him the *ignominious sort of death* that he was to suffer, for his chief horror seemed not so much the fear of *death itself* as the *mode* of it, beseeching as a favour to be shot as a soldier, but not hanged as a felon.

But I knew too well the sternness and peremptoriness of Lord Wellington to presume to interfere. I could not, however, bear to see him executed, and turned away from such a dreadful sight.

'When in Spain, robbing the beehives of the peasants was a frequent offence, in spite of Lord Wellington's repeated admonitions and reprimands, and honey was one of the chief profits of the Spanish farmers in some parts. All Lord Wellington's endeavours to stop these depredations having failed, he had recourse to the expedient of keeping whole divisions of the army in which such thefts had been committed "under arms" for days together, sometimes till the delinquent was discovered. This fell hard upon the innocent, and did not often lead to the detection of the thieves.

11. 'The Duke of Wellington, when in the south of France, anxious to conciliate the inhabitants, and particularly the authorities, received a letter from the Mayor of a village complaining of his favourite horse having been stolen by some of our people. The Duke returned a most polite and lengthy reply, which is recorded in his despatches. The French criticised and ridiculed the Duke for paying so much attention to this Mayor's complaint, and called it *une niaiserie*. But the Duke never wrote a letter which proved of greater advantage to him. This Mayor was a most influential person in his district, and was so pleased at the Duke's condescension, that in times of great scarcity of provisions and forage he exerted himself, in a most extraordinary way, to feed our troops.

The Duke's politeness rewarded.

12. '*Amateurs*, as they are called in the field, are a description of animal voted a great bore by real soldiers. They consisted of idle gentlemen who must needs try to show their pluck by poking their noses into danger in action (where they had no business to be), till it became too serious to be pleasant, when they immediately decamped, and became objects of derision. They had failed to ascertain the extent of their nerves. In this respect these said *amateurs* differed essentially from the *volunteers* we had with the army, who always reck-

Amateurs in action.

lessly exposed themselves, in order to render themselves conspicuous, as their object was to get commissions given to them without purchase. The largest proportion of these volunteers were killed, but those who escaped were well rewarded for their adventurous spirit. I have said that these *amateurs* were great bores in the field, and I will mention one glaring instance in the person of a distinguished sailor, though, in one respect, he was a *brilliant* exception to the faults of other amateurs, for when I mention his name, every one will know that he is the bravest of the brave—Admiral Sir Charles Napier. This distinguished man joined me whilst I was commanding the skirmishers of the rearguard, the day before we retreated on to the position of Busaco. He was most fantastically dressed in his sailor's dress, with a cutlass by his side, a brace of pistols at his waist, etc., looking a strange fellow. I need not say that he was all day amongst the foremost, but not content with this, he was urging me every moment to charge everything before me; but as I knew my business well, and as doing what he wished would have been contrary to all rule, and orders of Lord Wellington, I told him good-humouredly that he was a bore, and that if he were to go on this way we should realise the Irish story of the "Kilkenny cats," who fought so desperately and perpetually amongst themselves that they were all killed on both sides. Luckily he just after was wounded, and went to the rear, and to my great delight I got rid of him, but was glad he was not seriously wounded. There were four Napiers in the field at the battle of Busaco. Sir Charles, the one I have just spoken of, and his three cousins:—Sir Charles, the lamented man just dead; the present Sir William; and last, the present Sir George, three brothers *heroically* brave. The bravery of this family has always been proverbial, and seems hereditary. It was one of their ancestors who invented logarithms. But we unfortunately had a far different race of *amateurs* from the rare specimen I have just adduced in Charles Napier, the sailor, and though one would wish to be charitable towards these would-be heroes, yet I must say they were a great nuisance to the army. These foolhardy gentlemen—who could never be pitied if they got killed or wounded, for having "run into any kind of danger" which they were not bound to encounter, and which appeared to us a sort of vainglorious vanity, whereas with *us* it was a positive *duty*—were a numerous class who only contributed to *eat our provisions* when scarce, but who suddenly disappeared when things became serious and unpleasant. Mr. L——, whose posthumous memoirs have lately, most injudiciously, been published, was officially employed with the army in the Peninsula, as Deputy-Judge-Advocate. His functions were confined strictly to the closet, and he had no business whatever ever to poke his nose in danger, yet this civil functionary relates the "dangers he had passed," and seems to be proud of them ; but he was also a specimen of the genus *amateur*. As this gentleman has thought proper, cursorily, to introduce my name in his memoirs, and in a manner *à propos* to the amateur mania, I will just elucidate this subject by a case in point, in which Mr. L—— is concerned :—

'He thought fit to come out to the front when some fighting was going on, and happened to hear me exhorting some noble fellows of the Basque peasantry, who were enthusiastic in our favour, but only armed with sticks and scythes, not to expose themselves in this reckless manner, as the French cavalry, with which I was engaged, might occasionally drive us back, and cut up these brave peasants. Mr. L—— very readily took the hint as applying equally to his own precious person, and retreated. But I could not resist the temptation of giving this "quill-driver" my opinion, not only as to the folly of such persons coming, ostentatiously, to the front, but as to the mischievous effect on the *morale* of troops, by their precipitate retreat when they became frightened, for, though the soldiers heartily laughed at these *amateurs*, yet, at such moments, anything that diverts the attention of troops from what ought to be their only object, viz. "the enemy," is hazardous.

'Mr. L——'s stories of his own adventures are perfectly

absurd, except where his *stomach* was concerned, and he is then most in earnest.

Value of guerillas.

13. 'The guerillas were certainly of great assistance to us, and of great annoyance to the enemy, harassing them continually, but in *battles* they were of little use. They are a very similar force to the Cossacks, and it may be said of them, as Ségur said of the latter, "Ils ont beaucoup d'audace mais point de bravoure."

French matrimonial ideas.

14. 'French officers are not very scrupulous as to matrimony. A Spanish lady, at Valladolid, who had a beautiful niece living with her, told me that the aide-de-camp of a French general who was quartered in her house, courted this niece, and proposed marriage. The aunt inquired of the general the character of his A.D.C., when the former said:—"I can assure you, madam, that my A.D.C. cannot fail to make a good husband, having had much experience in that line, for, to my certain knowledge, he has married in every town we came to, in Spain, for the last three years." ! ! !

Parisian puns, 1815.

15. 'Some fulsome French flatterer, who had probably just before licked the dust off Napoleon's feet, said to the Emperor of Russia, just after he entered Paris as a conqueror, on showing him the letter "N," which in compliment to Napoleon had been put on all the public monuments and buildings, "Sire, il avait des *ennemis* partout" (i.e. "*des N mis*" partout), "Vous, Sire, aurez *des amis* partout" (i.e. "*des A mis*" partout).

'The famous antique horses plundered from Venice by the French were placed on the triumphal arch at the Tuileries, and there was a chariot attached to them, but no human figure in it, and it was currently understood that, some day, when he expected to have completed his conquests, he (Napoleon) intended his own statue to have been placed there. Another fulsome French flatterer said to the Emperor of Russia, who observed there being no figure in the chariot, "Napoleon, *Le charlatan*" (i.e. "*Le char l'attend*").

Single combat.

16. 'In reference to the event described in General Macdonald's letter, I must observe that, although it has been my lot to have been engaged in several single combats in the field,

during the Peninsular War, yet they were never of my own seeking, but that I was provoked to them by the braggadocio manner of the French officers; for I hold it to be very bad taste and feeling to engage *designedly* in these single combats, which in a *mêlée* are unavoidable. War would indeed be more horrible than it is if such individual hostilities were encouraged, which were only suited to a barbarous age. For the sake of example, an officer cannot well decline a challenge to single combat offered in the bullying manner with which the officer commanding the French regiment of Dragoons bore himself towards me on this occasion. Generally, the skirmishing of the cavalry in the Peninsula used to be carried on in the most chivalrous manner, I had almost said *amicable* manner, sometimes even the officers of both parties *shaking hands before* commencing; and often have we drank a glass of wine together *after* the day's fight was over!

'In this instance there was a coarse, bullying manner in the French officer which made me wish to chastise him, but I was on a very small Spanish horse, not much higher than fourteen hands, whilst he was, in all points, a formidable antagonist, *in appearance*, mounted on an immense horse—a very large, powerful man himself, with an immense fur cap—in short, *looking* as savage as a dragoon could look. My own men and General (then Captain) —— tried to dissuade me from encountering this *Goliath*, but I could not stand his taunts, and rode at him on my little charger, intending to equalise the combat through the agility of my little horse, in compensation for the great weight of my antagonist. He did not, however, wait for me, but, just before I came up to him, he turned his horse and retired amidst the yells and hisses, not only of *my* own men and the British officers present, but of *his* own men also, and, although I found myself alone amongst the latter, not only did they not attempt to cut at me or even to interrupt my return to my own troops, but showed me every mark of respect and approval of my conduct, by cheering me and waving their swords.

'Sir Andrew Leith Hay, in his book, describes a very different affair of this sort, in which I was engaged, and in which I met a most chivalrous, fine young fellow of a French officer, and I have never ceased to lament having wounded him so seriously as to have caused his death, but in this instance, likewise, I have the consolation that it was not of my own seeking, but a determined challenge on the part of my opponent in front of the line of battle.

'This alone can, in my opinion, excuse a single combat in war, or its being accidental in a *mêlée*.

Horses shot under me.

17. 'I had nine horses shot under me during the Peninsular War. This is not a great number considering the constant exposure to fire. When a horse has once been hit by musketry, ever after he trembles under fire, as if by instinct. It may be my fancy, but I have always thought I perceived this.

Shifts of cowards.

18. 'In a regiment which has been long in the field, and frequently in action, not only are the officers well known by the men, but every individual is known to each other, reciprocally. Thus I knew to a nicety what each officer and private in the 14th Light Dragoons was worth, and what he was fit for. One of the indifferent ones in action, well known to his comrades as such, in my squadron, being one day in the front rank, when we were advancing to charge a French squadron opposite to us, cut his horse's head almost in two at the moment he thought we were about to come in contact with the enemy. The enemy, however, turned just before we reached him, thus exposing the man's trick.

Making an example in action.

19. 'It was once reported to me, when we were formed up and exposed to a severe fire, that a man in my squadron had exclaimed—"This is murder, it is too bad to leave us here." As this speech was calculated to dishearten the men, I thought it advisable to make an example of the man at the moment, instead of bringing him to trial afterwards, and I called him out of the ranks and took him to the most exposed spot I could find, where I thought a cat with nine lives could not live for a minute. I left him there, and told him to stop there as a

vedette. The balls fell thick about him, but he escaped them all as if by a miracle, whilst his comrades in the ranks fell fast. Such is the fatality in action, verifying the soldier's saying, "Every bullet has its billet." To my great disappointment this cowardly rascal escaped unhurt. The proceeding, however, had its desired effect, for the men seemed delighted at it.

20. 'In one of our sharpest skirmishes it was reported to me, by one of the men, that a sergeant was showing a very bad example, and, instead of exposing himself, like others, to the fire of the enemy, he was skulking repeatedly behind the large cork trees, and on my notice being directed towards one of these trees I saw the sergeant hiding himself behind. In the moment of indignation I rode up to him, and licked him as long as I could stand over him, with the flat of my sabre, and ordered him out of the field. Chastising a skulker.

'*After* this chastisement I could not have tried him by a court-martial for cowardice, but I might myself, if he had chosen to come forward against me, have got into a scrape, for the proceeding on my part was certainly hasty and irregular, but he knew too well that if he complained, although I might have been reprimanded, he would have been the more shamefully exposed.

'He asked for his discharge, and left the regiment.

'After leaving the service, one might have supposed that this man would have shown spite and revenge, but he felt conscious that the infliction of the punishment was merited, for I met him often, after our return to England, in the streets of London, when he always saluted me most respectfully. He had become a schoolmaster, and was well qualified for the situation, as he was an excellent scholar.

21. 'On the occasion of erecting a monument to one of my old comrades, Colonel Townsend[1] of the 14th Light Dragoons, and introducing on it the Prussian Eagle on the helmet, which the Fourteenth wore in those days, because the regiment was the Risking life by wearing the Eagle.

[1] Colonel Townsend died in Ireland in 1845.

"Duchess of York's Own" (who was a Prussian princess), I am reminded of a curious little incident which occurred to me through wearing this regimental badge.

'When attached to the Spanish General, the Marquis de la Romana, in Galicia, just after he had separated from Sir John Moore, I was one day traversing a narrow defile, with my orderly dragoon behind me, when I was suddenly assailed by a volley from half a dozen muskets, which wounded my orderly's horse, and one ball went through my helmet without hurting me. Being in a part of the country where I knew the peasantry to be friendly to our cause, and hostile to the French, I was at a loss to account for this extraordinary proceeding, seeing at the time the peasants on the top of a rock, who had fired at me. In a moment after, however, I was surrounded by several hundreds of these armed peasants and ferociously assailed by them, seemingly with a determination to despatch myself and my orderly.

'I was at this period but little acquainted with the Spanish language, as this occurred at the end of 1808, and we had not then been long in the country, but I could understand sufficiently to know that the eagle on our helmets was the cause of it all. They mistook it for the French eagle, to which it is very similar, and had it not been for a friar who was amongst these peasants, and armed as they were, and apparently their leader, we should probably in one more instant have been sacrificed by this infuriated rabble. Fortunately he spoke French, and on my explaining to him that I was an English officer, and that the eagle was the Prussian eagle and not the French, he, though with some difficulty, persuaded the peasants of the fact, and we were released and allowed to proceed.

Never had I a narrower escape than on this occasion. Guns and pikes, etc. etc., were put close to my body with a view of putting an end to me, and my orderly dragoon, a fine, brave fellow, was so exasperated at seeing me thus in peril, that he made the matter worse, and nearly caused my instantaneous death, by drawing his sabre and cutting right and left at my assailants.

'I was taken to the prison at Lamego till my passport was sent to Oporto to be verified; and it was returned with an order from the Bishop for my release. Maréchal Beresford (now Lord Beresford), on being informed of the following conduct of a little American settled as a watchmaker at Lamego, had him sent for a long period to prison, till I interceded for him. This little wretch was employed by the mob, who doubted my being an Englishman, to ascertain, by conversing with me, whether I was or not; and he, with all the low spite of a Yankee, swore that I was not, and that I could barely speak English, when the mob again wanted to put me to death as a spy, and were only prevented by the courageous friar who had before saved my life. I must explain, in elucidation of the above story, that the Bishop of Oporto at this date was a very powerful personage in the north of Portugal, and governed it like a king. He was an intriguing, artful priest, and hated the English as heretics.'

The two following anecdotes of General Brotherton's do not refer to his service in the 14th Light Dragoons, but to the time when he served in Egypt as a young officer in the Coldstreams, in the army under General Sir Ralph Abercromby, in 1801, but their great military interest must excuse their insertion here. Anecdotes of service in Egypt, 1801.

1. 'The first time I was under fire was when carrying one of the colours of the Coldstreams in Egypt, the battalion marching in line. The sergeant behind me (called the covering sergeant), seeing me a raw youth, then only sixteen years of age, said in a respectful but half-joking way, "How do you feel, sir?" to which I replied, "Pretty well, but this is not very pleasant": the men were falling fast. The sergeant, who was a seasoned veteran, liked the reply, for he seemed to take me under his special protection and care ever after. His name and appearance I shall never forget. It was First time under fire, 1801.

Sergeant Stuckey—I often went to Chelsea to see him, where he died at the age of eighty-four, about the year 1840. Probably, had I pretended to feel quite at my ease, and to despise the danger altogether, the shrewd old sergeant would have put me down as a "humbug." It was this same sergeant who, along with poor Beckett, assisted me in the water, when a shell sunk the boat, to the westward of Alexandria.

Adventure in Egypt, 1801.

2. 'When part of the army moved to the west of Alexandria, we had to embark in flat-bottomed boats on a lake. I was carrying one of the colours of the Coldstreams, and Beckett (afterwards killed at Talavera) the other. He, poor fellow, was a very tall man—six feet two or three, I think. We were shelled by the enemy, and one shell fell in the boat, took off the legs of two men, and sunk it. The lake was very shallow, so much so that Beckett stalked along, walking with his colour in his hand, with water just up to his chin. I had to swim and carry my colour, which was no easy job; but I would have died, of course, sooner than let it go. Beckett, who was a delightful, good-natured fellow, stuck to me and assisted me all he could, but shell and shot fell thick around and many men suffered from them and were drowned. I was fortunate enough, however, to reach the shore with my colour, and we drove the enemy before us. It is singular that Beckett met with nearly a similar adventure on the landing in Egypt, two months before, and then had to swim for it, with the colours in his hand.'

Services of General Brotherton.

General Sir T. W. Brotherton, G.C.B.

P General Sir Thomas William Brotherton, G.C.B., the author of the above anecdotes of events in the Peninsular War, entered the army as an ensign in the Coldstreams in January 1800, became Lieutenant and Captain, July 1801; Captain in the 14th Light Dragoons in 1807; Major, November 1811; Lieutenant-Colonel by brevet, 19th May 1814; served in the Fourteenth till 1820, retired on half-pay same year; obtained the rank of Colonel, 22nd July 1830, and was appointed aide-de-camp to the King. On 8th February 1832 he became Commandant of the Cavalry Depôt at Maidstone. He was promoted to the rank of Major-General, 23rd November 1841, and on 17th August 1842 was appointed to the staff of the Northern District at York. On 1st January 1847 he became Inspecting-General of the Cavalry in Great Britain; was promoted Lieutenant-General on 11th November 1851, and General on 1st April 1860. He was made K.C.B. in 1855, and G.C.B. on 28th June 1861. He served under General Sir Ralph Abercromby in Egypt in 1801; in Germany under Lord Cathcart in 1805; and in Portugal, Spain, and France during the whole of the Peninsular War, from 1808 to 1814. For his services he received the war medal (P) with seven clasps, for the battles of Busaco, Fuentes d'Onor, Salamanca (where he was wounded), Vittoria, Pyrenees, Nivelle, and Nive, in which last battle he was again wounded and taken prisoner. Besides these battles he was present at all the cavalry affairs and skirmishes in which his regiment, the 14th Light Dragoons, was engaged, and was at the action on the Coa. He was several times wounded in these skirmishes. He died in January 1868.

Expedition to Gulf of Mexico, North America, August 1814.

The contest in North America, which had been recommenced in 1813, was still going on; and on the 31st August, 2 troops of the Fourteenth, followed by 2 troops on the 10th October, dismounted, sailed from Portsmouth and Plymouth respectively, one party arriving at Jamaica on 24th November. Here an expedition was assembled under Major-General (afterwards Lord) Keane for an attempt on New Orleans, on the river Mississippi, 110 miles from the Gulf of Mexico. The expedition arrived off the coast of Louisiana on the 10th December, where the troops had to be landed in light vessels and open flat-bottomed boats to navigate Lake Borgne, and then to traverse a difficult morass before approaching the enemy's works.

The Americans opposed in such overwhelming numbers, having extensively fortified lines and batteries as well as armed vessels on the river, that the enterprise failed and had to be relinquished.[1]

The first squadron of the Fourteenth arrived in the Mississippi river on the 27th December 1814, and the second squadron on the 5th January 1815. On the 2nd January 1815 a third squadron embarked for North America, but afterwards rejoined the regiment at Hounslow, and remained in England.

Major T. W. Brotherton became Brevet Lieutenant-Colonel on the 19th May 1814, and also received the Companionship of the Bath; and Major C. Baker became Brevet Lieutenant-Colonel on the 4th June 1814. At this period the uniform still remained blue, with orange facings and silver lace.

1815

Two squadrons engaged at New Orleans, 8th Jan. 1815.

In the second attack on New Orleans under Sir A. Cochrane, on the 8th January, the two squadrons served dismounted. Major-General the Honourable Sir E. Pakenham, K.C.B., was killed. Major-Generals Gibbs and Keane were dangerously wounded, when the command of the troops devolved upon Major-General Lambert, who in his despatch to Earl Bathurst, dated off Chandeleur Island, 28th January 1815, states: 'The conduct of the two squadrons of the 14th Light Dragoons, latterly commanded by Lieutenant-Colonel Baker, previously by Major Milles, has been the admiration of every one, by the cheerfulness with which they have performed all descriptions of service.

P.S. I regret to have to report that during the night of the 25th, in very bad weather, a boat containing Lieutenant Brydges and Cornet Hammond with 37 men of the 14th Light Dragoons unfortunately fell into the hands of the enemy, off the mouth of the Rigolets. I have not been able to ascertain correctly the particular circumstances.'—(*London Gazette Extraordinary*, 9th March 1815.)

The troops returned to the fleet, and this was the occasion

[1] Cannon's *Record*, p. 52.

Walter L. Colls, Ph. Sc.

Lieutenant-General James Dormer,
Founder of the Regiment 22nd July 1715,
Colonel of the 14th Dragoons. 1715-1720.

Walter L. Colls, Ph. Sc.

Field-Marshal John Campbell,
5th Duke of Argyll,
Colonel of the 14th Dragoons,
1757-1765.

Walter L. Colls. Ph. Sc.

Major-General Sir John Burgoyne, Bart.
(Colonel of the 23rd Light Dragoons.)
Commanded 14th Dragoons, 1773-1781.

General Sir John O. Vandeleur, G.C.B.,

Colonel of the 14th (Duchess of York's Own) Light Dragoons,

1823-1830.

Walter L. Colls. Ph. Sc.

Colonel John Townsend, A. D. C.,

Commanding 14th (King's) Light Dragoons,

1829-1845.

Walter L. Colls, Ph. Sc.

His Majesty King William IVth

Royal Patron of the Regiment,

30th JULY, 1830.

Walter L. Colls Ph. Sc.

Lieutenant-General Wm. Beckwith, K.H.,
Colonel of the 14th (King's) Light Dragoons.
1860-1871

Walter L. Colls, Ph. Sc.

General the Hon. Sir Henry Murray, K.C.B.,
Colonel of the 14th (King's) Light Dragoons,
1853-1860.

Walter L. Colls, Ph. Sc.

Colonel (afterwards Lieut-General)
Charles Steuart, C.B.,
Commanding 14th (King's) Light Dragoons,
1857–1861.

Walker & Cockerell, Ph. Sc.

Colonel (afterwards Major-General) P. S. Thompson, C.B.,
Commanding 14th (King's) Hussars,
1864-1875.

Walter L. Colls, Ph. Sc.

Lieutenant-Colonel F. P. Campbell,

Commanding 14th (King's) Hussars.

Walter L. Colls. Ph. Sc.

Colonel J. A. Knox,

Commanding 14th (King's) Hussars,

1881-1885.

Photo: Fradelle & Young. Walter L. Colls. Ph. Sc.

General C. W. Thompson,

Colonel of the 14th (Kings) Hussars,

1882-1896.

Walter L. Colls, Ph. Sc.

Colonel Charles Falkiner Morton,
Commanding 14th (King's) Hussars,
1885-1887.

Walter L. Colls. Ph. Sc.

Colonel Henry Blackburne Hamilton,
Commanding 14th (King's) Hussars,
1887-1891.

Photo: Devereux. Walter L. Colls, Ph.

Lieutenant-Colonel A. J. English.
Commanding 14th (King's) Hussars.
1896-1899

on which the boat with 2 officers and 37 men of the regiment fell into the hands of the enemy, as related above.[1]

The four troops left the Isles Dauphins for England on the 23rd March. They anchored at Spithead on the 13th May, landed at Portsmouth on the 14th, 15th, and 16th May, and marched to join the remainder of the regiment at Hounslow, whither the depôt had moved from Radipole Barracks, Weymouth, two months previously, detaching 50 rank and file with officers and non-commissioned officers to Hampton Court Barracks.

Two squadrons from North America land at Portsmouth, May 1815.

On the 6th April the Fourteenth received an intimation from H.R.H. the Commander-in-Chief stating that H.R.H the Prince Regent had been pleased, in the name and on behalf of H.M. King George III., to grant them permission to bear the word 'Peninsula' on their guidons and appointments, in commemoration of their distinguished services in Portugal, Spain, and France, from 1808 to 1814.

'Peninsula' granted, 6th April 1815.

Napoleon Buonaparte having quitted Elba on 26th February 1815, returned to France and once more regained the throne as Emperor. War accordingly began afresh, and ended in the glorious victory at Waterloo. As two squadrons had been employed in the Mississippi, and a third had embarked on 2nd January for the same destination, the Fourteenth were unfortunately prevented taking any part in that action and in the operations on the continent which preceded it; but two of their officers, Colonel Sir F. B. Hervey, Bart., and Major the Honourable H. Percy, served on the personal staff of Field-Marshal the Duke of Wellington in the battle of Waterloo.[2]

The *London Gazette*, dated 22nd June 1815, states as follows: 'Major the Honourable H. Percy arrived late at night on 21st June, in London, with a despatch from Field-

[1] The Regimental Digest of Services has the following entry :—'1815. January 19th and 21st. At the disembarkation near New Orleans to land at Cat Island, Lieutenant Brydges, Cornet Hammond, 1 troop sergeant-major, 39 rank and file captured by enemy.'

[2] Cannon's *Record*, pp. 50 to 53.

Marshal the Duke of Wellington, K.G., to Earl Bathurst, His Majesty's Principal Secretary of State for the War Department, about the victory gained at Waterloo (dated 19th June 1815).' In this the Duke of Wellington says :—'I send with this despatch two eagles taken by the troops in this action, which Major Percy will have the honour of laying at the feet of His Royal Highness. I beg leave to recommend him to your Lordship's protection.—I have the honour, etc.

(Signed) WELLINGTON.'

The third squadron alluded to above, which sailed from Portsmouth, 2nd January, for North America, proceeded first to Cork to join the reinforcements, for same destination, collected there under Major-General Johnstone. The order, however, was countermanded, and the two troops 14th Light Dragoons returned from Cork to England, landed at Deal on 30th March, were detained a short time at Canterbury, and thence went to Hounslow.

Regiment proceeds to Ireland.

The regiment left Hounslow on 30th December for Bristol, under Major and Brevet Lieutenant-Colonel Baker, where they embarked in various parties at intervals up to 31st January 1816, and proceeded to Ireland, landing at Waterford and Cork.

Establishment reduced.

Eight farriers were for the first time appointed to be borne on the strength of the regiment in December, and the troops were reduced from ten to eight: total strength, exclusive of officers, being 677, and troop-horses, 581. The eight troops were designated by the numbers 1 to 8, according to the seniority of their captains.

1816

Regiment lands in Ireland, January 1816.

On the 1st January and following days the regiment landed at Cork and Waterford, and proceeded to Dundalk, the last detachment disembarking in February.

All the front peaks of the troop saddles of Light Cavalry were cut down on the 24th September, and a further reduction in the establishment of the Fourteenth also took place, the total strength of non-commissioned officers and men being 509, with 333 troop-horses.

One lieutenant-colonelcy was reduced, and in the Army List for 1817, Colonel Sir F. B. Hervey, Bart., is the only Lieutenant-Colonel of the Fourteenth, Major-General S. Hawker having been removed from that position, which he had occupied since 1800, with the exception of a short interval only (1802-3) when he was temporarily on half-pay.

1817

Further small reductions took place in the 8 troops; 1 lieutenant only was allowed per troop from 27th January. The Fourteenth remained stationed at Dundalk.

1818

Still further reductions were made in the numbers of the various ranks of the regiment, but the number of the troops remained eight. By an order dated Dublin Castle, 23rd October 1818, each troop was to have only 2 sergeants, 3 corporals, 1 trumpeter, 1 farrier, 42 privates, 34 troop-horses.

Further reductions in establishment, 1818.

On the 25th December, Captains Townsend and Badcock were promoted to the rank of Major in the army for their services in the field during the war in the Peninsula, and a second Assistant-Surgeon ceased to be borne on the establishment of the regiment.

In July the Fourteenth moved to Portobello Barracks, Dublin, to be quartered. During its stay in Ireland the regiment received on several occasions the thanks of both Major-General White and Major-General Sir T. Sidney Beckwith for its good conduct, discipline, and efficiency.

On the 27th May, General Sir George Beckwith, G.C.B., Commander-in-Chief in Ireland, formed up the regiment, in column of half squadrons, in Dublin, and personally complimented all ranks for their excellent conduct and discipline during the period of their stay under his command in Ireland.

1819

Establishment altered, 1819.

The total establishment of the Fourteenth was now fixed at

8 Troops.
28 Officers (5 Staff Officers).
404 Non-commissioned officers and men.
273 Troop-horses.

Regiment lands in England, June 1819.

In June they embarked at Dublin, and landed at Liverpool 11th June, having crossed the Channel in vessels named the *Duke of Leinster*, *Duke of Richmond*, *Shamrock*, and *Dauntless*. From Liverpool there was a long march to the south of England.

Major-General Sir Robert Bolton, Inspector-General of Cavalry, inspected the regiment in June at Camberwell, a halt being made for two days on the march. After 200 miles' march along the roads there were no sore backs, and the Major-General reported favourably on the well-regulated and established discipline which pervaded all ranks. The Fourteenth proceeded thence to Canterbury.

Whilst stationed at Canterbury the various troops of the regiment were much scattered through Kent, at Deal, Hythe, Dover, Ringwould, Folkestone, Romney, Lydd, Sandgate, and Highgate, small parties being detached to these various places and employed in assisting the Riding Officers of the Revenue.

On 26th August, 2 squadrons went from Canterbury to Chatham, returning on 7th September.

On 22nd September, 5 troops left Canterbury and proceeded as follows:—

1 Troop to Ipswich.
1 Troop to Bury St. Edmunds.
1 Troop to Lynn.
2 Troops to Norwich, and detachment to Yarmouth in relief of the 9th Lancers.
1 Troop and headquarters remained at Canterbury.
1 Troop at Hythe.
1 Troop at Deal.

In December the 2 troops at Deal and Hythe joined headquarters at Canterbury.

Blue-grey kersey overalls introduced.

On 31st July, blue-grey kersey wove overalls were taken into wear by all ranks.

On 24th September, Colonel Sir Felton B. Hervey, Bart., C.B., Commanding the Fourteenth, and Secretary to the Master-General of the Ordnance, died. His loss was most deeply deplored by the whole corps.

Lieutenant-Colonel—C. Massy Baker.

Brevet Lieutenant-Colonel Baker succeeded Sir F. B. Hervey as Lieutenant-Colonel of the Fourteenth on 30th September, and Brevet-Major T. P. Milles became Major in succession. Lieutenant-Colonel Baker had practically had command of the regiment for some time already in the absence of Sir Felton Hervey on staff employment.

1820

Five 'honours' granted to the regiment, 1820.

On 17th January, His Majesty's royal permission was granted to the Fourteenth to wear on its guidons and appointments the words

'Talavera,'
'Fuentes d'Onor,'
'Salamanca,'
'Vittoria,'
'Orthes.'

The letter containing the intimation to the regiment of the royal permission for these additional honours was signed 'HARRY CALVERT, *Adjutant-General*, Horse-Guards, 15th March 1820.'

On 24th June the Fourteenth were inspected at Canterbury by Major-General Lord Edward Somerset. The men had lately been employed in the flat marshy ground near Romney, and many were sick with ague and similar complaints. Vaccine inoculation was regularly practised in the hospitals at this period. The review report states that there was a Riding-Master now in the regiment, but his name does not appear in the Army List amongst the roll of officers until the year 1823. Four men per troop were taught how to shoe horses on an emergency.

In June the 5 troops which were in Norfolk and Suffolk rejoined headquarters at Canterbury, and the Fourteenth again furnished detached troops to Deal and Hythe for revenue services.

On 19th July a squadron was sent from Canterbury to Dover on civil duty.

On 31st July, 4 troops and headquarters left Canterbury for Brighton.

On 14th August a squadron left Dover for Lewisham, and at the same time a squadron moved from Brighton to Reigate, *en route* to Richmond.

It appears from the general marching orders of this period that 6 troops were at Richmond, Putney, Mortlake, etc., early in September, and were ordered to march about 10th September to Brighton in two divisions, and that they arrived at Brighton, 13th September, where the 85th Regiment (Duke of York's Own Light Infantry) arrived from Richmond, Twickenham, etc., on 15th, 16th, and 18th of the same month. It also appears from same marching orders that the 6 troops of the 14th Light Dragoons left Brighton again on 30th September for the above-named places in the vicinity of London, and part of the regiment was employed at Richmond in September and October, and at Wimbledon in October, when it was again inspected by Major-General Lord Edward Somerset, Inspector-General of Cavalry.

There appears to have been a new system of military equitation introduced about this period, and the Fourteenth were temporarily cantoned during the autumn months about Richmond and Wimbledon, having come there from Kent and Brighton, and afterwards returned about October to Brighton, sending detachments to places along the coast of Sussex. Major and Brevet Lieutenant-Colonel T. W. Brotherton went on half-pay on 25th September, and Brevet Lieutenant-Colonel the Honourable Henry Percy, C.B., who had served on the Duke of Wellington's staff at Waterloo, became Major on the 12th October.

1821

George IV.

George IV., 1821.

The establishment was reduced from 8 to 6 troops, consisting of 23 sergeants, 6 farriers, 6 trumpeters, 18 corporals, 282 privates, and 253 troop-horses, and on 25th April the officers' clothing was regulated by His Majesty's warrant.

Establishment reduced to 6 troops, 1821.

On 16th June, Major-General Lord R. E. H. Somerset inspected the regiment at Brighton, under the command of Lieutenant-Colonel Baker. There were detachments at—

Hastings, under Captain Babington.
Arundel, under Captain Townsend.
Eastbourne, under Captain Charlton.

Pulmonic complaints had been very prevalent in the Fourteenth during the last six months in the regimental hospital at Brighton.

There were 83 soldiers' wives in the regiment at this time, and 110 children, 50 of whom attended school. The boys were taught trades, and the girls were taught to knit and sew in the regimental school.

Up to this period, corporal punishment appears to have been much resorted to in the army generally, and was very frequent—in fact, it seems to have been almost the only punishment awarded by regimental courts-martial for the rank and file, reductions for non-commissioned officers being the only alternative.

Veterinary-Surgeon A. Black and the farriers had made a very successful use of Colman's patent shoe in cases of disease in the horses' feet, for which they were commended in the Major-General's report.

The Fourteenth proceeded during the year from Brighton

to Hounslow, and were inspected there again by Major-General Lord E. Somerset on 23rd October.

Captain (Brevet-Major) J. Townsend became Major on 13th September.

1822

On 1st June the Fourteenth took part in a review on Hounslow Heath before H.R.H. the Duke of York, Commander-in-Chief, together with the three Household Cavalry Regiments, the 10th and 15th Hussars, and a brigade of Royal Horse Artillery. A very high meed of praise was bestowed by H.R.H. on all the troops present, for their general appearance, equipment, and movements. This was expressed in a letter written by the Adjutant-General of the army, Sir Henry Torrens, and forwarded through General Earl Cathcart and Major-General Lord Edward Somerset, to those concerned.

In the summer the Fourteenth marched to Coventry, Birmingham, Nottingham, and Abergavenny. At this period the total number of officers of all ranks in the regiment was 28, and the number of troops 6.

1823

In June the Fourteenth moved to Dorchester, sending detached troops to Christchurch and Truro. General the Earl of Bridgewater died, and was succeeded as Colonel by Lieutenant-General Sir John O. Vandeleur, K.C.B., on 28th October. He came from the 19th Light Dragoons, which had been disbanded in 1821. Mr. Collyer, of Park Place, St. James's, had been regimental agent up to this time, since 1799.

Colonel—Sir J. O. Vandeleur.

1824

Headquarters and 2 troops moved from Dorchester in the autumn to Exeter.

A hospital sergeant was first appointed to the regiment in this year, and one private taken off the establishment.

Messrs. Greenwood, Cox, and Hammersley were this year appointed the regimental agents.

1825

In April the Fourteenth left Exeter, etc., and embarking 18th April at Bideford, landed at Waterford on the 20th and 21st of April, and were quartered at Cork, Fermoy, Bandon, etc. **Regiment lands in Ireland, 1825.**

1826

The headquarters remained at Cork till the 8th May, when the regiment marched to Dublin, and on the 12th June was inspected by Major-General Sir Colquhoun Grant, K.C.B., and also by him again on the 14th September, when the Major-General was pleased to express his entire approbation of its appearance, discipline, and interior economy.

1827

On the 11th and 12th January the Fourteenth marched to Newbridge, *en route* to Athlone, Ballinrobe, Gort, and Loughrea.

1828

On the 4th and 5th March the Fourteenth moved to Dublin under Lieutenant-Colonel C. M. Baker, where it was inspected on the 22nd March by Lieutenant-General Sir G. Murray, G.C.B., Commander of the Forces in Ireland, from whom it received a very high meed of praise for its appearance, discipline, and general good conduct. He either spoke or wrote the following words to Lieutenant-Colonel Baker after his inspection of the regiment:—'Colonel Baker, it gives me the greatest pleasure to express the great satisfaction this inspection has afforded me. The high state of discipline, the efficiency and interior economy of the 14th Light Dragoons were never more conspicuous, and coincide with all the previous reports I have had from Major-General Sir Colquhoun Grant. Now that the regiment is about to embark, I will take this opportunity of stating that my long acquaintance with it and its distinguished services in the Peninsula, will always cause me to feel the most lively interest in its welfare and prosperity.'

Recruiting.

The Fourteenth of late had been recruited from time to time by means of a recruiting party stationed at Parsonstown, King's County, together with recruits raised at headquarters.

Regiment lands in England, 1828.

On the 26th March the Fourteenth embarked in two divisions for Liverpool, disembarked there the 28th, and marched thence on the 3rd April to Coventry and Birmingham.

Captain William Beckwith became Major on the 14th February, *vice* Milles.

1829

On the 16th April, Lieutenant-Colonel Baker retired after ten years in command of the regiment, and Major J. Townsend, who had served almost since he was a boy in the Fourteenth, succeeded him as Lieutenant-Colonel, and Captain E. L. Parry became Major.

Lieut.-Colonel—John Townsend.

On the 9th February, lapels for Light Cavalry were abolished, and the colour of the overalls, which for the last ten years had been blue-grey, was changed to Oxford mixture on the 15th April.

Changes in uniform, 1829.

The strength of the regiment at this period was only 326 men and 250 troop-horses, and during the year 5 men died and 4 deserted, according to the annual return.

On the 8th May, 3 troops left Birmingham, 2 for Burnley and 1 for Rochdale (Lancashire). On the 9th May, headquarters and 2 troops left Coventry for Leeds, and on the 5th June, 1 troop left Coventry for the same destination.

Major William Beckwith was awarded the dignity of a Knight Companion of the Hanoverian Order of the Guelphs ('K.H.').

By an order dated 9th December, forge carts were withdrawn from cavalry regiments in Great Britain.

1830

William IV.

William IV., 1830.

In April the Fourteenth marched from Leeds to Brighton and Chichester, 4 troops and headquarters to Brighton, 2 troops to Chichester. This move was completed by the 20th May, and on the 24th May, Major-General Sir Hussey Vivian, K.C.B., inspected the headquarters and 2 squadrons at Brighton, and highly commended the regiment for its appearance, discipline, and interior economy.

Recruiting was temporarily suspended about this time.

On the 19th June, at Brighton, the full Colonel, Lieutenant-General Sir J. O. Vandeleur, G.C.B., made a farewell inspection of the regiment, and was very complimentary in his remarks afterwards to Lieutenant-Colonel Townsend. He particularly noted the excellent management of the school, and the great progress of the pupils.[1]

Colonel—Sir Edward Kerrison, Bart.

On the 18th June, Major-General Sir E. Kerrison, Bart., K.C.B., was appointed Colonel in succession to Lieutenant-General Sir J. O. Vandeleur, who went to the 16th Light Dragoons.

On the 16th July an order was issued directing that only one major be borne on the establishment of cavalry regiments at home.

On the 22nd July the Fourteenth marched to London, and were quartered in the neighbourhood of Paddington and Lambeth from 24th to 29th July.

14th (King's) Light Dragoons, 30th July 1830.

On the 26th July, His Majesty King William IV. reviewed the regiment in Hyde Park. His Majesty was much pleased

[1] See Cannon's *Record*, p. 56, etc. It is a curious coincidence that at Brighton again, nearly sixty years afterwards, the excellent management of the regimental school was very highly commended by the District Inspector in 1888-90, as recorded in his official report to the Commanding Officer.

with the regiment, and commanded that it should in future bear the distinguished title of 'The Fourteenth,' or 'The King's,' instead of the 'Duchess of York's Own Regiment of Light Dragoons.' His Majesty expressed his great approval of the regiment through Major-General Sir E. Kerrison, Bart., K.C.B., who was present at the inspection. This order was dated 30th July 1830, and in consequence of becoming a Royal regiment the facings should have been changed from orange to scarlet, but this change does not appear to have been carried out immediately.

On the 29th July the Fourteenth proceeded from London to Birmingham and Coventry.

Regiment employed in quelling riots in the Midlands, October 1830.

On the 5th October, Major-General Dalbiac inspected at Birmingham, and was much pleased with the efficiency of the regiment. On the 11th November, 1 troop was sent from Coventry to Banbury, where it apprehended 17 rioters the same night, and from this date up to the 24th December all the troops of the Fourteenth were frequently on the move in the neighbouring towns and counties owing to riots.

On 2nd December headquarters moved from Birmingham to Oxford.

On the 3rd December, 1 troop moved from Kidderminster to Oxford.

On 4th December, 1 squadron moved from Northleach to Barford.

On 18th December, 1 troop went from Wellingborough to Northampton.

On 21st December, 1 troop went from Northleach to Coventry.

On 19th December headquarters moved from Oxford to Coventry.

Cavalry uniforms ordered to be scarlet.

On the 2nd August 1830, all cavalry, excepting the Royal Horse Guards (Blue), were ordered to be clothed in red, as His Majesty King William IV., being a sailor, preferred blue for the navy, and red for the army. All authorities, including the Army Lists, give the facings of the Fourteenth at this period as orange, but Cannon records that they were scarlet.[1] He does not, however, notice the general change of uniform from blue to red which took place throughout the cavalry from 1830 to 1832. The silver lace was ordered to be changed to gold this year.

1831

On the 1st January the number of troop-horses was increased from 253 to 271, and an orderly-room clerk, ranking as sergeant, was added, one private being at the same time reduced from the establishment. **Establishment.**

The headquarters were at Coventry till February, and from the 9th February till the 24th November the various troops of the Fourteenth were moved about on duty during the riots in the Midland and Southern counties, in aid of the civil power. Some troops went as far as Dowlais and Llandrillo, in Merionethshire.

Headquarters moved in February to Birmingham, sending 1 troop to Kidderminster.

In April, owing to elections at Coventry, 3 troops moved hence, 2 to Nuneaton and 1 to Leamington, returning to Coventry on the 9th May.

On 7th July headquarters and 2 troops marched from Birmingham for Worcester, thence *via* Tewkesbury to Gloucester to be quartered. Four troops and headquarters were inspected at Worcester on the 9th July by Major-General Sir C. Dalbiac. These 4 troops moved to Gloucester on the 11th July to be stationed, but on the 6th August, owing to the assizes taking place there, headquarters and 3 troops moved to Tewkesbury, and 1 troop to Upton-on-Severn.

The headquarters and 4 troops from Tewkesbury and Upton-on-Severn returned to Gloucester on the 19th August.

Regiment employed at riots in South Wales, August 1831.

On the 27th June a troop which had been on detachment was sent from Coventry to Abergavenny. A troop went from Dowlais to Abergavenny on the 29th August, having been sent

[1] Cannon's *Record of 14th Light Dragoons*, p. 57.

to aid the civil power at the Dowlais iron works, where there were riots, and 1 troop had moved on the 20th August from Abergavenny to Llandilo, owing to riots at Carmarthen, but returned on the 1st September to Abergavenny.

On the 26th September, 2 troops moved from Gloucester to Abergavenny, 2 troops at Abergavenny went to Gloucester, and a detachment went from Gloucester to Newport on the 1st November. On the 2nd November a troop went from Gloucester to Tewkesbury, on the 3rd November 1 troop went from Bristol to Gloucester, and on the 6th November 1 troop was moved from Abergavenny to the Tredegar iron works till the 8th November, and owing to Bristol riots, 3 troops were at Sudbury and Clifton in October, and in the same month 2 troops were employed at Bristol in the riots there, under command of Major William Beckwith, K.H.,[1] who had with him also a troop of the 3rd Dragoon Guards. The energetic conduct of Major William Beckwith on this occasion helped materially to suppress the riots, and to save the city from wholesale plunder by the rioters. Sir Charles Wetherall was the recorder of Bristol, and it was his arrival to hold the Sessions which brought the riots to a head, on Saturday, the 29th October. The troops at Tewkesbury returned to Gloucester by the 18th November.

Regiment employed at the Bristol riots, 1831.

General Lord Hill, Commander-in-Chief, gave the Fourteenth very high praise for their services during the suppression of the riots in various places, but particularly at Bristol, on which latter occasion he specially commended Major Beckwith, K.H., and Captains Gage and Musgrave.

Lieutenant Van Straubenzee[2] was one of the officers of Captain Musgrave's troop of the 14th Light Dragoons present at these Bristol riots, and the following description of what took place is taken from a letter written by him some years after to the editor of the *Daily Telegraph*:—'In the year 1831 it was my misfortune to be present at several riots as

[1] Afterwards Lieutenant-General William Beckwith, K.H., Colonel of the 14th (King's) Light Dragoons.

[2] Afterwards Colonel Van Straubenzee, commanding 2nd West York Militia.

well as at Bristol. In the latter, I may mention that after being pelted with stones and harassed in every sort of way, the captain in command of the troop of the Fourteenth, during a temporary absence of the commandant,[1] went to the mayor and his brother magistrates and expressed his opinion that if he was allowed to act the riots might soon be stopped. He obtained leave to act, but unfortunately as he was leaving the Mansion House he met the commandant, who told him he must do nothing without his orders. The consequence was the troops were pelted and harassed as before. There was one particular passage where the rioters had collected a great amount of ammunition, and from which issued a volley of stones every time the troops passed. The officer in command told them if the same thing happened the next time, he would fire into the passage. Before returning he got a pistol from one of the men, and as the same thing occurred, he fired instantly and killed a man. This stopped their stone-throwing, but they instantly went to the colonel in command of the troops and promised him if he would send away the "bloody blues,"[2] as they pleased to designate the 14th Light Dragoons, they would be quiet. He unfortunately believed this, and ordered the Fourteenth out to Keynsham. After they were sent away fearful excesses commenced: it was then that the three gaols were broken open, all the prisoners released, and the buildings destroyed by fire. The bishop's palace was broken into, sacked and burned down, and many of the rioters, being intoxicated by the contents of the cellar, were buried in the ruins. The toll-houses were burned down, and the Mansion House was set fire to, and many of the rioters perished by the fall of the walls. Fortunately the aide-de-camp to the Commander-in-Chief happened to be staying in the neighbourhood, and he came over and took upon himself to order the return of the troop of the Fourteenth,

[1] The commandant was disposed to be very lenient towards the rioters.

[2] This nickname seems to prove that the scarlet uniform with gold lace was not yet adopted in the Fourteenth.

and just when it arrived at Bristol, Major Beckwith, K.H., with another troop arrived from Gloucester, and he, taking command of the squadron, with a few vigorous charges, in which several men were cut down, struck terror amongst the populace and crushed the riot at once. Our squadron was very weak, and I don't think the officers, non-commissioned officers and men amounted to more than eighty altogether. Yet the infuriated and intoxicated mob, that had for many hours been in possession of the town, were subdued in an incredible short space of time.'

Major-General T. Van Straubenzee, C.B., late Royal Artillery, son of the late Colonel Van Straubenzee, says in a letter to the author:[1] 'As far as I remember, my father was in Captain Musgrave's troop at the time (as Lieutenant), and in charging the rioters he was knocked off his horse and stunned by a brickbat, and pulled into a house to save him from further violence.'

The riots at Bristol took place from the 29th to 31st October, and in addition to the 14th Light Dragoons, the Royal Horse Artillery, the 3rd Dragoon Guards, and the 52nd Light Infantry took part in their suppression, under the command of Major-General Sir R. Jackson, K.C.B.

Orange facings changed to blue, 1831.

On the 12th April it had been ordered that the facings should be changed from orange to blue.[2] The orange facings had been worn by the Fourteenth since 1798.

1832

Regiment at the assizes and riots in South Wales.

On the 13th January the troops at Abergavenny went to Chepstow and Usk till 1st February, and the troops temporarily at the Hot Wells, Clifton, came in to Gloucester on the 5th February. Owing to the assizes and riots several further movements of troops took place in March and April, and 1 troop from Abergavenny went to the Beaufort Ironworks on the 14th April till the 9th May.

Headquarters moved in March to Stroud, and back to

[1] Dated 23rd October 1900. [2] *War Office Circular-Book, E.*, p. 130.

Gloucester in April. In March, 1 troop went from Clifton to Sudbury.

A very handsome pecuniary reward of £40 was voted by the authorities at Gloucester to the non-commissioned officers and men of the Fourteenth who had been employed there when a great fire took place on the 23rd April, at the Lunatic Asylum. The money was paid by the Lunatic Asylum Protector Insurance Company, and was divided among the non-commissioned officers and men of the Fourteenth who had rendered such valuable aid on the occasion, under the orders of Lieutenant-Colonel Townsend and his officers. The Mayor of Gloucester, Mr. Alexander Walker, wrote a most complimentary letter, dated 4th June, to Lieutenant-Colonel J. Townsend as to the valuable service thus rendered by the troops.

Regiment employed on Royal escorts at Hounslow, 1832.

In June, 5 troops of the regiment were moved from Gloucester, Clifton, and Coventry, to Hounslow, where they were employed on King's duty, in furnishing escorts for their Majesties and the royal family.

One troop went to Hampton Court, and a detachment to Kensington, consisting of 1 subaltern, 1 sergeant, and 32 rank and file. On the 25th June, 4 troops were at Kensington for the night.

King's crest authorised for the appointments, and the Prussian Eagle continued on the guidons, 1832.

The King's crest (the royal cypher within the garter) was, by His Majesty's royal favour, permitted to be borne on the appointments; and the Prussian eagle, which had been hitherto used as the regimental badge since 1798, was now authorised to be continued in the second and third corners of the regimental guidon.[1] The regiment being a royal one was entitled to scarlet facings from 1830, but as the uniform was changed from blue to scarlet about the same date, or at all events not later than 1832, and the facings to blue, it is not probable that scarlet facings were worn until the uniform was changed again to blue in 1840.

Establishment.

During this year, according to the annual return, 6 men died and 13 deserted; the establishment of the regiment being 335 men and 270 troop-horses.

[1] Cannon's *Record*, pp. 59-60.

Recruiting.

The Fourteenth had been recruited lately in the vicinity of its quarters, in the counties of Gloucester and Warwick, and by means of a recruiting party at Worcester.

1833

Major-General Sir E. Kerrison, Bart., K.C.B., inspected the Fourteenth at Hounslow on the 18th March, and expressed himself very pleased with what he saw.

Regiment proceeds to Ireland, 1833.

The troops at Kensington and Hampton Court had come to Hounslow on the 16th March, and on the 21st March the Fourteenth moved from Hounslow to Bristol, where they embarked for Dublin, the third and last division disembarking on the 2nd April. They were quartered in Dublin until the end of the year, going on the 5th July from Portobello Barracks for temporary duty to Newry, Dundalk, Armagh, Castleblaney, and Monaghan, in aid of the civil power, but returning from Armagh to Dublin by the 20th July. Major-General Sir E. Blakeney, K.C.B., inspected the regiment in Dublin at Portobello Barracks, on the 17th May; he expressed himself very much pleased, and especially commended the comfortable messing arrangements for the rank and file, and the admirable state of the regimental school. He also said that the steadiness of the men and the celerity of their movements at his review in the Phœnix Park could not be surpassed.

Lieutenant-General the Right Honourable Sir Hussey Vivian, Bart., G.C.H., K.C.B., Commander of the Forces in Ireland, held a review of the regiment in the Phœnix Park on the 27th May. At the conclusion of the parade he addressed the regiment, and said: 'The 14th Light Dragoons are, as I have always found them, and I have no doubt they will ever continue to be, as perfect as any regiment of cavalry can be, and you have my unqualified approbation.'

In Dublin, on the 14th August, Lieutenant-General Sir Hussey Vivian again had a review, when the Cavalry Brigade consisted of the 5th Dragoon Guards, 10th Hussars, and 14th Light Dragoons.

On the 18th April, Cornet Surtees died at Clifton, and Major William Beckwith, K.H., left the Fourteenth on appointment to an unattached lieutenant-colonelcy.

On the 8th September a slight reduction in the troop-horses and rank and file of the regiment was ordered, and on the 29th November recruiting was suspended. Establishment reduced. Recruiting suspended.

1834

On the 1st April headquarters and 3 troops moved from Portobello Barracks, Dublin, to Longford, 1 troop went to Gort, 2 troops to Athlone, and a small detachment went to Birr on the 23rd April from Athlone, owing to 4 Baronies in King's County being proclaimed under the Coercion Act, and in June another detachment went from Longford to Ballinasloe.

On the 24th May, His Majesty King William IV. commanded that the use of guidons in regiments of light dragoons should be discontinued.[1] Those in regiments of lancers and hussars had been already discontinued two months earlier. Guidons abolished, 1834.

In April the establishment was reduced to 274 men and 253 troop-horses. Establishment reduced.

On the 4th and 7th June, General Sir J. Buchan, K.C.B., inspected the Fourteenth at Longford, and expressed his entire satisfaction. He also inspected again on the 8th November, and spoke in very flattering terms of the satisfactory state of the regiment.

1835

During the parliamentary elections from the 8th to the 28th January the Fourteenth were employed in aid of the civil power, to preserve the peace, in Galway, Ennis, Castlebar, Ballinasloe, Tuam, Roscommon, Loughrea, and adjacent districts, without any casualties to man or horse. In April, 1 troop went to Carrickmacross. From the 5th to the 19th May the regiment was moving to Dundalk, 4 troops and headquarters; detaching Regiment employed on election duty, 1835.

[1] The Fourteenth thus lost the White Horse of Hanover granted by Royal Warrant in 1751.

1 troop to Belturbet with a subaltern's party at Ballyshannon, and 1 troop to Monaghan. Headquarters were inspected at Dundalk by Major-General Macdonell, C.B., on the 29th May, and again on the 25th September following, and on both occasions the Major-General was pleased to express his approval.

Regiment employed in aid of civil power for collection of tithes.

During June, July, and August the Fourteenth were constantly moved about in aid of the civil power. In December the troops at some of the out-quarters were exchanged for others from Dundalk, to which place they came in; the subaltern's party at Ballyshannon was not continued, and the Monaghan troop moved to Belturbet. In the previous month (November) the subaltern's party was sent from Ballyshannon to Monaghan to aid the civil power in collection of tithes.

Cornet C. Jones died at Longford on the 6th April, and Captain C. Delmè died at Dundalk on the 14th November.

Band to be dressed like the trumpeters.

Captain J. W. Sim Smith became Major *vice* Beckwith on the 17th July. The dress of the band was now assimilated to that of the trumpeters by an order of the 11th April applying to all cavalry regiments.

1836

On the 3rd February the Monaghan troop came to Dundalk and was not relieved. During the months of March, April, and the beginning of May the various troops of the Fourteenth were moved about in aid of the civil power at Newry, Carrickmacross, Newtown-Hamilton, Kingscourt, etc. There was a squadron now at Belturbet which joined headquarters at Dundalk on the

Regiment lands in Scotland, 1836.

11th May. On the 12th May the headquarters and 5 troops went from Dundalk to Belfast, and, commencing on the 17th May, the whole regiment embarked in 8 divisions in small steam vessels for Glasgow, arriving there on the 25th May.

Recruiting.

At this time the recruiting of the regiment was mostly kept up by means of recruits raised at headquarters and detachments. Major-General the Honourable Patrick Stewart inspected at

Glasgow on the 7th June, and gave the regiment the greatest credit for its high state of efficiency.

The second annual inspection was made by Major-General Sir C. Dalbiac, K.C.H., on the 8th October, which also passed off in the most satisfactory manner. The effective strength of the regiment was 305 non-commissioned officers and men, and 232 troop-horses. A detachment was stationed at Hamilton.

1837

Queen Victoria

Queen Victoria, 1837.

On the 4th April the Fourteenth commenced the march to Piershill Barracks, Edinburgh, from Hamilton and Glasgow.

'Douro' granted on the appointments, 22nd July 1837.

On the 22nd July, twenty-eight years after the event, in reply to an application made by Lieutenant-Colonel Townsend, permission to bear the word 'Douro' on the appointments was granted—an honour all the more precious since it is not shared by any other British cavalry regiment.

The Fourteenth were employed at Musselburgh and the adjacent localities from the 22nd July till the 5th August during the general election; and on the 22nd August 3 troops were employed at Dalkeith till the 28th, during the election of peers for Scotland.

1838

In May the regiment left Edinburgh and was at Langholm on the 1st June, afterwards proceeding, 3 troops and head-quarters to Birmingham and 2 troops to Coventry.

One troop went from Carlisle to Bath the same month, and in the following month moved to Clifton, when the squadron at Coventry moved to Weedon; the dismounted party and the baggage went by Glasgow and Liverpool to Birmingham.

When the Fourteenth left Edinburgh, Major-General Lord Greenock, Commanding the troops, issued a most complimentary farewell order in a letter dated Edinburgh, 27th May 1838.

He referred in the highest terms to the conduct of the regiment whilst stationed under his command, and said how gratifying it was to him to observe that a corps so distinguished as the 14th (King's) Light Dragoons had ever been, was careful to maintain its great reputation earned in the Peninsula, and how rejoiced he was that the corps still preserved unchanged after so many years of peace the same character to which it owed its former reputation, and that it had lost nothing of that spirit which animated it during the brightest period of the late wars. His lordship continued :—'This is mainly to be attributed to the devoted zeal and indefatigable exertions of its present commanding officer, who, having been brought up himself in the regiment and a participator in all the actions in which it was engaged in the Peninsula, knows so well how to keep alive both by precept and example the *esprit de corps* which pervades all ranks, and enables him, with the assistance of Captain Leary, who was trained in the same school and likewise fought gallantly in the Peninsular campaign, so successfully to carry out the excellent system of discipline and instruction which now renders the 14th (King's) Light Dragoons, in the Major-General's estimation, for all the purposes of Light Cavalry service, the most efficient corps in the British Army.

(Signed) GREENOCK, M.G.,
Commanding in North Britain.

'*To* Lieut.-Colonel TOWNSEND,
Commanding 14th (King's) Light Dragoons,
Piershill Barracks.'

On the 1st June, Captain C. Barton became Major *vice* Smith. Captain C. Royds, when on leave, died of consumption at Avignon, France, on the 27th March, and Assistant-Surgeon J. Huey died of laryngitis at Edinburgh on the 28th April.

Establishment. Recruiting.

Early in the year a slight increase in the rank and file took place, and recruits were partly raised in London, Birmingham, and Saxmundham. The number of troop-horses was raised from 253 to 271.

On the 31st July the 2 troops at Weedon proceeded to Northampton, and on the 19th May they marched for Coventry and arrived on the 20th.

On the 1st August, Major-General Sir Charles Dalbiac, Inspector-General of Cavalry, attended by his Brigade-Major, Lieutenant-Colonel the Honourable William de Ros, inspected the Fourteenth at Birmingham, and expressed his entire satisfaction. On the 31st December, Sir Charles Dalbiac became Lieutenant-General, and was succeeded as Inspector-General of Cavalry by Major-General Sleigh.[1]

1839

In April the Fourteenth marched to Hounslow and Hampton Court, a subaltern's party of 30 men went to Kensington, and a squadron to Hampton Court. The dismounted party went by railway from Birmingham to Harrow for Hounslow on the 11th April. Headquarters and 4 troops were at Hounslow.

Percussion carbines issued in place of pistols, and wallets in place of holsters.

On the 4th May pistols for the ranks were discontinued, and percussion carbines were issued on the 23rd October. Each officer, sergeant-major, and trumpeter still retained a pistol, and by an order issued by General Lord Hill, G.C.B., G.C.H., Commanding-in-Chief, dated the 4th May, leathern wallets took the place of holsters for Light Dragoons and Hussars.

On the 29th May, at Windsor, the 14th (King's) Light Dragoons, together with the Royal Horse Guards (Blue), were inspected by H.I.H. the Hereditary Grand Duke of Russia, who was attended by General Lord Hill, G.C.B., G.C.H., Commanding-in-Chief. His Imperial Highness was much pleased with the performances of the regiments.

On the 3rd June, Major-General Sleigh, Inspector-General of Cavalry, inspected at Hounslow, and highly commended the field movements and general state of the regiment. General Lord Hill, G.C.B., G.C.H., Commanding-in-Chief, inspected the regiment on Hounslow Heath on the 16th July,

[1] Afterwards General Sir J. W. Sleigh, K.C.B.

and expressed himself very pleased with the appearance of the men and horses, and the celerity of their movements, and he referred with much gratification to the many favourable reports he had received of the corps from the Inspecting-Generals of Cavalry at their several inspections.

Her Majesty Adelaide, the Queen Dowager, through Lord Howe, in a letter addressed to Lieutenant-Colonel Townsend, expressed her grateful thanks for the assistance so promptly rendered on the occasion of a fire at Bushey Farm by the Hampton Court detachment of the Fourteenth on the 25th August, and offered to give the men a dinner, or to reward them in any other manner suggested by the Colonel.

Queen Victoria reviews the regiment in the Home Park, Windsor.

On the 1st November, Her Majesty Queen Victoria held a review in the Home Park, Windsor, at which the 14th (King's) Light Dragoons were present, together with the 2nd Life Guards, and the 2nd battalion of the Rifle Brigade. Her Majesty was graciously pleased to forward a most flattering acknowledgment of her approbation of the regiment to Lieutenant-Colonel Townsend, to be communicated to the officers and men who took part in the review.

1840

Straight swords issued.

On the 1st January the regiment received straight swords in place of the curved light cavalry swords hitherto in use. All heavy cavalry and dragoons had straight swords as far back as 1812.

Escort for H.R.H. Prince Albert, 8th February 1840.

The following extract from the *Times* newspaper (taken from a Sunday print) for Monday, 10th February 1840, has reference to the distinguished honour which the Fourteenth had conferred upon them by being selected to supply an escort for His Royal Highness Prince Albert, on Saturday the 8th February 1840, the day he arrived in London previous to his marriage with Her Majesty Queen Victoria, on the 10th February:—'From Chatham His Royal Highness and suite set out for Gravesend and so on to Dartford, where they were met by one of Her Majesty's carriages attended by servants in Royal livery, and thence proceeded direct to town. On the arrival of the cavalcade at New Cross a detachment of the 14th (King's) Light Dragoons was in waiting, and escorted His Royal Highness from thence through Peckham, Camberwell and Stockwell, over Vauxhall Bridge to his happy destination. As early as three o'clock, notwithstanding the unfavourable state of the weather, groups began to assemble in St. James's Park in anticipation of His Royal Highness's arrival. Some stationed themselves at the Horse Guards from a belief that His Royal Highness would have come over Westminster Bridge, but the greater proportion congregated in the immediate vicinity of Buckingham Palace. About four o'clock the Earl of Albemarle, the Earl of Errol, the Earl of Uxbridge, Lord Melbourne and several of the Cabinet Ministers arrived at the Palace in readiness to receive the expected guest, and soon after, carriages, in which were the Duchess of Richmond, the Marchioness of Anglesey, Lady Byng, and other ladies of distinction, drew up in front of the principal entrance to the Palace through the Marble Arch, while the Palace Guard were drawn out in front of the guard-house prepared to offer the usual salute. All suspense was shortly afterwards terminated by the rapid approach of the vanguard of the 14th (King's) Light Dragoons, who galloped forward and took their station in front of the Marble Archway, the gates of which had been previously opened by one of the royal servants. The main body of the escort soon followed, surrounding the carriage, in which were His Royal Highness Prince Albert, the reigning Duke of Saxe-Coburg-Gotha, Prince Ernest, and Lord Torrington. A second carriage came immediately after containing Colonel Grey, Count Kelowrath, Baron Alvensleben, and another gentleman.'

Escort for H.M. Queen Victoria and H.R.H. Prince Albert, 10th Feb. 1840.

The Fourteenth also had the high honour of escorting Her Majesty Queen Victoria and His Royal Highness Prince Albert on the day of their wedding, 10th February, on the occasion of their departure for Windsor in their travelling chariot from Buckingham Palace as far as Colnbrook, and here they were relieved by an escort of the 2nd Life Guards, under command of Lieutenant Tottenham.[1]

In April and May the Fourteenth marched from Hounslow, Hampton Court, and Kensington as follows:—

2 Troops and headquarters to Dorchester.
1 Troop to Cardiff. 1 Troop to Trowbridge.
1 Troop to Weymouth. 1 Troop to Exeter.

Uniform blue, facings scarlet.

On the 29th June an order was issued that the facings of the Fourteenth be changed from blue to scarlet, and the uniform was changed at the same time from scarlet to blue.[2]

On the 5th August, Major-General Sleigh, Inspector-General of Cavalry, made his inspection of the regiment, and expressed himself much pleased with all he saw.

Messrs. Cox and Company were now appointed the regimental agents.

Recruiting done at headquarters since 1838.

The recruiting of the Fourteenth was now, and had been for the last two years, kept up by enlisting at headquarters.

1841

Establishment raised to 9 troops. Recruiting at Worcester.

In January a recruiting party was sent to Worcester. On 3rd January the Fourteenth were placed under orders for India to relieve the 4th Light Dragoons in Bombay, and the establishment was augmented to 9 troops, 55 sergeants, 12 trumpeters, 8 farriers, 40 corporals, 627 privates, and 701 troop-horses. The recruiters at Worcester and at headquarters were actively engaged enlisting men to complete the strength.

On the 30th March the Fourteenth proceeded to Canterbury, and arrived there on the 10th April preparatory to embarkation for India. Major-General J. W. Sleigh, C.B., Inspector-General of Cavalry, made a farewell inspection of the regiment

[1] *Times*, Tuesday, 11th February 1840.

[2] See *War Office Circular-Book, E.*, p. 118.

at Canterbury on the 20th April, and afterwards wrote a very complimentary letter to Lieutenant-Colonel J. Townsend, commanding the regiment, of which the following is a copy:—

'My dear Townsend,—It gives me much pleasure to be able to beg of you to express to the officers and men of the 14th (King's) Light Dragoons the gratification I have had in making my inspections of the regiment since I have held the appointment of Inspector-General of Cavalry. I am aware I can add nothing to establish the character of a corps which has ever borne so distinguished a place for gallantry in the field, and good conduct in quarters; yet it is a pleasing task for me to bear record on your departure for India that the same good conduct still exists which has heretofore gained you the high opinion of all those officers under whom your regiment has served.—Believe me, my dear Colonel, faithfully yours, J. W. Sleigh.'

After this inspection on the 20th April the Fourteenth were dismounted, and the horses were drafted to several other cavalry regiments during the four following days.

Regiment gives over horses, April 1841.

On the 30th April, Major William Havelock, K.H., eldest brother of the renowned Major-General Sir Henry Havelock, K.C.B., was appointed second Lieutenant-Colonel from the 4th Light Dragoons on augmentation.

Captain C. Harvey was promoted Major on augmentation, and Lieutenant C. W. Thompson[1] (afterwards General) joined the 14th Light Dragoons from the 81st Regiment.

On the 15th May, Lieutenant-General Sir Edward Kerrison, Bart., G.C.B., G.C.H., full Colonel of the 14th Light Dragoons, inspected them at Canterbury, and afterwards issued a very gratifying and most highly complimentary farewell order to Lieutenant-Colonel Townsend, the officers, non-commissioned officers, and men. The 13th Light Dragoons

[1] General C. W. Thompson was Colonel of the 14th (King's) Hussars, 1882-96.

being at this time quartered at Canterbury, the Fourteenth renewed their old friendship with them, and as related by Cannon in his *Record of the 13th Light Dragoons*, 'the friendship of the Ragged Brigade, which had begun with and had continued throughout the eventful careers of the two regiments in the Peninsula, was cemented afresh, and on this occasion the Fourteenth presented to the Thirteenth their handsome mess-tables to perpetuate in the latter corps a kindly remembrance of their old companions in arms. Sic, VIRET IN ÆTERNUM.'

The regiment presents mess-tables to 13th Light Dragoons.

Embarks for India, May 1841.

The Fourteenth marched from Canterbury to Herne Bay on the 24th May. The 1st division under Lieutenant-Colonel J. Townsend embarked on the same day at Gravesend in the *Repulse* freight steamship of the East India Company, and arriving at Bombay on the 7th September, disembarked there on the 8th in boats, and marched to Kirkee, arriving on the 19th September.

At Kirkee, October 1841.

The 2nd division under Major Barton embarked at Gravesend on the 14th June in the East India steamship *Reliance*, landing at Bombay on the 5th October, and marched to Kirkee, where it arrived on the 13th October, and the whole regiment was concentrated there under command of Lieutenant-Colonel Townsend.

On the 1st October, 150 volunteers from the 4th Light Dragoons were transferred to the Fourteenth.

Moustaches generally worn in cavalry.

It was about this period that moustaches were generally worn in the cavalry, and it is commonly said that H.R.H. Prince Albert (the late Prince Consort) introduced them into our service.[1] On the 12th November the regiment was inspected at Kirkee by His Excellency Lieutenant-General Sir Thomas McMahon, Bart., K.C.B., Commander-in-Chief in Bombay. On the 23rd November, Lieutenant-Colonel J. Townsend was appointed A.D.C. to Her Majesty the Queen, with the rank of Colonel in the army.

[1] As far back as 1812 Hussars wore moustaches, and other cavalry (Dragoons and 'Heavies') shaved the upper lip.

1842

The regiment was inspected at Kirkee by Major-General Sir Charles J. Napier, K.C.B., commanding the Poona Division of the Bombay Army, on the 7th January, and again on the 13th June by the same General Officer. The Commander-in-Chief at Bombay (Sir Thomas McMahon, Bart.) also inspected the regiment at Kirkee on June 23rd, and gave the highest credit to all ranks. On the 18th August, His Excellency Lieutenant-General Sir T. McMahon, Bart., K.C.B., again inspected at Kirkee in the presence of the Honourable the Governor of Bombay (Sir George Arthur, Bart.), and they both made the most complimentary remarks as to the high state of efficiency in which they found the regiment.

A draft arrived from England on the 9th November, consisting of 1 sergeant and 99 men, under Cornet Brown, *per* steamship *Coromandel.*

Cornet William McMahon, son of Lieutenant-General Sir Thomas McMahon, Bart., K.C.B., joined the Fourteenth on the 25th November; and Captain F. H. Stephens became Major on the same date *vice* Barton, who retired. On 20th December, Major-General Macneil inspected at Kirkee, and reported that he found everything in the highest state of efficiency.

This year a Hospital-Sergeant was added to the establishment of the regiment, and one private was reduced. A Hospital-Sergeant had previously been appointed to the regiment in 1824. Establishment.

1843

On the 3rd February, Major C. P. Ainslie joined the Fourteenth as Major *vice* Stephens, and Lieutenant R. P. Apthorp became Adjutant in succession to Captain William Clarke, promoted. On 5th June, Major-General Macneil made his half-yearly inspection at Kirkee, and expressed himself very much pleased with all he saw, and His Excellency Lieutenant-General Sir Thomas McMahon, Bart., again inspected on the

Two squadrons proceed in October to the Mahratta campaign (Kolapore), October 1843.

7th November. In October 2 squadrons left Kirkee and proceeded on field-service to Kolapore under Major Harvey. They took part in the southern Mahratta campaign during this and the following year, when the forts of Monshin and Munsomtosh were taken. The 2 squadrons on field-service numbered 15 officers, 289 non-commissioned officers and men, with 289 troop-horses.

1844

On 20th January, Major-General Macneil made his first half-yearly inspection, and his second on the 29th of June, both at Kirkee. His Excellency the Commander-in-Chief of the Bombay army (Lieutenant-General Sir Thomas McMahon, Bart.) also made an inspection on the 27th June.

1845

The Fourteenth were inspected by Major-General Macneil at Kirkee on 21st January in light marching-order, for outpost and picquet duty, and in review-order on the following day; and Lieutenant-Colonel Havelock was directed to convey to the regiment the Major-General's entire approbation of the results of both days' work.

Colonel J. Townsend, A.D.C., had gone home on leave in the spring, and he died at Castle Townsend, County Cork, Ireland, on 22nd April, after nearly forty years' service in the Fourteenth and sixteen years in command.

Lieut.-Colonel —William Havelock.

Lieutenant-Colonel William Havelock, K.H., succeeded to the command of the Fourteenth, and Lieutenant-Colonel E. Harvey became second Lieutenant-Colonel on 23rd April.

On 16th May, Captain W. H. Archer became Major.

Two squadrons return from Kolapore, 19th March 1845.

The 2 squadrons from Kolapore rejoined headquarters on 19th March, having been present at the investment and capture of the fortresses of Panulla and Powrghur.

On 10th June, His Excellency Lieutenant-General Sir Thomas McMahon, Bart., K.C.B., Commander-in-Chief at

Bombay, made an inspection of the regiment in 'watering-order,' and was very pleased.

On 20th June, Major-General Macneil made his half-yearly inspection, and the Honourable the Governor-General of Bombay, together with His Excellency the Commander-in-Chief, made an inspection a few days later and expressed their entire satisfaction.

On 27th November, His Excellency General Sir Thomas McMahon, Bart., G.C.B., inspected the Fourteenth for the last time, in marching-order, previous to its departure from Kirkee, which station it left on 15th December to march to the upper provinces of Bengal.

Regiment leaves Kirkee, 15th December.

After the inspection Sir Thomas McMahon made a most flattering report of the state of the Fourteenth, and he said he should convey the same to the Governor of Bombay, the Commander-in-Chief of the forces in India, and to the authorities in England. He said he considered the 14th (King's) Light Dragoons in as high a state of efficiency as any regiment in Her Majesty's service.

On 14th December the Major-General commanding the division promulgated a very complimentary farewell order. He particularly remarked on what an excellent school he considered the 14th (King's) Light Dragoons must be to produce good officers, when such as Lieutenant-Colonel Talbot, Colonel Sir Felton Bathurst Hervey, Colonel T. W. Brotherton, and the late lamented and gallant Colonel Townsend had been brought up in it.

On 15th December the regiment left Kirkee, marching towards Agra.

Regiment marches to Agra.

1846

The Fourteenth had now to undergo a long monotonous march of nearly three months, during which time the cholera was raging in the Malwa Jungle.[1] Lieutenant F. D. Gray,

[1] Malleson relates that Lord Elphinstone (Governor of Madras, 1837-42), who was revisiting India in 1845-46, marched in company with the 14th Light Dragoons

1 sergeant, 1 corporal, and 15 privates died in passing through the jungles between Ahmednuggur and Mhow. Lieutenant Herbert Gall had a most wonderful recovery. At some halting-place *en route* he was left behind for dead, to be interred, but it is said that some champagne brought him round. He was actually being placed in a coffin, when he uttered some ejaculation, and the hospital apothecary gave him champagne, having been told by the surgeon to let him have anything he asked for!

Lieutenant Gall's recovery from cholera.

Stampede at Indore.

A very remarkable circumstance occurred during the four days' halt of the Fourteenth at Indore. Owing to some display of fireworks by the native prince, 250 of the troop-horses broke loose in a stampede, and were not recovered for some days, and then with the greatest difficulty. The terrified animals broke away from their picquet-ropes in a dark night, and scampered far and wide over the surrounding country. In about 5 days, however, all were brought back, excepting 6 or 7, and these were afterwards discovered at Kirkee, over 400 miles away.

Arrival at Agra, 27th February. Arrival at Meerut, 16th March 1846.

On 22nd and 27th February the Fourteenth entered Agra; on 4th March they proceeded to Meerut, and reached that station on 16th March.

Arrival at Umballa, 17th April.

In April the march was continued towards Umballa, which was reached about the 17th. On 2nd May, at Umballa, Major-General Gilbert, Commanding Sirhind Division, made a minute inspection of the Fourteenth in marching-order, and expressed complete satisfaction at their turn-out and general state.

On 8th September, Major J. W. King, previously in the 5th (Princess Charlotte of Wales's) Dragoon Guards, joined the Fourteenth as Major *vice* Archer.

The Resident of Indore, Mr. R. N. C. Hamilton, wrote a very complimentary letter, dated Camp Cheetawond, 7th March 1846, in which he expressed his gratification at the

under Lieut.-Colonel Havelock, his former Military Secretary, from Bombay through Central India to the headquarters of the British army before Lahore.

excellent conduct of all ranks of the regiment when passing through that district.

On 1st December, Major-General Sir W. R. Gilbert, K.C.B., Commanding Sirhind Division, made his second half-yearly inspection, and was very much pleased.

1847

On 7th January, at Umballa, Lieutenant-General Lord Gough, Commander-in-Chief in India, inspected the Fourteenth, and on the 19th he visited the Cavalry Brigade, to which the 14th Light Dragoons belonged, and was pleased to express his entire approbation of their appearance in the lines and in barracks. On the 7th May, Major-General Sir W. Gilbert inspected the regiment in watering-order, which, he said, was a most satisfactory turn-out, and he inspected again on the 16th December in complete marching-order.

On 30th March, Lieutenant Arthur Need was appointed to the Fourteenth, and joined at Umballa, and on 23rd November Captain H. E. Doherty became Major.

1848

Arrival at Ferozepore and Lahore (Anarkullee).

The Fourteenth left Umballa on the 15th February for Ferozepore, where they arrived on the 29th February, and remained there till about 13th August; but the left wing, consisting of the 3rd and 4th squadrons, had been previously detached to Anarkullee (Lahore) on the 8th May, and arrived there on the 12th May. Strength—10 officers, 213 men and 213 troop-horses.

The regiment was inspected at Ferozepore on 24th April by Major-General Sir Walter Raleigh Gilbert, K.C.B., who gave it great credit and was much pleased with its appearance.

A squadron sent into the Sikh country.

On the 1st June the 3rd squadron, 4 officers, 91 men and horses, left Lahore under Major King on field-service, and proceeded across the river Ravee to operate against Bahre Maharajah and his adherents. The squadron returned to Lahore on the 7th June. Lieutenant-Colonel Patrick Grant,[1] Adjutant-General of the Army in India, by direction of the Commander-in-Chief, wrote a very complimentary letter from Simla as to the success achieved by this detachment in the expedition across the Ravee, in which he spoke very highly of the zealous exertions of both officers and men engaged.

The headquarters arrived in Lahore on the 19th August, having left Ferozepore on the 14th.

The second Sikh campaign, which ended in the fall of the Sikh power and the annexation of the Punjaub, shortly after this broke out, and during the time the regiment was on field-service with the army of the Punjaub (3rd November 1848 to 1st May 1849) a depôt was formed at Anarkullee (Lahore) for the baggage, regimental stores, and the women and children of the regiment left behind.

Establishment, 701 troop-horses.

The establishment of the Fourteenth at this period was 9 troops, 56 sergeants, 12 trumpeters, 8 farriers, 40 corporals, 627 privates, and 701 troop-horses. There were 2 Lieutenant-Colonels and 2 Majors, viz. Lieutenant-Colonels William Havelock, K.H., and J. W. King (who had succeeded Lieutenant-Colonel Harvey recently), and Majors H. C. Doherty and Charles Steuart, the latter having been appointed to succeed Lieutenant-Colonel J. W. King, promoted to the second Lieutenant-Colonelcy.

Charge of Ramnuggur, 22nd November 1848.

The 14th Light Dragoons now took the field with Lord Gough's Army of the Punjaub. They numbered 442 sabres besides officers, and their first engagement with the enemy took place at Ramnuggur, on the banks of the Chenab, on the 22nd November 1848. Here the regiment, led by their gallant Colonel, William Havelock, made those brilliant charges, against overwhelming numbers of the enemy, which have since become matters of history. The memory of that glorious day has ever since been held sacred by the

Memory of the day celebrated annually in the Fourteenth.

[1] Afterwards Field-Marshal Sir Patrick Grant, G.C.B., G.C.M.G.

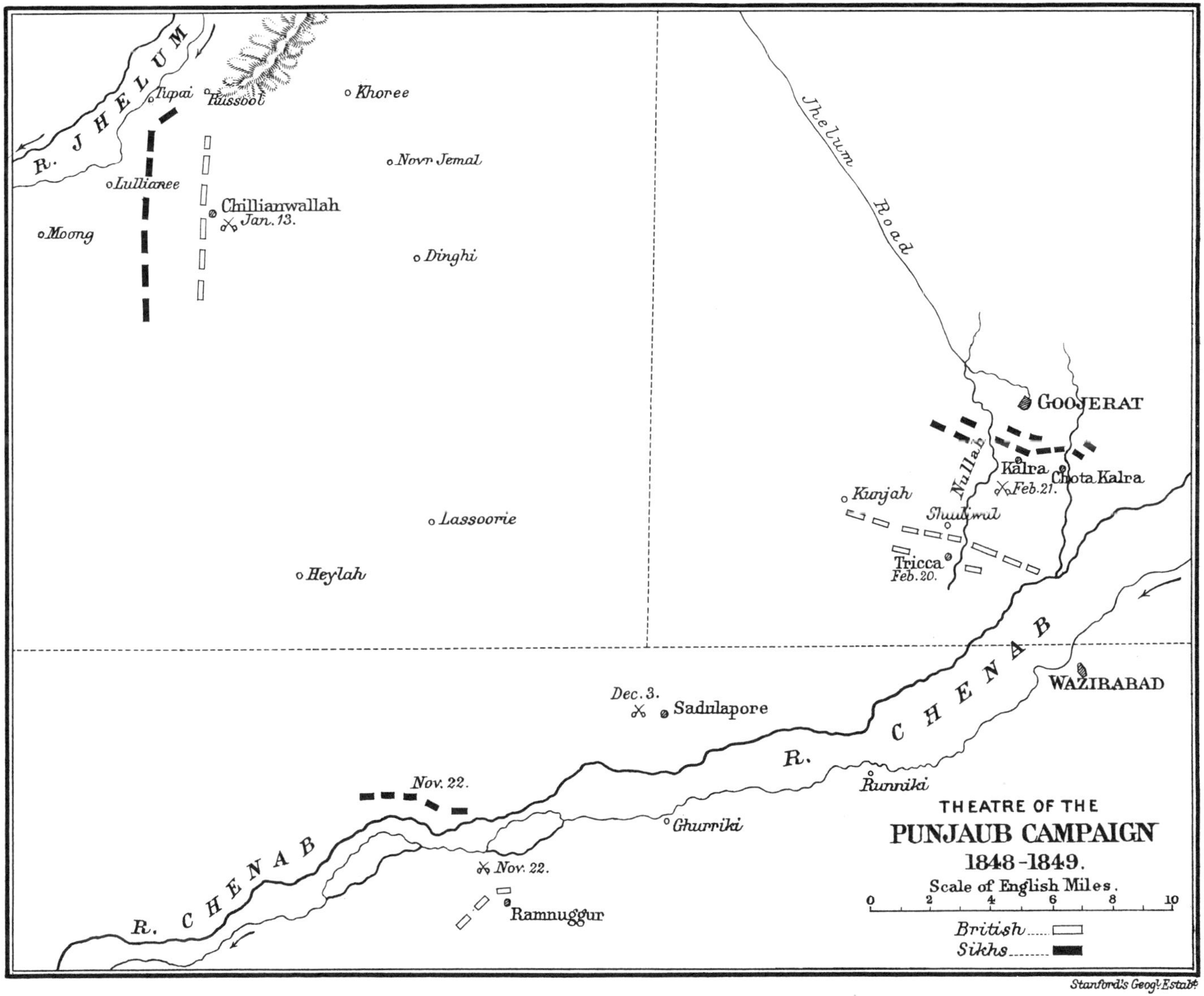

London: Longmans, Green & Co.

Fourteenth, and as year by year comes round the anniversary is celebrated, especially in the sergeants' mess, where a ball or other entertainment takes place. Past and present officers with friends and guests are invited, and on these occasions it is the time-honoured custom to revive old memories and stirring scenes long past, by toasts and libations freely taken from the celebrated 'Ramnuggur Cup,' a handsome silver bowl presented to the sergeants' mess, many years ago, by the gallant 5th Light Cavalry who shared in the glories of that day, and charged along with the Fourteenth as brave and trusty comrades. The cup is emblematic of the occasion and has a suitable inscription engraved upon it, with a glorious list of the actions in which the regiment has been engaged with the enemy, in their campaigns in the Peninsula, Punjaub, Persia, and Central India, numbering no less than 28, as follows :— The Ramnuggur Cup.

Douro.	Chillianwallah.	Betwa.
Talavera.	Goojerat.	Jhansi.
Fuentes d'Onor.	Persia.	Koonch.
Salamanca.	Dhar.	Golowlee.
Vittoria.	Mundesor.	Calpee.
Orthes.	Rathgur.	Morar.
Peninsula.	Barodia.	Gwalior.
Kolapore.	Muddenpore Pass.	Jowra-Alipore.
Punjaub.	Chanderi.	Ranode.
Ramnuggur.		

Ramnuggur was essentially a cavalry affair, and was brought about by a reconnaissance in force under the Commander-in-Chief in India, General Lord Gough, G.C.B., who intended merely to reconnoitre the enemy and explore the fords of the river in that locality. It was probably between the hours of 1 and 2 P.M. in the afternoon, or perhaps earlier, when Lieutenant-Colonel Havelock received his orders from a Staff Officer sent by General Lord Gough, who was not far off, to charge the Sikh cavalry that had crossed the river in large masses and were Description of the charge.

clearly visible to our front. Havelock was thirsting for glory, and to use his own expression, he felt the opportunity had come to 'win his golden spurs.' He led off with 2 squadrons of the Fourteenth in column of troops across the plain, and for half a mile at least these squadrons were exposed during their advance to an incessant fire from the Sikh guns posted on the banks of the river. They then approached a steep bank leading down to the nullah, a partially dry channel of the river, beyond which lay the Sikh forces who had crossed the river from the right bank and were posted on a sort of island formed by the main channel of the river and a small stream. Here they undoubtedly had guns in position and infantry entrenched, hitherto unseen and quite unknown to us. As the Fourteenth came up to this steep bank, each squadron in turn paused, thus causing some temporary confusion in the ranks, but they soon plunged down, following their gallant Colonel, forming squadrons and line on the move as best they could, and galloped rapidly across the nullah, charging the Sikhs, horse and foot. They were now exposed to the close matchlock fire of the enemy, and as the ground near the island was of a boggy nature, ill suited to cavalry, water having recently subsided, many horses floundered about and frequently got into difficulties, but the men rode gallantly forward against the opposing Sikhs, sabring hundreds of them and driving the 'Gorchurras' (Sikh irregular cavalry) back helter-skelter into the river and numerous channels which ran up the creeks and banks on all sides. It was now that Havelock perceived for the first time the large bodies of infantry concealed in these dry channels running along the bed of the river, and although his first attack had been eminently successful in driving back the Gorchurras as desired by Lord Gough, his characteristic dash and headlong pluck seem to have overswayed his prudence and better judgment, for without hesitation, brave leader that he was, he determined, notwithstanding the overwhelming odds against him and the adverse circumstances in which he was placed, to make another charge, and assail the heart of the Sikh position.

Accordingly, he retired his men a little and re-formed the squadrons, which by this time had been reinforced by another squadron of the Fourteenth, as well as by a considerable body of the 5th Light Cavalry under Brevet Lieutenant-Colonel Alexander, and placing himself in front of the line, Havelock once more boldly advanced to meet the foe. The squadrons were exposed to a murderous fire from the batteries on the right bank of the river, as well as from the guns posted on the island, yet nothing daunted they charged right into the centre of the Sikhs, driving them back towards the river, and sabring right and left in a desperate *mêlée* which ensued. They were also confronted with the fire of the matchlock men, both horse and foot, who slowly retired disputing every inch of the ground. Both regiments behaved splendidly, and it was not to be wondered at that both lost heavily, the total number of casualties in each regiment being about equal in proportion to the numbers of each engaged. The gallant but too daring chief of the Fourteenth fell in this attack, and not less than 74 men, including officers, with 96 horses were placed *hors de combat*. Of these numbers the 14th Light Dragoons had 44 men (including 6 officers) and 56 horses either killed, wounded, or missing, whilst no less than 30 men (including 3 officers) and 40 horses belonging to the 5th Light Cavalry bit the dust. The numbers of the Fourteenth engaged were about 350 sabres. It was a short but very sanguinary business. Colonel Havelock's body was found and fully identified, though headless, about twelve days after the engagement. It was lying with the bodies of 9 troopers of the Fourteenth heaped on it, showing that his men had rallied round and fought for their chief. His left arm and leg were nearly severed, as well as the thumb of his right hand. Captain Fitzgerald of the Fourteenth was mortally wounded in the *mêlée* and died subsequently: one of his sword-cuts penetrated the brain and another the spine. Major Doherty brought the charging squadrons out of action, and Lieutenant-Colonel King, who had been ordered to command the support when the Fourteenth advanced, came up just at the right moment with

Death of Lieut.-Colonel William Havelock, K.H.

a squadron which formed a welcome nucleus for the other squadrons to form on. Alas! the brave Havelock was not amongst them. He was last heard of wounded and hacked at by several Sikhs in the *mêlée*.

As Ramnuggur was considered merely an affair of outposts and a purely cavalry fight, no honorary distinction was conferred for it, but none the less most will concede that these charges of the 14th Light Dragoons and 5th Light Cavalry deserve a high place amongst cavalry charges delivered under unfavourable circumstances. The Sikhs were immensely superior in cavalry, besides being assembled in great force with infantry and artillery posted and partially entrenched beyond a nullah, having the further advantage of a sandy river-bed and boggy ground between them and the attacking squadrons. Hence it is that the memory of Ramnuggur has always been held very dear by all in the Fourteenth, and both Havelock and his brave companions-in-arms have ever been reckoned heroes by succeeding generations of those serving in the regiment.

Casualties at Ramnuggur.

The casualties of the regiment in this affair were as follows:—

Killed—
- 1 officer (Lieutenant-Colonel William Havelock, K.H.).[1]
- 1 sergeant (John Harwood).
- 1 corporal.
- 12 privates.
- 37 horses.

Wounded—
- 5 officers—Captain J. F. Fitzgerald,[2] very severely.
  - Captain R. H. Gall, severely.
  - Captain A. Scudamore, slightly.
  - Lieutenant William McMahon, severely.
  - Cornet the Hon. R. W. Chetwynd, slightly.
- 4 sergeants.
- 18 privates.
- 15 horses.

[1] Lieutenant-Colonel Havelock was officially reported 'Missing.'
[2] Captain Fitzgerald died of his wounds a few days afterwards.

Missing { 2 rank and file.
4 horses.

The 5th Light Cavalry had

1 non-commissioned officer and 12 privates killed.
2 officers and 15 privates wounded.
40 horses killed and wounded.

Brevet Lieutenant-Colonel Alexander lost his arm by a round-shot which killed the Quartermaster-Sergeant and wounded Lieutenant Reilly in the foot. The 3rd Light Dragoons and 8th Light Cavalry, who had made some charges against the Sikh Gorchurras at an earlier part of the day, but had not crossed the nullah, suffered very slight casualties. The 3rd Light Dragoons had 5 privates wounded and 3 or 4 horses struck by round-shot; the 8th Light Cavalry lost a subadar-major killed; and the 12th Irregular Cavalry had Captain Holmes wounded. The troop of Horse Artillery, under Lieutenant-Colonel Lane, which lost a gun and a couple of wagons about the time of the advance of the 3rd Light Dragoons and 8th Light Cavalry, only had 1 private wounded and 4 horses killed. Amongst the staff, Brigadier-General C. R. Cureton, C.B., commanding the Cavalry Division, who was a very distinguished cavalry leader with brilliant war services, was shot in the chest and killed, and Lieutenant Hardinge, Aide-de-Camp, was wounded.

It will be seen by the above list of casualties that the losses of the Fourteenth were much heavier than those of any other regiment engaged, and there is no doubt that they and the 5th Light Cavalry had the brunt of the affair all to themselves. It is true the 3rd Light Dragoons and 8th Light Cavalry charged across the plain several hours before the charges of the Fourteenth took place, and the Sikh Irregular Cavalry (Gorchurras) retreated before them, but neither regiment penetrated into the position of the Sikh forces posted beyond the nullah, as it is believed Brigadier White did not consider the ground suitable for cavalry, and was induced to retire his brigade from the ground subsequently crossed by Havelock, in the exercise, as was afterwards proved, of a very wise and soldierly judgment. This would naturally account for their lighter list of casualties.

Brigadier Charles Robert Cureton, C.B., 16th Lancers.

The lamented loss of Brigadier Charles Robert Cureton, C.B., A.D.C., which occurred about the time of Colonel Havelock's second attack, deserves more than a passing notice, not only on account of his great reputation as a brilliant cavalry officer, already distinguished in former campaigns, but also because of his old connection with the 14th Light Dragoons in which he enlisted when a youth under the name of 'Charles Roberts,' serving with them in the Peninsula. The probably true and authentic account of Cureton's death is as follows:—He was riding at the time along with the Commander-in-Chief (Lord Gough) and the rest of his staff, among whom were Lieut.-Colonel J. B. Gough,[1] officiating Quartermaster-General, (in India) and Major F. P. Haines,[2] Military Secretary. Following closely was the chief's personal escort, a troop of the 5th Light Cavalry. Cureton, perceiving that Havelock was forming up his squadrons with the evident intention of making another charge, and not approving of this under the circumstances, rode off hurriedly to stop him, but was almost immediately shot dead by a bullet which struck him in the chest. This version of Cureton's death is strongly corroborated by an original letter in the handwriting of the late General Sir J. B. Gough, G.C.B. (then Lieutenant-Colonel J. B. Gough), who says in the letter, dated 23rd November 1848, 'Poor Cureton was shot dead within five yards of me and close to Lord Gough.' This is certainly at variance somewhat with the statement in Brigadier-General C. Campbell's despatch quoted in Appendix A, page 569, but it agrees with the well-known water-colour picture by Henry Martens of the 'Charge of the 14th Light Dragoons at Ramnuggur,' and the engravings of that picture by Harris, published by Rudolph Ackermann of 191 Regent Street, in

[1] Afterwards General Sir John B. Gough, G.C.B.

[2] Now Field-Marshal Sir Frederick P. Haines, G.C.B., G.C.S.I., C.I.E.

1851,[1] where Cureton is depicted as shot in the throat and chest as he was galloping over the plain towards the charging squadrons, apparently in the endeavour to deliver some message or order. We can only conclude therefore that the allusion in Brigadier Campbell's despatch was written under some slight misapprehension or confusion of details.

General Brotherton gives the following account of Brigadier Cureton as he knew him in the Peninsula, amongst his interesting episodes. The General writes as follows :—

'I went out to the Peninsula in command of a troop, and that troop produced besides numerous brave and good soldiers an extraordinary man, notorious throughout the service for bravery and intelligence. He afterwards became Brigadier-General Cureton, so highly extolled in Major-General Sir Harry Smith's despatch on the battle of Aliwal. He served all through the Peninsular War in the 14th Light Dragoons, and there learned the rudiments of outpost duty, for his consummate knowledge of which he is so highly praised by Sir Harry Smith.[2] Baron Osten perfectly recollects a dialogue that passed between Corporal Cureton and me, in Portugal, he under the name of "Roberts," in which he had enlisted.

'Cureton, the son of most respectable parents, tradespeople in Lancashire, became first a corporal in my troop, and then a sergeant. He distinguished himself daily in the field, but being, moreover, a well-educated young man, and a particularly good penman, Lord Fitzroy Somerset asked for him to go to headquarters, to write in his office. I reluctantly parted with him, and even tried to dissuade him from going, representing to him that as we were then daily engaged with the enemy, he would

[1] At the foot of these coloured engravings a quotation from the *Illustrated London News* of 27th January 1849 is usually printed, giving some account of the supposed third charge of the Fourteenth at Ramnuggur, which it is clearly proved never took place. See p. 221.

[2] 'The manner in which this famous officer handles his cavalry, under the hottest and most galling fire, ranks him amongst the first cavalry officers of his age.'—(Sir Harry Smith's despatches. Battles of Aliwal and Sobraon, 28th January and 10th February 1846.)

have such frequent opportunities of distinguishing himself as would infallibly lead to promotion, and bring him conspicuously into notice. His better star, however, prevailed, and he did go to headquarters, where he was not only employed in writing, but acted as mounted orderly to Lord Fitzroy Somerset in all the subsequent battles. His merit being very conspicuous, he soon got a commission, ultimately became Adjutant-General in the East Indies, witnessing all the late operations in that part of the world, was in the unfortunate Cabul expedition, shared, most conspicuously, in many glorious fights against the Sikhs, and fell at last on the battlefield of Ramnuggur, when in the act of distinguishing himself. It was by a most extraordinary chance he fell gloriously with his old corps, the 14th Light Dragoons, which he had left many years before.

'His modesty was equal to his bravery, for though he had become a most distinguished person, and had been mainly instrumental by his daring and judicious management of the cavalry in gaining the critical and glorious battle of Aliwal under Sir Harry Smith, yet his modesty, and I may say humility, were remarkable.

'On his return once to England, Lord Fitzroy Somerset invited him to meet me at dinner at his house, and though he had perfectly the manners of a gentleman, though risen from the ranks, he never presumed, and showed me the same deference and respect as when he was a corporal in my troop. When the clasps for the battles in the Peninsula were issued, instead of sending in his application for his share of them (which was *every battle* that had been fought, as he had been present at all of them) through the regular channel, and direct to Horse Guards, by way of paying me a compliment he sent it through me, saying that "I knew his services better than any one."'

The affair of Ramnuggur is very graphically described in the *Illustrated London News* of 27th January 1849, where two letters from correspondents are published, one from 'Eyewitness,' dated

25th November 1848, and one from 'An Officer of Bengal Horse Artillery,' dated 1st December 1848. In Appendix A, pages 561-567, extracts from these are published, as they appear reliable and are in accordance with the recollections of officers now living who were present and took part in the events of the day. They also coincide in most respects with the narrative of Ramnuggur given at pages 205-209 of *The Sikhs and the Sikh Wars*, by General Sir Charles Gough, V.C., G.C.B., and A. D. Innes, M.A. (published in 1897), with one obvious discrepancy, and this demands some mention here. In 'Eyewitness's' letter in the *Illustrated London News* there is a detailed account given of a supposed third attack and charge made by the 14th Light Dragoons and led by Lieutenant-Colonel King after Havelock fell. There appears to be no corroboration whatever of this charge, and absolutely no other authority for it. The incident has therefore been altogether omitted in the extracts published in the Appendix. If such a charge had taken place, surely some of the survivors with whom the author has had the advantage of both personal conversation and correspondence would know of it. Again, the 'Regimental Digest of Services' kept in the orderly-room of the regiment states distinctly 'the regiment made two charges'; and in *The Sikhs and the Sikh Wars* there is no allusion to such an event, for it is stated at page 209, after mentioning Cureton's death, 'Before any fresh order could be despatched, Havelock had made the second fatal charge, and the horses began to flounder in the sandy river-bed. The Sikhs swarmed down on them; and though the brigade succeeded in cutting its way back, Havelock was slain.' Now, Sir Charles Gough, one of the authors of this work, was one of the officers who on that day charged with the 8th Light Cavalry at Ramnuggur, and was also an eyewitness of the charge of the Fourteenth. The official account given in the Regimental Records above alluded to is as follows:—

'22nd November 1848. Ramnuggur.—The regiment made two successive charges, in the latter against an overwhelming

Extract from the Regimental Digest of Services of the 14th (King's) Hussars. 'Ramnuggur.'

force of cavalry and infantry, and exposed to the fire of three batteries.' Surely the above is sufficient to prove that there never was a third charge as described in the letter in the *Illustrated London News*, which was probably written hastily and without full and reliable information of the actual facts, so misleading on a field of battle to observers from different points of view,—indeed an *advance* of cavalry at a rapid pace might easily be mistaken for a *charge* by any one witnessing it from a distance.[1]

In the *Illustrated London News*' account there is one other point, but of less importance, which appears at variance with facts. It is the statement in the 'Bengal Horse Artillery Officer's' letter, which seems to connect closely the loss of the gun and wagons of Lieutenant-Colonel Lane's troop of Horse Artillery with the charges of the 14th Light Dragoons. The real version of this incident appears to be that the gun stuck in the sand, and had to be abandoned at an earlier part of the day than that at which the Fourteenth made their two charges, in the second of which they were supported by the 5th Light Cavalry. The *contretemps* of the gun sticking in the sand probably happened at the time the charge under Brigadier-General Michael White of the 3rd Light Dragoons took place, and when the latter regiment with the 8th Light Cavalry advanced across the plain, driving back the Sikh cavalry to the bank of the nullah, but thence retired in good order, not deeming a further advance advisable.

Lieutenant-Colonel—J. W. King.

On 23rd November, Lieutenant-Colonel J. W. King succeeded to the command of the Fourteenth; on the 28th November, Major H. E. Doherty became second Lieutenant-Colonel, and Captain William Clarke, who had risen from the ranks, became Major. Subsequently Lieutenant-Colonels King and

[1] Even Captain Apthorp in his remarks (published in Appendix A, p. 569, etc.) speaks of three charges, but he explains that he calls the advance of Havelock before his first change of front a charge. As Captain Apthorp was Adjutant of the 14th Light Dragoons at the time, he would be the person to make the entry in the 'Digest of Services,' under the orders of the commanding officer of the regiment, and the entry is, 'The regiment made *two* successive charges.'

Doherty were each awarded a Companionship of the Order of the Bath.

After the affair of Ramnuggur the whole army remained there watching the enemy till the 30th November, when a body of troops was despatched at midnight, under Major-General Sir Joseph Thackwell (a Waterloo veteran), to Wazirabad ford, 22 miles up the river, and effected a passage of the Chenab at that point. There were two safe fords nearer than Wazirabad, namely Ghurriki, 7 miles from Ramnuggur, but 4000 Sikhs held this; another higher up at Runniki was also guarded. There was yet another, higher up still, at Ali-Shor-Ke-Chuk, but it was dangerous, the stream being too rapid. The division that was sent under Sir Joseph Thackwell consisted of White's Cavalry Brigade, 3rd Light Dragoons, 5th and 8th Light Cavalry, and 2 Irregular Native Cavalry Regiments, in place of the Fourteenth, with Colin Campbell's Infantry Brigade, consisting of Her Majesty's 24th and 61st Foot, and 5 Regiments of Native Infantry; also 30 field-guns and 2 heavy guns. The whole force crossed over from the left bank safely on the 1st and 2nd December, and commenced to march down the right bank in the direction of Ramnuggur, with a view to driving out the Sikhs from their position opposite our camp, for our evacuation of Ramnuggur was out of the question whilst they remained there, as such a step would leave the way to Lahore unguarded. By the evening of the 2nd December, Major-General Sir Joseph Thackwell had reached within 9 miles of the Sikh position on the right bank, and having driven off the outposts from the fords at Ghurriki and elsewhere, had established communication with Lord Gough at Ramnuggur. On the 3rd December, Thackwell moved to Sadulapore, sending a sufficient force to guard the Ghurriki ford, whence reinforcements were expected under Brigadier Godby. Shere Singh, at this juncture, late in the afternoon of the 3rd December, came forward with his troops, occupying a strong position, covered by sugar-cane fields lying in front of him, and the British accordingly fell back somewhat to gain better ground, but the Sikhs did not come

Movements of the British after Ramnuggur.

Sadulapore, 3rd December 1848.

Retreat of the Sikhs from Ramnuggur to the Jhelum, 3rd December 1848.

on. A cannonade on both sides was maintained till evening, and when darkness set in Shere Singh retired from his entrenchments, carrying with him his artillery, setting fire to his magazine, and withdrawing his whole army northwards. He afterwards took up a strong position covered by jungle near the banks of the Jhelum. The losses incurred at Sadulapore by our force were slight, being about 80 men killed and wounded. The Sikhs suffered heavily: it is believed their losses were ten times as great as ours, owing to the greater precision of our artillery fire. On hearing of the retreat of the Sikh army, Lord Gough at once pushed forward across the Chenab as a pursuing force the 9th Lancers and 14th Light Dragoons, and these troops joined Sir Joseph Thackwell's main body. The Sikhs, it appears, retreated in the greatest disorder, leaving in the villages numerous wounded men. They subsequently divided into three divisions, which became more a rout than a retreat. The 14th Light Dragoons arrived at Camp Heylah on 4th December, and were at once sent by orders of Major-General Sir JosephThackwell along the road to Dinghi. The 9th Lancers took another route, to try and come up with the fugitive Sikhs, but they were reported to have gone over the Jhelum, so the cavalry returned to Heylah without having effected anything decisive.

1849

The beginning of the new year found the regiment in camp at Heylah, the depôt troop being at Maidstone with a strength of 103 men.

Battle of Chillianwallah, 13th January 1849.

On 12th January, Lord Gough's army advanced on Chillianwallah, and next day was fought that sanguinary battle in which, after several hours' desperate fighting, the British forces held their ground, having gained a somewhat indecisive victory, and bivouacked on the battlefield. Our losses were 2357 men and 176 horses, of whom 1000 were Europeans, including 89 officers.

The killed alone were about 700 of all ranks. On the other hand the Sikhs, whose men and guns considerably outnumbered ours, lost heavily in killed and wounded—1000 killed, 2000 wounded. Of the latter many died subsequently. The enemy was not absolutely driven off the field, but remained in an entrenched position about 2 miles from the British camp, having his centre at Russool village and his right flank resting on the Jhelum.

Lord Gough's army at the battle of Chillianwallah was composed as follows:—

Cavalry Division.—Major-General Sir Joseph Thackwell, K.C.B., commanding.

1st Brigade.—Brigadier M. White:
3rd Light Dragoons,
5th and 8th Light Cavalry.

2nd Brigade.—Brigadier Pope:
9th Lancers,
14th Light Dragoons,
1st and 6th Regiments Light Cavalry.

The European cavalry regiments numbered about 400 men each, the Native cavalry 300, effective in the field.

Infantry (*2nd Division*)[1] under Major-General Sir Walter Gilbert, K.C.B.

1st Brigade.—Brigadier Mountain:
Her Majesty's 29th Regiment,
The 30th and 56th Regiments Native Infantry.

2nd Brigade.—Brigadier Godby:
2nd European Regiment,
The 31st and 70th Regiments Native Infantry.

3rd Division.—Brigadier General Colin Campbell, commanding.

1st Brigade.—Brigadier Pennycuick:
Her Majesty's 24th Regiment,
The 25th and 45th Regiments Native Infantry.

[1] The 1st Division of the army of the Punjaub was employed in the siege of Mooltan.

2nd Brigade.—Brigadier Hoggan :

Her Majesty's 61st Regiment,

The 36th and 46th Regiments Native Infantry.

3rd Brigade.—Brigadier Penny :

The 15th, 20th, and 69th Regiments Native Infantry. *Artillery Division*, under Brigadier Tennant :

Six troops of Horse Artillery under Brigadier Brooke, with Colonels Brind and C. Grant; the troops respectively under Lieutenant-Colonel Lane, Majors Christie, Huish, Warner, Duncan, and Fordyce.

Two batteries of four 18-pounders and two 8-inch howitzers each, under Majors R. Shakespeare and Ludlow; Major Horsford commanding.

Three field batteries : No. 5, commanded by Lieutenant Walker in the absence of Captain Kinleside, sick; No. 10, commanded by Lieutenant Robertson, in the absence of Captain Austin, wounded; and No. 17, commanded by Major Dawes.

The Foot Artillery, under Brigadier Huthwaite.

The European Infantry numbered about 900 each, and the Native Infantry about 700. Her Majesty's 24th, an exceptionally strong regiment, turned out 1000 men in the field. The Sikhs were lying with the Jhelum behind them, on the west, occupying a group of villages protected by jungle, with their left reaching to Russool, where there was a belt of hills. Chillianwallah, in advance of the line, was held only by their outposts. Lord Gough's intention was to march from Dinghi to Chillianwallah, drive in the outposts, and defer the attack till next day, when he had gained more accurate information of the position of the enemy.

At 7 A.M. on 13th January he advanced from Dinghi with his army formed in line of contiguous columns. By 12 o'clock he drove in the enemy's outposts. The Sikh position, which was entrenched, stretched from Russool on their left to Moong on their right, and Lord Gough decided to encamp, reconnoitre the position, and give battle on the following day. Whilst preparations for laying out the camp were in progress the Sikhs opened fire with their guns, and advanced beyond their entrenchments, whereupon the British general gave orders to prepare for immediate action. For the first hour the battle was an artillery duel, in which the enemy had the advantage both of position and of guns. At about half-past three in the afternoon our advance commenced.

The cavalry were on our flanks—White's on the left, Pope's on the right. Hoggan's Brigade was on the left of the line, with Pennycuick's next, both under Colin Campbell; then came Mountain's, and then Godby's, both under Gilbert, with Penny's Brigade (the 3rd of Campbell's Division) in reserve. The artillery were ranged—*in the centre* the heavy batteries; *attached to Campbell's Division*—3 troops Horse Artillery (Brind), and Walker's and Robertson's field-batteries, both under Major Mowatt; *attached to Gilbert's Division*—3 troops Horse Artillery (Grant) and Dawes's field-battery.

On the enemy's side, the Sikh line extended for about 6 miles, covered by thick jungle, and with their right considerably overlapping the British left.

At the commencement of the general advance, Campbell, seeing the intricate nature of the ground to his front, covered as it was by thick jungle, directed his Brigadier Pennycuick to act independently with his Brigade, and he himself accompanied Hoggan with the other (left) brigade of his Division. Robertson's field-battery, which should have acted with Hoggan's Brigade, was sent off early in the action, by orders of a staff-officer, to the left of our line, to help the Horse Artillery who were trying ineffectually to silence the Sikh guns on their left front; at the same time Pennycuick's Brigade advanced at such a rapid pace that it got in front of Mowatt's field-battery, which should have accompanied it, and consequently the latter could not give the required assistance to that brigade, but was enabled to render most important services to the other brigade (Hoggan's), which had lost its own field-battery (Robertson's) as above stated. This field-battery had silenced the Sikh guns which were engaged with Brind's Horse Artillery, and then tried to rejoin Campbell's Division. Pennycuick's Brigade in their rapid advance soon got into very difficult ground, and came under a terrible fire, first of roundshot and then of grape, to which the centre regiment, Her Majesty's 24th, were principally exposed, and the enemy keeping behind trees and thick bushes was completely screened from view, so that the losses of that gallant regiment were most severe. Still, nothing daunted, the brigade pushed on, and at length came in view of the Sikh guns. There still remained some swampy and very difficult ground to be traversed before they made the final charge right up to the cannons' mouth. Brigadier Pennycuick and many officers and men fell in this last attack, and although the 2 Native Infantry Regiments were advancing with the 24th Regiment, the Brigade lost so heavily that it had to retire to a position clear of the jungle. The 24th Regiment had lost their Colonel (Brookes) close to the guns, 13 of their officers were killed and 10 were wounded, while no less than 231 men were killed and 266 wounded. The Native Regiments also suffered heavy losses.

Meanwhile Mowatt's field-battery had covered the advance of Hoggan's Brigade, led by Campbell, who had advanced with great care and regularity through the difficult ground he had to traverse. Her Majesty's 61st had scattered the Sikh cavalry on their front; the 36th Native Infantry were attacked on their flank by a large body of Sikhs, but rescued by 2 companies of the 61st Regiment, who captured 2 of the enemy's guns. The 46th Native Infantry were opposed on the left by a large body of Sikh cavalry, but repulsed them under their leader, Major Tudor. Later the Sikhs brought up fresh cavalry and 2 more guns against the right of Hoggan's Brigade at the time it had first been driven back, but the Brigade rapidly changed front and drove the enemy opposed to them off the ground. They captured in all 13 guns, after some very severe hand-to-hand fighting, in which the Brigadier himself received a severe sword-cut on his right arm.

On our extreme left, White's Cavalry Brigade had advanced at same time as Campbell's Division. The ground was wholly unsuitable to the action of cavalry, and they were exposed to a heavy fire of round-shot during their advance, and consequently no great success resulted. After remaining in support of the Horse Artillery for some time, Major-General Sir Joseph Thackwell had ordered an attack. It was at the time that the Sikh battery, as already related, had been silenced. The Grey Squadron of the 3rd Light Dragoons with the 5th Light Cavalry advanced, led by Captains Unett and Wheatley. The 5th Light Cavalry came upon a large body of Sikhs, and being exposed to heavy musketry fire, were repulsed, and had to retire; but the 3rd Light Dragoons, Unett's squadron, swept on gallantly right through to the enemy's rear, and after a splendid charge, had to cut their way back, suffering severe loss—23 men killed, 2 officers and 15 men wounded, 15 horses killed and missing, 7 horses wounded.

After this affair, Sir Joseph Thackwell detached a troop of Horse Artillery (Warner's) and a squadron of the 8th Light Cavalry to support Hoggan's Brigade, and he himself with the remainder of his cavalry gave what support he could to Campbell's Division, moving on its left for the rest of the day. Now to turn to the right and right-centre of the British line: we find that Sir Walter Gilbert was preparing to advance shortly after Campbell's Division had done so. On the extreme right was Pope's Cavalry Brigade with 3 troops of Horse Artillery under Colonel C. Grant; then Godby's (infantry) Brigade, then Mountain's, with Dawes's battery between them. Brigadier Pope observing a considerable body of Sikh cavalry on the heights about Russcol, detached a wing from each of the 1st and 6th Light Cavalry, a wing from the 9th Lancers, and 8 Horse Artillery guns, retaining the other wing of each regiment, and the whole of the 14th Light Dragoons with the rest of the guns; he sent the detached troops to his right flank under the command of Lieutenant-Colonel Lane, with orders to protect that flank and to act as occasion demanded.[1] The Brigadier then advanced with his 9 squadrons and 10 guns on the right of Gilbert's Division in one line, without any supports or reserves, thus:—

Advance of Pope's Cavalry Brigade.

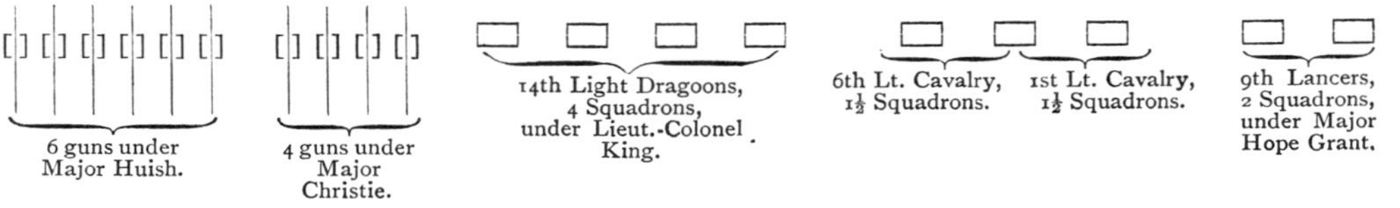

Very soon a body of Sikh horse was observed in front, and the guns were pushed forward, but, whilst they were getting into action, Pope came on with the cavalry, passed in front of them, and so masked their fire. The ground was particularly unsuited to the advance of a long line of cavalry, owing to its jungly nature and from being so full of obstructions, which made it very difficult for the squadrons to keep in touch and act in concert. There was no real charge made by the cavalry, but they were advancing as best they could and as the difficult nature of the ground permitted, when all of a sudden there was a distinct order given to retire by the command 'Threes about,' which was heard proceeding from the centre of the line where the Native cavalry was posted. The centre of the Brigade went about, the command was quickly taken up by the whole line, and a retrograde movement took place which was carried out at an increased pace for a considerable distance before the line was halted and re-formed. This withdrawal of the cavalry left the Horse Artillery guns (commanded by Majors Christie and Huish) wholly unprotected, and they retired with the cavalry, but the Sikh horsemen coming up cut down several of the gunners, and captured 4 guns, 2

[1] This body of Cavalry and Horse Artillery under Lieutenant-Colonel Lane acted quite independently of the remainder of Pope's Brigade, and no connection was kept up with the latter during the battle.

wagons, and 53 horses, besides rendering the 6 other guns useless for the day. The whole occurrence is shrouded in mystery: it may have been partly caused by a sudden panic to which the best of troops are at times liable, but it seems far more likely that the manner in which the brigade was handled was the real cause and origin of what happened. The Horse Artillery suffered the greatest loss: Major Christie was badly wounded and died soon afterwards, 7 of his men were killed, and 2 were wounded; in Huish's troop 5 were killed and 5 wounded; in the Cavalry Brigade, Brigadier Pope was badly wounded and had to be conducted off the field; and the cavalry regiments lost 2 officers, 2 native officers and 11 men killed, 2 officers and 30 men wounded. Owing to the brigade having advanced without any supports or reserves whatever, there was no point for the retiring squadrons to rally on, which added considerably to the confusion, and rendered the matter far more serious than it otherwise would have been.[1] This incident occurred just as Sir W. Gilbert was leading his Division to the attack, Dawes and his field-battery being in line with the skirmishers in the front; and Gilbert, when he perceived his right flank suddenly uncovered by the absence of the cavalry, had to refuse his right (Godby's) Brigade in order, to some extent, to protect it. The troops of this Division behaved most steadily, and Mountain's Brigade (on the left) continued their advance covered by the fire of No. 17 Field Battery, charging and taking a Sikh battery in front of the village of Lullianee. Godby's Brigade had also charged and carried some guns to its front, when their rear was suddenly attacked: this was caused by the unguarded flank left after the withdrawal of Pope's cavalry, hence Godby had to face his men to their rear, being literally surrounded by the enemy. At this juncture Dawes's battery came up and poured in a heavy fire on the Sikh horsemen who were about to charge, which drove them back. The

[1] See Appendix A, p. 573, etc.

2nd Bengal Europeans[1] under Major Steele then charged a body of the enemy who were still threatening the rear, and after a sharp struggle successfully repulsed them. The same occurred in Mountain's Brigade, which was also compelled to face about and drive off the enemy from its rear; and soon afterwards Colin Campbell with Hoggan's Brigade and White's cavalry, as well as the whole of the artillery of the left, moved up and joined in the fight on the right.

Penny's Brigade, which had been ordered up from the reserve when the disaster happened to Pennycuick's Brigade, lost its way in the jungle, and afterwards came up on the right of Godby's Brigade about the time when it was attacked in rear by the enemy. This brigade was threatened on its right flank and in front by large bodies of Sikhs, but was rescued by the fire of Dawes's field-battery, which happened to be there at the right moment, and promptly drove the enemy off. The whole of the British artillery now opened fire on the masses of the Sikhs who were retreating in disorder upon Tupai, and Lieutenant-Colonel Lane, who had not taken any active part in the battle since he was detached, as already related, to the right flank, by Brigadier Pope's order, with his Horse Artillery and 5 squadrons of cavalry,[2] now appeared, and poured in a heavy fire where he observed the enemy in retreat, thus rendering valuable aid at the close of the battle.[3]

Casualties of 14th Light Dragoons at Chillianwallah.

The losses of the Fourteenth at Chillianwallah were as follows:

Killed { 1 Officer (Lieutenant A. J. Cureton).
1 Man.
2 Horses.

[1] Now the 2nd Battalion Royal Munster Fusiliers.

[2] Two squadrons 9th Lancers and three squadrons from the 1st and 6th Regiments of Light Cavalry (one and a half squadrons from each). Lane also had his own 6 guns, and 2 guns detached from Christie's troop of Horse Artillery.

[3] This account of the battle of Chillianwallah is mostly taken from Gough's and Innes's *Sikhs and Sikh Wars*, some of it almost *verbatim*.

Wounded { 1 Officer (Major C. Steuart).
14 Men.
2 Horses.

Missing { 2 Men.
4 Horses.

During the whole of the two days following the battle, being the 14th and 15th January, rain fell incessantly, and during this time the melancholy task of burying the dead was carried out by the troops. Afterwards the two armies lay at Russool and Chillianwallah watching one another, Lord Gough being determined to wait for the fall of Mooltan, so as to get the reinforcements from the south, especially as the enemy had already received large accessions of force under Chutter Singh. The Sikhs began to evacuate Russool on 11th February and fell back on the Jhelum. They subsequently turned and made an endeavour to cross the Chenab near Wazirabad, but Major-General Whish's column coming from Mooltan forestalled them there, and they had to take up a position on the plain between the two rivers near Goojerat. By the 14th the whole of the Sikhs had left their encampments at Russool. Accordingly on the 15th Lord Gough moved to Lassoorie, whence, pushing forward a column to guard the ford at Wazirabad, he moved on 16th to Pukee Nuggar and Sadulapore. By the 20th, Major-General Whish's troops from Mooltan (which city fell on 25th January) having come up, the army, now complete, was concentrated under Lord Gough at Shadiwal and Kunjah, the Sikhs being at Goojerat a little to the north of that position.

Movements of the two armies after Chillianwallah.

Our forces numbered 25,000 men with 96 guns. The combined Sikh armies under Chutter Singh and Shere Singh are believed to have numbered from 50,000 to 60,000 men, and they had 60 guns. Of the latter we subsequently captured or destroyed 53 guns at Goojerat.

The battle of Goojerat was fought on the 21st February 1849, and the arrival of the Mooltan army gave Lord Gough

Battle of Goojerat, 21st Feb. 1849.

that preponderance in artillery which he desired. The forces under his command were as follows :—

There were in the first place the same regiments which had been present at Chillianwallah. The divisional commanders were as before; but Brigadier Penny was now in command of what had been Godby's Brigade in Gilbert's Division, while Penny's and Pennycuick's Brigades, in Campbell's Division, were commanded by Carnegie and M'Leod.

To these were added the 1st Infantry Division, under Whish, with Brigadier Markham in command of the 32nd Foot and the 49th and 51st Native Infantry, and Hervey in command of the 10th Foot and the 8th and 72nd Native Infantry; also Dundas's Bombay column, 60th Rifles, 3rd Bombay Native Infantry, Bombay Fusiliers, and 19th Native Infantry. The Scinde Horse and 4 regiments of irregulars were added to the cavalry. The artillery, under Brigadier Tennant, now numbered 96 guns, 18 being of heavy calibre. The engineers and sappers were under command of Major-General Cheape, who had returned with Major-General Whish from Mooltan.[1]

The Sikhs were drawn up in the form of a rough crescent, facing nearly due south. Their right flank lay across a deep dry nullah which curved round the city behind them and covered part of their front, then took a sharp turn south, and passed through the centre of the British encampment. Their left reached to a smaller nullah full of water, which ran south into the Chenab. Their cavalry, Afghan horsemen chiefly, extended beyond the nullahs on right and left. The villages of Kalra and Chota Kalra, lying between the nullahs, had been occupied, fortified, and loop-hooled with great skill by the Sikhs.

Position of Lord Gough's army.

The British faced them, looking north, their line divided by the great nullah. Gilbert's Division was on the right next to this nullah, having Mountain's Brigade on the left and Penny's on the right. Beyond Gilbert on the right was

[1] The above is taken almost *verbatim* from *The Sikhs and the Sikh Wars*, as well as much that follows in the account of the battle.

Whish's Division, with Hervey's Brigade in the front line and Markham's in the second. On the right flank were the Cavalry Brigades of Lockwood and Hearsey.

On Gilbert's left was the heavy battery of 18 guns; Whish was supported by 3 troops Horse Artillery, with Dawes's battery and 2 troops Horse Artillery for the time in reserve. The cavalry were supported by Warner's troop of Horse Artillery.

On the left of the nullah was Campbell's Division, with Carnegie and M'Leod in the front line and Hoggan in second line, supporting Dundas and the Bombay column on the left. White's Cavalry Brigade, with Sir J. Thackwell, was on the left flank, supported by 2 troops of Horse Artillery. The Bombay column was supported by Blood's Bombay Horse Artillery; Campbell by the 2 light field batteries of Ludlow and Robertson. A reserve, consisting of the 5th and 6th Light Cavalry, the 45th and 69th Native Infantry, and the Bombay field battery, was in charge of the rear.

By half-past 7 the troops had started. The Sikhs opened fire, which showed the position and range of their guns. Our line halted, and the whole force of artillery was moved to the front covered by infantry skirmishers. By 9 o'clock the long line of guns was in position, about 800 yards from the Sikhs, and then the battle opened in earnest. For two and a half hours the artillery duel was continued. Both sides suffered considerably, but the enemy had by far the worst of it. Then at half-past 11 a general advance was ordered, the artillery still leading. Now followed some very stubborn fighting, when Penny's Brigade, chiefly the 2nd European Light Infantry under their Brigadier and Major Steele, stormed the village of Kalra. Chota Kalra, too, offered a desperate resistance to the 10th Foot in Hervey's Brigade under Colonel Franks. Markham's and Hervey's Brigades on our right had very hard fighting, but the fire of the Horse Artillery guns overwhelmed the enemy in that direction. The artillery on our left was most effective, and dominated the nullah, so that

Campbell's infantry had little opposition. On our extreme left the Afghan horsemen tried to turn our flank, but Thackwell brought his Horse Artillery to bear on them, and to cover a brilliant charge executed by the Scinde Horse, supported by the 9th Lancers, who drove back the enemy's squadrons in precipitate flight, and Thackwell was enabled to turn their flank. The ground prevented further action on the part of the cavalry on that flank, but the guns were brought forward and aided the rout of the enemy which was now beginning. On our right, the Afghan horsemen kept us constantly on the alert, but the ground there was very unfit for cavalry action owing to the wet nullah and the villages lying to our front. By half-past 12 the whole Sikh army was in full flight, and by 1 o'clock Goojerat itself, the Sikh camp, their baggage, and most of their guns were in possession of the British. On the left of the town, Dundas and Thackwell followed in pursuit, and on the right the cavalry under Lockwood and Hearsey, aided by infantry. The cavalry pursued for upwards of 12 miles, till it was dark, and the flying Sikhs had to drop most of the few guns they had carried off the field in their hurried flight. It was a thorough rout, and every branch of the army, horse, foot, and artillery, did its work thoroughly and with complete success.

The Fourteenth took a glorious part in the pursuit following upon this crowning victory of the campaign, and their steady conduct during the battle had the desired effect of holding in check and driving off those goles of Afghan horsemen which seriously threatened the right flank of our army. The Fourteenth were in the 2nd Cavalry Brigade, with the 1st Light Cavalry and the 11th Irregular Cavalry, commanded by Brigadier-General G. H. Lockwood, C.B., supported by Warner's troop of Horse Artillery, and the duty of this brigade was to defend the right flank. The Sikhs and Afghan horsemen were perpetually menacing this flank during the battle, and did their best to surround us by endeavouring to get to our rear. One troop of these audacious warriors did actually get round our rear, and threatened the Commander-in-Chief and his staff, but they were driven off most gallantly and cut to pieces by Lord Gough's escort, a troop of the 5th Light Cavalry, ably led by Lieutenant Stannus,[1] who was severely wounded on the occasion.

The successful manner in which these various attacks were met by our cavalry on the right is best described in the following extracts from despatches relating thereto. Brigadier-General G. H. Lockwood, C.B., writes:— Extracts from despatches. Brigadier-General Lockwood, C.B.

'CAMP GOOJERAT,
February 22, 1849.

'On the morning of the 21st I deployed my Brigade in the following manner:—In the first line,

5 troops of Her Majesty's 14th Light Dragoons,
2 squadrons 1st Light Cavalry,

with 1st troop 3rd Brigade Horse Artillery on the left, escorted by a troop of the 14th Light Dragoons and a ressalah of the 11th Irregular Cavalry; in support the remainder of the 11th Irregular Cavalry; and in reserve (under Lieutenant-Colonel Doherty),

1 squadron 14th Light Dragoons,
1 squadron 1st Light Cavalry.

The enemy's horsemen appeared in great force upon our right, threatening to turn our flank, so I changed front to the right, directing the reserve to retain its front and communicate with the infantry on its left. Captain Warner's guns opened with great effect upon the horsemen and turned them, but they only retired a short distance, and then a regiment of their regular cavalry moved round by a circuitous route and got completely into our rear. I immediately detached towards them three guns with a squadron of the 14th Light Dragoons, who, in conjunction with Major Christie's corps of Irregular Cavalry, drove them off. About this time a large gole of horsemen

[1] Afterwards General Henry James Stannus, C.B.

came on towards me and I prepared to charge, but as they turned at once from the fire of the guns, and as there was a nullah in our front, I refrained from advancing after them. The reserve also now advanced in support of Colonel Hervey's Brigade of infantry. I then received orders to bring on the brigade, and followed in the pursuit of the enemy. In the pursuit the 14th Light Dragoons and 1st Light Cavalry cut down and shot a considerable number of the Sikh Infantry, and Corporal William Pain of the 14th Light Dragoons captured a red silk standard, killing in single combat the horseman who bore it. I beg to state to the Major-General (Commanding the Cavalry Division) that I had the greatest satisfaction in witnessing the steadiness of the troops composing the brigade in performing several manœuvres under a heavy fire of artillery. My best thanks are due to Captain Warner, Horse Artillery; also to Lieutenant-Colonel Bradford, commanding 1st Light Cavalry; Lieutenant-Colonel King, commanding 14th Light Dragoons; and Lieutenant-Colonel Doherty, 14th Light Dragoons, who commanded the reserve.' —*London Gazette*, April 19, 1849.

Corporal Pain captures a Sikh standard.

Brigadier-General Hearsey.

Brigadier-General Hearsey (commanding 4th Cavalry Brigade), who was commanding in the pursuit in which Brigadier Lockwood's (2nd) Brigade took part, says in his despatch, dated

'CAMP, near KOREA,
23rd February 1849.

'The distance the cavalry under my command went over in pursuit was fifteen miles. I found Brigadier Lockwood, C.B., most zealous, and desirous of having an opportunity to charge at the head of the 14th Light Dragoons. The conduct of that regiment throughout the day was most exemplary and steady, and I have not the least doubt had an opportunity occurred it would have been most eagerly seized to the utter destruction of any body opposed to it.'

In his despatch to the Adjutant-General, dated Headquarters, Camp Goojerat, 25th February 1849, Major-General Sir Joseph Thackwell, K.C.B., commanding Cavalry Division, says :— Major-General Sir J. Thackwell, K.C.B.

'I am gratified to learn that both officers and men of Brigadier Lockwood's Brigade behaved greatly to his satisfaction, and that the 14th Light Dragoons, under Lieutenant-Colonel King, and the 1st Light Cavalry, under Lieutenant-Colonel Bradford, conducted themselves gallantly, and evinced every anxiety to close with the enemy. I am happy to observe that the Brigadier has mentioned with great approbation the conduct of Lieutenant-Colonels Bradford and King in command of their regiments, and I cannot avoid here stating for the information of his Lordship that I observed with much satisfaction the zeal and judgment evinced by both officers when in command of considerable bodies of cavalry detached from the camp at Chillianwallah on important duties.'

In his despatch, dated Headquarters, Camp Goojerat, 26th February 1849, His Excellency the Commander-in-Chief in India (Lord Gough) says :—'The determined front shown by the 14th (King's) Light Dragoons and the other cavalry regiments on the right, both regular and irregular, completely overawed the enemy and contributed in a very large measure to the success of the day : the conduct of all in following up the fugitive enemy was beyond all praise.' This despatch was subsequently published by the Right Honourable the Governor-General of India in his general orders, for the information of the army after the battle. General Lord Gough, G.C.B.

The following reference to the pursuits at Goojerat occurs in Thackwell's *Narrative of the Sikh Wars*, p. 227 :—'The brigades of Hearsey and Lockwood captured several guns in their advance, and committed awful havoc amongst the flying Sings. The 14th Light Dragoons were conspicuous in this bloody work of retribution. Captain Scudamore of that regiment was on the point of despatching a flying Sing when the cry of "Mercy!" arrested his arm. No sooner had the gallant officer passed him than he turned quickly round and shot his generous preserver. The zeal and activity of Lieut.-Colonel King in this pursuit were surpassed by no one.'

Casualties at battle of Goojerat.

The casualties of the 14th Light Dragoons at Goojerat were as follows :—

Killed	Lieutenant Ambrose Lloyd. 4 horses.
Wounded	Captain J. H. Goddard, severely. Captain A. Scudamore, dangerously. 4 rank and file. 2 horses.

The pursuit after Goojerat.

In the pursuit on the day of the battle our cavalry and several troops of Horse Artillery tried hard to catch the Sikh cavalry and Afghan horsemen, but although the 9th Lancers, 3rd Light Dragoons, 14th Light Dragoons, and all the light and irregular cavalry of our army did their best, the enemy's cavalry was too fleet for them. However, the Khalsa Infantry were not so lucky: they were early overtaken on our right by Hearsey's and Lockwood's Brigades, in which were the Fourteenth, who simply drove the stubborn Sikhs from every garden, ravine, or hedge where they attempted to make a stand, and shot and cut them down without mercy. Shere Singh never halted in his flight till he had passed the Jhelum; guns, standards, the whole of his camp—left standing with all it contained—all the baggage of his army, ammunition, and several thousand head of cattle were the spoils of the victors.

Losses in both armies at Goojerat.

The total losses of the British at Goojerat were 5 officers, 92 men, killed; 24 officers, 682 men, wounded.

In killed, wounded, prisoners, and missing, the Sikhs probably lost from 15,000 to 20,000 men, a great portion of their army having disbanded themselves.

Goojerat saw the Khalsa army completely shattered, and the victory was a glorious triumph for Lord Gough and the Army of the Punjaub. After the news of Chillianwallah reached England, popular feeling was very strong against Lord Gough and his 'Tipperary tactics,' as they were called, and the heavy loss of life incurred in his battles made him unpopular in the country. The Government of the day determined to recall him, and on the 7th March 1849, Lieutenant-General Sir Charles Napier, K.C.B., was appointed Commander-in-Chief in India, with the local rank of General, but long before he arrived to assume command, the gallant old soldier he came to supersede had, by his brilliant victory of Goojerat, amply vindicated his reputation, and was restored once more to the confidence of his Queen and country.

Immediately after Goojerat, Major-General Sir Walter Gilbert, K.C.B., a noted horseman and a very energetic officer, was despatched with 12,000 men of all arms, amongst whom were the Fourteenth, with orders from Lord Gough to pursue across the Jhelum, and to seize at once Rhotas, Attock, Peshawur, and the whole frontier up to the passes leading into Afghanistan. This force experienced very inclement weather, and had to make a temporary halt in consequence, on the 2nd March. Sir Walter Gilbert's orders were to grant no terms save unconditional surrender. By the 6th March the Sikhs, so hotly pressed by our troops and worn out with flight, without supplies of food or ammunition, restored all their prisoners, and finally, on the 12th March, at Rawul Pindhi, laid down their arms and surrendered at discretion to Sir Walter Gilbert. The Fourteenth were present on this occasion, also at the capture of Attock, 17th March; whence, continuing their pursuit of the flying Afghan horsemen, they crossed the Indus, were at Peshawur on 21st March, and taking part in the expulsion of the Afghans beyond the Khyber Pass, encamped under Jumrood Fort till the 2nd April. The result of the victorious campaign now closed was the annexation to the British Crown of the whole of the Punjaub, from the Khyber Pass to the banks of the Sutlej, and this event took place on the 1st April 1849. The whole campaign, from the time the Commander-in-Chief actually took the field to the victory at Goojerat, occupied three months.

Pursuit to the Khyber Pass.

Annexation of the Punjaub, 1st April 1849.

On 2nd April the Fourteenth left Jumrood on their

Regiment arrives at Lahore.

march back; they left Peshawur 4th April, and arrived at Lahore 1st May, where, before the troops were dismissed to their lines, the 'muster' was held. On 7th May, Major-General Sir Walter Gilbert, K.C.B., inspected the regiment at Lahore, and again on the 17th October at the same station, and expressed satisfaction with all he saw. In the month of November, 109 men joined from England, making the strength present at headquarters up to 700 men.

Depôt at Maidstone.

The depôt at Maidstone under Colonel C. Middleton, commandant of the Cavalry Depôt there, consisted of 76 men on 1st December.

On 6th December the Marquis of Dalhousie, Governor-General of India, inspected the Fourteenth at Anarkullee (Lahore), on which occasion the new Commander-in-Chief in India, General Sir Charles James Napier, G.C.B., was present, and expressed his approbation of the movements of the Cavalry Brigade, in which the Fourteenth took part.

On 17th December the Commander-in-Chief inspected the regiment again at Lahore, and was pleased to express his satisfaction with what he saw.

1850

During the whole of this year the Fourteenth were stationed at Anarkullee,[1] Lahore. On 25th January, Lieutenant William Featherstonehaugh became Paymaster *vice* Rofe, who had served in that capacity ever since the year 1812, having then succeeded Mr. Flanegan, the first Paymaster ever appointed to the 14th Light Dragoons.

On the 21st February and on 5th April the regiment was inspected by Major-General Sir. W. Gilbert, G.C.B.

On the 2nd March, Quartermaster G. Shenton died at sea.

On the 25th March, His Excellency the Commander-in-Chief in India, General Sir Charles J. Napier, G.C.B., made an inspection of the garrison at Lahore, seeing all the troops there, including the 14th Light Dragoons.

[1] The military cantonment is now called Mean-Meer.

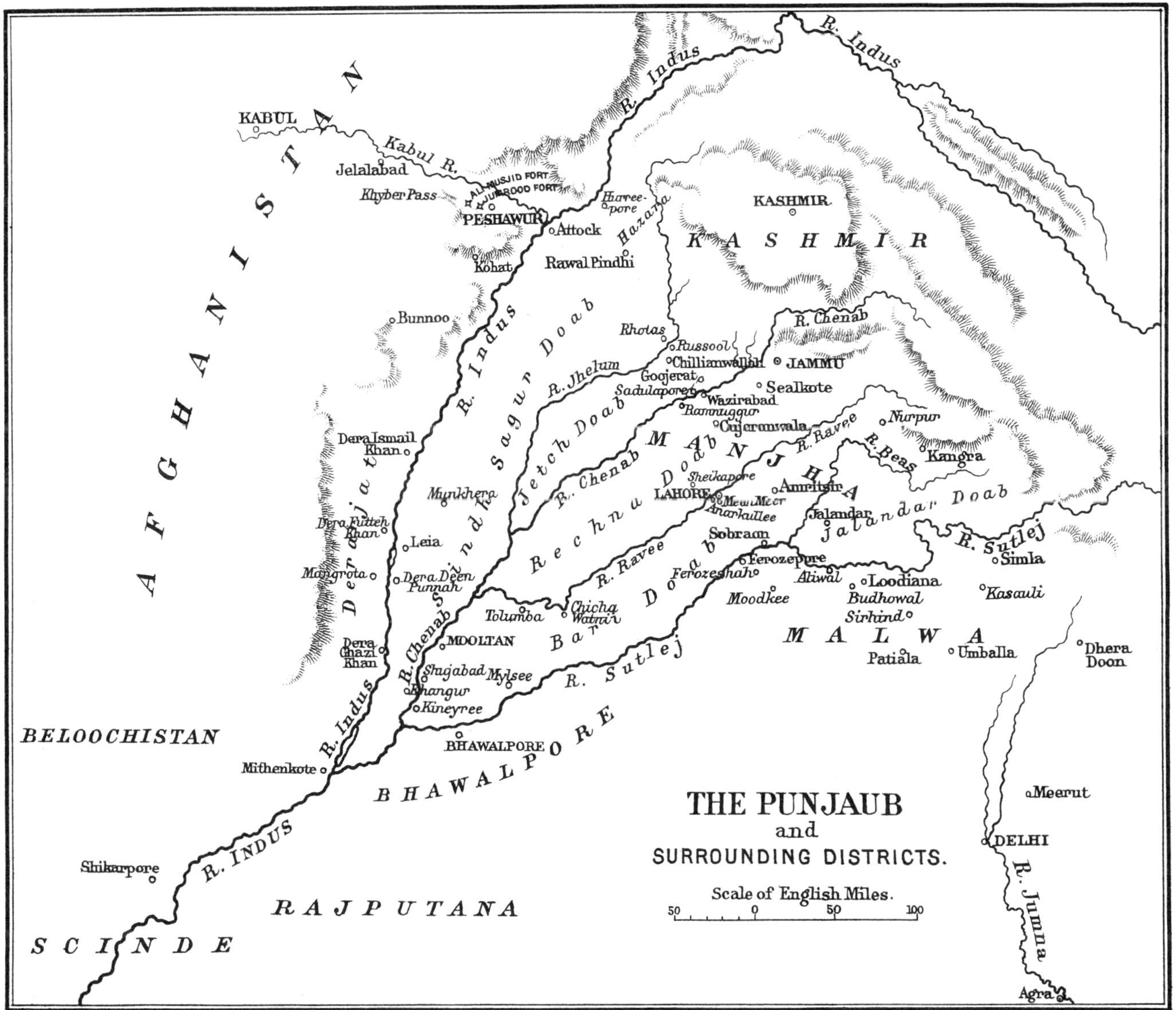

London: Longmans, Green & Co.

On the 6th July, Lieutenant-Colonel J. W. King, C.B., died at Lahore, and was succeeded in command of the regiment by Lieutenant-Colonel H. E. Doherty, C.B.; on the 7th, Major C. Steuart became 2nd Lieutenant-Colonel; and on the 17th September, Captain J. H. Goddard became Major. Death of Lieut.-Colonel King. Lieut.-Colonel—H. E. Doherty.

On the 29th October, Major-General Sir Walter Gilbert, G.C.B., inspected the regiment, and expressed himself very much pleased with the result. He saw the Fourteenth at 5.30 A.M. in 'marching-order,' and at 5 P.M. in 'watering-order.'

1851

The Fourteenth left Lahore 19th January, and marched to Meerut, arriving there 6th March. At Meerut.

On 1st February, headquarters were at Camp Loodiana; on 1st March at Camp Mangal.

On 17th March the Fourteenth took part with the other regiments stationed at Meerut in a review held by Lieutenant-General Sir Joseph Thackwell, G.C.B., K.H., at which His Excellency General Sir William Gomm, K.C.B., Commander-in-Chief in India, was present and was highly pleased; and on the same day His Excellency visited the barracks, stables, regimental hospital, and the men's dinners, and expressed to Lieutenant-Colonel Doherty, C.B., how satisfied he was with everything he had seen.

On 22nd March, Lieutenant-General Sir J. Thackwell inspected the Fourteenth very minutely, and he inspected a second time on 17th November, and on both occasions expressed satisfaction with all he saw.

Lieutenant H. C. Reader was appointed Adjutant *vice* Lieutenant R. P. Apthorp invalided home. Lieutenant M. C. Smith died at Meerut 14th March; and Cornet G. A. Francklyn died at the same place on 18th December of pleurisy and lung disease. There is a very handsome silver claret-jug of a most uncommon pattern in the officers' mess which was presented by the family of this young officer in his memory.

1852

The Fourteenth remained all this year at Meerut. On the 27th March and 29th October, Lieutenant-General Sir J. Thackwell made his inspections of the regiment, and on both occasions expressed himself satisfied with the appearance of the men, the condition of the horses, and the interior economy of the corps.

On the 23rd July, Veterinary Surgeon A. W. Caldwell died.

Authority to bear words 'Punjaub,' 'Chillian-wallah,' and 'Goojerat' on appointments granted, 1852.

On 14th December authority was granted for the regiment to bear on its appointments the words 'Punjaub,' 'Chillian-wallah,' and 'Goojerat,' in consideration of its services in the second Sikh campaign in the Punjaub, 1848-49.

1853

The Fourteenth remained at Meerut all this year. The strength of the service troops under Lieutenant-Colonel Doherty, C.B., at Meerut was 770 men; the depôt troop at Maidstone, under the Commandant, Colonel F. Griffiths, consisted of 43 men.

Colonel—Hon. H. Murray.

Lieutenant-General the Honourable Henry Murray, C.B., was appointed Colonel of the regiment on 18th March, in succession to General Sir E. Kerrison, Bart., G.C.H., K.C.B., deceased.

On 20th July, Lieutenant and Riding-master John Holliday died at Meerut.

On 5th August, Cornet Charles Wemys Thesiger[1] joined the Fourteenth on transfer from the 5th Madras Light Cavalry.

On 30th December, Captain W. Wilmer became Major *vice* Clarke.

The Fourteenth were inspected at Meerut by Lieutenant-General Sir Joseph Thackwell on 30th March, and by Brigadier-General J. Scott, C.B., on 5th November. Both officers expressed their satisfaction at the efficient state in which they found the regiment.

[1] Now Lieutenant-General the Hon. C. W. Thesiger, Colonel of the 14th (King's) Hussars.

1854

During the whole of this year the Fourteenth remained at Meerut. On 17th March, Brigadier-General Scott, C.B., inspected, and on 7th November, Brigadier-General George Brooke, C.B., and both officers seemed very pleased with what they saw. Regiment at Meerut.

Three officers died this year, viz. Lieutenant W. D. Boyd, at Bromley, Kent, on 14th January; Major J. H. Goddard, who had been wounded at the battle of Goojerat,[1] at Simla, on 31st May; and Assistant Surgeon R. Wigstrom at Mazagon, Bombay, on 8th September.

On 1st June, Captain A. Scudamore became Major *vice* Goddard. The establishment was now— Establishment.

9 Troops.	10 Farriers.	626 Privates.
57 Sergeants.	40 Corporals.	703 Troop-horses.
14 Trumpeters.		

A Saddler-Sergeant was included in the establishment this year for the first time.

His Excellency General Sir William Gomm, K.C.B., Commander-in-Chief in India, expressed his great satisfaction at receiving so good a report of the regiment from Brigadier-General G. Brooke, C.B., after his inspection held on 7th November at Meerut, as stated in a letter from Colonel Frederick Markham, Adjutant-General of Her Majesty's army in India, addressed to Brigadier-General Brooke, C.B., commanding at Meerut, which was notified to the officer commanding the regiment by His Excellency's desire. In an extract from General Orders, dated Headquarters, Shunishabad, 30th December 1854 (No. 91), His Excellency remarks with reference to the 14th Light Dragoons having been placed under orders to proceed to Bombay with all despatch, preparatory to embarkation for the seat of war in Turkey, as follows:— Regiment placed under orders for Turkey and the seat of war.

'The Commander-in-Chief heartily congratulates the corps on its brilliant destination, and on the prospect this opens to it of early opportunity occurring for further ennobling its Sir William Gomm's farewell to the regiment.

[1] Major Goddard was shot in the ankle at Goojerat.

standards, rich already in the records of Peninsular and Indian victory. Her Majesty's Fourteenth will bring an accession, as we confidently trust, nothing short of 667 British sabres, with which it will quit Meerut, to the ranks of the heroic army, which in brief space of time has achieved exploits unsurpassed in the annals of British intrepidity and constancy. Sir William Gomm has pleasure in publicly recording the tenor of the half-yearly report made to him of the state of the regiment on 7th November last by an Inspecting Officer so competent in every respect to the duty as that by whom it was then visited. Brigadier Brooke reports in his "Summary of Remarks" as to the 14th Light Dragoons—"The regiment appears to be in a state of complete efficiency. The horses are in good working condition, and the appointments of both men and horses in excellent order."

(Signed) 'WILLIAM GOMM, General,
'*Commander-in-Chief, East Indies.*'

1855

Orders for the Crimea countermanded.

On 8th January, at Meerut, the Fourteenth received orders for the Crimea, where the war between the Russians and the allies was now in progress. On the 16th the march commenced from Meerut, but on 17th an order came countermanding the embarkation for the Crimea, and the regiment marched back to Meerut, and re-entered the station on the 19th January.

Ordered to Kirkee.

On the same day orders were received from the Governor-General direct by electric telegraph for the Fourteenth to march to Kirkee, and on 24th January the march to Kirkee, 75 marches and 884 miles distant, was commenced. The strength on leaving Meerut was 20 officers, 705 men, and 701 horses.

Regiment arrives at Kirkee, 21st April 1855.

The Fourteenth entered Kirkee on 21st April, losing 2 men by death on the march. On the 28th April, Brigadier-General Trydell, commanding the Poona Brigade, made his inspection. He was much pleased, and said he could scarcely have expected to find the regiment in such an efficient state after their long march, as he did.

Lieutenant-General the Honourable H. Murray, C.B.,

became General on the 16th February. On 28th June, His Excellency General Sir H. Somerset, Commander-in-Chief of the Bombay army, reviewed the Fourteenth, and was very pleased with the manner in which the field movements were executed. The Right Honourable the Governor of Bombay, Lord Elphinstone, was present, and told Colonel Doherty that, as an old cavalry officer, he never saw a regiment 'trot past' better than the Fourteenth had done that day.

On 2nd November, Major-General F. Schuler, commanding the Poona Division of the army, made an inspection of the regiment, and was highly pleased.

1856

The Fourteenth remained at Kirkee this year.

Colonel Doherty, C.B., left India for England on leave of absence on 30th August, after commanding the regiment for six years, and he never rejoined it, as he retired on half-pay the following year.

On 7th November, Riding-master Joseph Raiker was appointed to the regiment.

His Excellency General Sir Henry Somerset, K.C.B., inspected on 20th March, and again on 17th December; Major-General Schuler inspected on the 28th October, and both expressed themselves perfectly satisfied.

On 31st October, Colonel Charles Steuart arrived from England and assumed command of the regiment during the absence of Colonel Doherty.

1857

Regiment proceeds to Persia on service, 20th Feb. 1857.

On 20th February the regiment, under Colonel Steuart, left Kirkee for Persia on field-service with the expeditionary force commanded by Lieutenant-General Sir James Outram, K.C.B., and reached Bushire in March, where it disembarked and was encamped. The Fourteenth had moved by squadrons from Kirkee to Bombay, the last squadron marching on 24th February. The strength of the service troops proceeding to Persia was 25 officers, 614 men, and 649 horses. The headquarters embarked at Bombay 8th March, and landed at Bushire 31st March, and they left Bushire 30th April, landing at Bombay 15th May. Captain Need's troop had embarked at Bombay in the ship *Raby Castle* with a strength of 76 horses on the 1st March, and landed on the 21st March at Bushire, where was an open roadstead with a very rough sea, in consequence of which many of the small boats used for landing were capsized and several horses drowned. Another troop sailed in the *Golden Era*, but was much delayed through grounding on a sandbank. Assistant-Surgeon J. H. Sylvester of the Indian Medical Service and Veterinary-Surgeon H. Dawson were attached to the regiment, and served with it in the Persian expedition.

The Persian expeditionary force proceeded by sea from Bombay in the months of February and March to the Persian Gulf, and landed at Bushire: it consisted of 2 divisions of infantry with artillery, sappers, and a cavalry division.

General Stalker commanded the 1st Division; Brigadier-General Henry Havelock, C.B., the 2nd Division; and Brigadier-General Jacob, C.B., the Cavalry Division; the two brigades of cavalry being commanded by Colonel Tapp, and Colonel Steuart of the 14th Light Dragoons.

One troop of the Fourteenth under Captain Prettejohn formed part of the force which started from Bushire towards the end of March and proceeded up the gulf with Lieut.-General Sir James Outram towards Shatta-el-Arab. Several of the vessels containing troops grounded on the bar at Shatta-el-Arab, the name given to the place at the mouths of the Tigris, Euphrates, and minor tributaries in the Persian Gulf, and thus considerable delay and inconvenience was caused, and most of the cavalry was late to be of efficient service partly in consequence of this, and partly owing to the small supply of boats for landing the troops. The forts of Mohamra were situated along the banks of the river, but the town of the same name lay 40 miles higher up, on the north side of the river Karoon, which at this point, near its junction with the Shatta-el-Arab, is from 600 to 800 yards wide. Four armed steamers and two sloops of war, under Captain Young of the Indian navy (acting commodore), had preceded the

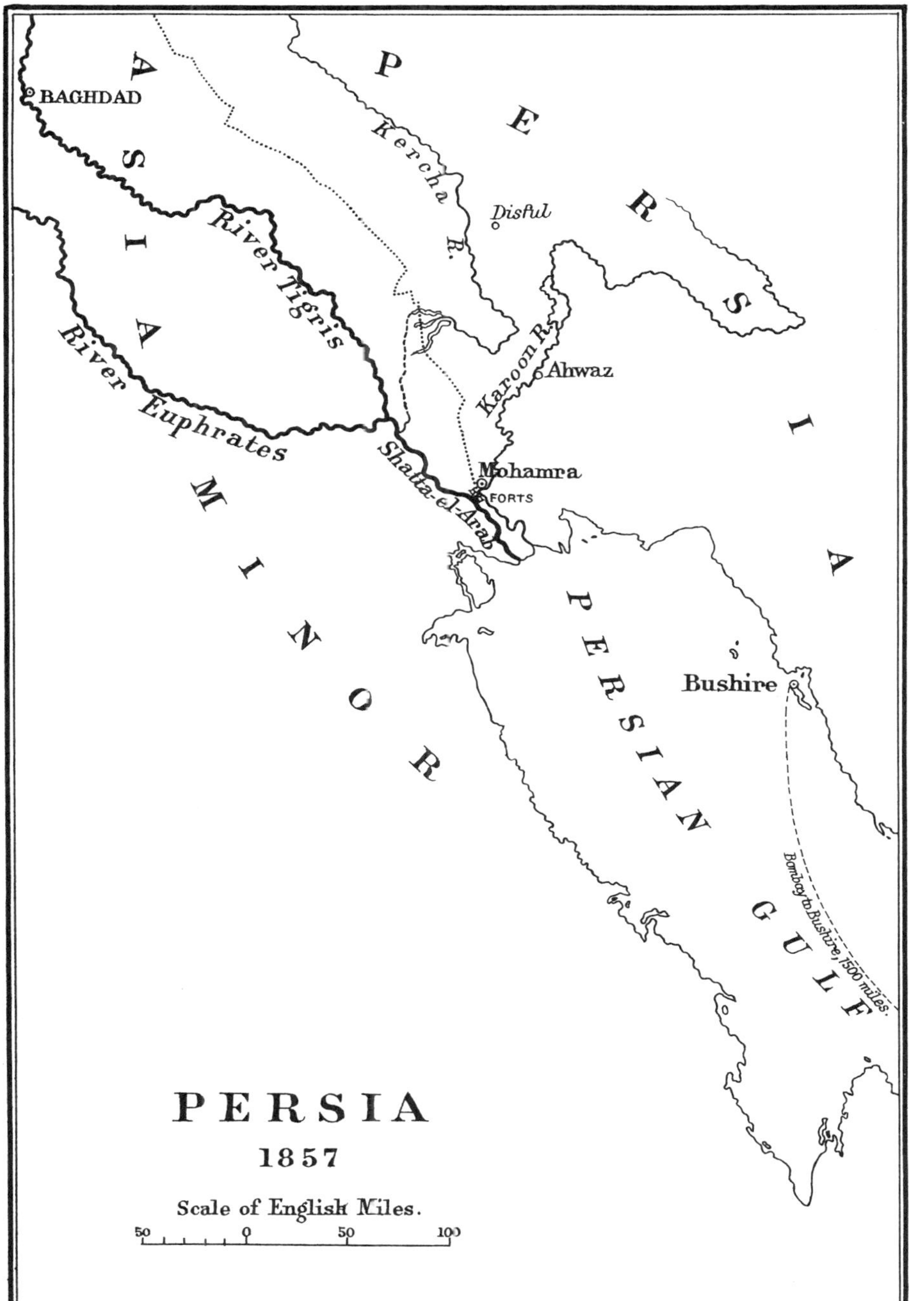

Stanford's Geog! Estab!

London: Longmans, Green & Co.

flotilla of vessels carrying the troops, and these successfully bombarded the forts and entrenched positions of the Persian army within gun-fire, which had the effect of completely demoralising the enemy, as the fire was well directed; and subsequently, when the force landed under Sir James Outram, who was himself in the leading vessel, the *Feroze*, the whole Persian army evacuated their camp, leaving tents standing, and all their property, together with ammunition and 17 guns, which fell into our hands. The expeditionary force had to land in small boats, running the gauntlet of both gun and musket fire from some of the Persian batteries on the river bank, but so soon as ever the landing was effected and the advance on the enemy's camp 2 miles off actually made, the Persians fled, having first exploded their principal magazine. Owing to the want of cavalry just when it was needed, no effective pursuit took place, although a detachment of the Sinde Horse under Captain M. Green did attempt it, but without much result, and the fugitives made good their escape, losing 200 killed. The only portion of the military forces engaged was the mortar battery under Captain Wogan, and a few European riflemen who were employed on board the war vessels when they bombarded the Persian earthworks and batteries lining the banks. There were 23 men killed and wounded altogether in the squadron during the bombardment, and the victory was really a naval one.

Flight of the Persians.

The strength of the Persian army was 13,000 men, whilst the British and Indian troops numbered under 5000 men, as follows:—

	Men.
14th Light Dragoons,	89
Sinde Horse,	303
64th and 78th Regiments,	1534
2 Native Regiments,	1465
The Light Battalion,	920
Sappers and miners,	233
12 guns Horse Artillery and No. 2 Light Field Battery,	342
Total,	4886

Mohamra captured, 26th March 1857.

The capture of Mohamra took place on 26th March. Sir James Outram had himself left Bushire on 18th March, reaching Shatta-el-Arab, 174 miles, on 21st March. He left a large portion of the expeditionary force, to the number of 3000 men, under the command of Brigadier-General Jacob, C.B., at Bushire, where General Stalker had unfortunately just died, and Commodore Ethersey of the Indian navy, commanding the squadron, died in the Persian Gulf about the same time. The climate is a very pestilential one, and dysentery and fever are prevalent and very fatal to Europeans.

It was only Captain Prettejohn's, the 'H' troop of the Fourteenth, which was present at the taking of the town and fortress of Mohamra, as this troop happened to arrive first at Bushire and so was taken on first, as Jacob's Horse had not arrived, being delayed by the gales. The officers with Captain Prettejohn were Lieutenant Mackenzie, Cornet Ridley, and Assistant-Surgeon Fox.

The other troops of the Fourteenth and headquarters remained some time at Bushire, and some who left Bombay in sailing-vessels were sent back without even landing or doing any service in the expedition, much to their disappointment. The only losses suffered by the Fourteenth in Persia were 22 troop-horses by death, the result of accident or disease. After the affair of Mohamra some infantry with an armed flotilla were despatched up the river Karoon as far as Ahwaz, under command of Captain Rennie, Indian navy, the detachment of the 78th Highlanders being commanded by Captain Hunt; but though the Persians numbered 7000 they retreated in hot haste before a body of 300 British infantry, 3 small steamers, and 3 gunboats.

Thus ended the war. The 64th Regiment and the troop of the 14th Light Dragoons were shortly afterwards sent back to Bushire, whilst the remainder of the force was kept some time longer in the unhealthy vicinity of Mohamra and Shatta-el-Arab, exposed alike to Persian sun and gulf fever. Sir James Outram himself did not finally quit Persia till 17th June, by which time the mutiny in India had broken out.

The Lieutenant-General says in his despatch, referring to the services of the military portion of his forces employed in this expedition :—'Although not actively engaged with the enemy, I am not the less indebted to all ranks for their exertions and zeal, and especially for the great order and despatch with which the landing of the troops was effected under Brigadier-General Havelock, C.B. The highest spirit prevailed, and had the large Persian army only waited our approach out of the range of the ships' guns, I feel confident it would have received a lasting lesson.'[1]

The Fourteenth reached Kirkee in May and had scarcely settled down in their former quarters on return from the successful expedition to Persia, when they had to respond once more to the old familiar trumpet-call of 'Boot and Saddle.'

Regiment returns to Kirkee, May 1857.

The 8th of June found the left wing of the regiment, 'B,' 'C,' 'D,' and 'E' troops, *en route* to Ahmednuggur and Nusseerabad; another troop, probably the 'A' troop, followed on the 11th, making the strength of those on field-service up to 10 officers, 300 men, and 305 troop-horses.

Left wing leaves Kirkee, 8th June 1857, for the Deccan.

Major-General Schuler inspected the headquarters at Kirkee on 15th and 16th July, after the remainder of the regiment had left for service. 'G' and 'H' troops had been moved from Kirkee into Poona to be quartered there when the service troops left; but on 16th June, 'H' troop proceeded temporarily to Sattara in the southern part of the Bombay Presidency, followed by some more detachments of the regiment to the same town, their presence being urgently required there in consequence of an outbreak at Kolapore,[2] situated further south.

On 10th July the right wing began to take the field, and on that date 120 sabres (probably 'H' and 'K' troops)[3] marched to join the other portions of the regiment already on service in the Deccan, followed on 16th July by some more men of the right wing, so that by this time it may be said the whole

Right wing proceeds to the Deccan, 16th July 1857.

[1] The above is quoted from the *Biography of Sir James Outram*, by Sir F. Goldsmid.

[2] Kolapore, or Kolhapur, is in the Mahratta country.

[3] Prettejohn's and Need's.

regiment was fairly launched upon the campaign against the mutineers in the great Sepoy revolt, which lasted till May 1859.

Since their arrival in India the Fourteenth had always worn the puggrie round their forage-caps, excepting when in full dress, and then the shako was worn; but now the latter was given up altogether, and the authorities gave permission for the men to wear turbans instead, which was a far more sensible head-dress for an Indian sun. Before the mutiny broke out the men of the Fourteenth were always called

'Puggrie Wallahs.'

'Puggrie Wallahs' by the natives, having, it is believed, initiated the wearing of forage-caps with puggries round them.[1] Gloves and stocks were also discarded by the men at this time when going on field-service, as being useless encumbrances for actual fighting and hard work.

Major-General Woodburn's column.

It was owing to the disturbed state of affairs throughout India generally just at this time, May 1857, that Lord Elphinstone, the Governor of Bombay, designed to form a column to secure and hold the great line of road between Bombay and Agra. This column was only a small one. It was placed under the command of Major-General Woodburn, C.B., and was intended by Lord Elphinstone to be used to open out communications with Central India and the North-West Provinces. It consisted of the following troops:—

5 troops of the 14th (King's) Light Dragoons,
The 25th Bombay Native Infantry,
Captain Woolcombe's battery of Horse Artillery,
A pontoon train.

The column started from Kirkee and Poona on 8th June, with orders to march at once to Mhow, in order to save that place and prevent the spread of the insurrection which had already taken place in the Malwa district and along the northern portion of the Bombay Presidency. The state of affairs at Mhow and Indore was very serious. Malleson[2] says:—'It was just possible that, making forced marches, General Wood-

[1] See *Scraps from my Sabretache*, by Stent, 1882, p. 144, from which several quotations have been made in this Record.

[2] Vol. v. p. 7.

burn might approach so near to Indore[1] as to baffle the plans of the discontented. Circumstances, however, occurred which baffled the hopes expressed by Lord Elphinstone, when, acting on his own unaided judgment, he pressed upon the military authorities the necessity for General Woodburn to advance.' The circumstances alluded to by Malleson were the revolts in Malwa and the Deccan, especially in Aurungabad, which latter city was once the capital of the kingdom of Ahmednuggur, and later the favourite residence of the Emperor Aurungzeeb. It is situated in the north-western corner of the Nizam's dominions in Hyderabad, Deccan. On the second day's march of the column from Poona, it was joined by the 25th Bombay Native Infantry, under the command of Major Follett. The 25th Bombay Native Infantry was destined afterwards, in company with the 14th Light Dragoons, to perform the most brilliant services in Central India under Major-General Sir Hugh Rose, K.C.B., but just at the present crisis doubts were actually entertained of its loyalty. Subsequent events proved how utterly unfounded these were. The whole of this column, which has been called first the Deccan Field Force, secondly the Malwa Field Force, and thirdly the Nerbudda Field Force,[2] entered the city of Ahmednuggur at the same time, Captain Woolcombe's battery of Horse Artillery (native drivers and European gunners) having joined it simultaneously with the 25th Bombay Native Infantry. Aurungabad.

The Nizam's army was now in a very doubtful state of loyalty, and it was on the third day's march of Major-General Woodburn's column that intelligence came of the disaffection of the 1st Hyderabad Cavalry at Aurungabad, a day's march from the halting-place at Seeroor,[3] as well as of the alarming danger which threatened Mhow.

[1] Malleson's Hunterian nomenclature is not adhered to in this Record, but the older spelling is generally preferred.

[2] The designation of the column was altered as it advanced and grew by reinforcements. When in the Deccan it was so called; leaving and entering Malwa it became 'The Malwa'; after quitting that territory and entering the valley of the Nerbudda it changed its name again.

[3] Or Sírur.

The monsoon had just burst, and naturally caused much inconvenience to the marching of the troops.

As the reports from Mhow were so serious, the order of march was altered to Malligaum, leaving Aurungabad to the right; but almost directly after the route was again changed, and the column recalled to proceed to Aurungabad to help the Europeans who were in such dire straits there, as the troops had mutinied. On the 19th June, Captain H. O. Mayne arrived in camp from Aurungabad with the ladies and children of that station.

Aurungabad Mutiny quelled, 23rd June 1857.

After a wet march the column reached Aurungabad at 10 A.M. on 23rd June. Captain Abbott, commanding the 1st Cavalry of the Hyderabad Contingent, with the other officers and such of the men as had remained trustworthy, were posted in the mess-house, but the rebellious portion of the garrison, which consisted of the 1st and 3rd Hyderabad Cavalry, 2nd Hyderabad Infantry, and a battery of the artillery of the Hyderabad Contingent, had encamped themselves on high ground beyond the cantonment, on the Jaulna road. It was the 1st Cavalry who were mutinous. Malleson [1] says:—'General Woodburn's column marched at once to the ground occupied by the mutineers, and ordered the men to give up their arms. With the exception of one troop of the 1st Cavalry, all obeyed. The General gave the men of that troop six minutes to consider the course they would pursue. When the time elapsed, the men, instead of submitting, put on a bold front, and attempted to ride away. In this attempt most of them succeeded. The next morning, some three or four, convicted of attempts at assassination, were hanged, and order was restored.'

The formation of the column at the rebels' camp was, 14th Light Dragoons on the left, guns in the centre, 25th Native Infantry on the right. The 1st Cavalry fell in on foot, facing the column, with the native officers only mounted. The guns were previously loaded with canister, and pointed at the rebels; but during the six minutes' grace allowed, the men

[1] Vol. v. p. 9.

got on their horses and made off. Now, when too late, the guns were fired, knocking over a few horses left behind, and one native syce. Then the 14th Light Dragoons, with tired horses after a long forced march, were sent after the runaways, but they for the most part, as was to be expected under the circumstances, made good their escape. Captain Gall of the 14th Light Dragoons led his troop after those who were escaping by the Jaulna road. Captain Barrett pursued across the open country, and some of the loyal Native Cavalry under Captain Abbott joined the Fourteenth in their almost fruitless pursuit. One squadron of the regiment under Lieutenant Leith,[1] with 2 guns, some of the 25th Native Infantry, and some sappers and miners, had been very judiciously posted near the bridge over the river leading to the cavalry lines, so as to prevent any disturbance occurring in the cantonment. In the end some few rebels were taken and brought back prisoners. Courts-martial were held, and several rebels were convicted and executed, either by hanging or else shot or blown away from the cannon's mouth; but there is no doubt that as a whole the mutineers on this occasion were treated too leniently, and without the necessary decision which is of paramount importance on such critical occasions, especially when dealing with the natives in India.

Mutineers pursued by Gall, Barrett, and Abbott.

The next day a squadron of the 14th Light Dragoons, accompanied by 2 guns of Woolcombe's battery and some Native troops, were placed under command of Captain Gall, 14th Light Dragoons, and marched off at dusk for Boldana, in Berar, as there was a large sum of money in the treasury there under the guard of a troop belonging to the mutinous cavalry of the Nizam. This column reached Boldana, 100 miles off, in three days, whence, after taking over from Mr. Bullock, the acting resident, the whole of the mutinous troop, previously disarmed, marched back under Captain Gall to Aurungabad, bringing the prisoners with them. They passed *en route* through Jaffarabad and Jaulna, and arrived on 6th July at Aurungabad. On 9th July news came to Aurungabad con-

Captain Gall's column marches to Boldana and arrests mutineers.

[1] Afterwards Major Leith, V.C.

Left wing marches for Central India, 12th July 1857.

cerning the Mhow and Indore mutinies, and on the 12th the left wing of the Fourteenth and the rest of the column marched for Central India, Major Follett having assumed the command in consequence of Major-General Woodburn being incapacitated by illness. A detachment of 50 sabres of the Fourteenth was left behind in Aurungabad for the protection of the place, as well as 2 guns. This troop of the regiment appears to have gone back to Kirkee in October when no longer required in the Deccan, and it remained there for a time with other details, forming a sort of depôt under command of Brevet-Colonel C. P. Ainslie,[1] who had been reappointed on 26th August from half-pay to serve once more in the Fourteenth as second Lieutenant-Colonel. On 17th June, Lieutenant C. W. Thesiger[2] exchanged into the 6th (Inniskilling) Dragoons, which regiment he afterwards commanded. On 20th June, Captain Gaussen died in England, and on 26th August, Colonel H. E. Doherty, C.B. (in England), retired on half-pay, being succeeded in command of the Fourteenth by Colonel Charles Steuart.[3] On 18th September, Captain R. H. Gall became Major *vice* Wilmer.

Lieut.-Colonel—Charles Steuart.

The march of the column which started from Aurungabad for the relief of Mhow on 12th July, under the temporary command of Major Follett, 25th Bombay Infantry, and with which was the left wing of the Fourteenth under Captain Gall's command, lay through the Asseerghur jungles, which from June to November are very unhealthy and often prove fatal to Europeans, as there is much danger of malaria in that season.

At Edulabad, Major Follett fell a victim to cholera, and was buried at Burhampur.[4] When Major-General Woodburn fell ill at Aurungabad, Colonel C. S. Stuart,[5] of the Bombay Infantry, had been appointed as Brigadier to command the force, and he assumed command at Asseerghur, which was

[1] Colonel Ainslie commanded the station at Kirkee.

[2] Now Lieut.-General the Hon. C. W. Thesiger, Colonel of 14th (King's) Hussars.

[3] Afterwards Lieut.-General Charles Steuart, C.B.

[4] See Sylvester's *Campaign in Central India*, p. 15.

[5] Afterwards General Sir Charles Shepherd Stuart, G.C.B.

reached on 22nd July. Here also Lieutenant-Colonel Durand, R.E.,[1] Officiating Agent to the Governor-General for Central India (in the absence of Sir Robert Hamilton), Mrs. Durand, Dr. Henry Wilson, as well as several fugitive officers and ladies who had escaped from the mutiny at Indore, joined the column on the march.

On 27th July the troops crossed the Nerbudda in the vicinity of Hoosingabad, the 14th Light Dragoons swimming the river,[2] the infantry, guns, and baggage passing by the ford of Mokka-ka-Turr, near Burwai. As the force ascended the Vindhya range at a height of 1650 feet, all traces of cholera and malaria quickly disappeared, but the heavy rains continued almost incessantly during the greater part of the march.

On the 28th July the column was joined by the 3rd Regiment of Cavalry, Hyderabad Contingent, commanded by Major S. Orr; on the 31st it ascended the Simrol pass, and the following morning entered Mhow. That night the monsoon came on with great violence, and very heavy rains set in and continued for the next two months. It was a most critical time: Holkar's troops at Indore, only 13 miles off, were in open mutiny; and there was no European infantry at Mhow, where the force, under Brigadier Stuart, consisted of 5 troops 14th Light Dragoons, 3rd Cavalry, Hyderabad Contingent, 1 horse battery of European Artillery, the 25th Bombay Native Infantry, and a pontoon train. A few days later, as related in Kaye and Malleson's *History of the Indian Mutiny*,[3] 4 companies of the 86th Regiment[4] arrived to reinforce the column at Mhow; but no operations could be satisfactorily carried on during the heavy rains, so it was not till October that anything further was undertaken. On the 20th of that month Brigadier Stuart led forth his column, now the Malwa[5] Field Force, with a view to clear the rebels

Advance of the Malwa Field Force from Mhow under Brigadier C. S. Stuart, Oct. 1857.

[1] Afterwards Sir Henry Marion Durand, K.C.B.

[2] Sylvester's *Campaign in Central India*, p. 16.

[3] Vol. v. p. 42.

[4] Now 2nd Battalion Royal Irish Rifles.

[5] Malwa, in Central India, comprises the states of Gwalior, Indore, and Dhar.

out of the surrounding districts where they had occupied some of the most important forts and towns of Malwa.

Mundesor. The first point of intended attack was Mundesor. This town is situated on a tributary of the river Chambal, about 120 miles from Indore. Malleson[1] says:—'In the month of July this place had been occupied by some of Sindhiá's revolted troops, and these had been joined, and were being constantly further strengthened, by Afghan, Mekráni, and Mewatí levies. In August the insurrection at Mundesor threatened not only to embrace all Western Malwa, but Neemuch as well. The leader of the Mundesor insurgents was Ferozshah, a sháhzáda or prince connected with the imperial family at Delhi. It was estimated in September that some 15,000 men, with 16 or 18 guns, had rallied round his standard, and this estimate was subsequently found to have been below the actual number.' The force under Brigadier Stuart at the time only consisted of about

200 of 14th Light Dragoons.
170 Artillery.
230 of 86th Regiment.
350 of 25th Bombay Native Infantry.
350 of 3rd Nizam's Cavalry.
Madras Sappers and Miners.

Lieutenant-Colonel Durand, the political officer, saw the importance of striking a blow as quickly as could be; and accordingly, as Mundesor was too strongly held to be attacked immediately, it was determined first to march on the fort of Dhar. On the 14th of October an advanced party under Major Robertson, 25th Bombay Native Infantry, consisting of a troop of the 14th Light Dragoons, 3 companies Native Infantry, 2 guns, and some Hyderabad Cavalry, had been sent on from Mhow in the direction of Dhar to form a kind of advance-guard to the column, with orders to proceed *via* Gujri. The main body, which left Mhow on 20th October, was divided into 2 columns, one under Major Keane, 86th

[1] Vol. v. pp. 44, 45.

Regiment, and the other under the Brigadier. Each of these took a somewhat different direction, in order to try and clear the whole of Malwa of its rebels, and on the 22nd the entire force arrived in the vicinity of Dhar. Captain Mayne, 3rd Irregular Cavalry, who was attached to the force, had ridden forward with some sowars, and reported he had been fired on at a village about 4 miles from Dhar. On coming in view of the fort and town the 86th Regiment advanced in skirmishing order, and it was ascertained that a large body of the garrison had sallied out from the fort to attack our force. Malleson says they had planted 3 guns on a hill south of the fort, from which point they were extended along its eastern face in skirmishing order, and advanced boldly against the British.[1] The guns were charged and captured by some of the Fourteenth, acting with the 25th Native Infantry, led by Major Robertson, whose men promptly turned them on the rebels. 'Almost simultaneously the four companies of the 86th Regiment and the sappers, flanked by Woolcombe's (Bombay) and Hungerford's (Bengal) batteries, advanced against the enemy's centre, whilst the cavalry threatened both flanks—the dragoons under Captain Gall, the left, the Nizam's cavalry under Major Orr, the right. Baffled in their advance by the action of the 25th Bombay Native Infantry and the play of the British guns on their centre, the enemy made a rapid movement to their left, and attempted to turn the British right. But the dragoons, led by Gall, and the Nizam's cavalry led by Orr and Macdonald, Deputy Quartermaster-General of the force, charged them so vigorously that they retired into the fort, leaving 40 dead bodies of their companions on the field. On the British side 3 dragoons and 1 native trooper were wounded, a jamadar and a native trooper were killed. The fort was now invested, but the British force had to wait for their siege-guns, expected on the 24th. They arrived on the evening of that day ; the next morning they were placed in position.'[2]

Action at Dhar, 22nd October 1857.

During this action, as above related, the rebels' battery had

[1] Kaye and Malleson's *History of the Indian Mutiny*, vol. v. pp. 47-48. [2] *Ibid.*

Three guns captured by Captain Gall at Dhar.

been gallantly charged by a small body of the 14th Light Dragoons, led by Captain Gall and Cornet Giles, as well as by the 25th Bombay Native Infantry under Major Robertson. The 3 guns (brass 9-pounders of English manufacture) were captured and then turned on the enemy by the Native Infantry, who worked them, whilst our line advanced and the cavalry made their final charge. The enemy were thus driven from the outworks on all sides, pursued by the cavalry. Three standards were captured with the guns, and several acts of bravery were performed by the men of the Fourteenth, who behaved with great gallantry on this occasion. Sergeant G. Gardiner attacked, with only a few light dragoons, a party of the enemy who had fired at him from an ambush, killing several of them. A division of 'D' troop charged a body of the rebels who threatened an attack on the baggage, when Troop Sergeant-Major Grainger displayed great resolution and courage, cutting down two of the rebel horsemen, one of whom was pressing a light dragoon very hard. In this *mêlée*, Troop Sergeant-Major Grainger received a spear-wound in the wrist, and his horse was wounded by two sabre-cuts. Most of the rebels took refuge inside the fort and town, and it was not till the 31st, at night, that the fort of Dhar was actually captured. The garrison for the most part had cleverly eluded us and made good their escape that night, quitting the fort between 9 and 10 P.M., and going to the north-west, in the direction of Mundesor. This occurred at the time the breach was entered by our men, the rebels escaping by the main gate. The outlying picquet of the 3rd Hyderabad Cavalry had a skirmish with the rearguard of the retreating enemy, but the main body had passed it and the dragoons unobserved; and although the cavalry pursued when the alarm was given, the rebels had got too far away, and only a few stragglers were captured. It had unfortunately happened that the European picquet, which had been stationed near the spot where the garrison escaped for some days, and which knew the ground well, had been changed that morning, and when the men of the Fourteenth belonging

Sergeant Gardiner behaves gallantly.

Troop Sergeant-Major Grainger distinguishes himself in action.

Fort of Dhar captured, 31st October 1857.

to Captain Barrett's troop galloped up to where they heard firing going on, not knowing the ground well, several of them got hopelessly bogged in marshy ground. Moreover, the sowar sent by the jamadar[1] of the native picquet to give the alarm fell with his horse on the way, and was disabled.[2] A quantity of treasure and handsome elephant gear was found in the fort, and it was forwarded to Mhow under escort, the troops subsequently sharing in prize-money.

The Field Force, Hyderabad Contingent, joins the Malwa Field Force at Dhar, October 1857.

A large accession of troops came up for the British at this juncture in the shape of what was designated the Field Force, Hyderabad Contingent, consisting of troops of all arms sent by the Nizam. These were placed under the command of Major Orr, and did good service now with the Malwa Field Force and subsequently with the Central India Field Force under Sir Hugh Rose's command.

These combined forces continued the march through Western Malwa in the direction of Mundesor and Neemuch, on the track of the rebels. There was nothing heard now but incessant reports of mutinies everywhere, and repeated massacres of Europeans took place. The rebels from Dhar had attacked Mehidpoor cantonment on the 8th November, and Captain Mills of the Native Cavalry fell in a charge against the mutineers made by half a troop of loyal sowars. The commandant, Major Timmins, with the adjutant, Lieutenant Dysart, escaped and reached the camp of the Malwa Field Force on 9th November, escorted by the loyal half-troop of Native Cavalry. On the 19th November the force crossed the Chambal river and reached Hernia.[3]

Execution of Mehidpoor rebels.

Whilst encamped on the banks of the Chambal, a parade was ordered to witness the execution of about 70 of the Mehidpoor rebels who had been taken, after a severe fight with a large body of mutineers, by Major Orr and his Hyderabad

[1] Jamadar or jemadar.

[2] Lowe's *Central India.* Malleson, vol. v. p. 49; and Sylvester's *Central Indian Campaign*, p. 31.

[3] Lowe's *Campaign in Central India* gives an animated account of the troops crossing this river. See Malleson, vol. v. p. 52.

Cavalry, at the village of Rawal. A drumhead court-martial was held, and the rebels were shot. Owing to reports received of the siege of Neemuch by the rebels, the march of the Malwa Field Force was hurried on, and by the 21st November it encamped within 4 miles' distance of Mundesor.

Mundesor, 21st to 24th November 1857.

Whilst the men were at breakfast that day, an alarm was given that the enemy meant to attack. A strong force had previously been posted by us on the heights immediately to our front, which hid the city from our view, and Major Robertson was in command of it. Seeing that we did not mean to attack them that day, the enemy were emboldened to attack us, and sallied forth in great numbers, crossing a small river in their passage, and began by attacking Major Robertson's position.

Charge of Lieutenant Dew and picquet of the 14th Light Dragoons.

He opened his guns on them, and directed Lieutenant Dew of the 14th Light Dragoons to charge with the small body of cavalry he had: this he did most gallantly, and with a most successful result. Lieutenant Dew had only about 20 light dragoons of his outlying picquet with him, but he made a furious onslaught with these upon 300 footmen (Velaitees),[1] and the latter were so taken aback by this sudden attack that they made a hasty retreat to the river they had previously crossed in their advance.[2] They were, however, pursued by the cavalry to its brink, and many were shot or cut down in attempting to cross. This success of the cavalry had an excellent effect on the attacking force, as the rebels all retired and left us quiet for the rest of that day. It may fairly be said that this resolute charge of Lieutenant Dew's picquet was the main cause of the defeat of the enemy on this occasion, and contributed in no small degree to the success of Major Robertson's force in warding off from our camp the determined attack which would otherwise have been made upon it. Lieutenant Dew's personal bravery was so conspicuous that he was recommended, though unsuccessfully, for the coveted distinction

[1] Velaitees were matchlockmen.

[2] Lieutenant Dew was well supported by some of the Hyderabad Cavalry on this occasion.

of the Victoria Cross, and several men of the Fourteenth were brought to notice for their gallant conduct on the same occasion. Regimental Sergeant-Major Clark, who was severely wounded, did excellent work on the right flank, taking with him a party of skirmishers, following up the enemy and pursuing a considerable distance till they were finally driven off. Lieutenant Gowan was sent with a troop of the Fourteenth to support the skirmishers and cut off the rebels, and he reported very highly as to the admirable coolness and bravery of Private Buchanan when in presence of the enemy. It is believed 100 of the rebels were killed in this affair, and large numbers also were wounded.[1]

Regimental Sergeant-Major Clark and Private Buchanan, 14th Light Dragoons, distinguish themselves.

Next morning, 22nd November, the British force advanced, hoping to take Mundesor. The cavalry and artillery were in advance, and as we passed the large straggling town, surrounded by trees, a few shots reached our column. The rebel picquets fell in as we advanced, our long stream of baggage and native followers skirting round the town so as to reach the Neemuch road. About midday we halted in some jowarree fields, and the natives here told us that a large body of the rebels had left the town, but that the fort was still held by a garrison of 2000. Shortly afterwards reports came that a body of rebels was advancing from a distance on Mundesor, and to meet these the wing of the Fourteenth under Captain Gall, together with the Nizam's cavalry, were sent off. These crossed the river and dashed forward at a smashing pace across-country, through fields of standing grain,[2] but did not come in sight of the enemy for several miles, when they were seen making rapidly for the cover of the village of Peeplia. The Fourteenth and some of the Hyderabad Cavalry managed to get amongst the rebels' infantry, and cut up quite a hundred, pursuing them within matchlock range of the mud walls of the village, where they had a strong force under cover of the buildings; accordingly, as the cavalry had neither guns nor infantry with them, it

[1] Sylvester, p. 38, etc. [2] *Ibid.*

was decided to rejoin the main body, especially as two officers and some men were already wounded, and the horses much fatigued after a long gallop. During the retirement to camp, numbers of the enemy who had hidden themselves in trees and standing corn were killed by the cavalry. On the 23rd the force began to cross the Mundesor river, near which the camp lay on the preceding day, and whilst the baggage was being transported, an alarm was raised amongst the native followers. It soon appeared that the garrison of the fort had sallied forth and was advancing on our line of march. A troop of the Fourteenth was sent back across the river to endeavour to repulse this attack, which it apparently did most successfully, as the column now effected the passage of the river to the right bank, and then proceeded along the road towards Neemuch, 22 miles distant from Mundesor, for the relief of the Europeans there, who were in imminent danger, was the real object the Brigadier had in view. Soon, however, was heard a report of cannon to the right front, and an alarm was also raised from the rear that an attack was imminent on our rearguard and baggage, where Lieutenant Fenwick of the 25th Bombay Native Infantry was in charge. The guns heard to the front were from the village of Goraria, about 4 miles distant, where the rebel force which had come from Neemuch had taken up a strong position, 'their right resting on the village, their centre on a long hill, and their left well covered by fields of uncut grain, with broken ground and nullahs in their front, full of water and mud.'[1] To meet the attack from the Mundesor garrison on our rear a troop of the Fourteenth, with Lieutenants Leith, Redmayne, and another subaltern was immediately detached, and endeavoured to keep the enemy at bay for a time, as the rearguard had not actually been engaged, but was only threatened. The guns of our force meanwhile opened on the enemy in front: the 14th Light Dragoons under Captain Gall were on the right, the Hyderabad Cavalry under Major Orr

Battle of Goraria, 23rd November 1857.

[1] Malleson, vol. v. p. 54.

on the left, Hungerford's and Woolcombe's batteries formed our right centre, the Hyderabad bullock-battery the left centre, the 86th Regiment and 25th Bombay Native Infantry the centre, with the Hyderabad Infantry and Sappers on the left of the Nizam's battery. Our guns and the fire of the infantry—especially that from the Enfields of the 86th Regiment—soon weakened the fire of the enemy's field battery, which had been carefully placed near his centre, masked by date, palm, and other trees;[1] and behind it, sheltering the mutineer Mehidpoor Cavalry and rebel infantry, were some ruined huts on the edge of the village, whilst on the left of the village was a gharry road constructed in a deep cutting.

It was now that Lieutenant Martin placed himself at the head of 19 troopers of the Fourteenth, who were escort to Woolcombe's battery,[2] and with these gallant fellows charged across the deep cutting into the midst of the rebels' guns and actually took them, but so hot was the musketry-fire from the infantry posted in the huts in rear of the battery that the party had to retire, Lieutenant Martin being himself severely wounded. Captain Gall with a squadron of the Fourteenth, composed of 'B' and 'D' troops, quickly came to the rescue, making a gallant and most successful charge, and captured the battery of 5 guns. He cut down the gunners, and then pursued and cut up 200 of the rebel infantry. Not long after this the enemy retired, though slowly, still clinging tenaciously to the village of Goraria, which they continued to occupy when night fell.

Captain Gall captures 5 guns, 23rd November 1857.

Malleson states that the British loss was considerable on this occasion, amounting to not less than 60 killed and wounded. The enemy lost on a much larger scale, especially in the outskirts and in the surrounding country, for our cavalry pursued round the village in every direction and cut up hundreds of them.

Captain Gall went to the right with the Fourteenth, Major

[1] Sylvester, p. 40.

[2] This battery opened fire at 900 yards' range, and then made a forward movement to its right to enfilade the enemy's line. Sylvester, page 41.

Orr with the Hyderabad Cavalry went to the left. They found a large quantity of loot and plunder on the rebels, which the latter had brought from Neemuch, where they had raised the siege and were hoping to rescue Ferozshah and the garrison of Mundesor, when they were thus so opportunely caught and defeated at the right moment by our column at Goraria.

To return now to the movements that had been going on in rear of the column whilst this fight had been carried on in front. A party of infantry, some Hyderabad Cavalry under Captains Abbott and Murray, as well as 2 guns, had been sent back to the rearguard to reinforce the troops of the Fourteenth under Lieutenant Leith already there. The baggage and rearguard had been attacked by a large force, about 2000, from the Mundesor garrison. Our guns immediately opened fire on the rebels, the cavalry charged—both the Fourteenth and the Hyderabad Horse—and drove them back, pursuing and sabring a large number. The pursuit was continued up to a point where a small pond of water and some shallow pits or stone quarries joined : here the enemy drew on our cavalry into broken ground, and Lieutenant Redmayne, who was leading in front, followed by a few men, wheeled round the pond, and was shot down in a tremendous volley fired by some men hidden in the gravel quarries.[1] Poor Redmayne met a soldier's death. His body was mercilessly cut up by the rebels, his charger and accoutrements instantly carried off, and several of his men belonging to the Fourteenth were wounded at the same time.[2] Private O'Neill behaved most gallantly in this affair, and was shot through the chest by a jingal-ball of great size; he lived and recovered from his wound. At this critical juncture Captain Abbott luckily arrived by a different route, on the other side of the water and gravel pits, which caused the enemy to retire within the fort.

Death of Lieutenant Redmayne.

[1] Sylvester, pp. 43, 44.

[2] The charger was subsequently recovered at the capture of Rathgur Fort, 28th January 1858, by Sir Hugh Rose, who bought the charger. See page 276.

During the night our cavalry surrounded Goraria, and the remainder of the force encamped, the enemy keeping up a continuous matchlock fire even after darkness set in.

On the morning of the 24th our heavy guns were brought to bear on the village, and at last, about noon, some 200 of the rebels, the Velaitees, came out under a flag of truce and surrendered themselves prisoners. Those that still remained inside Goraria were the brave Rohillas, and they stuck to the last brick in the place.[1] At about 4 P.M. the 86th Regiment and the 25th Bombay Native Infantry stormed the village at the point of the bayonet, when those who rushed out of the houses were cut down by the cavalry, and several hand-to-hand fights took place in the sugar-cane plantations outside.[2] The village was set on fire, and the Madras Sappers and Miners fought nobly with the rebels amongst the fire and smoke.

Operations at Goraria, 24th and 25th November 1857.

On the morning of the 25th November not a living man remained in Goraria. Neemuch was now successfully relieved, and Captain Mayne was able to ride in safely, returning there with an escort of loyal sowars.

Malleson says:—'The stern defence of the Rohillas did service to their cause. Whilst the British force was dealing with them, the Sháhzáda and his 2000 Afghans and Mekránis, as panic-stricken as they had been bold, evacuated Mundesor and retreated on Nangárh, whither the cavalry, worn out by four days of unremitting exertion, was unable to pursue them. They [the rebels] fled through the country, endeavouring to seek refuge in the jungles.' Some of these were afterwards killed by loyal natives at Partabgharh, in Rajputana, and the remnant escaped across the Chambal river towards the east. Mundesor having thus fallen into our hands, the fort was subsequently dismantled and the guns destroyed. Thus our troops had been most successfully engaged in the western district of Malwa; Dhar, Neemuch, and Mundesor were cleared of rebels, and the Malwa Field Force had finished its appointed work. The column was henceforth to be designated the 1st

[1] Malleson, vol. v. p. 55. [2] Sylvester, p. 46.

Brigade of the Nerbudda Field Force under the same brigadier, Colonel C. S. Stuart, Bombay army, and still accompanied by the Officiating Agent to the Governor-General, Lieutenant-Colonel Durand. The march of this brigade was accordingly continued to Indore, passing through Mehidpoor on the banks of the Sipra in Western Malwa, and itself the scene of a recent mutiny, where Major Timmins had been defeated by the mutinous troops and several European officers and sergeants murdered, the whole native garrison, with few exceptions, going over to the rebels. After passing through Oujain,[1] another town on the Sipra, Indore, Holkar's capital, was reached about 14th December.

Here Holkar's disaffected regular cavalry were disarmed in the presence of Lieutenant-Colonel Durand and the Nerbudda Field Force, the disarmed soldiers being placed under the care of the Sikh cavalry of the late Bhopal contingent. One thousand six hundred men of Holkar's infantry were also disarmed the same evening by the Maharajah's chief minister at the request of Lieutenant-Colonel Durand. On the following day, 15th December, Sir Robert Hamilton, Agent to the Governor-General, arrived and relieved Lieutenant-Colonel Durand, and at the same time the 1st Brigade occupied the cantonments.

Sir Hugh Rose takes command of the army in Central India, 17th December 1857.

Major-General Sir Hugh Rose, K.C.B.,[2] having on 17th December assumed command of the forces in Central India, they were designated the Central India Field Force and the Hyderabad Contingent Field Force, the former consisting of the two Brigades of the late Nerbudda Field Force, under the same Brigadiers,[3] and Major Orr being commandant of the Nizam's Contingent. Sir Hugh, shortly after assuming command, ordered the 1st Brigade from Indore to Mhow, where it remained till 30th December. 'Long after these events,' says Sylvester, 'the despatches appeared in which Lieutenant

[1] Or Ujjain.

[2] Sir Hugh Rose marched from Poona to Mhow in company with 'G' troop of the Fourteenth under Captain W. McMahon, who brought up some horses left there sick by other portions of the regiment when they started on service.

[3] Colonel C. S. Stuart, 1st Brigade, and Colonel Charles Steuart (14th Light Dragoons), 2nd Brigade.

Martin's gallantry was noticed, and he well deserved it. Brigadier Stuart was made a Companion of the Bath, Lieutenant-Colonel Durand became colonel, and Captains Gall, Robertson, Woolcombe, and Hungerford, received the rank of brevet-major for their services.'

This brings to an end the campaign of 1857, in which the left wing of the Fourteenth under Major Gall played such a prominent part whilst with the Deccan Field Force, and afterwards in the 1st Brigade of the Malwa and Nerbudda Field Forces, commanded by Brigadier C. S. Stuart of the Bombay army.

Movements of the right wing, 14th Light Dragoons.

The right wing of the Fourteenth had meanwhile marched from Aurungabad and the Deccan under Major Scudamore, on 31st October, in order to reach Sehore, a town in the Bhopal state, Central India, where a mutiny had broken out amongst the Bhopal contingent. They marched by way of Asseerghur and Hoosingabad, Lieutenant Travers acting as staff officer. The town at Sehore is a large one, and is situated 22 miles to the west of Bhopal. After a long march Scudamore's wing reached Sehore the 23rd November, and formed part of the 2nd Brigade of the Nerbudda Field Force under Brigadier Charles Steuart. On the 17th December this wing became a part of the 2nd Brigade, Central India Field Force, commanded by Major-General Sir Hugh Rose, K.C.B., as already related.[1]

Active preparations for the coming campaign in Central India were carried on during the remaining few days of 1857, both at Mhow and Sehore: the siege-train was hurried on to completion, and the transport improved, while camels in abundance were easily procured in the surrounding districts of Malwa and Rajputana.

The small portion of the Fourteenth, representing the depôt or headquarters of the service troops, remained at Kirkee till the end of the year, under command of Colonel C. P. Ainslie, and on 29th November sent some detachments from Kirkee to Sattara in the southern part of the Bombay

[1] See page 268.

Presidency, where they were required for temporary duty. During 1857 the depôt troop in England remained under the commandant of the Maidstone Cavalry Depôt, Colonel C. M. Balders.

Establishment augmented from 9 to 10 troops.

The establishment of the Fourteenth had been augmented in June from 9 to 10 troops, and consisted of—

59 Sergeants.	40 Corporals.
14 Trumpeters.	626 Privates.
10 Farriers.	703 Troop-horses.

1858

This proved a very memorable and eventful year in the history of the Fourteenth, owing to a succession of most brilliant services performed by them in Central India, Gwalior, Bundelcund, and the North-west Provinces, during the suppression of the Indian Mutiny. 'The Fighting Fourteenth,' a sobriquet gained nearly fifty years earlier in the Peninsular campaign, was most amply justified in this campaign by the successors of the men who gained it; and these by their gallantry and valour, displayed in many an action, siege, and pursuit in which they took part, during these operations conducted against the rebels, have added to the long list of honourable distinctions already earned by the regiment yet another name—that of 'Central India.'[1] During the whole year of 1858 and for several months in 1859 the Fourteenth were continuously in the field, either in brigade or movable column, hunting up the rebels, and most of these services were performed under the command of those two able leaders, Major-General Sir Hugh Rose, K.C.B.,[2] and Major-General Sir Robert C. Napier, K.C.B.[3]

Central India Field Force, January 1858.

The Central India Field Force consisted of 2 brigades, as follows :—

[1] 'Central India' is borne on the appointments of three other cavalry regiments, viz. 8th Hussars, 12th Lancers, and 17th Lancers.

[2] Afterwards Field-Marshal Lord Strathnairn, G.C.B., G.C.S.I.

[3] Afterwards Field-Marshal Lord Napier of Magdala, G.C.B., G.C.I.E.

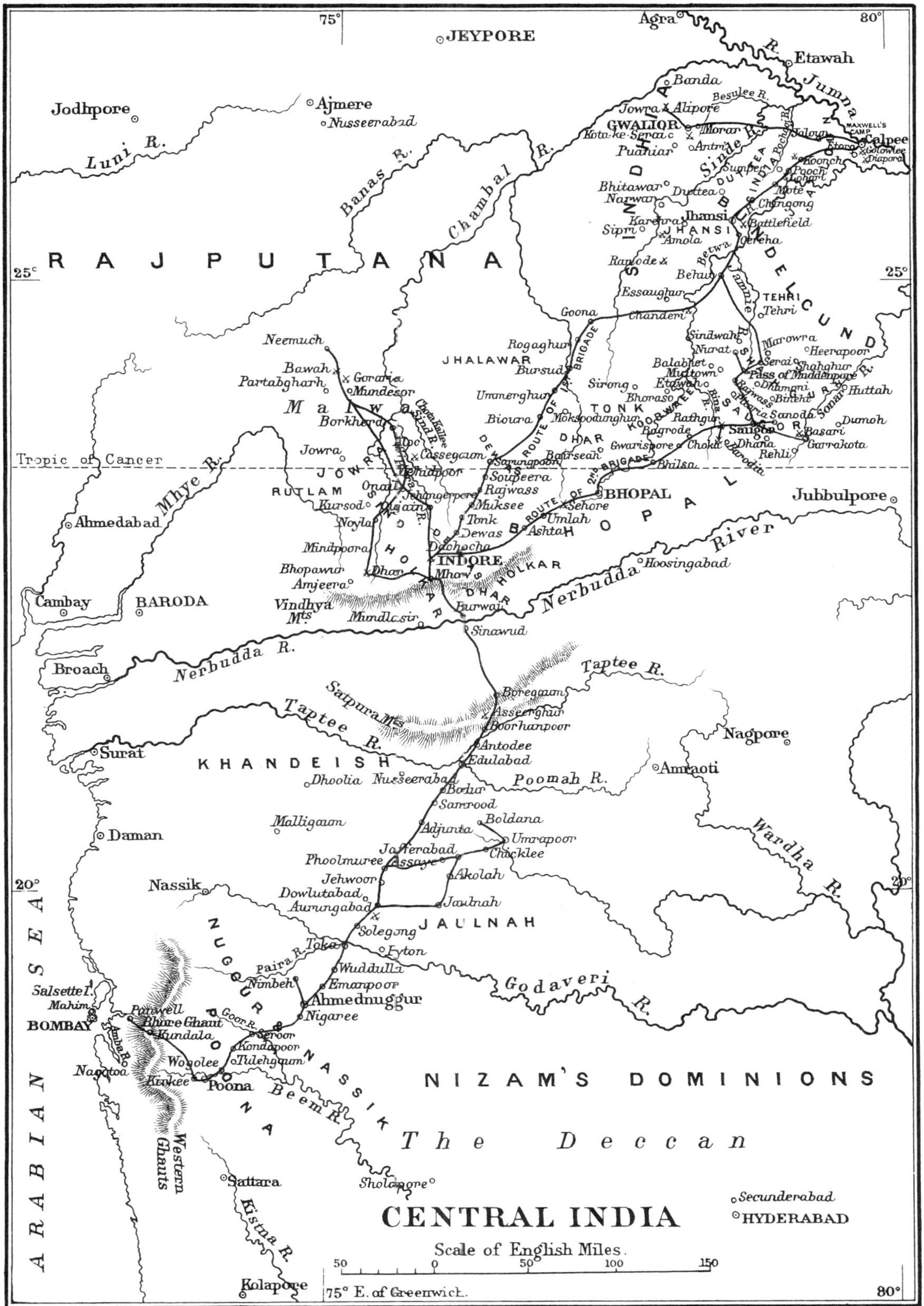

To Bangalore 330 miles

Stanford's Geog! Estab! London.

London: Longmans, Green & Co.

At Mhow, 1st Brigade, commanded by Colonel C. S. Stuart, Bombay Army, as Brigadier, consisting of—

Left wing, 14th Light Dragoons under Major Gall.
1 Troop 3rd Bombay Light Cavalry.
2 Regiments of cavalry, Hyderabad Contingent.
2 Companies 86th Regiment (Royal County Down).[1]
25th Regiment Bombay Native Infantry.
1 Regiment of infantry, Hyderabad Contingent.
3 Light Field Batteries { 1 Royal Artillery. 1 Bombay Artillery. 1 Hyderabad Contingent.
Detachment of sappers.

At Sehore, 2nd Brigade, commanded by Colonel C. Steuart, 14th Light Dragoons, as Brigadier, consisting of—

Right wing and headquarters 14th Light Dragoons, under Major Scudamore.
Headquarters 3rd Bombay Light Cavalry.
1 Regiment of cavalry, Hyderabad Contingent.
3rd Bombay European Regiment.[2]
24th Bombay Native Infantry.
1 Regiment of infantry, Hyderabad Contingent.
1 Battery Horse Artillery.
1 Light Field Battery.
1 Battery Bhopal Artillery.
1 Company Madras Sappers.
Detachment of Bombay Sappers.
Siege-train, with guns worked in action by drafts from field batteries.

The recent hard work and privations to which most of these troops had been exposed during their services in the Malwa and Nerbudda campaigns necessitated some repose before starting on a fresh campaign. We have recounted what they

[1] The remainder of the 86th Regiment joined this brigade later, on 16th March, the day before the attack and capture of Chanderi.

[2] Now the 2nd Battalion of the Prince of Wales's Leinster Regiment (Royal Canadians), late 109th Regiment.

did at Aurungabad, Boldana (in Berar), Dhar, Mundesor, Goraria, and Mehidpoor; it now remains to record their gallant deeds in Central India.

Malleson bears ample testimony to the good services of the Fourteenth, and particularly of that gallant officer, Major R. H. Gall, afterwards Major-General Gall, C.B., who commanded the left wing of the regiment at this period.[1] He says, when referring to the Malwa campaign and operations carried on towards the close of the year 1857:—'Many officers distinguished themselves in this campaign. One of these, who for his daring, his gallantry, and his brain power, was especially noticed by Colonel Durand, requires mention here. "Much of the success in quelling this insurrection," wrote Durand[2] to Lord Canning at the end of November 1857, "is due to the judicious daring, the thorough gallantry with which, whenever opportunity offered, Major Gall, his officers and men, sought close conflict with the enemy—a bold one, who often fought most desperately. I feel it a duty to Major Gall and Her Majesty's 14th Light Dragoons, men and officers, thus especially to beg your lordship's influence in favour of officers and men who have merited, by conspicuous valour, everything that Her Majesty's Government may be pleased to confer. They deserve most highly."'

Malleson's and Durand's tribute to Major Gall and the 14th Light Dragoons.

After a rest of three weeks at Mhow and Indore, Major-General Sir Hugh Rose, K.C.B., took the field with the Central India Field Force and the Hyderabad Contingent Field Force. He left Mhow on 6th January escorted by a troop of the Fourteenth and some Hyderabad Contingent Artillery, and accompanied by the Agent to the Governor-General at Indore, Sir Robert Hamilton, who was the political officer for the whole of Central India. On arriving at Sehore on 8th January, the Major-General found there the 2nd Brigade of the Central India Field Force, under command

[1] *History of the Indian Mutiny*, by Kaye and Malleson (Longmans, Green, and Co., 1898), from which work many quotations have been made in these records. See vol. v. p. 59.

[2] Colonel Henry Marion Durand, R.E., Officiating Agent to the Governor-General for Central India.

of Brigadier Charles Steuart, C.B., 14th Light Dragoons, in which was the right wing of the Fourteenth under command of Major Scudamore. Brigadier Steuart had just been awarded the Companionship of the Order of the Bath for his services in Persia in 1857.

On 15th January, Lieutenants Leith and Dew with 'E' troop of the 14th Light Dragoons arrived at Sehore from Mhow, escorting the much-needed siege-train for the 2nd Brigade. Sir Hugh Rose on arrival at Sehore lost no time in bringing the mutineers of the Bhopal contingent to punishment. There were about 150 of them, who were put to death after having been found guilty by a drumhead court-martial. They were all shot. Stent, in his *Scraps from my Sabretache*, relates how that the number to be shot was 149, but that after the execution 150 bodies were counted, and he accounts for it by saying that 'a brother of one of the prisoners came to see the last of him, and in the *mêlée* must have shared his brother's fate, whether by accident, or purposely, no one knew.' It is said by the same authority, that as the rebels knew our words of command perfectly, when the word 'Fire!' was given many threw themselves down uninjured on the ground, the shots passing harmlessly over them, and these had to be afterwards despatched by men placed near for the purpose with muskets and carbines.

Bhopal mutineers shot.

The 2nd Brigade and the Hyderabad troops left Sehore on 16th January. They had been reinforced by 800 men of Bhopal levies, contributed by the loyal Begum of that principality, and all marched for the relief of Saugor, where the garrison was besieged by rebels; but Rathgur had first to be reduced.

Malleson says:—'The 1st Brigade left Mhow on the 10th January, and then marched in a line parallel with the 2nd Brigade upon Chánderi, a very famous fortress in the territories of Sindhiá.' The operations of the 1st Brigade will be referred to further on.[1] We will now follow the fortunes of the 2nd Brigade, with regard to which Malleson[2] writes:—'Rathgur,[3] distant only 25 miles from Saugor, is situated on the spur of a

[1] See page 280. [2] Vol. v. pp. 95, 96. [3] Or Ratghur.

Fort of Rathgur besieged, 24th January 1858.

long high hill and commands the country surrounding it. Near its base runs a deep and rapid river, the Bíná. Altogether it was a most formidable position. Sir Hugh Rose arrived before this place on the morning of the 24th January. He at once, with small loss, drove the enemy from the outside positions they had occupied in the town and on the banks of the river, and then completely invested the place. Fronting the eastern face he posted the Bhopal levies, facing the northern the 3rd Bombay Light Cavalry and the Hyderabad Contingent Cavalry. With the remainder of the force he occupied the plain across which runs the road to Saugor. He then reconnoitred the ground preparatory to selecting sites for his breaching batteries. Early on the morning of the 26th Sir Hugh Rose made a move forward. Crossing the Saugor road with the 3rd Europeans, followed by the 18-pounders, howitzers and mortars, and the guns of the Hyderabad Contingent, he entered the jungle. But no sooner had these troops reached a point well within its thick covering than the enemy, who had been working near, fired the jungle-grass on all sides. For a few moments the position was perilous, but Sir Hugh, turning back beyond the range of the flames, sent his sappers to cut a road for the guns up the height to the north of the town. This operation and the bringing up of the guns occupied the greater part of the day. Meanwhile the remainder of the force had occupied the town of Rathgur, and driven the enemy within the fort.'

During the 26th, 27th, and 28th, Sir Hugh besieged the fort with his mortar battery and other artillery, whilst the 3rd Europeans employed their Enfield rifles to suppress the matchlock fire of the enemy. By 10 P.M. on 28th, a large breach had been made in the fort. That same night, as related by Malleson[1]: 'The Rajah of Bánpúr advanced on the rear of the besieging force with a considerable body of revolted Sepoys and other levies. He came on with great boldness, his standards flying and his men singing their national hymns. Instead of ceasing his fire against the fort Sir Hugh Rose

[1] Vol. v. p. 97.

redoubled it, but at the same time detached a small force, consisting of a detachment of the 14th Light Dragoons, the 3rd Bombay Light Cavalry, the Horse Artillery, and the 5th Hyderabad Infantry, to deal with the Rajah of Bánpúr and his followers. They did not wait to be charged, but throwing away their arms and ammunition, made off with such celerity that, though hotly pursued, a few only were cut up.'

After this relieving party of the rebels had been driven off those within the fort apparently lost heart. The garrison was said to number from 500 to 600 men—Velaitees, Mekranies, and Bundeelas, all fighting-men from their boyhood; but they silently evacuated the fort during the night. When Sir Hugh discovered the precipitate flight of the garrison he gave orders for a pursuit, but the rebels had gone too far, and no great results came of it. About noon on the 30th January information came that 'the Rajah of Bánpúr, reinforced by the garrison, had taken up a position near the village of Barodiá, about 15 miles distant.'[1] Sir Hugh, with a considerable portion of his force of all arms, including the greater part of his cavalry (3 troops Fourteenth and a squadron of Bombay Cavalry), went in pursuit. 'About four o'clock in the afternoon he came upon them posted on the banks of the Bíná, and prepared to dispute his passage. Sir Hugh at once attacked, and, though the rebels fought well, he forced the passage of the river. The country on the other side was thick and bushy, and the rebels took every advantage of it. From the river to Barodiá, Sir Hugh had to fight his way step by step. He did not do this without loss.'[2] Captain Neville, R.E.,[3] was killed, and numerous other casualties of officers and men occurred in killed and wounded. The Rajah himself escaped, but his followers were completely defeated, and the column returned to Rathgur about 2 o'clock on the morning of the 31st January. Malleson remarks: 'The fall of Rathgur had effected two most important objects. It had cleared the country south

The fort evacuated, 28th January 1858.

Engagement at Barodia, 30th January 1858. Three troops of the 14th Light Dragoons present.

[1] Malleson, vol. v. p. 98 *et seq.* [2] *Ibid.*

[3] This officer had only just joined the force. He had served with distinction in the Crimean campaign.

of Saugor of rebels, had re-opened the road to Indore, and had made it possible for the General to march to the relief of Saugor, now beleaguered for nearly eight months.'[1]

Amongst the rebel chiefs who fell into the hands of the British at the capture of Rathgur was Mahomed Fazil Khan, one of the Delhi royal family. He was captured, when hiding, by a native servant in the employment of Captain Need, 14th Light Dragoons. This fortunate native servant obtained a handsome pecuniary reward which had been set on Mahomed Fazil Khan's head by the Government, and the luckless chieftain was doomed to be hanged over the gate of the fort, where his body was seen suspended by the side of Khamdar Khan, a former adherent of the British in Kolara, who had recently thrown in his lot with the mutineers. The charger of Lieutenant Redmayne, 14th Light Dragoons, who fell in action at Mundesor on 23rd November 1857, was recovered here when the British captured the fort.[2] The animal had received a severe shell-wound over the eye, and was purchased by Sir Hugh Rose. The Shahzadah of Mundesor took possession of it when Lieutenant Redmayne fell, and had brought it to Rathgur, but he abandoned it in his flight.

Our cavalry pursued the fugitive garrison from Rathgur, and succeeded in cutting up several chieftains as well as about 70 rebels, and nearly 100 followers were taken prisoners.[3] 'The standards taken here, as at Mundesor, were marked with the crescent and bloody hand. Before leaving Rathgur the fortifications and defences were laid waste by the Engineers, and Sir Hugh Rose led his troops in triumph to Saugor.'[4] The 31st Native Infantry had remained loyal, so the houses and property in the cantonments of Saugor were intact. Stent remarks:—'As we passed under the walls of the fort we were greeted by the ladies, who thronged the battlements, with the waving of hands and handkerchiefs (I will not be positive that they did not even cheer us), and we were proud

Saugor relieved, 3rd February 1858.

[1] Malleson has 'Sagár,' 'Indúr,' 'Rátgarh' or 'Ráhatgarh.'

[2] See p. 266 *n*. [3] Sylvester, pp. 60-62. [4] Sylvester.

to think that our timely arrival had saved them from the clutches of the rebels. Saugor had been from six to seven months beleaguered, and the poor creatures shut up there had heard with intense delight the pounding of our guns at Rathgur, which is only 22 miles distant.'[1] The relief of Saugor took place on the 3rd of February, and the column rested there several days. On the 8th a small force was sent under command of Captain Hare, Hyderabad Contingent, to destroy the fort of Sanoda. On the 9th, Sir Hugh Rose marched with his troops towards the fort of Garrakota,[2] standing on an elevated plateau 25 miles east of Saugor, with the wide river Sonar flowing past it. According to Malleson the fortifications were strong. He says: 'It was held by the revolted Sipáhís (Sepoys) of the 51st and 52nd Native Infantry and other rebels, well supplied with ammunition.'[3]

'G' troop of the Fourteenth, under command of Captain William McMahon, together with 2 companies of the 24th Bombay Native Infantry, was left for the time in Saugor to protect the station when the remainder of the Central India Field Force marched on Garrakota.

Capture of the fort of Garrakota, 13th February 1858.

Sir Hugh Rose with his force arrived in sight of Garrakota on the afternoon of the 11th, and made a thorough reconnaissance of the place. He drove in the rebels from the village of Basári,[4] where they occupied a position, and next day commenced his attack. A steady fire from our guns and mortars was kept up all day: this silenced the enemy's guns, but under cover of darkness the garrison slipped away by the Paunch Ghat towards Dumoh and made good their escape, as, owing to the smallness of our besieging force, it was unable to guard this part of the fort. As soon as the flight of the garrison was reported, early on the morning of the 13th, Captain Hare with the Hyderabad Cavalry, 2 Horse Artillery guns under Lieutenant Crowe, and 2 troops Fourteenth under Captain R. J. Brown and Captain Arthur

[1] *Scraps from my Sabretache*, p. 188.

[2] Garhákóta (Malleson).

[3] Vol. v. p. 99.

[4] Or Bassaree.

Need, followed in pursuit. They came up with the rebels at the Biás[1] river, a tributary of the Sonar, near the village of Biás. The cavalry and guns crossed the river, and the latter opened fire, then the cavalry charged and pursued for a considerable distance. Captain Need highly distinguished himself by his bravery and swordsmanship. The cavalry slew nearly 100 men, and of these Captain Need himself killed 5.[2] The 2 troops of the Fourteenth engaged were 'A' and 'K' troops, the captains of which were Need and Brown, with Lieutenants Leith and Dew. In Captain Hare's report of the affair he speaks of Captain Need, 14th Light Dragoons, as 'a good and dashing cavalry officer,' and he specially mentions in terms of the highest praise the gallant conduct of Captain Need and his troop, adding, 'Captain Need pursued with his gallant troop until dark.' Sir Hugh Rose recommended Captain Need to the favourable consideration of His Excellency the Commander-in-Chief in India for his services on this occasion.

Gallant charge of the 'A' and 'K' troops under Captains Need and Brown at the Biás river, near Garrakota, 13th February 1858.

A large quantity of supplies fell into our hands at Garrakota, as the place was well stored with provisions of grain, flour, etc. There was also a collection of all sorts of loot, evidently derived from English sources, and plundered by the natives from the houses of the English. Sir Hugh had a portion of the fortress destroyed under direction of Major Boileau and the Madras sappers and miners, and then marched back to Saugor, arriving there 15th February. Captain McMahon's troop of the 14th Light Dragoons then rejoined the right wing of the regiment under Major Scudamore.

Malleson[3] says:—'Jhánsí, 125 miles distant to the north, was the next point to be aimed at. But between Saugor and Jhánsí lay the passes of Malthone and Muddenpore, the forts of Suráhi and of Maráura, the towns of Sháhgarh and Bánpúr.[4] After overcoming the obstacles which these places

[1] Or Beas. [2] See Appendix B, p. 592, etc. [3] Vol. v. p. 100.

[4] 'Maráura lies 37 miles north of Ságar, and 22 west by north of Sháhgárh. Sháhgárh lies 40 miles north-east of Ságar. Bánpúr is in the Lálitpúr district' (Malleson).

would probably offer, Sir Hugh would have, before marching on Jhánsí, to effect a junction with his 1st Brigade under Brigadier Stuart. He could scarcely move from Saugor until he should hear that Brigadier Whitlock's column had started from Jubbulpore[1] for that place. Meanwhile he would have time to repair damages and to store supplies.

'At length news came that Whitlock had left Jubbulpore. Sir Hugh's preparations were now as complete as they could be made.' He had, according to Malleson, caused to be collected large supplies of sheep, goats, oxen, grain, flour, tea, soda-water, an additional supply of elephants, and summer clothing for his European soldiers. In addition he had transferred the sick and wounded to the Saugor Field Hospital, and re-supplied the siege-train with ammunition, strengthening it by the addition of heavy guns, howitzers, and large mortars from the Saugor arsenal. On the 26th February a start was made; on the 27th the fort of Barodiá was shelled and taken. On 3rd March the pass of Malthone was sighted, but it was resolved to force the pass of Muddenpore, making simultaneously a feint on Malthone Pass, which was fortified and held in force by the rebels. For this purpose Major Scudamore, with a squadron of the Fourteenth, some guns, some native infantry and other details was detached, whilst with the main body of his force Sir Hugh moved rapidly on the pass of Muddenpore. 'He soon came under the fire of the defenders, supposed to number 9000 men, chiefly composed of Bundeelas and Velaitees, as well as some Bengal Sepoys.'[2] For a time the British advance was checked and the fire was so heavy that their guns had to be retired some distance, and Sir Hugh's horse was shot under him. Later on the superior tactics of the British prevailed, and a determined bayonet charge by the 3rd Europeans,[3] aided by the Hyderabad Infantry, completely carried the day. The enemy fled in disorder into the town, but our howitzers drove them out of that, and they

Barodiá fort captured, 27th February 1858.

Muddenpore Pass forced, 4th March 1858.

[1] Jabalpúr (Malleson).

[2] Sylvester, p. 67.

[3] Now the 2nd Battalion Leinster Regiment.

fled to the jungles pursued by the cavalry, amongst which were 3 troops of the Fourteenth, who drove them up to the fort of Sarálic. Major Scudamore received the thanks of the Major-General for the able and successful manner in which he had conducted the feint on Malthone.

'The effect of this victory,' writes Malleson,[1] 'was very great. It so daunted the rebels that they evacuated without a blow the formidable pass of Malthone, the fort of Narhat to the rear of it, the little fort of Saráhí, the strong fort of Maráura, the fortified castle of Bánpúr (the residence of the rebel Rájah called after it), and the almost impregnable fortress of Tal-Bahat on the heights above the lake of that name. They abandoned also the line of the Bíná and the Betwá, with the exception of the fortress of Chandérí, on the left bank of the latter river.'

Operations of the 1st Brigade, Central India Field Force.

It will now be necessary to return for a time to the operations of the 1st Brigade, Central India Field Force,[2] which we left at Mhow in January of this year, under command of Brigadier Stuart, Bombay army, and in which Major Gall with the left wing of the Fourteenth was serving. Stuart left Mhow on the 10th January, and marched along the Agra Trunk road as far as Goona. About 70 miles to the east of Goona are situated the fort and important town of Chanderi. The former is situated on a lofty hill, and during the month of February 1858 numbers of the rebel Sepoys already defeated by the 2nd Brigade of the Central India Field Force, as above related, flocked thither for refuge, and with a firm determination to make a last stand there against the British troops. On the 5th of March, Stuart arrived in the vicinity of Chanderi, and the men encamped near a small village 6 miles short of the town, in order to give time for a thorough reconnaissance of the position. A party of the 14th Light Dragoons and Irregular Horse,[3] accompanied by Major Gall of the Fourteenth, Captain Fenwick, R.E., Major Keatinge[4] of the Bombay Artillery, the political agent, and

Chanderi attacked by 1st Brigade, 5th March 1858.

[1] Vol. v. p. 103.
[2] See page 273.
[3] Sylvester.
[4] Afterwards General Keatinge, V.C.

Assistant-Surgeon J. H. Sylvester,[1] attached to the 14th (King's) Light Dragoons, were sent forward through the dense jungle to reconnoitre. After proceeding a considerable distance the party was fired upon by a volley of musketry, and as it would have been the height of folly to proceed further, the presence of the enemy having been ascertained, the party returned to camp.

Next day, 6th March, with a strong advance-guard under Major Robertson, Bombay army, we advanced on this once famous place. It is said that Chanderi[2] was very important in the prosperous days of the Moghul Empire. Malleson says there was a proverb in the time of Akbar, 'If you want to see a town whose houses are palaces, visit Chanderi.' It has been described as being in those days a city possessing 14,000 houses built of stone, 384 markets, 360 caravansaries, and 12,000 mosques. The oppression of the Mahrattas subsequently brought it to a very different state, added to which its manufactures had suffered from competition with Manchester. Although its former splendour had departed, much that was picturesque remained at the time it fell into our hands.

There was a good deal of fighting outside the fort on the 6th as our advance-guard moved forward. The infantry advanced in skirmishing order, 2 companies 86th Regiment and 25th Bombay Native Infantry, whilst the artillery opened with round-shot and shell, driving the rebels from the outlying ruins and summer-houses, out of which they had been firing upon us as we passed through an intervening gorge. The enemy then took shelter behind a loop-holed wall, which defended the fort and town at the point where we advanced to the attack. After a while this position was taken by our men, and the enemy retired within the town and fort, half a mile distant. Our brigade subsequently encamped on one of the hills which commanded the fort on the west side.

For the next few days our men were employed in clearing

[1] Mr. Sylvester was in medical charge of the left wing of the Fourteenth throughout this campaign. He belonged to the Indian Medical Service, and served later in Beatson's Horse and Mayne's Horse (2nd Regiment).

[2] Chanderi is in the Gwalior State.

out the surrounding villages, in reconnoitring, and in placing guns in favourable positions. The cavalry were kept busy reconnoitring, but the ground was not suitable for horsemen to act in. The 24-pounder guns were dragged up by elephants on the 10th March, and by the 13th of that month the breaching batteries commenced to fire. On the next day a breach was effected, and on the 17th the fort was stormed by the men of the 86th Regiment (now the 2nd Battalion Royal Irish Rifles, formerly the County Down Regiment), who had been augmented by the arrival of their other wing the day before, as well as the 25th Bombay Native Infantry. The assault was eminently successful: the rebels for the most part hurled themselves over the parapets and made a hasty retreat. Our cavalry was not numerous enough to prevent the escape of the garrison, as the fort and town were so extensive and of such great size. A magazine exploded during the capture, by which several men of the 86th Regiment were killed and others badly injured and burned. Most of the enemy effected their escape through the town; any that remained were shot down or bayonetted. The 3 troops of the Fourteenth which were present under Major Gall had been chiefly employed in patrolling and reconnoitring, but the country at Chanderi was not in any way suitable for the action of cavalry. On the night before the assault Lieutenant Gowan with 'C' troop of the Fourteenth was posted in a selected position, and at the signal for the assault at daybreak on the morning of the 17th, made a very successful feint by firing some rounds of blank ammunition so as to draw away the attention of the garrison from the real point of attack. This duty was most satisfactorily carried out, and Lieutenant Gowan and his troop received the thanks of the Brigadier for its complete success. The fort of Chanderi alone was 4 miles in circumference, and occupied a very commanding situation, with a fine view over the surrounding country. All the guns as well as stores of grain and salt fell into our hands, but nothing of value, and only about 100 of the rebels were killed.[1] The British loss was 29, including 2 officers.

Chanderi taken, 17th March 1858.

Lieutenant Gowan with 'C' troop, 14th Light Dragoons, makes a successful feint at the storming of the fort of Chanderi.

[1] Sylvester.

On the 15th, Lieutenant Dowker and 30 sowars of the Hyderabad Contingent Cavalry had arrived with despatches for the Brigadier from Major-General Sir Hugh Rose, who was anxiously looking for our junction with him and the 2nd Brigade in front of Jhansi.

As it was the 17th March, the bands all played 'St. Patrick's Day' whilst the troops marched through the now deserted town to their camp, and of course the men of the 86th (Royal County Down) Regiment were the heroes of the hour. As soon as the fort had been dismantled the 1st Brigade moved away to join Sir Hugh Rose in the vicinity of Jhansi.

On 19th March the left wing of the Fourteenth was ordered from Chanderi to Jhansi to join the right wing and headquarters of the regiment, which were with the 2nd Brigade, and had arrived there on the 20th inst. Accordingly Major Gall made all haste, and by forced marches his 3 troops covered 70 miles in 2 days, arriving in front of Jhansi on the 21st March, in good time to take part with the rest of the regiment in the investment and ultimate capture of that important city and fortress.

According to Malleson, both Lord Canning, the Governor- Jhansi. General of India, and Lord Elphinstone, the Governor of Bombay, attached the very greatest importance to the fall of Jhansi. Jhansi was regarded as the stronghold of rebel power in Central India, the main strength of the formidable rebel force on the Jumna. Here, too, English men and women about 9 months earlier had been slaughtered under circumstances of peculiar atrocity.

It was on the 20th March that Brigadier Steuart with the cavalry and artillery of the 2nd Brigade of the Central India Field Force had arrived and invested Jhansi. With this brigade were 5 troops of the Fourteenth, being the right wing of the regiment, under Major Scudamore's command, numbering 325 rank and file; and it was on the following day that the other 3 troops of the regiment, composing the left wing, under Major Gall, numbering about 200 rank and file, also arrived. On the 22nd March the city and fortress were completely invested by our cavalry.[1] When Sir Hugh Rose arrived in front of Jhansi with the 2nd Brigade, the latter was halted for a time in the plains at some distance from the town, and the Major-General with his staff and escort proceeded to reconnoitre the position thoroughly. This was not done without attracting the fire of the enemy's batteries from all sides. Malleson[2] says:—'It was at 9 A.M. on 21st March when Sir Hugh Rose arrived at Jhansi, and he did not finish his reconnaissance of the place till 6 P.M., so completely did he do the work. The city was walled in; the fortress, standing on a high granite rock, was due north of the city, overlooking it. The fortress commands the city and surrounding country; it is built of excellent and most massive masonry; it is difficult to breach because, composed of granite, its walls vary in thickness from sixteen to twenty feet. It has extensive and elaborate outworks of the same solid construction, with front and flanking embrasures for artillery-fire, and loop-holes, of which in some places there were five tiers, for musketry. Guns placed on the high towers of the fort commanded the country all around. On one tower, called the "white turret," then recently raised in height, waved in proud defiance the standard of the high-spirited Ranee.[3] The fortress is surrounded on all sides by the city of Jhansi, the west and part of the south face excepted. The steepness of the rock protects the west; the fortified city wall springs from the centre of its south face, running south-east, and ends in a high mound or mamelon, which protects by a flanking fire its south face. The mound was fortified by a strong circular bastion for five guns composed of solid masonry, round part of which was drawn a ditch twelve feet deep and fifteen broad.

'The city of Jhánsi is about four miles and a half in circumference. It is surrounded by a fortified and massive wall, from six to twelve feet thick, and varying in height from eighteen to thirty feet, with numerous flanking bastions armed

[1] See Appendix B, pp. 582-83. [2] Vol. v. p. 108 *et seq.* [3] Ráni (Malleson).

as batteries, with ordnance, and loop-holes, and with a banquette for infantry.'[1]

According to Malleson, the town and fortress were garrisoned by 11,000 men, composed of rebel Sepoys, foreign mercenaries, and local levies, and they were led by a woman who believed her cause to be just, and who, classified according to Channing's definition of greatness, was a heroine, though of the third order.

In his long reconnaissance of the 21st March, Sir Hugh Rose had noted all the strong points of the defence, and had examined the nature of the ground. He noted the many difficulties presented to the attack by the fort perched on a lofty granite rock, with its three lines of works, its flanking fire, its thick and solid walls. He had discovered that it would be necessary to take the city prior to assailing the fortress, a work involving double labour and double danger. In this reconnaissance, however, he had decided on his plan of attack. That night he was joined by the cavalry of the 1st Brigade: the next day he completely invested the city and fortress with his cavalry.[2] Jhansi invested.

The cavalry investment was carried out with great tact and discretion. There were seven flying camps of cavalry established with their chain of outposts and vedettes on duty round the city day and night. Stent's[3] description of the part taken by his troop gives a good idea of what the work was:—'My troop [of the 14th Light Dragoons], the "K," under Captain Brown, numbered about 60 men. We were expected to cover a certain portion of the city, to see that none escaped, or to turn out at any moment, and on any emergency. Consequently we were never out of harness, sleeping in front of our horses, which were always ready saddled and bridled—never having the bits taken out of their mouths, night or Flying camps of cavalry round Jhansi.

[1] The above description, Malleson says (vol. v. p. 109, footnote), is taken from Sir H. Rose's despatch of 30th April 1858, where it is stated, 'a remarkable feature in the defence was that the enemy had no works or forts outside the city.'

[2] See Appendix B, p. 582. [3] *Scraps from my Sabretache*, p. 197, etc.

day, except a few at a time for feeding purposes, or to give them a drink in comfort, so that it came harder on the horses than it did on us. As for ourselves, I don't think we were able to change our clothes, or have a wash, for about a fortnight, and it may be imagined that we were rather dirty, and that a bath would have done the whole of us good; but we couldn't even wash our faces, to say nothing of the elaborate luxury of a bath. Yet somehow, in spite of this and the dreadful heat, none of us fell sick, and all of us seemed to enjoy the life we led. One day, six privates and myself were out in charge of a young Irish officer of the regiment, belonging to "K" troop, named Beamish, and we had caught a party of Sepoys in a small building. They had retreated up a narrow staircase which was only wide enough for one to go up at a time, and could easily have kept us at bay if they had not been apparently panic-struck at our appearance. We had all dismounted, and our leader was soon busily engaged pulling the Sepoys one by one down the stairs by their "hind-legs" (as a comrade observed), and handing them over to our tender mercies. This amusement highly delighted Cornet Beamish, who, when he had finished, declared it was much better fun than "drawing badgers."

Cornet Beamish catches Sepoys.

'One cavalry flying camp was commanded by Major Gall with a squadron of the Fourteenth from the 1st Brigade. One was commanded by Captain Thompson, 14th Light Dragoons, and it was posted near the water palace and lakes of Jhansi. Another was under Captain Forbes, 3rd Bombay Light Cavalry. Three more were commanded by Captains Abbott, Murray, and Clarke, Hyderabad Contingent Cavalry, and Major Scudamore commanded the seventh, and he was placed in command of the whole of the investing force of cavalry, and on the night before the battle of the Betwa (31st March) he had command of the whole of the troops employed in the investment of the city and fortress. Brigadier Stuart with his 1st Brigade from Chanderi arrived about 24th

March and was encamped about 2 miles from the 2nd Brigade, and 1 mile from the fortress of Jhansi. The siege commenced on the 22nd March, and what was called the besieging force was divided into two attacks. The right attack was near the water palace and was carried on by the 2nd Brigade. The left attack was placed opposite the mamelon, and being carried on by the 1st Brigade, it was not thoroughly commenced until 25th March. From this date the siege was carried on with great vigour, and the system of investment by the "flying camps" of cavalry was most admirably conducted—it was said a cat couldn't pass their lines. Day after day the same routine was rigidly enforced. No quarter was given: those attempting to escape from the city were cut off by our vedettes and sentries, or attacked by our ambush parties posted at night. There was not a night passed but a large number of prisoners were taken by our cavalry picquets, and many of these were summarily disposed of.'

Siege of Jhansi, 22nd March 1858.

Malleson says:—'For 17 days the fire from the besieging batteries and from the walls of the city and fort was incessant. Shot and shell were poured into the city, and the enemy's guns never ceased to reply. The labour imposed upon the small force of the besiegers was tremendous. During the period of which I have spoken, the men never took off their clothes, and the horses were not unbridled except to water. Nor were the exertions of the besieged less determined. Women and children were seen assisting in repairing the defences of the walls, and in carrying water and food to the troops on duty, whilst the Ranee constantly visited the troops and animated them to enthusiasm by her presence and her words.'

As we had only 2 18-pounders for breaching purposes, the progress made against the massive masonry of the walls was somewhat slow, but by the 29th March the mamelon guns were silenced by the fire of our left attack, and on 30th and 31st our cannonading was continued with renewed vigour so that a breach, not, however, yet practicable, had been made.

Unluckily, just at this juncture Sir Hugh received in-

The army of the Peishwa attempts the relief of Jhansi under Tantia Topee.

telligence of the advance of a relieving army, which was the so-called army of the Peishwa, advancing on Jhansi from the north. Tantia Topee, the agent of Nana Sahib, was leading this army at the entreaty of the Ranee, and it was said that he had in his ranks Sepoys from numerous mutinous regiments, as well as levies from several rebel rajahs, and some of the finest regiments of the disaffected Gwalior contingent. His numbers were estimated at 22,000 men and 28 guns.

Such a sudden and unexpected danger placed Sir Hugh Rose in a most perilous position: he saw that to withdraw the troops then investing the fortress for the purpose of attacking this new enemy would be a most unwise and dangerous step. So he determined to gather together all the men he could who were not actually on duty in the siege, and face the foe with these, whilst at the same time the siege should be continued with unabated vigour by the others. Under this arrangement only 1500 men, including 500 British, were available to march against Tantia Topee's thousands. The force selected was furnished by detachments taken from both brigades. The detachments from the 1st Brigade were led by Brigadier C. S. Stuart, whilst Sir Hugh Rose himself led those supplied by the 2nd Brigade. On the night of the 31st March the men bivouacked in their clothes with everything ready at hand for immediate action. At 4 o'clock on the morning of the 1st April, Tantia Topee advanced towards the British encampments around Jhansi, hoping to sweep us from the face of the earth. Half an hour later, according to Malleson, our picquets fell back and gave the British general warning of the approach of the enemy, who came up to within 800 yards, completely overlapping our small line, and apparently hoping to envelop our flanks. The rebel guns immediately unlimbered and opened fire. The immense line of Tantia Topee's looked as if it would completely hem in and crush not only the small force brought out against it by Sir Hugh, but also the whole investing force round Jhansi, in which case we should have been placed between two fires—Tantia Topee's on the one side, and the guns of Jhansi on the other. Luckily Sir Hugh was able to grasp the situation, and in an instant he took the necessary steps to ward off this impending danger. He placed Captain Lightfoot's field battery on his left with a squadron of the Fourteenth under Captain Prettejohn: these were ordered to attack the enemy's right. In the centre he placed his heavy guns and infantry: the latter consisted of the 3rd Europeans [1] under Lieutenant-Colonel Liddell, the 24th Bombay Native Infantry, and the Hyderabad Contingent Infantry. Of these the 24th were formed as a support, which materially weakened the first line, but was unavoidable. The 1st Brigade detachments under Brigadier C. S. Stuart had been intended for Sir Hugh's second line, but he had subsequently ordered them off by a circuitous route to our left to watch some fords of the river and to prevent any portion of the rebels' forces from doubling back on Jhansi, in which case they would have cut off Colonel Scudamore's flying camps of cavalry.

Battle of the Betwa, 1st April 1858.

The infantry in the centre of our first line were placed behind some rising ground, and were lying down to avoid the heavy fire poured forth by the enemy at his first attack, and their orders were to advance so soon as the cavalry and artillery attacks on both our flanks were well developed.

On his right flank Sir Hugh placed Captain Need with a troop of the Fourteenth, as well as a troop of the Nizam's Cavalry under Clarke, and the Eagle Troop of Horse Artillery (the 1st troop of Bombay Horse Artillery), commanded by Lieutenant-Colonel Turnbull.[2] These were to attack the enemy's left. Sir Hugh Rose himself took command of the right of his line. At first it looked as if our guns would prove powerless to check the onward rush of the enemy, who so greatly outnumbered us, but Sir Hugh's tactics, as it happened, turned out to be exactly suited to the exigencies

[1] The 3rd Bombay European Infantry, afterwards the 109th Regiment, now the 2nd Battalion Leinster Regiment.

[2] Now 'N' Battery, Royal Horse Artillery.

of the moment. In order to enfilade the enemy's left he sent forward 2 guns from the Horse Artillery on our right under Lieutenant T. C. Crowe,[1] who moved diagonally to his right, and although one of the guns was disabled, the fire of the other was so rapid and so correct that the enemy's left was shaken.[2] The enemy's centre advanced very steadily, and poured in a heavy fire upon our centre which began to tell on our men around the heavy guns, and it was at this time that the infantry, who were lying down, had been ordered to advance, when the cavalry charges were being executed. Speaking of the infantry advance, Malleson says:—'The infantry sprang to their feet, advanced a few yards, then poured in a volley, and charged. The result was magical. The first line of the enemy at once broke and fled in complete disorder towards the second line, abandoning several of their guns.' The charges of the cavalry brought the battle to a speedy and successful issue, concerning which Sylvester says:—'The Horse Artillery gun having been disabled was more than Sir Hugh Rose could bear: the Major-General, at the head of Captain Need's troop of dragoons,[3] dashed into the enemy's left, while Prettejohn and McMahon, with Lieutenant Dew, led their troops[4] against the enemy's right, and doubled them up. This was a magnificent sight, and in a moment the enemy's ranks were a mass of confusion: they were shaken and disorganised, and commenced a disastrous retreat. They were hurled back on the Betwa by the irresistible attack of the dragoons.'[5] The Nizam's troop of Hyderabad Cavalry on our right also did good service in supporting Captain Need's charge, and Sylvester says that before this charge the same troop under Clarke had been ordered by Sir Hugh Rose to charge the battery which had disabled Lieutenant Crowe's gun, but that Clarke was thrice driven back by

[1] Now Major-General Crowe, late Royal (Bombay) Artillery.

[2] Malleson, vol. v. p. 113, etc., where 'enemy's right' is printed for 'enemy's left.'

[3] The 'A' troop 14th Light Dragoons.

[4] The 'H' and 'G' troops 14th Light Dragoons.

[5] Sylvester, p. 96, etc.

showers of grape and volleys from the Velaitee[1] matchlock-men, losing some men and horses and receiving a wound himself. Stent, who was present as a non-commissioned officer in the regiment, gives the following account of the charges of the Fourteenth on this occasion[2]:—

'It was a glorious sight to see them thundering along, headed by the General and Captain Prettejohn, the latter of whom was bareheaded, and who fought and shouted like a demon. One minute, and they were among the enemy, and all that was to be seen was a confused mass of flashing swords and bayonets, struggling men and horses, and hoarse shouts of rage. From this seething, struggling mass our men emerged victorious, for the result of the charge showed that an act of daring and personal bravery on the part of a leader (an act not often done—a Commander-in-Chief to lead a charge) will sometimes change defeat into victory as it did in this case.

'The rebels were thoroughly routed in this charge, and turned and fled; were rallied, formed up again, to be again charged and routed; and yet again, only to undergo the same infliction, losing all their guns, and finally bolting in the greatest confusion, pursued by our men, who cut up great numbers of them, stopping only at the river Betwa from sheer exhaustion. Many of the enemy who escaped our swords were drowned in attempting to cross the river; the whole of the ground passed over by our men was strewed with their bodies, and at the lowest estimate it was calculated that 1500 of them must have been slain, and no doubt the wounded were at least as many more. Our cavalry and artillery bore the brunt of this severe engagement, my regiment[3] suffering the most, from the nature of the conflict being a succession of charges and hand-to-hand fights.'

We must now leave for a moment the routed first line of

[1] The word Velait or Welait is Hindustani for 'foreign'—a Velaiter is a foreigner. The Velaitees were native mercenaries, generally armed with matchlocks and tulwars. They were the best fighting-men amongst the rebels, and were composed of cut-throats, 'budmashes,' and first-class scoundrels.—(Sylvester.)

[2] *Scraps from my Sabretache*, p. 215, etc.

[3] 14th Light Dragoons.

Tantia Topee, pursued by their victors, to glance at the movements of those detachments of the 1st Brigade under Brigadier C. S. Stuart, C.B., to whom allusion has already been made, and to show what an important effect their movements had on the fortunes of the day. 'Brigadier Stuart had moved round the hill into the plain on the right of the enemy, in order to check a large body of them who were taking advantage of the battle raging in front of the line to move off towards Jhansi. Stuart attacked, defeated them, and drove them back, hotly following them. So close indeed was the pursuit that they had no time to re-form, but fled in confusion, leaving gun after gun in the hands of the victors, and numbers of their men dead or dying on the field.'[1]

Gallant conduct of Lieutenant Giles and part of a troop of 14th Light Dragoons with the 1st Brigade.

This affair of the 1st Brigade was most opportune, for had this body of rebels not been cut off it is highly probable they would have worked round to Jhansi, and made the position of Colonel Scudamore's flying camps of cavalry very perilous. Lieutenant Giles, 14th Light Dragoons, with 30 or 40 of his troop, distinguished himself very highly in this pursuit, and cut up a large number of the enemy; but his force was too small and quite inadequate against the dense masses of matchlockmen that confronted him. The ground, too, was most unsuitable for cavalry and utterly impracticable for artillery, and the infantry could not keep up from sheer fatigue after their long march. The rebels sheltered themselves behind the adjacent rocks and caused several casualties in Lieutenant Giles's troop, killing 1 man, wounding 5, and placing 10 horses *hors-de-combat*, so that, notwithstanding the dauntless courage and gallantry exhibited by this officer, he could not do more than content himself with cutting up outsiders. As it was, however, 250 of the rebels had been killed by Stuart's attack, and many more wounded, besides which 6 guns, 2 elephants, some camels, ammunition and treasure fell into his hands. Malleson says :— 'Tantia beheld in dismay the men of his first line rushing helter-skelter towards him, followed by the three arms of the British in hot pursuit; but he had scarcely realised the fact

[1] Malleson, vol. v. p. 114.

when another vision on his right flank came to add to his anguish.' This vision was the rout of a large body of the enemy, caused by the very successful flank movement executed by Brigadier C. S. Stuart, and in which Lieutenant Giles, with his small troop of the Fourteenth, played such a conspicuous part. The second line of the rebel army was commanded by Tantia Topee in person. That crafty leader had occupied a selected position about 2 miles in rear of his first line, and here he placed his second line and reserve upon rising ground, with its front covered by jungle. Thus he was able to descry in an instant the reverse that had happened to his first line. Malleson says :—'It had the effect of forcing upon him a prompt decision: the day, he saw, was lost, but there was yet time to save the second line and his remaining guns. The jungle was dry and easily kindled: Tantia Topee at once set fire to it, and under cover of the smoke and flames commenced a retreat across the Betwa, hoping to place that river between himself and the pursuers. His infantry and horsemen led the retreat, his guns covered it. Right gallantly and skilfully they did it, and he did succeed in crossing the Betwa with his reserve and guns and some of the fugitives of the first line.'

When the first line of the rebels was routed, as we have already described, the pursuit became general. Our cavalry and artillery, with Prettejohn[1] and Need[2] at the head of the Fourteenth, dashed at a gallop through the burning jungle, cutting up hundreds of the rebels in their onward course, and determined to capture every gun of the enemy that had opened fire upon them. The flying enemy often rallied in a mass or 'gole,' and many hand-to-hand fights between them and our cavalry took place, as in the swamps about Dhar; but the further the pursuit continued the thinner and fewer these rallying masses became, till at last little squares and groups, and then only single fugitives, dotted the plain. The pursuit did not cease till 2 troops of the Fourteenth and a troop of the Nizam's Cavalry

[1] Afterwards Major-General R. B. Prettejohn, C.B.

[2] Afterwards Lieutenant-Colonel Sir Arthur Need, Lieutenant of the Yeomen of the Guard.

had actually crossed the Betwa, and here they became exposed to the heavy fire of the enemy, both in crossing the ford and also in ascending the steep road on the opposite bank. The rebels made great efforts to carry off their guns, elephants, and ammunition, but the greater part fell into our hands. Eighteen guns, 2 standards, and large stores of ammunition were captured, and after the capture of the last gun, an 18-pounder, about a mile and a half on the other side of the Betwa, Sir Hugh withdrew the cavalry from pursuit. They had been marching and fighting incessantly for many hours, and both men and horses were quite exhausted. Nine miles remained to be traversed back to Jhansi, but the troops returned to camp amply compensated at having achieved so brilliant a victory over their vaunting foes. Malleson says 1500 rebels were killed and wounded that day. The British losses were 81 men and 29 horses, of which 15 men and 13 horses were killed. The Fourteenth suffered in proportion more heavily than any other regiment or corps. Their losses were: killed, 5 men, 11 troop-horses; wounded, 25 men, 16 troop-horses.

Guns, stores, and standards captured from the rebels.

Casualties of the Fourteenth at the Betwa.

In Appendix B, pages 593-97, extracts from Sir Hugh Rose's despatch, dated 30th April 1858, are inserted *verbatim*, and in these the reader will find officially recorded the gallant part taken by the regiment in this important victory over the Peishwa's army. The names of three officers of the Fourteenth —Captain Need, Captain Prettejohn, and Lieutenant Leith— are brought to notice for their important services on this occasion, as well as that of one non-commissioned officer, Sergeant Gardiner, who also, on a former occasion at Dhar, behaved so gallantly. Lieutenant Leith, who was in charge of the 'A' troop, led by Captain Need, performed an act of bravery and devotion for which he was recommended by Sir Hugh Rose for the Victoria Cross, and received that much-coveted decoration subsequently.[1] In that charge Need, who was a great swordsman, dashed up rocks where no horse could keep its footing, and was in great danger; but Leith saw his

Lieutenant Leith earns the Victoria Cross.

[1] See Appendix B, page 598.

perilous position and came to his captain's help at the right time and saved his life. Need was not actually unhorsed, but was almost surrounded by the enemy's infantry. His saddle, and even his jacket, which was loose and large, was slashed in several places, and his reins were cut.

Telegram sent by Sir Hugh Rose to Government about the battle of the Betwa.

The following copy of the telegram forwarded by Sir Hugh Rose to the Government gives a very clear and concise account of this battle, and is taken from Stent's *Scraps from my Sabretache* :—'This morning, at daybreak, the force under my orders fought a general action with the so-called Peishwa's army, and, by the blessing of God, gained a complete victory. The rebels are stated to have numbered from 20,000 to 25,000 men: they were under Tantia Topee, Nana Sahib's relative, and their object was to relieve Jhansi. I did not discontinue the siege or investment of Jhansi, consequently the force with which I fought was extremely weak. The rebels, amongst whom were the grenadier regiment and another regiment of the Gwalior contingent, fought, except the cavalry, desperately; but I turned their left flank with artillery and cavalry, and after making two stands they broke and fled, after defending themselves individually to the last. I pursued them to the river Betwa, taking all their guns, eighteen in number, and an English 18-pounder of the Gwalior contingent drawn by two elephants; an 8-inch mortar, and quantities of ammunition, including shells, 18-pounder shot, ordnance park, and two elephants. Two standards were also taken. The enemy tried to stop our pursuit by setting the jungle on fire, but nothing could check the ardour of the artillery and cavalry, who galloped in pursuit across the country in flames. I cannot calculate at present the enemy's loss in killed, but it must have been very great, as the country is strewed with dead bodies, chiefly those of sepoys. As I now shall be free from the attacks of a numerous attacking army I hope to conclude speedily the siege of Jhansi.'

The remnant of the Peishwa's army, with Tantia at their head, made their escape towards Calpee,[1] and the victors

[1] Kalpí (Malleson).

resumed their former positions round Jhansi the same evening. During the time this battle had been raging the besieged had redoubled their fire; they manned the walls, poured down volleys of musketry, shouting and yelling hideously, and seemed to threaten a sortie. It seems incredible that no effectual attempt was made by the garrison to come out and attack the attenuated lines of besiegers, but perhaps they were deterred by the increased vigour displayed by our batteries, and also by a feigned attack which was purposely made by a body of troops under Major Gall, 14th Light Dragoons, and Lieutenant Lowrie, Royal Artillery, upon a distant part of the city wall, where a party of rebels did make some slight attempts at a sortie, but were effectually driven back by a 9-pounder and howitzer, together with some native infantry and dragoons acting under Major Gall.

Major Gall makes a feigned attack on city wall during the battle of the Betwa.

Sir Hugh Rose determined to take advantage of the discouragement which it was well known the defeat of the relieving army would undoubtedly produce on the minds of the Ranee and her garrison. Accordingly the siege was prosecuted with renewed vigour, and when it was known that the breach in the city wall was just practicable, he determined to storm the place on 3rd April. His plan of attack, according to Malleson, was this: to make a false attack on the west wall with a small detachment under Major Gall, 14th Light Dragoons. As soon as the sound of his guns should be heard, the main storming-party was to debouch from cover and enter the breach, whilst on the right of it attempts should be made to escalade the wall. His dispositions were as follows:—

Sir Hugh's plan of attack on Jhansi, 3rd April 1858.

1. The right attack, composed of the Madras and Bombay Sappers, 3rd Bombay Europeans, and infantry of the Nizam's Hyderabad Contingent, was divided into two columns and a reserve:—

The right column under Lieutenant-Colonel Liddle, 3rd Europeans; the left column under Captain Robinson, 3rd Europeans; the reserve under Brigadier C. Steuart, C.B., 14th Light Dragoons.

This attack was to attempt to gain the town by escalade.

2. The left attack, composed of the Royal Engineers, the 86th Foot, and the 25th Bombay Native Infantry, was similarly divided into two columns and a reserve:—

The left column, commanded by Lieutenant-Colonel Lowth, 86th Regiment, was to storm the breach; the right column, led by Major Stuart, 86th Regiment, to escalade the rocket-tower and the low curtain immediately to the right of it. The reserve was commanded by Brigadier C. S. Stuart.

Storming of Jhansi, 3rd April 1858.

At 3 o'clock on the morning of the 3rd April, the storming-parties marched to the positions assigned them, to await there the signal from Major Gall's party. No sooner was it given than the stormers dashed to the front.

The left attack was successful, but on the right the escalading party suffered very heavily, and their ladders were too short and too weak. Lieutenants Dick and Meiklejohn of the Bombay Engineers, while nobly leading, were killed, and Lieutenant Bonus of the same regiment was hurled down from the wall struck in the face by a log or stone, Lieutenant Fox of the Madras Sappers was shot in the neck on the wall; but at length the stormers got a footing on the rampart, and owing to an opportune rescue made by a party from the left attack under Captain Brockman, who with some men of the 86th Regiment took the defenders in flank and rear and drew them off, the right attack was enabled to hold its own and to join the left attack in dispersing the defenders, so that eventually, after a sanguinary contest, the city and palace were successfully captured. The resistance in the stables of the palace, as well as in the apartments, was very determined, and severe conflicts ensued here and in some of the streets leading through the city, where desultory fighting continued during the night and following day, as well as in the suburbs. Whilst the palace was being captured a body of rebels, about 400 in number, driven from the city, took refuge on a hill to the west of the fortress, where they were surrounded by our cavalry.

Major Gall storms a hill in the suburbs of Jhansi.

On hearing of this Sir Hugh ordered a mixed force, under command of Major Gall, 14th Light Dragoons, to storm the position and capture the rebels. Major Gall, with his usual skill and courage, took the position with the 24th Bombay Native Infantry, and all the defenders were either killed or blown up, with a loss of 1 officer and a few men of the 24th Bombay Native Infantry. Another body of 1500 men held out in a suburb, but these also were driven out with a loss of 300. Sir Hugh now bethought himself to make plans for capturing the fortress, but the Ranee saved him all trouble on that score. She had fled there for safety when the palace was taken, but on the night of the 4th of April, despairing of success, she evacuated the fortress and fled to Calpee, intending to join Tantia Topee there.[1] She was reported to have escaped on horseback with her child before her, and a drawn sword in her hand, accompanied by her few remaining followers and a cavalry escort of sowars, along the Banda road. The following description of the Ranee of Jhansi is from Stent: 'She was a very handsome woman about 24 years of age, a perfect Amazon in bravery, heading her troops, mounted like a man—just the sort of dare-devil woman that soldiers admire. She was finally killed, fighting against us at Gwalior.'[2] It is believed she was wounded by a carbine bullet in the fighting at Morar, June 1858, and was carried to the rear, where she expired, and was burned according to the Hindoo custom. At the time of her death she was attired as a cavalry soldier.[3]

Escape of the Ranee of Jhansi, 4th April 1858.

Pursuit of the Ranee.

As soon as the flight of the Ranee was known, a cavalry force was sent in pursuit, but she had got too great a start and made good her escape to Calpee. Lieutenant Dowker with some sowars went in hot pursuit, and actually got near her at one time, but he was wounded by her escort whom he engaged, killing a large number of them and capturing the Ranee's tent. Sir Hugh's force occupied the fortress on 5th April. Our losses in and around Jhansi, including the

Fortress of Jhansi occupied by British on 5th April 1858.

[1] Malleson, vol. v. p. 115 *et seq.*

[2] *Scraps from my Sabretache.*

[3] Sylvester, p. 183.

action fought on 1st April, amounted to 343 killed and wounded, of whom 36 were officers. The enemy's loss was computed at 5000: 1000 bodies were actually burned or buried in Jhansi itself.[1] Losses at Jhansi.

During the last few days of the siege, and for some time after the capture of Jhansi, Major Gall, 14th Light Dragoons, had been given the command of 'The North Outpost,' with a force under him of 2 and sometimes 4 pieces of ordnance, a squadron of the Fourteenth, and about 200 native infantry. He had constantly to move out of camp to his right or to his front, either to support the picquets near or to attack the enemy in the city lying to his front, and to prevent them from escaping. During the day of the battle of the Betwa, on 1st April, he was employed with this force in making an attack on the city wall, and preventing an expected sortie of the garrison. Major Gall's 'North Outpost' at Jhansi.

On the day the city was stormed, 3rd April, he made the feigned attack on the bastion of the north-east wall, so as to divert attention from the real attack on the south-east, and thus gave the signal to the storming-party to enter the breach.

On the 5th April, taking 150 of his infantry, Gall moved out of camp, and swept the ground along the city walls on the north side. Lieutenant Gowan, 14th Light Dragoons, with a small party of his men was also employed on the same duty. These parties sabred or shot down a large number of fugitive rebels who had secreted themselves in the cornfields and gardens about, not having been able to escape through the cavalry picquets when the city was taken. Most of the cavalry were thus employed on the day after the Ranee escaped, when the fort was evacuated. By this means six or seven hundred of the enemy are said to have been cut up or shot. Cornet Beamish, 14th Light Dragoons, accounted for about 300 in this way with his party of dragoons, and a very large number, about 220, were disposed of by the Hyderabad Cavalry under Captain Abbott. Major Gall's infantry killed 100, Lieutenant Gowan's men, 30. Desultory fighting outside Jhansi by 14th Light Dragoons and Nizam's Cavalry.

[1] Malleson.

After the fall of Jhansi, Sir Hugh Rose remained there nearly 19 days, partly to rest his troops and partly to make the necessary preparations for another campaign. His next object was to march on Calpee: this was the arsenal of the rebels. Rao Sahib, the nephew of Nana Sahib, was there, with large supplies of artillery and warlike stores, and a numerous garrison. Calpee was situated on the river Jumna, 102 miles north-east of Jhansi, and 46 miles south-west of Cawnpore.[1]

Calpee.

Lieut.-Colonel Gall's column marches from Jhansi on Koonch, 22nd April 1858.

On the night of the 22nd April a column under Major, now Brevet Lieutenant-Colonel Gall, 14th Light Dragoons, consisting of 3 troops (B, C, and E) 14th Light Dragoons, Lightfoot's battery of artillery, some companies of 3rd Europeans, some companies of 25th Native Infantry, and some Hyderabad Cavalry, was sent off from Jhansi to a place on the Calpee road. This force reached Pooch, 14 miles from Koonch, on 1st May.

Arrives at Pooch, 1st May.

On 25th April, at midnight, Sir Hugh left Jhansi for Koonch with the 1st Brigade, leaving orders for the 2nd Brigade to follow in two days. Major Orr with the bulk of the Hyderabad Contingent troops had been sent previously to cut off some rebels from crossing the Betwa and so doubling back southwards, and he too was to march on Koonch. On 1st May Sir Hugh with the 1st Brigade came up to Pooch, and joined Lieutenant-Colonel Gall's column, and by the 5th May Sir Hugh was joined by his 2nd Brigade, now augmented by the arrival of the 71st Highlanders. The whole force now marched on Lohari, 10 miles nearer to Koonch, but on arrival there it was ascertained that the rebels had possession of the fort. Lieutenant-Colonel Gall with his column was detached to take it. This he successfully achieved after a stiff fight and some loss. The 3rd Europeans and the 25th Native Infantry had severe hand-to-hand fights, but the whole garrison was destroyed and the fort captured. Lieutenant-Colonel Gall and 4 other officers were wounded, and we lost 1 man killed and 13 wounded. One brass gun was captured in the fort. Lieutenant-Colonel Gall himself personally led the stormers,

The two brigades arrive at Pooch and march to Lohari.

Lieut.-Colonel Gall storms the fort of Lohari and captures it, 5th May 1858.

[1] Malleson.

and was afterwards highly complimented by the Major-General for the able manner in which he had conducted the capture.[1]

After the flight from Jhansi, Tantia Topee and the Ranee of Jhansi had both escaped to Calpee. Here Rao Sahib, at the Ranee's request, reviewed the whole of his army, and then placed Tantia Topee in command of it, and ordered him to take it forward at once against the British. Tantia accordingly proceeded to Koonch on the Jhansi road, 42 miles from Calpee, and there entrenched himself in a strong position covered by woods and gardens, with temples at intervals, and surrounded by a strong wall. Meanwhile, as we have seen, the main body of the Central India Field Force and Hyderabad Contingent Force was advancing on Koonch under Major-General Sir Hugh Rose to confront the rebels. The Major-General had decided to make a flank march, and so to turn the enemy's position by attacking him in flank instead of making the usual frontal attack so dear to Asiatics. Accordingly on 6th May he broke up his camp and proceeded towards Koonch, so as to gain a position facing the unfortified side of the town, and hoping thus to cut off the enemy's line of retreat in the direction of Calpee.[2] On 7th May Sir Hugh's force came in sight of Koonch by 7 A.M., having marched 14 miles. Here he halted and drew up his line 2 miles distant from the town: the 1st Brigade formed on the left, the 2nd Brigade were in the centre, and the Hyderabad troops, under Major Orr, were on the right. Preparatory to the attack the troops were served out with a dram of grog and some biscuit. At 8 A.M. Lieutenant-Colonel Gall was sent forward with some cavalry to reconnoitre the wood, gardens, and temples, and his advance was covered by artillery fire. At the same time the siege-guns were moved to a position whence they could effectually play upon the town.

Battle of Koonch, 7th May 1858.

Lieut.-Colonel Gall sent to reconnoitre.

Lieutenant-Colonel Gall soon returned and reported that

[1] See Appendix B, p. 597.

[2] Malleson, vol. v. p. 120. In Sir Hugh Rose's despatch the date of the battle at Koonch is 7th May. Malleson's *History* has 6th May. Sylvester, who was present, confirms Sir Hugh's as the correct date.

the enemy had retreated through the wood nearer to the town, that they had cavalry in their rear, that the fire of the siege-guns had driven the rebels from the right of the wood into the town, but that some outworks were still held by them. The Major-General determined to clear the wood and the outworks with his infantry, and then to storm the town. This was well and successfully carried out, and the fort was occupied by the men of the 1st Brigade, led by Sir Hugh, the fighting and skirmishing in the wood, however, causing several casualties. The enemy were ultimately driven from the wood, gardens, and outworks, through the town, out of the town, and along the plain on the road leading towards Calpee. This was on the north side. Steuart with the 2nd Brigade, arriving from the west and occupying the centre of the line, had attacked a body of rebel infantry strongly posted to his front, directly in his intended line of advance. He met with a very determined resistance, and it was not till a body of the 1st Brigade came to his assistance by a well-delivered flank attack on the rebels that he was enabled to disperse them. In this attack of the 2nd Brigade the 2 troops of the 14th Light Dragoons, under Captains Thompson and Gordon, took a prominent part, and at the critical moment Captain Gordon led his troop forward and made a gallant charge against a mass of the enemy, whom he broke and cut up, for which he was specially mentioned by Sir Hugh Rose in his despatch of the 24th May 1858.[1] Steuart then endeavoured, but ineffectually, to cut off the rebels from their line of retreat, but instead of working on through the town as he had been expected to do, his Horse Artillery and Cavalry moved round to the south side of it and joined in the general pursuit. Major Orr, with the Nizam's troops, arriving from the south, had acted on the right of the line in the attack, and also joined in the attempt to cut off the fugitives in the direction of Calpee. The action was over in about an hour, but the heat of the sun was so terrific that the infantry were not in a fit state to be sent in pursuit. The fugitives at once commenced a well-organised and orderly retreat. The greater

Sir Hugh Rose's plan of attack.

Captains Thompson's and Gordon's troops of 14th Light Dragoons, engaged.

[1] See Appendix B, p. 587.

part of the cavalry, including the Fourteenth and some of the Hyderabad Cavalry, together with the Horse Artillery and light field guns, were launched forth in hot pursuit, the guns pouring volleys of grape on the retiring masses. Malleson says:—'The manner in which the rebels conducted their retreat could not be surpassed. They well remembered the lessons they had learned from their European officers. There was no hurry, no disorder, no rushing to the rear; all was orderly as on a field-day. Their skirmishers fired, then ran behind the relieving men and loaded, the relieving men then fired and ran back in their turn.'[1]

Pursuit at Koonch.

Captain William McMahon's squadron and Captain Blyth's troop of the Fourteenth charged the enemy's skirmishers magnificently. The former officer had to lead his men over a piece of very heavy ploughed land under a heavy fire: several of his men and horses were killed and a large number of saddles emptied. When the squadron got to close quarters the enemy fought fiercely. After firing they used their bayonets and native knives, with which they cut and slashed desperately at our men and horses. Captain McMahon received 3 sabre wounds, but continued to lead his squadron through the pursuit. These 3 troops of the Fourteenth cut up almost to a man the rearguard, which the enemy had thrown out in skirmishing order. Captain Blyth charged on another occasion and captured a gun under a very heavy fire; and Captain Prettejohn, by Sir Hugh's personal order, gallantly led a troop of the Fourteenth against an enfilading line of the enemy's skirmishers, also under a very heavy fire—'an order,' says Malleson, 'carried out by that most daring officer with great gallantry and success.'

Three troops of the Fourteenth charge the enemy.

Captain Prettejohn's charge.

This pursuit was continued up to 7 miles from Koonch in the burning sun, and the men and horses suffered intensely from the exposure. Even the Major-General was so much overcome at one time by the heat that he had to dismount and seek shade, where Dr. Vaughan attended him, but he subsequently resumed his duties in the pursuit.[2] Sylvester, who

[1] Vol. v. p. 124.

[2] Sylvester.

was present on the occasion, says: 'The heat was terrible; I never recollect suffering so much from thirst, and there was little water to be had. The Fourteenth were never better than on this day; they charged like a body of demons straight into the huge masses of revolted Sepoys until Scudamore screamed "Halt! it is perfect madness." I counted 34 dead Sepoys in one spot. The enemy got away with an 18-pounder and a 24-pounder gun along the Jaloun road to Calpee.' The rebels never gave in till nearly all their guns were captured, and they then crowded along the road a helpless mass of fugitives. At length our tired troops were recalled from pursuit and brought back to bivouac at Koonch, having been 16 hours marching, fighting, and in pursuit. It was 8 o'clock before the cavalry returned in the evening. We captured 9 guns and a large quantity of ammunition. The enemy lost about 600 men in the action and pursuit. Our losses were 3 officers and 59 men killed and wounded, besides a large number struck down by the sun.[1]

Casualties of 14th Light Dragoons at Koonch.

The Fourteenth lost heavily:—

Killed { 5 Men.
3 Troop-horses.

Wounded { Captain McMahon, severely. (Sabre-cuts on leg and right hand.)
17 Men.
6 Troop-horses.

Missing 4 Troop-horses.

'Sunstroke.'

In addition to these killed and wounded, the 14th Light Dragoons had 2 officers (Captain Need and Lieutenant Travers) and 16 men struck down by the sun, of whom 2 men died subsequently. In his despatch about Koonch,[2] Sir Hugh Rose brings the gallant services of the regiment into very prominent notice, and specially mentions the following officers for the favourable consideration of the Commander-in-Chief: Captain Gordon, Captain McMahon, Captain Prettejohn, Captain Todd (Staff Officer to Brigadier Steuart, C.B.), and Captain Blyth.

[1] Malleson.

[2] See Appendix B, p. 585 *et seq.*

The Ranee of Jhansi, present during the action, had fled from Koonch to Calpee, attended by her cowardly horsemen, who turned tail, having never faced us at all; and Tantia Topee, always foremost in flight, escaped to a place near Jaloun, where his parents dwelt.[1] Our men had nothing to eat till 8 P.M. at night, except what they carried in their haversacks.

The 8th of May was a day of rest for both brigades. Their tents were pitched at daylight, and men and horses pulled themselves together for the coming march on Calpee. The men of the Fourteenth were in a splendid state of physique at this time, being well inured to the climate, so they suffered less than the other Europeans, of whom the newly arrived 71st Highlanders lost more men by sunstroke than any other corps present, but even in the Fourteenth there were nearly 150 men, including sick and wounded, belonging to the left wing only, under medical treatment on the day after the battle of Koonch.

At 2 A.M. on 9th May the Major-General moved off with the 1st Brigade, followed next day by the 2nd Brigade. The march to Calpee was a most trying one, owing to the intense heat and the want of water. The rebels had thrown up elaborate fortifications along the direct road leading from Koonch to Calpee, hoping to make short work of us as we passed; but Sir Hugh got timely notice from spies, and from reports brought in by Lieutenant-Colonel Gall, 14th Light Dragoons. As a result of his reconnaissance it was ascertained that the enemy was entrenched in large numbers to our left. Sir Hugh took a circuitous route, and moved towards Golowlee, on the right bank of the Jumna, 6 miles from Calpee, avoiding the direct road, and thus outflanking the enemy, who found himself completely outwitted and his great defences on the main road turned and useless.[2] March on Calpee.

There was another advantage too in this plan of campaign, for Sir Hugh had received an intimation that a column under Colonel G. V. Maxwell, with the 88th Foot and other details, would co-operate with him, and had already reached the left

[1] Malleson.

[2] Sylvester.

Lieut.-Colonel Campbell assumes temporary command of 2nd Brigade.

bank of the Jumna opposite Golowlee. On the 13th May, owing to the illness of Brigadier C. Steuart, C.B., 14th Light Dragoons, the command of the 2nd Brigade devolved for a time upon Lieutenant-Colonel Campbell, 71st Highland Light Infantry. On the 15th, Sir Hugh with the 1st Brigade reached Golowlee, and established communication with Colonel Maxwell's column on the opposite bank by means of pontoons, as no boats could be found. On the 16th, Lieutenant-Colonel Campbell brought up the 2nd Brigade, which had been menaced by the enemy on the march from Etora. This brigade encamped near Golowlee, on the left flank of Major Orr's Hyderabad Force, and opposite the village of Diapora. The 1st Brigade had also been harassed by large bodies of the rebels on their last day's march to Golowlee, when their rearguard was attacked by cavalry.

Enemy attacks the rearguard of 2nd Brigade near Golowlee.

On the 16th, after the 2nd Brigade had reached camp, the rearguard was attacked; it was commanded by Major Forbes, 3rd Bombay Cavalry, who had with him 170 of his own sowars, a weak troop of the Fourteenth under Lieutenant Beamish, 2 guns Royal Artillery, 200 Irregular Horse, a company of 3rd Bombay Europeans, a company of Bombay Sappers, and 116 men of the 24th Bombay Native Infantry, under Lieutenant Estridge. It seems that 6000 of the enemy, who had occupied the village of Etora as soon as the brigade moved out, had cut off some of our baggage in rear. They followed up to within 600 yards of camp, and fired on our men with artillery shots and volleys of musketry, whilst their cavalry seemed threatening a charge. We lost some men of the 24th Bombay Native Infantry and of the 3rd Bombay Cavalry, whilst an artilleryman had his pouch shot off his belt, and one of the Fourteenth had his turban shot off,—the latter was replaced with the most perfect nonchalance! Major Forbes handled his force most skilfully, and, notwithstanding his critical position, brought it safely into camp. The whole force was eventually turned out under the Major-General, and desultory firing went on till dark, when the enemy, tired out like ourselves, gradually ceased firing and left us quiet.[1] The position of the 2nd Brigade camp was 6 miles from the Jumna, and was much exposed to attacks from the enemy; the 1st Brigade was encamped within a mile of the river's bank nearer Golowlee. There was a village called Muttra near the camp of the 2nd Brigade, and it was here that the enemy was constantly endeavouring to attack us and to get round the left rear of the position. The determined fire of Captain Field's battery, however, kept him at a respectful distance, though his round-shot would occasionally roll in among the tents of the Fourteenth, and in an attack made by the rebels on the 17th May several troops of the Fourteenth, with infantry and artillery, had to be employed to drive them off and to keep up connection between the 2nd Brigade and village of Muttra, which lay to our left.[2] There were incessant skirmishes going on and threatened attacks by the enemy on the 18th, 19th, and 20th May. The heat of the sun was terrific, and in consequence our men suffered dreadfully from the exposure and harassing work, for the enemy knowing this persisted in making attacks. A mortar battery in front of the 1st Brigade was established by the 19th, and on the 20th, Colonel Maxwell, from the other side of the Jumna, sent over 2 companies of the 88th Regiment and 120 Sikhs to reinforce Sir Hugh. By the 21st, Maxwell's batteries opened on the town of Calpee. There was a move forward of all the camps on 19th, after which both brigades and the Nizam's troops rested on the right bank of the river Jumna, encamped one behind the other, with Golowlee in front, and a mile of ravines lying between them and the river bed.[3] In the forthcoming attack on Calpee, Sir Hugh intended Maxwell to shell the fort and city, whilst he would clear the ravines and other obstacles and attack the fort on the left side.[4] The left of the British force nearly touched the road running from Calpee to Banda. Malleson describes the fortress

Fortress of Calpee.

[1] The above account is from Sylvester's *Recollections of the Campaign in Malwa and Central India under Major-General Sir Hugh Rose, G.C.B.*, p. 146, etc.

[2] Appendix B, p. 601 *et seq.* (Lieutenant-Colonel Campbell's report).

[3] Sylvester.

[4] Malleson.

of Calpee as being protected by five lines of defence to its front, and by the river Jumna to its rear. These five were—

1st. A series of entrenchments with flank defence;
2nd. 84 temples of solid masonry, with walls round them of the same;
3rd. An outwork of ravines;
4th. City of Calpee;
5th. A second chain of ravines;

and lastly came the fort itself, built on a precipitous rock.

The rebels at Calpee were commanded by the Nawab of Banda, a fugitive from the defeat inflicted on him by General Whitlock at Banda, and he was aided by the brave Ranee of Jhansi, driven from Koonch. They had a large number of sepoys as well as 2000 horsemen from Banda, besides guns and numerous followers.

It would be difficult to describe all the suffering our men and horses went through at this period of the campaign, especially on the march from Koonch, and in the operations about Calpee, both from the fierceness of the sun and the great scarcity of water, which prevailed until the camps were moved within reach of the Jumna. The sick-list was enormous, and the cases of sunstroke which occurred incessantly had a most depressing effect on the whole force. A foraging-party of 33 men belonging to the Fourteenth left camp one day, and of this number only 19 returned on their saddles. Of the infantry, the 71st Highlanders suffered most. They had only recently come to India, before which they had been in the Crimea. The rebels well knew how fatal the exposure to the sun's rays was to Europeans, and so they invariably arranged their fights to take place at the hottest time of the day. On the 22nd May, the date of the capture of Calpee, the enemy commenced the attack by marching in masses along the Banda road about 10 A.M. and threatening the British left near Golowlee. Another body opened fire at the same time against our centre.[1] According to Sylvester,[2] the disposition of the British forces was as follows:—

Defeat of the Calpee army near Golowlee, 22nd May 1858.

[1] Malleson, vol. v. p. 127 *et seq.*

[2] P. 154 *et seq.*

Brigadier C. S. Stuart, commanding 1st Brigade, posted himself by 8 A.M. at the mortar battery in front of the camps, taking with him half Woolcombe's battery and a party of the 3rd European Regiment. On his right, extended as a line of skirmishers reaching along the ravines to the Jumna, was the 86th Regiment under Colonel Lowth; on the left of the Brigadier was a wing of the 25th Native Infantry; in rear as a support was a troop of the Fourteenth, as well as a troop of the 3rd Bombay Light Cavalry. Towards the centre of the line was Colonel Robertson, with the other wing of his corps (the 25th Bombay Native Infantry), the remaining half of Woolcombe's battery under Lieutenant Strutt, and the 21st company Royal Engineers; on the left centre Lightfoot's troop of Horse Artillery, and 2 troops of the Fourteenth, our heavy guns, the Royal Artillery field battery, 71st Highlanders, and main body of the 3rd Europeans. On the extreme left were the Hyderabad Contingent force, Maxwell's camel-corps riflemen and Sikhs. Here, too, as the ground was more suitable for cavalry than on our right, was a squadron of the Fourteenth under Lieutenant-Colonel Gall, and 3rd Hyderabad Cavalry under Captain Abbott.

Such was the disposition of Sir Hugh Rose's little army when the enemy commenced their attack, as already stated above. By 9 A.M. our vedettes began to fire and fall back, but owing to the nature of the ground, especially in the direction of the ravines, a good half of the approaching rebels came on without being visible to us. The mass of their cavalry and artillery were on the right of their line, the ravines being impracticable to their left. Our infantry began to advance, and our guns opened fire at the same time.

The enemy's attack on our left was only intended as a feint. It was led by the Nawab of Banda and Rao Sahib, the nephew of the Nana. It proved, however, in the end serious. It was pushed well home, and our left became heavily engaged. Sir Hugh, notwithstanding, being fully persuaded in his own mind of the real intention of the enemy, did not move a man from his right, and his forecast was correct. Suddenly, as if by magic, the whole line of ravines became lighted up by a mass of fire, both artillery and musketry, which was brought to bear with overwhelming force on the British right. As Malleson says: 'The suddenness of the attack, the superior numbers of those making it, and the terrible heat of the day, gave the rebels a great advantage.' The sun had struck down an unusual number of Europeans, and some of our men's Enfield rifles having become clogged by constant use in all weathers, were difficult to load, so that at one time things looked very critical when the enemy, starting up in great numbers from the ravines, pressed forward with loud yells and caused the British to fall back to the position where our light field-guns and mortar battery were posted. It was here that Brigadier Stuart made a gallant stand; himself dismounted, and standing by the guns, he bade the gunners defend them with their lives.[1] The 86th Regiment and 25th Native Infantry in a thin extended line made a good resistance, and disputed the advance step by step; 'still,' says Malleson, 'the rebels pressed on, and it seemed as though from their very numbers they must prevail, when Sir Hugh, to whom news of the attack had been conveyed, brought up the camel-corps, which had opportunely crossed the river that very morning, at their best pace; then, dismounting the men, and leading them forward himself at the double, charged the advancing foe, who was then within a few yards of the British guns. For a moment the enemy stood, but only for a moment. A shout, a dash forward from our whole line, and they went headlong into the ravines below. Not only was the attack on our right repulsed, but the battle was won. The attack on our left collapsed when it was seen that that on the right had failed, and our guns, gaining the rebels' flank, inflicted great loss on them as they fled. Sir Hugh followed them up so closely that he cut off a number of them from Calpee. The fire from Maxwell's batteries made those who reached that fort feel that it was no secure place of refuge. They evacuated it accordingly. The

[1] Malleson.

main bulk of the enemy, pursued by the Horse Artillery and cavalry, lost their formation and dispersed, losing all their guns and baggage. Even the Ranee of Jhansi, who fled with them, was compelled to sleep under a tree!' The numbers of the defeated army are computed by an eye-witness[1] as being 'ten times our number.'

Calpee taken by the British, 23rd May 1858.

Next morning Sir Hugh marched into Calpee. The 1st Brigade under Brigadier C. S. Stuart went through the ravines, following the course of the Jumna, whilst Sir Hugh led the 2nd Brigade himself along the Calpee road, and Colonel Maxwell's batteries shelled the fort and villages in front. There was, however, no resistance offered, and both brigades entered the town and fort, whence the rebels had fled, leaving their great arsenal replete with all the munitions of war entirely in our hands.[2]

Pursuit by Lieut.-Colonel Gall with cavalry and horse artillery.

Lieutenant-Colonel Gall with all the available men of the Fourteenth, Horse Artillery, and Hyderabad Cavalry was immediately sent in pursuit, and, as related in Sir Hugh Rose's despatch of 22nd June 1858, extracts of which are printed in Appendix B,[3] this duty was most ably carried out. The enemy was pursued as far and as closely as possible: the whole of his guns and 6 elephants were captured, and large numbers of the rebel sepoys, especially the Sind Velaitees and mercenaries of the Nawab of Banda, were cut up by the Fourteenth and the Nizam's Cavalry.

Regiment mentioned in despatches, also Lieut.-Colonel Gall.

In Sir Hugh Rose's despatch of 13th October 1858, the services performed by the regiment, the privations undergone from exposure to the sun, want of water, and scarcity of forage, are fully recorded, and Lieutenant-Colonel Gall's name is specially brought to notice for his gallant conduct in the pursuit of the rebels after their defeat at Golowlee and Calpee, in which 5 troops of the Fourteenth took part. In this pursuit the cavalry, who had been 13 hours in the saddle, pursued 8 miles on the Jhansi road. The names of Captain Barrett,

Charges made by the Fourteenth in the pursuit at Calpee.

[1] Dr. Lowe, author of *Central India During the Rebellion of* 1857-58.

[2] Malleson, vol. v. p. 130. [3] See p. 600 *et seq.*

Captain Need, Lieutenant Giles, Surgeon Stewart, Acting Regimental Sergeant-Major Clark, and Private Winton of 'B' troop were specially mentioned by Lieutenant-Colonel Gall in his report of the 25th May as having performed gallant services on the same occasion. Captain Need led a charge against the rear of an infantry column, in which the Fourteenth captured 3 guns and cut down 200 rebels. On this occasion Lieutenants Giles and Beamish used their revolvers with great effect. A charge led by Lieutenant-Colonel Gall with Captain Barrett, 14th Light Dragoons, and Captain Abbott, Nizam's Cavalry, at the head of their respective troops, went straight into a line of the retiring enemy and cut up about 300 rebels, losing only 2 men wounded, with 3 horses killed and wounded, and also captured 4 elephants.[1] The Ranee and the Nawab of Banda both made a precipitate flight from Calpee, at midnight, after the battle. It was said that a shell from Maxwell's battery burst in the Ranee's room and killed two of her attendants, which somewhat hastened her departure.[2] Subsequently, on 1st June, she was in the vicinity of Gwalior when the Maharajah's troops deserted him, and, after his flight to Agra, took possession of the city and fort of Gwalior together with Rao Sahib and Tantia Topee.

Flight of the Ranee and the Nawab of Banda.

On the 25th May, Sir Hugh despatched a 'pursuing column,' in consequence of the information brought by Lieutenant-Colonel Gall on his return from the pursuit, that he was certain a very large body of rebels had retreated along the Jaloun road to reach a ford across the Jumna, heading to the north. With this column a troop of the Fourteenth was despatched at first, and shortly afterwards, on the 29th May, 2 squadrons of the Fourteenth and a wing of the 86th Regiment followed from Calpee as reinforcements.[3] Lieutenant-Colonel Robertson reported soon after that there was no doubt the main body of the Calpee rebels had taken the road to Gwalior, and the 1st Brigade under Brigadier

Lieut.-Colonel Robertson's 'pursuing column' to Gwalior. The left wing of 14th Light Dragoons forms a part of this force.

[1] See Appendix B, pp. 602, 603. [2] Sylvester. [3] *Ibid.*

Stuart set off on the 1st June in hot haste after them: 2 troops of the Fourteenth went with this brigade.[1] On the 6th June the Major-General himself started for the same destination, with a part of the 2nd Brigade, in which were 2 more troops of the Fourteenth. Thus the whole of the Fourteenth once more set out on the march for further fighting and campaigning in the very hottest time of the whole year, and as the rains were nearly due, every endeavour would have to be made to reach Gwalior before they regularly set in.

The 1st and 2nd Brigades march on Gwalior.

This march of the Central India Field Force on Gwalior was mostly carried out at night, to avoid the terrific heat of the day. The men usually left camp at 11 P.M., reaching their camping-ground by sunrise. Brigadier-General R. Napier, C.B., arrived from Lucknow about the 14th June, and assumed command of the 2nd Brigade. Notwithstanding the great heat,[2] Sir Hugh Rose made forced marches, and reached the vicinity of Morar cantonments by 16th June. Here the rebels were ready to receive us in force, but, tired though his men were, the Major-General, with his usual dash, engaged them, and took the cantonments after some severe fighting. He first reconnoitred the position carefully, and ascertained the enemy's strength, then decided to attack. The Fourteenth and Nizam's Cavalry were placed on either flank, guns and infantry in the centre; the 2nd Brigade supported, the 1st was in front. Sir Hugh moved somewhat to his right and turned the enemy's left. The enemy being taken thus unexpectedly, gave way on all sides, but a considerable number took shelter in a dry nullah with ravines, in rear of the cantonments and a village adjoining, and here the 71st Highland Light Infantry, under Lieutenant-Colonel Campbell, lost an officer and several men in clearing out the rebels. The enemy was eventually driven by our men right through the cantonments, and the rout was completed by a very successful charge of a wing of the

Brigadier-General R. Napier, C.B., commands 2nd Brigade.

Capture of Morar cantonments, 16th June 1858.

Charges of 14th Light Dragoons at Morar.

[1] See despatch of 13th October 1858, p. 603 *et seq.*

[2] It was 130° Fahrenheit in the shade at times.

Fourteenth led by Captain P. S. Thompson. He caught the fugitives in the plains and cut up a large number of them. One troop of the Fourteenth under Lieutenant Gowan also charged a body of rebels as they fled from the ravines, and destroyed a considerable number.[1]

On the 17th, Brigadier Smith, coming from Rajputana with a column to join the Central India Field Force round Gwalior, had arrived near Kota-ke-Serai, 5 miles to the south-east of Gwalior, and had been fighting all day with strong bodies of the rebel infantry. He had reported his need of assistance, and Sir Hugh sent off immediately, as a reinforcement, the 25th Bombay Native Infantry, 4 guns of artillery, and 3 troops of the Fourteenth, all under command of Colonel Robertson.

Death of the Ranee of Jhansi.

During a cavalry charge made by a squadron of the 8th (King's Royal Irish) Hussars belonging to Brigadier Smith's column, which was led by Colonel Hicks and Captain Heneage of the 8th Hussars, the Ranee of Jhansi met her death. She was slain by a trooper, her horse having fallen, and he being quite ignorant of her rank or sex. She was dressed as a cavalry soldier. Her body was picked up and burned that night by some of her devoted followers, in accordance with the Hindoo custom. On the following day, 18th June, Sir Hugh started with the 1st Brigade from Morar cantonments (which were left in charge of Brigadier-General Napier and the 2nd Brigade), and marched in the afternoon to join Brigadier Smith. It was a long, circuitous march, very trying, especially to the infantry, who knocked up in considerable numbers. At night they bivouacked on the Morar river, and in the morning Scindiah had arrived from Agra to witness the result of that day's fighting, which was destined to restore to him his lost principality with the city and fort of Gwalior. After an early reconnaissance, Sir Hugh Rose decided to attack as speedily as possible. During the whole of the previous day, the 18th, the

[1] Brigadier-General R. Napier, C.B., mentions in his report (see Appendix B, p. 607) the gallant part taken on this occasion by the right wing of the Fourteenth.

enemy's guns, posted on a ridge to the left of Brigadier Smith's camp, had been pounding our troops. A canal lay between the two forces, and this had to be bridged over by us. This duty was performed with alacrity by the Madras sappers and miners, whilst all the time the enemy's battery on the ridge kept up a heavy fire on the working parties, and masses of their infantry were seen moving to take up positions from which they could act with the best advantage to oppose our advance. Sir Hugh lost no time in sending his infantry, 86th and 95th British Regiments, as well as the 10th and 25th Bombay Regiments, across the canal to seize the ridge and attack both flanks of the opposing rebel infantry. He directed Brigadier Smith to be ready to attack the enemy's positions at the Phool Bagh[1] and beyond it with a troop of Horse Artillery and a squadron of the Fourteenth, whilst he sent a light field battery, escorted by 2 troops of the Fourteenth, to the heights as a covering party for his advanced line, intending these guns to reply to the enemy's batteries in position in front of Gwalior.[2]

The infantry attack was admirably carried out. Several of the enemy's guns were taken in position, and turned on themselves by the men of the 86th and 95th Regiments, whilst the Bombay Regiments also captured some guns. 'The day was won,' says Malleson; 'the heights were gained, Gwalior lay, as it were, at the feet of the British.' To their right, looking down from the heights above across the plain, lay the Phoolbagh and the old city, surmounted by the fort; to their left lay the Lushkar (literally 'encampment,' but really the 'New City'), with the Maharajah's palace, and its spacious streets and houses. On the plain were visible the enemy, who had been driven from the heights, endeavouring to seek refuge in one or other of the fortified places or walled enclosures at hand. Sir Hugh at once ordered a general advance. The 1st Bombay Lancers were sent round the rear of the heights by a circuitous

[1] 'Flower-garden.' A garden and palace at the foot of the fort.
[2] See Appendix B, pp. 605, 606.

road, and were thence to make an attack on the New City (the Lushkar). The 3rd Bombay troop of Horse Artillery and a squadron of the Fourteenth protected the right flank of the troops attacking the grand parade, and also turned the enemy's left. The 86th Regiment advanced on the left, the 95th on the right. The rebels appeared paralysed. It is true their guns replied, but their infantry fell back. The charge of the Bombay Lancers added to their dismay, after which our infantry came up and completed the capture of the Lushkar. Meanwhile, on the right, Brigadier Smith had taken the Phoolbagh, killing large numbers of the rebels, and, as a considerable body of them, including some cavalry, fled towards the British Residency, which lay about 8 miles from Gwalior in the direction of Agra, he launched forth after them his cavalry and guns in pursuit, which was carried on till long after dark, inflicting great loss on the fugitives, and capturing most of their guns. Brigadier Smith speaks very highly of the steadiness with which the 2 troops of the Fourteenth, escorting the 3rd troop Bombay Horse Artillery, stood the enemy's artillery fire of 'shot and shell, and of the ardour with which they afterwards fell on the guns and the retreating enemy.'[1] Sylvester says:—'The enemy was taking away a large number of guns and ammunition wagons, but Brigadier Smith with the mounted part of his brigade gave pursuit, overtook, fought, and captured several guns, and only allowed the remnant to escape because the exhausted men and cattle could go no further in the darkness, for night had not closed the work. Here, as ever heretofore, did the 14th Light Dragoons and Bombay Horse Artillery add to their now weighty crown of laurels.'

Two troops of the Fourteenth had been held in support on the lower slopes, during the final attack on the two cities of Gwalior. The fort still remained unconquered. It was owing to the gallantry of a young officer, Lieutenant Rose, 25th Bombay Native Infantry, assisted by Lieutenant Waller of the

[1] See Appendix B, p. 606.

same regiment, who had under them a detachment of their own men, and who were posted on duty near the main gateway of the rock fort, that this important and formidable fortress fell into our hands on the 20th June, the day following the battle of Gwalior. It was 'a deed of unsurpassed daring,' says Malleson. Lieutenants Rose and Waller with a party of sepoys and some of Scindiah's police (Kotwallahs) forced the gate, passed through the other six gates, charged a gun at the archway of the fortress, and closed with the rebel garrison opposed to them. Here they were engaged in a hand-to-hand contest. A desperate fight ensued, and many fell on both sides, when, just in the hour of victory, Rose fell mortally wounded by a musket-ball fired from behind the wall. The man who fired it was despatched by Waller, but Rose died a hero's death. The rock fortress was gained. Sir Hugh Rose thus mentions the circumstance in his despatch: 'But the gallant leader, Lieutenant Rose, who has been twice specially mentioned by me for good and gallant conduct, fell in the fort mortally wounded, closing his early career by taking the fort of Gwalior by force of arms.'

Fort of Gwalior captured on 20th June 1858.

Tantia Topee, as usual, had deserted his post and fled when the battle was at its height, drawing off numbers of the rebel cavalry with him as a guard, 'and setting to his men a base and cowardly example, which went unpunished till a later period, when the halter encircled his "Bunniah"-born neck at Sipri.'[1] Thus Morar, the two cities and the fort of Gwalior were all in our hands by the 20th of June, and the total losses of the British arms were less than 100 men killed and wounded.

A squadron of the 8th Hussars and a squadron of the 14th Light Dragoons, 'honourable representatives of my force,' says Sir Hugh Rose,[2] 'escorted His Highness the Prince of Gwalior to his palace in the Lushkar.' This ceremony was attended by the Major-General and his staff, who received Scindiah with every possible mark of respect upon his arrival, in company with Sir Robert Hamilton, Agent to the Governor-General for

[1] Sylvester, p. 190. [2] See Appendix B, p. 606.

Central India, with a large and imposing retinue. Scindiah was everywhere greeted with enthusiasm by the populace, and after eighteen days' occupation by the rebels the city, fort, cantonments and states of Gwalior were restored to their ruling prince. Subsequently, as a mark of his appreciation of the great services rendered to him, Scindiah presented to the officers and men of the Central India Field Force the Gwalior Star, an elegant design in frosted silver, bearing on it a snake of gold, the crest of the Prince of Gwalior, and the figures '1858,' worn with an orange ribbon; and the royal authority was afterwards granted for its issue to, and acceptance by, the troops engaged.

The Gwalior Star.

In Appendix B are extracts from the despatch of Sir Hugh Rose, dated 13th October 1858, giving full particulars of the recapture of Gwalior and Morar.[1]

Sir Hugh Rose's farewell order to the Force.

After the victory at Gwalior, Sir Hugh Rose proceeded to Bombay to assume command of the army of that Presidency. He issued the following farewell order:—'The Major-General commanding, being on the point of assuming the command of the Poona Division of the Bombay army, bids farewell to the Central India Field Force, and at the same time expresses the pleasure he feels that he commanded them when they gained one more laurel at Gwalior. The Major-General witnessed with satisfaction how the troops and their gallant comrades-in-arms, the Rajputana Brigade under Brigadier-General Smith, stormed height after height, and gun after gun, under the fire of a numerous field and siege artillery, taking finally by assault two 18-pounders at Gwalior. Not a man in these forces enjoyed his natural health or strength: an Indian sun and months of marching and broken rest had told on the strongest; but the moment they were ordered to take Gwalior for their Queen and country they thought of nothing but victory. They gained it, restoring England's true and brave ally to his throne, putting to rout the rebel army, killing many of them, and taking from them in the field, exclusive of those

[1] See p. 603 *et seq.*

in the fort, 52 pieces of artillery, all their stores and ammunition, and capturing the city and fort of Gwalior, reckoned the strongest in India. The Major-General thanks sincerely Brigadier-General Napier, C.B., Brigadier Stuart, C.B., and Brigadier Smith, commanding brigades in the field, for the very efficient and able assistance which they gave him, and to which he attributes the success of the day. He bids them and their brave soldiers once more a kind farewell. He cannot do so under better auspices than those of the victory of Gwalior.'

On the 19th June, at Gwalior, immediately Sir Hugh saw that success was certain, he sent off an express to Brigadier Napier at Morar requesting him to pursue the rebels as far and as closely as he could. Napier received this order between 5 and 6 o'clock on the morning of the 20th June, and started within two hours, taking the following troops of his 2nd Brigade with him[1]:—

	Men.
Lightfoot's troop of Bombay Horse Artillery, .	99
Captain Prettejohn's troop 14th Light Dragoons,	62
Captain Abbott's Hyderabad Cavalry, . .	245
3rd Light Cavalry, 2 troops, . . .	104
Meade's Horse, ...	180
Total, . .	690

This force, after a fatiguing march of 25 miles exposed to a terrific sun, rested for the night, and next morning, shortly after sunrise, came up with Tantia Topee and a strong body of the rebels at Jowra-Alipore.

Affair at Jowra-Alipore, 21st June 1858.

In his *Sepoy Revolt*, p. 281, General Innes, V.C., gives the following account of the engagement which ensued:—'The routed and flying enemy were forthwith pursued by a light column under Brigadier-General Robert Napier, C.B., who next day overtook them at Jowra-Alipore, drawn up 12,000 strong in two lines. Holding his cavalry in hand in their

[1] See Brigadier Napier's report, Appendix B, p. 608 *et seq.*

front, but sheltered from them by intervening high ground, he sent the horse artillery to enfilade their lines from their left flank. This speedily shook and began to roll up their lines, when the cavalry were let loose and made a frontal charge. On this the enemy broke and fled, losing 25 guns, all their equipment, and 300 killed. The rout was complete, and from that time, the end of June, Tantia Topee's force became a fugitive one: it gradually dwindled away, though it never surrendered.' This force, which was so effectively dealt with by Brigadier-General Napier's column, was composed of the remnants of the Calpee army, with considerable additions picked up at Gwalior, and was headed by Tantia Topee, Rao Sahib, and the Nawab of Banda. Malleson says:—'The result was decisive. Prettejohn's distinguished valour and Abbott's gallant leading were especially conspicuous. The dash of Lightfoot's Horse Artillery was superb to look at. "You cannot imagine," writes an eye-witness, a cavalry officer, "the dash of the artillery: it was wonderful. We could scarcely keep up with them." But, in fact, every man behaved like a hero; each vied with his comrade. After a brief resistance the rebels broke and fled, hotly pursued. They lost 25 guns, all their ammunition, elephants, tents, carts, and baggage, and had 300 to 400 men killed. Never was a rout more complete.' The Fourteenth had only one man (Private G. Staple) wounded in this affair. Captain Prettejohn, Captain Todd, and Surgeon Stewart were specially mentioned in the Brigadier's despatches,[1] and he recommended for the Victoria Cross Private Novell, of 'H' troop, for an act of conspicuous bravery in charging alone, under a heavy fire, into a village and killing one of the enemy. The pursuit was continued by the cavalry for 6 miles from the scene of their first attack.

The following is a brief recapitulation, taken from Malleson, of what the Central India Field Force had accomplished in less than six months:—

On 6th January, Sir Hugh Rose had left Indore.

[1] See Appendix B, p. 609.

On 24th January, he laid siege to Rathgur.
28th January, he defeated the Rajah of Banpur.
29th January, he took Rathgur.
3rd February, he relieved Saugor.
13th February, he took the strong fort of Garrakota.
4th March, he forced the Muddenpore Pass.
17th March, his 1st Brigade stormed the fort of Chanderi.
22nd March, he invested Jhansi.
1st April, he defeated Tantia Topee on the Betwa.
3rd April, he successfully stormed Jhansi.
6th May, he defeated Tantia Topee and the Ranee of Jhansi at Koonch.
23rd May, he beat the rebels at Golowlee near Calpee and occupied that fort the following day.
16th to 20th June, he recaptured from the rebels the Morar cantonments, the heights, cities and fort of Gwalior, and reinstated Scindiah on his throne.

The Central India Field Force was now to a great extent broken up. Brigadier-General R. Napier, C.B., took command of the Gwalior Division. The larger portion of the infantry remained at Morar and Gwalior with some cavalry and artillery, and a brigade of native infantry, with cavalry and artillery, was sent to hold Jhansi. Of Brigadier Smith's brigade a part was sent to Sipri, a part to Goona, and part remained at Gwalior. The distribution of the Fourteenth was as follows: 3 squadrons at Gwalior, 1 squadron at Jhansi.

Many of the officers and men of the Fourteenth had suffered from illness and sunstroke during the operations about Koonch, Calpee, and Gwalior. Colonel Charles Steuart, C.B., Brigadier of the 2nd Brigade, became ill at Koonch, and during the month of May he, together with Captain Brown, Lieutenant Leith, Lieutenant Travers, and Veterinary-Surgeon Dawson went home to England on medical certificate, owing to illness caused by the hardships and exposure they had undergone in the campaign. At Golowlee, on 22nd May, Major Scudamore, who was in command of the Fourteenth, was struck down by the sun's rays, and Brevet Lieutenant-Colonel Gall assumed temporary command of the regiment, and was senior officer with the five troops taking part in the pursuit on that day and at the capture of Calpee.

Casualties by sunstroke amongst 14th Light Dragoons at Calpee.

In the operations about Calpee the Fourteenth had lost, from sunstroke, 12 men: 8 of these were in the left wing with the 1st Brigade, and 4 were in the right wing with the 2nd Brigade.

The tired and enfeebled men of the late Central India Field Force were now looking forward to some rest and repose, and as the rains had begun to fall in torrents this gave them relief from the oppressive heat. But their rest was destined to be cut short. Tantia Topee had still to be pursued and the rebels must be hunted down: indeed, for several months to come there was plenty of work for the British troops in various directions over Central India and other provinces. Brigadier Napier operated with his detached parties first in Gwalior, then in the districts to the west and south-west of Gwalior, bordering on Rajputana, up to the end of November. In December a new enemy invaded Gwalior. The pseudo-prince, as Malleson calls Ferozshah, who was expelled by our troops from Mundesor in November 1857, had now determined to join Tantia Topee in his struggles against the British. Accordingly he crossed the river Jumna on 9th December, coming from Etawah, and proceeded in the direction of Jhansi. Previous to this, in the month of August, a column had been sent out from Jhansi under Captain P. S. Thompson, 14th Light Dragoons, in which was included a squadron of the Fourteenth consisting of 'C' and 'E' troops. This column was employed in various operations against the rebels, and had an engagement on 23rd September at Garotha. For his services on this occasion Captain Thompson subsequently received the thanks of His Excellency the Most Honourable the Viceroy and Governor-General of India in Council.

Affair at Garotha, 23rd September 1858.

From the 10th September to the 4th October, Major Scudamore had command of a column in the field against the rebels, in which a squadron of the Fourteenth was included, consisting of 'A' and 'H' troops; and again from the 5th October to the end of the year 1858, Major Scudamore was hunting down the enemy in command of another column, included in which was a squadron of the Fourteenth, composed of 'B' and 'D' troops. This column, like the previous one, took the field and operated in various directions throughout the districts surrounding Gwalior. Major Scudamore's columns.

In the month of December, owing to reports received by Brigadier-General Sir Robert Napier, K.C.B., about the advance of Ferozshah, as alluded to above, he sent off several detachments to watch the roads by which that chieftain would probably come.[1] On the morning of the 12th December he received from the commander of one of these, Captain William McMahon, 14th Light Dragoons, who was posted with a detachment of his men near the confluence of the Jumna, Chambal, and Sind rivers, information that the rebels had passed into the Lohar district of Kuchwaghar, a marshy spot, and were advancing up the jungles of the Sind river. The Major-General, hoping to intercept them, marched that day from Gwalior to Antri, on the Jhansi road, his force consisting of the following details:—

	Men.
1 squadron 14th Light Dragoons, . .	150
Mahratta Horse, ...	100
71st Highlanders, ...	117
25th Bombay Native Infantry, . . .	50
40 camels, Gwalior Camel Corps, . .	...
2 Bombay light field battery guns, . .	...

The squadron of the Fourteenth consisted of the 'A' and 'H' troops under Captain Prettejohn, and there were also 25 men of the Balandshar Horse, who joined Sir R. Napier's

[1] Malleson, vol. v. p. 251.

force at Narwar (for he had marched to the south from Antri, owing to fresh intelligence received of the rebels' movements), and by the 17th, marching with all speed, he reached Ranode, which is a large town 50 miles north-east of Goona.

Affair at Ranode, 17th December 1858.

As it happened, Ferozshah was marching from a different, though parallel, direction on Ranode that very same morning, with an intention of sacking the place. He was quite ignorant of the arrival of the English, and his army, a somewhat irregular mass, extended with a front of nearly a mile.[1] Sir R. Napier had scarcely time to form up the Fourteenth when the rebels were within a few yards of him, as the other portions of his little army were delayed by the camels in crossing a ravine. The troops actually engaged in the action were Prettejohn's squadron 14th Light Dragoons; Mahratta Horse, under Captain F. H. Smith; and 38 men of the 71st Highlanders under Captain Smith, mounted on camels, guided by Captain Templer.

Prettejohn, directly he saw the opportunity for charging, dashed with his squadron into the midst of the rebels, and the blow completely doubled them up. They turned and fled before the Mahratta Horse were upon them, but the latter were in time for the pursuit. Captain Prettejohn and 13 men of the Fourteenth were wounded in the charge, which was carried out with special dash and complete success. The pursuit was continued for 7 or 8 miles. The rebels, who as a mass made no resistance, fought bravely in individual cases. We captured 6 elephants, several horses and ponies, and a quantity of arms. There were 150 dead bodies of the enemy left on the scene of this charge at Ranode. As Captain Prettejohn was wounded, the command of the Fourteenth devolved upon Captain Need, who estimated the loss of the rebels in the subsequent pursuit at 300.[2]

[1] Malleson, vol. v. p. 253.

[2] See Appendix B, p. 610 *et seq.*, Captain Need's report and Sir R. Napier's despatch.

The following extract from a letter written by Brigadier-General Sir R. Napier, K.C.B., to Major Scudamore, then commanding 14th Light Dragoons (owing to the absence on sick-leave in England of Colonel C. Steuart, C.B., and Colonel Ainslie being employed elsewhere in India), has reference to the affair of Ranode:—

'This morning the enemy were beaten and pursued with slaughter for eight miles, your glorious Fourteenth going a mere handful into the mass without looking twice, Prettejohn leading, as you would expect.'

For their gallant conduct on this occasion the following officers and men of the Fourteenth were specially brought to notice in despatches:—

Captain Prettejohn.
Captain Need.
Captain Todd.
Lieutenant Giles.
Regimental Sergeant-Major Thomas Clark.
Corporal George Best ('H' troop).

Casualties of Fourteenth at Ranode.

Captain Prettejohn's wound was a severe sabre-cut on the outside and back of his left thigh, three inches above the knee. In addition to the 13 men wounded, the Fourteenth also had 1 officer's charger wounded and 1 missing, 5 troop-horses wounded and 13 troop-horses missing.

Lieutenant Hugh Gough[1] of the Mahratta Horse was present in this affair, and was mentioned in despatches for his gallant conduct on the occasion. He has very kindly allowed the following extract from a book written by him to be inserted here[2]:—

'We (2nd Mahratta Horse) arrived at our destination, Morar (Gwalior), just in time to take a share in Sir R. Napier's dashing pursuit of Ferozshah, ending with the action and complete dispersal of his followers at Ranode on 17th December. Sir R. Napier had received intelligence at

[1] Now General Sir Hugh Gough, G.C.B., V.C.

[2] *Old Memories*, by General Sir Hugh Gough, G.C.B., V.C. Blackwood, 1897.

Morar that Ferozshah, with about 2000 rebels, was endeavouring to make his way across Bundelkund towards Central India, endeavouring to join forces with the famous Tantia Topee, about the most energetic and restless of the rebel leaders. Hoping to cut off Ferozshah, Sir Robert organised a movable column, consisting of a squadron of the 14th Light Dragoons, 100 men of the 71st Highland Light Infantry (as a camel corps), the 2nd Mahratta Horse, and the Torvanna Horse (independent Punjaub Cavalry) under a native leader, Jehan Khan. With this small force, about 350 all told, Napier left Morar on, I think, December 13, 1858, starting down the Jhansi road. On arriving near the village of Ranode in the early morning of the 17th December, our little force was carefully distributed in ambush, and as the enemy approached, when thoroughly exposed to our attack on an open plain, the order was given to "Advance and charge." We were down on them like lightning, and in an instant all was confusion, slaughter, and flight. Ferozshah was the first to bolt, and being well mounted, he got away; but most of the leaders were killed, and little resistance was made. It was a case of *sauve qui peut* from first to last. Our newly raised recruits (Mahratta Horse) were not in it with the 14th Light Dragoons, who rode like mad and pursued for seven miles over a most break-neck country. We cut up numbers, and dispersed the whole gathering, besides capturing six or seven elephants. Prettejohn of the Fourteenth got a severe sabre-cut across the thigh, and on my way back I passed him, much distressed, not at being wounded, but at being unable to get at his cigars, which, being in his holster, had disappeared with his horse! After this affair at Ranode, the column marched for a time about the jungles in pursuit of Tantia Topee, and we passed through Goona and Augur where "Meade's Horse"[1] were, and subsequently we returned to Morar.'

[1] It was the commandant of 'Meade's Horse,' Major Meade, who subsequently captured Tantia Topee.

1858-59

In the subsequent pursuit of the rebels by Brigadier-General Sir R. Napier's column, Major Prettejohn's men endured some very harassing marches, continuing, as they did, for a prolonged period.

On 1st March 1859, the 'B' troop joined this column, and all were employed actively till 3rd April in hunting up bodies of rebels scattered through dense jungles, the men being frequently in the saddle from sunrise to sunset. On 2nd April, 'A' and 'H' troops were present at the attack of a village during some operations in the Gwalior district.

After the battle of Gwalior on 19th June 1858, the headquarters of the Fourteenth had remained stationed in the Morar cantonments for nearly a year; 'C' and 'E' troops were sent for a time to Jhansi, but they returned, under Major Thompson, on 3rd August to Morar.

Most of the various troops of the Fourteenth had been kept employed in different expeditions, on field service, during these eventful times.

In recognition of their services Majors Scudamore and Gall were promoted Brevet Lieutenant-Colonels, and nominated Companions of the Most Honourable Order of the Bath.

Captains Todd, Thompson, and Prettejohn became Brevet-Majors, and later on they received the brevet rank of Lieutenant-Colonel; and Captains McMahon and Need became Brevet-Majors.

Regimental Sergeant-Major Clark was awarded the medal 'for gallant conduct in the field.'

During the year 1858 the headquarters were at Kirkee in January.

On 1st February they were at Camp Rahala.
On 1st March ,, ,, ,, Camp Jhilwand.
On 1st April ,, ,, ,, Camp Goona.

On 1st May they were at Camp Jhansi.
On 1st June ,, ,, ,, Camp Calpee.
On 1st July ,, ,, ,, Morar till the end of the year.

Brevet-Colonel Ainslie had been for a time at Jhansi in command of a brigade, which operated against the rebels and advanced towards Ranode in December of this year (1858).

Description of Lieutenant-Colonel Gall, C.B., 14th Light Dragoons.

(Extract from *Scraps from my Sabretache*, by G. Carter Stent. W. H. Allen and Co., 1882.)

'Colonel Gall deserves something more than a passing notice, and though I disliked him personally, I cannot refrain from expressing my admiration of him as a soldier and a daring officer.

'He was a short, spare, sallow-visaged man; but in his little frame was an immense amount of courage and endurance. He, I believe, gloried in danger, and would face anything or everything—the devil himself. He had so much confidence in himself, and during the Sikh war was endeavouring to seize a standard, when he received a sabre-cut which rendered his right hand useless. He thought to get over that, however, by inventing a sword which could be fitted to his wrist; this, after a trial or two, he found did not answer, so he had to give it up and use his left hand. In leading a charge, either against the enemy or at a field day, he would turn round in his saddle and say, "Now, men, you are quite at liberty to gallop over me—if you can!" He was always so splendidly mounted that that was an impossibility.

'In riding, his light frame seemed to grow out of the saddle; as the old soldier constantly remarked, "He sticks to the saddle like a sick monkey on the yard-arm!" He was reported never to undress, but always to sleep booted and belted, and dressed ready for a turn-out on the instant; and this would appear to be true, for at the first blast of the trumpet he would appear riding down the lines fully equipped, as if he had been waiting for the trumpet to sound. I am not his biographer, or I might write a volume concerning him; but I will conclude with observing that, though he was not very popular among his own men as a commanding officer, every man among us admired him for his daring as a soldier.'

Sylvester, who lived in the same tent with him during the greater part of the campaign, says of Lieutenant-Colonel Gall: 'His energy wore him to a shadow; he was brave to a fault, but fussy to a degree. He often slept in his uniform, and when near the enemy would often get up in the night and ride round the picquets.'

Since the 1st April 1858, a schoolmaster-sergeant had ceased to be borne on the strength of cavalry regiments. Establishment altered, 1858.

During a portion of 1858, Captain J. Barrett had been in command of the headquarters, nearly all the other officers being with the various portions of the regiment engaged on service in the field—some with the Nerbudda Field Force, some with the Central India Field Force, and some with the Gwalior Division. Five troops were present at the capture of Garrakota, 5 troops at the forcing of the Muddenpore Pass, and 3 troops at the capture of Chanderi.

Towards the close of the year 1858, the service-troops in India numbered 660 non-commissioned officers and men. The depôt at Maidstone consisted of 134. Effective strength, 1858-59.

1859

From 1st January to end of April several of the troops were on field-service with the Gwalior Division, and some, as we have already related, were detached with flying columns through the provinces.

'Persia' granted, 1859.

On 21st January the Fourteenth received permission to bear the word 'Persia' on their cap-plates and appointments in consideration of their services in that country in 1857.

Regiment ordered to England, March 1859.

On the 4th March, orders were received for the Fourteenth to go to Bombay, preparatory to giving up their horses and embarking for England. There were 71 men who volunteered to remain in India, and they were transferred chiefly to the newly formed regiments, now the 19th and 20th Hussars, and the 21st (Empress of India's) Lancers, which were composed of volunteers from the old Bengal European Light Cavalry Regiments belonging to the late Honourable East India Company's service. On the 24th April the Fourteenth gave up their horses to the Commissariat Department.

The headquarters remained at Morar (Gwalior) till 26th April, when they left by bullock-train *en route* to Bombay, the rest of the regiment having proceeded there in advance by same route.

Farewell orders to the Fourteenth by Sir Robert Napier and Brigadier-General Stuart.

On the 20th April, at Camp Serony, when the Fourteenth were about to proceed to Bombay, the following farewell order was issued by Brigadier-General Sir Robert Napier, K.C.B., commanding Gwalior Division:—'I cannot part with the troops now leaving the Gwalior Division, after having been associated with them during a year's eventful service, without a few words to convey to them my admiration and regard for their excellent and soldier-like conduct. To Colonel Scudamore it is due that the State of Duttea and the central districts of Gwalior were protected from the rebels under the Rao Sahib, who were baffled by the movements of his small but undaunted column. The brilliant 14th Light Dragoons and their charges at Jowra-Alipore and Ranode will not easily be forgotten.'

The Brigadier's order was as follows:—'Brigadier-General Stuart has to express his best thanks and acknowledgments to the officers, non-commissioned officers, and men of that part of the 14th Light Dragoons that served under his command with the Malwa and Central India Field Forces, for the gallantry

HONI SOIT QUI MAL Y PENSE
DOURO
FUENTES D'ONOR
VITTORIA
PENINSULA
CHILLIANWALLAH
PERSIA
TALAVERA
SALAMANCA
ORTHES
PUNJAUB
GOOJERAT
CENTRAL INDIA
14TH (KING'S) HUSSARS

FIRST (OR KING'S) GUIDON.

1751.

SECOND GUIDON.

1751.

XIV
L.D
DUCHESS OF YORKS OWN
XIV
L.D

1798.

TALAVERA
FUENTES D'ONOR
VITTORIA
HONI SOIT QUI MAL Y PENSE
XIV
KING'S
L.D
SALAMANCA
ORTHES
PENINSULA

1832.

1742.

1715.

1776.

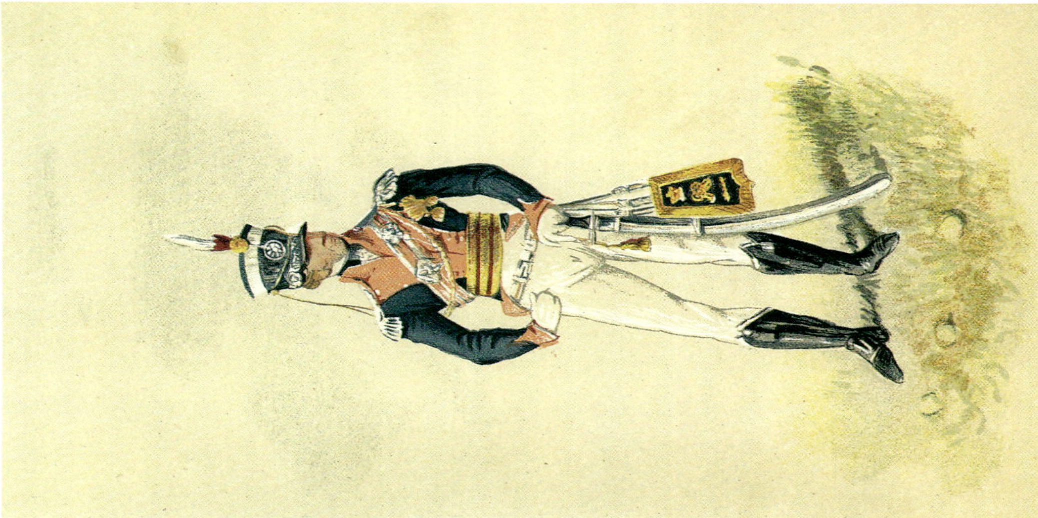

1831.

1815.

1819.

1836.

1841.

1851.

R. SIMKIN

1891.

1894.

1899.

1900.

Officer's Shako-Plate, from 1829 to 1860.
(Prussian Eagle worn in centre up to 1832.)

Officer's Button (XIV), from 1830 to about 1845.

Officer's Button (14), from 1845 to 1856.

Universal Button worn on Tartan Frocks since 1899.

and dash they have displayed on all occasions in which they were engaged with the enemy. Discipline and efficiency have been admirably maintained throughout under most trying circumstances. They have borne with great cheerfulness exposures and fatigues in all weather. In taking leave of this distinguished regiment, Brigadier Stuart feels that he is parting with tried friends and comrades, and most heartily wishes them all honour and happiness in their future career.'

By the 23rd May the regiment was assembled at Bombay preparatory to embarkation for England, but the order was countermanded, and it returned to Kirkee, where it arrived 26th May and remained for the rest of the year. Order for home countermanded, May 1859.

Being very short of horses, remounts were now quickly supplied to the regiment. On 6th June, 176 unbroken remounts joined, and by the 21st June, 14 days after these horses joined, 2 squadrons paraded in marching order for the inspection of Major-General Sir Hugh Rose, K.C.B., who complimented the commanding officer and all concerned at the promptitude with which the remounts had been trained. Regiment remounted, June 1859.

On 22nd June a further batch of 88 remounts joined, and on 6th July 88 more. This completed the regiment up to 55 horses per troop. They mostly consisted of Arabs and Persians.

On 27th July a squadron of the regiment in review order escorted Major-General Sir Hugh Rose, K.C.B., on the occasion of his being invested with the order of G.C.B.

On 12th August, Captain T. Barrett died of hepatitis at Kirkee.

On the 22nd August the Fourteenth were present at a brigade field-day at Poona, and although only six weeks had elapsed since the last batch of remounts had been received, the whole of the regiment was now quite fit again for active service.

On 12th November an order was received for the regiment to prepare for immediate embarkation for England. Regiment ordered to England, November 1859.

On 17th November, His Excellency General Sir H. Somerset inspected and was much pleased with the state of the regiment. He said that H.R.H. the Duke of Cambridge had repeatedly written to him expressing his admiration of the gallant conduct of the 14th Light Dragoons during the late campaign in Central India.

Volunteering was now opened, and 130 men volunteered to go to cavalry regiments serving in the Bombay and Madras Presidencies, and to the Bombay Horse Artillery.

1860

Depôt moved from Maidstone to Newbridge.

The Fourteenth were at Kirkee from the 1st January till they moved to Bombay in February for embarkation. On the 25th January the depôt troop was moved from Maidstone, where Colonel Gibson was then commandant, and proceeded to Newbridge, arriving there the 27th January, to await arrival of the service troops, having a strength of 8 officers, 188 non-commissioned officers and men, 12 women and 10 children.

Regiment embarks for England at Bombay, 11th February 1860.

On the 1st February the horses were handed over to the remount depôt at Kirkee, and on the 11th February the Fourteenth embarked at Bombay in 2 detachments for England, headquarters and 6 troops—'A,' 'B,' 'C,' 'D,' 'G' and 'H'—under Lieutenant-Colonel Scudamore in ss. *Sydenham*, being transhipped on the 1st June, four days after arrival at Plymouth, to ss. *Iberia* for Liverpool and Dublin. The remaining 2 troops ('E' and 'K') embarked at Bombay in ss. *Speedy* on the 13th February, under Major Thompson, for Gravesend, where they were on the 22nd June, and proceeded by rail to Liverpool, embarking there the 23rd for Dublin, where they landed the 24th June, and proceeded to Newbridge by rail.

Regiment at Newbridge, July 1860.

The headquarters arrived in Dublin on the 2nd June, and proceeded to Newbridge by rail on the 3rd June. The whole regiment was at Newbridge on the 1st July under the command of Colonel C. Steuart, C.B., and remained there till September 1861, having completed a service in India and Persia of nineteen years and one month.

During the Mutiny campaign in India, the Fourteenth

lost 1 officer and 72 men, and 6 men died on the voyage from India. Strength of the regiment embarked at Bombay: 13 officers, 391 non-commissioned officers and men, 19 women, 39 children.

Colonel Steuart received orders to complete the Fourteenth with troop-horses up to 428, which was done chiefly by purchases in the south of Ireland. Three hundred and thirty-four troop-horses had joined headquarters with the depôt, 46 of which were transferred from 1st Royal Dragoons, 5th Lancers, 3rd Light Dragoons, and 15th Hussars.

On the 3rd June the establishment of the Fourteenth was considerably reduced, and was fixed as under:— Establishment, 1860.

8 Troops (instead of 10).	40 Sergeants.
1 Lieutenant-Colonel.	32 Corporals.
2 Majors.	9 Trumpeters.
8 Captains.	8 Farriers.
8 Lieutenants.	537 Privates.
8 Cornets.	428 Troop-horses.

As the second lieutenant-colonelcy was reduced, Lieutenant-Colonel (Brevet-Colonel) C. P. Ainslie left the regiment.

The changes in establishment since 1854 (when a saddler-sergeant was added) had been few: in 1857 the number of troops was increased from 9 to 10, with

59 Sergeants,	40 Corporals,
14 Trumpeters,	625 Privates,
10 Farriers,	703 Troop-horses

and in 1858 the schoolmaster-sergeant was reduced At the present time the regiment was considerably below its authorised strength, and several recruiting-parties were sent out from headquarters, one to Norwich and one to Devizes, and in addition some of the infantry recruiting centres were given instructions to aid in obtaining recruits for the Fourteenth. The greater number of the recruits received came from London and Birmingham. Recruiting.

On 25th July, Major-General J. Lawrenson, Inspector-General of Cavalry, inspected the troop-horses and remounts.

On 12th October, Major-General William Parlby, commanding Cavalry Brigade, Dublin district, made an inspection of the regiment.

Colonel—William Beckwith.

On 12th November, Major-General William Beckwith, K.H., became Colonel of the Fourteenth, in succession to General the Honourable Sir Henry Murray, K.C.B., deceased. The latter officer had been promoted to the rank of Knight Commander of the Order of the Bath shortly before he died. Major-General Beckwith had previously served in the Fourteenth as major, up to the year 1833, when he left on being promoted to a half-pay lieutenant-colonelcy, and it was he who in 1831 commanded the squadron of the Fourteenth employed in the suppression of the Bristol riots, with such conspicuous ability and success.

On 12th December, Lieutenant G. M. Dew died at Liverpool of diphtheria.

1861

Major-General W. Parlby inspected the Fourteenth at Newbridge on 27th March.

On 16th April, General the Right Honourable Sir George Brown, G.C.B., K.H., Commander of the Forces in Ireland, came from Dublin and reviewed the Fourteenth, when he expressed great satisfaction with its efficiency in every respect.

Lieutenant-Colonel—A. Scudamore.

On the 19th June, Major (Brevet Lieutenant-Colonel) A. Scudamore, C.B., became Lieutenant-Colonel, and succeeded Colonel C. Steuart, C.B., in command of the regiment.

On 13th August, General H.R.H. the Duke of Cambridge, Commanding-in-Chief, came to the Curragh Camp and reviewed the Fourteenth, together with the whole of the Cavalry Brigade of the Dublin Division, and he expressed himself very pleased at the rapidity with which the Fourteenth, composed for the most part of recruits and young horses, had since their return from India been brought up to their present state of efficiency.

On 13th August, Lieutenant-Colonel A. Scudamore was awarded a good-service pension of £100 a year.

On 15th August, H.R.H. the Prince of Wales honoured Lieutenant-Colonel Scudamore, C.B., and the officers of the regiment with his company at dinner, in their mess at Newbridge Cavalry Barracks. H.R.H. the Prince of Wales dines with the officers.

On the 17th August the 14th (King's) Light Dragoons were converted into Hussars, with the designation of 14th (King's) Hussars, the uniform being blue, busby-bags yellow, and plumes white. 14th (King's) Hussars, uniform blue; busby-bags yellow.

On the 26th August the Fourteenth were present at a review of troops held by Her Majesty Queen Victoria on the Curragh of Kildare. Her Majesty was graciously pleased to express herself much gratified with the appearance of the troops, and desired Lieutenant-General the Right Honourable Sir George Brown, G.C.B., Commanding the Forces in Ireland, to make this known to them. Reviewed by H.M. Queen Victoria on the Curragh, August 1861.

On the 13th, 14th, and 16th September the headquarters and 6 troops (marching by squadrons) left Newbridge and arrived on the 17th and following days at Dundalk, where Major-General Lawrenson, Inspector-General of Cavalry, made his inspection on the 25th September. He expressed himself perfectly satisfied in every respect with the appearance of the regiment. 'H' troop had marched to Belturbet on the 12th September, and 'G' troop on the same day to Belfast, there to be stationed.

Brigadier-General G. W. Key, C.B., commanding the Cavalry Brigade, Dublin Division, made his inspection of the regiment at Dundalk on the 11th October, and expressed his satisfaction.

The approximate strength now was 33 officers, 570 non-commissioned officers and men, 390 troop-horses. A sergeant-instructor of musketry was added to the establishment. Establishment.

On the 8th October, Lieutenant-Colonel Scudamore, C.B., exchanged to the 7th Hussars, and Lieutenant-Colonel Sir William Russell, Bart., C.B., came from the 7th Hussars to command the Fourteenth. Lieutenant-Colonel—Sir William Russell, Bart.

1862

Regiment arrives at Manchester and Preston, May 1862.

On the 6th May the regiment left Dundalk, Belfast, and Belturbet, for Manchester and Preston *via* Dublin and Liverpool, the headquarters arriving on the 13th May at its new quarters. 'C' and 'K' troops went from Dundalk to Preston, and the remaining 6 troops went from Dundalk, Belfast, and Belturbet to Manchester.

On the 10th July, Major-General Lawrenson, Inspector-General of Cavalry in Great Britain and Ireland, inspected the Fourteenth at Manchester, and seemed very pleased with all he saw. He particularly remarked in flattering terms as to the excellent manner in which the field movements were performed under great disadvantages of ground.

On the 6th September, 'A' troop moved from Manchester to Bury, and 'C' troop from Preston to Manchester, and on the 16th September, 'B' troop marched from Manchester to Ashton-under-Lyne, and 'K' troop from Preston to join headquarters at Manchester.

Bandmaster appointed, 1862.

A bandmaster was added to the establishment this year.

On the 24th September, Major-General Lawrenson, Inspector-General of Cavalry, again inspected the regiment. He was very pleased. He praised the good riding of the men especially, and said that the horses were well bitted, well groomed, and in good condition, and the non-commissioned officers' ride he considered excessively smart.

1863

In January, 'D' troop proceeded from Manchester to Bury, 'A' troop from Bury to Manchester, and 'B' troop came from Ashton to Manchester, 'E' troop going from Manchester to Ashton-under-Lyne.

On the 21st March, owing to the riots at Staleybridge, the 'E' troop from Ashton-under-Lyne, under Captain Chadwick, was ordered there in aid of the civil power, and the 'H' troop, under Brevet Lieutenant-Colonel Prettejohn, went the same night from Manchester to Ashton-under-Lyne to support the troop already there. 'H' troop returned on the 25th to Manchester.

On 27th March the regiment, under command of Brevet Lieutenant-Colonel R. H. Gall, C.B., was ordered to proceed by march route to visit the surrounding manufacturing towns where more riots were expected. It marched by Ashton-under-Lyne to Staleybridge, and the 'G' troop, under Captain Mather, was detached to remain for a time at the latter place, the rest of the regiment returning same day to Manchester.

On 2nd May, 'H' troop from Manchester relieved 'E' troop at Ashton-under-Lyne. On 15th May, 'G' troop from Manchester went to Bury and relieved 'D' troop, which arrived at Manchester the same day.

On 2nd October, at Manchester, the Inspector-General of Cavalry, Major-General Lawrenson, made his inspection of the regiment and was very pleased. He said, 'The Fourteenth is a very smart regiment.'

On 3rd September, Her Majesty's gracious permission was granted to the regiment to bear the words 'Central India' on the appointments. 'Central India granted, 3rd September 1863.

1864

On the 15th March, Brevet Lieutenant-Colonel R. H. Gall, C.B., retired on half-pay, and Captain R. J. Brown became Major.

Commencing on the 29th April, the Fourteenth left Manchester, Bury, and Ashton, in 5 divisions, and marched to Aldershot, arriving there on the 13th May, and were located in the North Camp. On the 11th June they were inspected by Major-General E. C. Hodge, C.B., commanding the Cavalry Brigade, at Aldershot. He was highly pleased with everything he saw, and particularly praised the interior economy and the manner in which the regimental and troop books were kept.

In July the regiment was encamped for 14 days at Sandhurst, under Major-General Hodge, C.B., commanding Cavalry Brigade, and experienced intensely hot weather.

On the 19th September it moved into the West Cavalry permanent barracks, South Camp. General Sir J. L. Pennefather, K.C.B., was the General in command at Aldershot.

Major-General E. Hodge, C.B., made a second inspection of the regiment on the 6th October.

Lieutenant-Colonel—P. S. Thompson.

On the 29th November, Lieutenant-Colonel P. S. Thompson succeeded Colonel Sir William Russell, Bart., C.B., in command of the Fourteenth, and Captain F. B. Chapman became Major, after only six years' service.

Uniform of Hussars.

At this time the uniform of all hussars was blue, facings same colour, except the 3rd Hussars (red) and 13th Hussars (buff); the Fourteenth having yellow busby-bags, thus retaining the old regimental colour of the facings.

1865

The regiment remained in the West Cavalry Barracks, Aldershot, the whole year, but was encamped in June for 21 days on Cove Common.

On the 5th May, Major-General Hodge, C.B., made his inspection, and was very pleased with the result. The cleanliness and good order of the stables, barrack-rooms, and married people's quarters especially called forth his unqualified praise. On the 8th August, the Inspector-General of Cavalry, Major-General Lord George Paget, C.B., inspected and was perfectly satisfied.

1866

The Fourteenth remained at Aldershot till February, when a sudden order was received on Sunday night to proceed on the 24th February to Hounslow, Hampton Court, and Kensington, in relief of the 6th Dragoon Guards (Carabiniers), who had been suddenly hurried off under Colonel Charles Sawyer by train and steamer to Ireland, owing to the 'Fenian' disturbances. Captain Travers and Lieutenant Ashton with 'D' troop went to Kensington, and 'E' troop, under Captain Chadwick and Lieutenant Bolton, to Hampton Court; 'A,' 'B,' 'C,' 'G,' 'H' and 'K' troops were with headquarters at Hounslow. Later in the year the 'E' troop came to Hounslow, and 'G' troop, under Captain Hay Newton, was sent to Hampton Court.

Education, nationality, and religion of the rank and file, 1866.

At this period there were 30 men in the regiment who could neither read nor write; 21 who could read but not write; 479 who could read and write; and 46 men of superior education. Out of the total of 576 non-commissioned officers and men serving, there were in the regiment—

491	English.	481	Church of England.
62	Irish.	62	Roman Catholics.
22	Scottish.	22	Presbyterians.
1	Foreigner.	11	Dissenters.
576		576	

Establishment

In April of this year 1 Sergeant-cook and 1 Sergeant-Instructor of Fencing were added to the regimental establishment. There were 2 privates who underwent corporal punishment at Hounslow, being the last cases previous to its abolition: one was for the crime of stealing, and the other for desertion.

On the 27th April the regiment was inspected at Hounslow by Major-General Lord George Paget, C.B., Inspector-General of Cavalry, who said he should have much pleasure in reporting very favourably as to its state to Field-Marshal H.R.H. the Duke of Cambridge, Commanding-in-Chief. On the 6th August, Lord George Paget, C.B., inspected the regiment again.

Hyde Park riots, December 1866.

On the 3rd December the Fourteenth marched from Hounslow to Wellington Barracks, London, and the Royal Mews, Buckingham Palace, being ordered there 'to aid the civil power should its services be required.' This was in consequence of the 'Beale' Hyde Park riots. On the 4th December the regiment returned to Hounslow.

1867

On the 6th January, Lieutenant-Colonel P. S. Thompson became Brevet-Colonel. On the 4th February, Captain A. N. Bolton died at Kingstown, near Dublin. On the 2nd April the Fourteenth began to march to the north of England. On the 23rd April, 'A' and 'C' troops arrived at Newcastle, under Major Chapman, and remained there till the end of August, when they proceeded to Hamilton, arriving there on the 10th September. During their stay in Newcastle they did excellent service at a fire rescue. Each man received a new suit of uniform as a reward for his bravery and the damage done to his clothing, also a free gratuity of a week's pay. During April and May the Fourteenth were marching to Edinburgh and out-quarters: they were 32 days on the march from Hounslow. 'B' troop went to Perth under Captain William Arbuthnot, 'D' troop (Captain Knox's) and 'G' troop (Captain Hay Newton's) to Hamilton. 'H' troop (Captain the Honourable F. Amherst's), 'K' troop (Captain Degby-Boycott's), and 'E' troop with head-quarters went to Piershill Barracks, Edinburgh. Major R. J. Brown with the sick and dismounted men, also the women and children, proceeded from Hounslow by rail and steamer *via* London and Leith. On the 17th August, 'A' troop under Captain Boycott marched to Kelso for the purpose of escorting Her Majesty Queen Victoria during her stay at Floors Castle, on a visit to the Duke and Duchess of Roxburgh, returning on the 26th August to Edinburgh. On the 8th October the regiment was inspected by Major-General Lord George Paget, C.B., Inspector-General of Cavalry, who was very pleased with all he saw.

Regiment arrives in Scotland, May 1867.

On the 23rd October, Cornet J. L. Milligan died at York.

On the 6th December there was a change of detachments. 'A' troop under Captain Campbell came from Hamilton to Edinburgh, and 'K' troop under Captain Knox marched from Edinburgh to Hamilton. Whilst the latter troop was marching through Linlithgow they came in for very hard weather, and all had to dismount and lead their horses owing to the ice on the roads.

Captain R. Chadwick was promoted Major *vice* R. J. Brown, retired.

A detachment from 'C,' 'D,' 'G,' and 'K' troops went from Hamilton to Glasgow, under Major Chapman.

On the 14th December the party marched at 2 A.M., and was billeted in the cattle-market, men and horses being provided for by the corporation, they having been ordered to Glasgow in aid of the civil power to suppress a contemplated rising of the Fenians there.

On the 16th December this detachment returned to Hamilton.

1868

In May the Fourteenth left Scotland for Ireland, and embarked in four successive parties at Glasgow in the ss. *Duke of Abercorn* for Dublin:—

The first division, 'B' and 'D' troops, under Major Chapman, arriving at Newbridge the 9th May.

Second division, 'A' and 'H' troops, under Major Chadwick, arriving at Newbridge the 13th May.

Third division, 'C,' 'G,' and part of 'K' troops, under Captain Baskerville, arriving at Newbridge the 18th May; and the headquarters and 'E' troop, under Colonel Thompson, leaving Edinburgh on the 16th May, arrived at Newbridge on the 21st May. Some of the troops remained a night in Portobello Barracks, Dublin, after disembarkation, and proceeded by march next day to Newbridge, where the Fourteenth relieved the 10th Hussars, under Colonel V. Baker.

Regiment at Newbridge, May 1868.

On the 8th June, Major-General A. A. T. Cunynghame, C.B., commanding the Dublin Division, inspected the regiment and expressed himself pleased with the result.

On the 9th September, Major-General Lord G. Paget, C.B., inspected, and previously to him Brigadier-General A.

Little, C.B., made his inspection, and both these general officers reported most highly to H.R.H. Commanding-in-Chief on the efficiency of the regiment.

On the 10th October, Brigadier-General A. Little, C.B., commanding the Cavalry Brigade, Dublin Division, inspected again at Newbridge.

On the 19th October the Fourteenth marched to Dublin, and were quartered in the Royal Barracks, one troop ('A') being at Portobello Barracks.

During November and December they were much detached in the north and south of Ireland in aid of the civil power at the parliamentary elections, and various troops were employed at the following places:—Ballincollig, Cork, Youghal, Kinsale, Bandon, Bantry, Skibbereen, Belfast, Dundalk, Drogheda, Carrickmacross, and Duleek. The last detachment returned to Dublin on the 15th December.

In November a wing under Major Chapman, consisting of 'B,' 'E,' 'H,' and 'K' troops, went by rail from Dublin to Cork, and sent detachments to Ballincollig, Youghal, Kinsale, Bandon, Skibbereen, and Bantry. These troops returned by march route after the elections, reaching Dublin on the 15th December.

'D' troop, under Captain Knox, marched to Drogheda the 13th November, and returned in ten days to Dublin.

A squadron, 'C' and 'G' troops, under Major Baskerville, went by rail to Dublin on the 23rd November. 'A' troop, under Captain F. P. Campbell, went by rail to Dundalk on the 19th November, thence on the 22nd November to Drogheda, on the 24th to Carrickmacross, 27th to Duleek, 28th by rail back to Dublin.

The squadron detached to Ballincollig found everything in a very 'rough-and-tumble' state, as no troops had been stationed there for several years previously. The troops at small places were for the most part kept on the move, never halting in the same place more than three days.

Captain F. P. Campbell became Major *vice* Chadwick, retired.

1869

Captain William Arbuthnot was awarded a Brevet-Majority in recognition of his services in the Abyssinian War under Lord Napier of Magdala.

From the 1st April the squadron system was introduced into the cavalry service. As is well known, it proved a failure and was very short-lived, partly owing to the purchase-system, as the four junior captains of every regiment who had purchased the command of their troops lost all command, and partly owing to the faulty and incomplete manner in which the system was introduced. Squadron system introduced, April 1869.

The following was the new organisation of the regiment :—

1st Squadron—Commanded by Brevet-Major Arbuthnot.
Second in command, Captain Ashton.

2nd Squadron—Commanded by Captain Knox.
Second in command, Captain Henry.

3rd Squadron—Commanded by Captain Mather.
Second in command, Captain F. S. Russell.

4th Squadron—Commanded by Captain the Honourable F. Amherst.
Second in command, Captain the Honourable J. St. V. Saumarez.

Establishment of the regiment :—

4 Squadrons.
1 Colonel.
1 Lieutenant-Colonel.
2 Majors.
8 Captains.
8 Lieutenants.
4 Cornets.
1 Paymaster.
1 Adjutant.
1 Riding-master.
1 Quartermaster.
1 Surgeon.
1 Assistant-Surgeon.
1 Veterinary Surgeon.
1 Schoolmaster.
48 Sergeants.
8 Trumpeters.
24 Corporals.
1 Saddle-tree maker.
12 Shoeing-smiths.
4 Saddlers.
426 Privates.
344 Troop-horses.

On the 13th May the Fourteenth were inspected in Dublin by Major-General Sir A. Little, K.C.B., commanding the Cavalry Brigade, Dublin Division. On the 28th July the whole regiment, having marched by squadrons from Dublin, commencing on the 22nd July, was encamped on the Curragh in Donelly's Hollow, under command of Colonel P. S. Thompson; the 6th Dragoon Guards (Carabiniers) were also there, under command of Lieutenant-Colonel C. W. Bruce, and worked in brigade with the Fourteenth, and a very wet season they had under canvas. Both regiments were very smart: the Carabiniers were noted for the celerity of their movements, the Fourteenth for their steadiness at drill and for the care they took of their young horses.

On the 16th August, at the Curragh Camp, Major-General Lord G. Paget, C.B., Inspector-General of Cavalry, made his inspection of the regiment, and Brigadier-General R. Wardlaw, C.B., commanding the Cavalry Brigade, Dublin Division, made his inspection on the 18th October. After an unusually wet and inclement drill season the camp broke up in October, and on the 23rd one squadron, with headquarters, marched to Longford, one squadron to Newbridge, one squadron to Limerick, half a squadron to Castlebar, and half a squadron remained at Curragh Camp under Captain the Honourable F. Amherst for vedette duty, the remainder of the Fourteenth having left camp by the 25th October.

On the 27th November the squadron at Newbridge went by rail to Nenagh in aid of the civil power during the Tipperary elections, and returned to Newbridge on the 30th November.

On 7th December the half squadron on vedette duty at the Curragh Camp proceeded to Longford, being relieved in the vedette duty on the Curragh Camp by a half squadron of the 6th Dragoon Guards (Carabiniers) from Newbridge, under Captain H. B. Hamilton.[1] The squadron at Newbridge marched to Athlone on the 7th December, and a half squadron

[1] Afterwards Colonel H. B. Hamilton, commanding 14th Hussars.

went on election duty to Ballymahon on the 16th December, returning the same day to Longford. The elections were going on in Tipperary and Longford, and the Fourteenth were much detached in various places for short periods from November 1869 to April 1870. On the 18th and 20th December a squadron went from Limerick to Ballinrobe, and detachments were employed in aid of the civil power during the winter at Westport, Claremorris, Tuam, and Dunmore.

One and a half squadrons, with headquarters, were employed on election duty at Longford on the 31st December.

Colonel P. S. Thompson was awarded a Companionship of the Order of the Bath this year in recognition of his services.

1870

In March a squadron ('H' and 'K' troops) under Major Chapman went from Athlone to Tuam in aid of the civil power, and a squadron went to Ballymahon from the 12th to the 14th April on similar duty, owing to County Longford elections.

The squadron organisation, after one year's existence, was abolished on the 1st April. The troop system was reintroduced into the cavalry service from the above date. *Troop system reintroduced, April 1870.*

Consequent on the abolition of the squadron organisation system on the 1st April, one troop in the regiment was reduced, and the establishment fixed as follows :—

Seven troops with 25 officers (the Surgeon and Assistant-Surgeon being attached, and no longer belonging to the regiment as before, and one major being reduced from the establishment).

39 sergeants.	21 corporals.
8 trumpeters.	386 privates.
4 farriers.	300 troop-horses.

Total all ranks, 483.

All ranks hailed the return of the popular old troop system, for the squadron organisation had only been introduced in a faint-hearted manner. It was feebly supported, and the way it was worked proved utterly unsuitable to the requirements of our weak cavalry regiments, especially when split up into several small detachments as was the case with regiments in Ireland at this period.

During the month of April, from the 10th to the 16th, the headquarters at Longford, together with the depôt troop of the 8th Hussars,[1] were employed under Colonel Thompson in aid of the civil power during elections at Longford. On the 27th May, Major-General R. Wardlaw, C.B., made his inspection of the Fourteenth and was quite satisfied. He particularly noticed the fine condition of the horses. Thirty-four troop-horses were transferred to the 7th Hussars.

During July the move of the regiment from Longford and out-stations to Cahir and out-stations took place.

On the 20th July the headquarters and 'C' troop, under Major Chapman, with the depôt troop of the 8th Hussars under Captain H. E. Malet, arrived at Cahir; 'B' troop, under Captain Laing, went to Fethard; 'D' troop to Fermoy, under Captain Knox, and subsequently 'A' troop also; 'H' troop to Waterford; 'G' troop to Clogheen.

A detachment went to Carrick-on-Suir.

Headquarters and remainder of the regiment went to Cahir, where Major-General Wardlaw, C.B., Commanding Cavalry in Ireland, made an inspection on the 19th October, and was very pleased.

1871

Establishment augmented to 8 troops.

On the 1st February a considerable augmentation took place in the establishment of the regiment, which was altered

[1] Under the new regulations a depôt troop from Canterbury belonging to the 8th (King's Royal Irish) Hussars, under Captain Harold Malet and Lieutenant Hooper, joined the 14th Hussars to be attached to it, the Cavalry Depôt at Canterbury having been temporarily abolished, pending reorganisation.

from 7 to 8 troops, with the following numbers of the various ranks of non-commissioned officers and men and troop-horses:—

42 sergeants.	32 corporals.
9 trumpeters.	515 privates.
8 farriers.	384 troop-horses.

Total non-commissioned officers and men, 606.

On the 24th February, Major-General H. Richmond Jones, C.B., who had passed all his service in the 6th Dragoon Guards (Carabiniers), became Colonel *vice* Lieutenant-General William Beckwith, K.H., deceased. Colonel—H. R. Jones.

On the 27th March, 'D' troop under Captain Knox marched from Fermoy to Clonmel. On the 3rd May, Major-General R. Wardlaw, C.B., made his half-yearly inspection of the Fourteenth at Cahir, and during the same month they marched to fresh quarters as follows:—

'A' and 'B' troops, under Brevet-Major Arbuthnot, from Fethard to Cork.

'C' troop, under Captain Russell, from Carrick-on-Suir to Bandon.

'D' troop, under Captain Knox, from Clonmel to Fermoy.

'E' troop, under Captain Palmer, from Cahir to Limerick.

'G' troop, under Captain Peel, remained at Clogheen.

'H' troop, under Captain the Honourable F. Amherst, and 'K' troop, under Captain Beaumont, from Cahir to Cork.

The headquarters at Ballincollig, being apparently represented by a few details, together with the band, and the depôt troop 8th Hussars, which latter did not go to the now re-organised Cavalry Depôt, Canterbury, till the 1st June 1871, having been attached to the 14th Hussars since April 1870. The headquarters, under Colonel Thompson, left Cahir on the 9th May, and reached Ballincollig on the 13th of that month.

The troop at Bandon under Captain Russell marched in aid of the civil power, on the 21st August, to Kinsale, and returned on the 23rd to Bandon.

The regimental club at Ballincollig gave several dramatic

performances during their stay there. One of the best plays was the *Seven Bags of Gold*, which was listened to repeatedly by some very appreciative audiences.

On the 3rd October, Major-General Wardlaw, Commanding the Cavalry in Ireland, inspected the Fourteenth at Ballincollig, and was very satisfied with the result.

Upwards of 50 men were ordered to be transferred to the 18th Hussars at Bangalore, owing to the losses there by cholera, and the transfers took place either this year or in 1872.

Captain and Brevet-Major William Arbuthnot, who had received a brevet-majority for his services in the Abyssinian campaign of 1868, became Major *vice* Chapman, retired.

1872

On the 1st February a detachment went from Fermoy to Listowel.

On the 3rd February a detachment went from Ballincollig to Killarney.

On the 4th February a detachment went from Limerick to Listowel.

The County Kerry elections were now going on, and the Fourteenth were much employed on election duty from the 1st to the 16th February, during which time they furnished detachments for duty at various polling-places, such as Listowel, Castle Island, Killorglin, Cahersiveen,[1] Kenmare, Killarney, and Dingle, in aid of the civil power. Captain the Honourable F. Amherst and Captain Knox commanded at Listowel on separate occasions; Captain Peel at Castle Island; Captain the Honourable J. St. V. Saumarez at Cahersiveen; Captain Lefroy at Dingle; Lieutenant M'Taggart at Killorglin; Lieutenant Merry at Killarney; and Lieutenant Bray at Kenmare.

On the 27th March, one troop from Limerick, under Captain Lefroy, proceeded to Galway, and returned on the 30th March,

[1] Or Cahirciveen.

being employed in aid of the civil power during the trial of the County Galway election petition.

On the 30th March, Major-General Holdich, C.B., Commanding the Cork District, inspected the headquarters at Ballincollig under Colonel Thompson, and Major-General Wardlaw, C.B., inspected at the same place on the 3rd May.

On the 1st April the Regimental Hospital-Sergeant was reduced from the establishment under the new system of the Army Medical Department, and regiments lost their medical officers. Establishment.

In May the Fourteenth marched from Ballincollig and out-stations to Newbridge, where they arrived on the 25th May. Lieutenant Kentish marched in charge of the young horses of the regiment. Captain the Honourable F. Amherst's troop ('H') took up the vedette duty on the Curragh Camp on the 5th July in relief of a troop of the 8th Hussars.

On 2nd and 3rd October, Major-General Robert Wardlaw, C.B., Commanding the Cavalry in Ireland and the Curragh District, made a minute inspection of the Fourteenth at Newbridge, and expressed himself very pleased with the state of the regiment, which was particularly praiseworthy considering the very scattered state in which it had been during the past three years, owing to its being quartered in so many detachments, and so much broken up for election duties all over Ireland. The following is an extract from Horse Guards letter received on 24th January 1873, through the Cavalry Brigade office, Curragh Camp, containing observations by the Field-Marshal Commanding-in-Chief, H.R.H. the Duke of Cambridge, on the last annual inspection of the regiment:—
'The 14th Hussars are in excellent order, and ably commanded by Colonel Thompson. H.R.H. desires this corps and their commanding officer should be highly commended.'

1873

On the 1st January, 'D' troop under Captain Knox proceeded from Newbridge to the Curragh Camp, and took over the vedette duty from 'H' troop, under Captain the Honourable F. Amherst, which latter troop joined the remainder of the regiment at Newbridge on the same date. 'D' troop returned to Newbridge from vedette duty on the 16th July.

On the 5th May the depôt, with recruits and young horses under Captain Harpur, proceeded to Dublin to the Royal Barracks, to be stationed there whilst the regiment was at summer drills.

Colonel—John Wilkie.

On the 15th May, General John Wilkie was appointed Colonel *vice* General H. R. Jones, C.B., transferred to the Colonelcy of the 6th Dragoon Guards (Carabiniers).

On the 4th July the Fourteenth participated in the review of the troops on the Curragh by the Right Honourable Lieutenant-General Lord Sandhurst, G.C.B., Commanding the Forces in Ireland, who was highly pleased with the appearance and movements of all.

On the 1st September, Major-General Wardlaw, C.B., made his inspection at Newbridge. He was very pleased, and noticed the fine condition of the horses.

In October the Fourteenth marched to Dublin and took over quarters in the Royal Barracks.

1874

Regiment proceeds to Aldershot, 1874.

In May the regiment, after six years' sojourn in Ireland, of which the greater part had been passed in small and scattered stations, left Dublin for Aldershot. The march was commenced on the 20th May, when the first division, consisting of 'A' and 'B' troops under command of Major William Arbuthnot, left the Royal Barracks, and embarking at North Wall, Dublin, in ss. *St. Patrick* for Birkenhead, proceeded thence by march route to North Camp, Aldershot. The second division, consisting of 'H' and 'K' troops under Captain the Honourable F. Amherst, embarked in ss. *Windsor* at the same place on the 22nd May, and proceeded by similar route to Aldershot. The third division,

'C' and 'D' troops under Captain Knox, embarked in ss. *St. Patrick* on the 25th May, and likewise marched from Birkenhead to Aldershot. The fourth division, consisting of headquarters with 'E' and 'G' troops under Major Campbell, and the young horses in charge of Riding-Master Harran, went by ss. *Windsor* to Birkenhead, and by route march to Aldershot; and the fifth and last division, consisting of the women and children in charge of Captain M'Taggart, embarked in ss. *Cymba* for Portsmouth, and thence went by rail to Aldershot.

The Fourteenth arrived at Aldershot by the 16th June and were placed under canvas in the North Camp, the horses being located in temporary stables. After taking part in the first period of the summer drills under command of Colonel Thompson, during which time the regiment was attached to the 2nd Division and encamped at Woolmer from the 22nd to the 24th June, it occupied the South Cavalry Barracks for the winter on the 3rd July in relief of the Royal Scots Greys. Later on the regiment also took part in the second period of summer drills with the 1st Division, encamping at Colony End on the Chobham Ridges on the 7th July, and returning to the South Cavalry Barracks on the 10th July.

On the 20th July the Fourteenth encamped with the rest of the troops of the Division at Woolmer, and returned to Aldershot on the 29th July. On the 6th, 7th, and 8th October, Major-General Sir Thomas W. McMahon, Bart., C.B., Commanding Cavalry in Great Britain, inspected the regiment and was very satisfied. He was pleased with everything he saw, but especially noticed the admirable manner in which the regimental office work and clerking were conducted.

Captain F. Russell received a Brevet-Majority for his services in the Ashantee War on the west coast of Africa this year.

During the year 19 men deserted, and 4 were given up as deserters to other corps. On the 30th December, Captain A. E. T. Preston died of scarlet fever at Mentone, in the south of France.

1875

Lieutenant-Colonel—F. P. Campbell.

On the 29th April, Colonel P. S. Thompson retired on half-pay, and Lieutenant-Colonel F. P. Campbell succeeded to the command of the Fourteenth. One major was reduced from the establishment.

The regiment took part in the summer drills at Aldershot in July with the 2nd Army Corps, marching to Coldingly on the 13th July and encamping there, thence to Hartford Bridge Flats, and encamping there 20th July, returned to Aldershot the 22nd July.

On the 26th July, 'A,' 'B,' 'C,' 'H,' and 'K' troops marched to Colchester, arriving on the 31st July. 'C,' 'D' 'E,' and 'G' troops marched on the 27th July and arrived on the 2nd August.

At Colchester the Fourteenth received orders to prepare for embarkation for India, and came under the command of Major-General Sir E. Greathead, K.C.B., commanding the Eastern District.

On the 25th September, H.R.H. the Duke of Cambridge made his farewell inspection of the regiment at Colchester. There was a story current in the Fourteenth that on this occasion the favourite pork chops destined for His Royal Highness's luncheon were owing to some unforeseen occurrence eaten previously by the Riding-Master![1] His Royal Highness, in bidding farewell to the regiment, was graciously pleased to express his entire satisfaction with its discipline and with the good conduct it had ever shown during the time of its home service. His Royal Highness especially referred to the distinguished name for good horses the regiment had in recent years obtained, and this he credited to the late commanding officer, Colonel P. S. Thompson, C.B.

On the 28th and 29th September, Major-General Sir T. W. McMahon, Bart., C.B., Inspector-General of Cavalry, made an

[1] The author will not vouch for the absolute accuracy of this statement. When the Duke used to come to inspect regiments it was always a rule to have pork chops amongst the other delicacies provided for His Royal Highness.

inspection of the Fourteenth. He was very pleased, and especially with the outpost duty. In November the actual 'orders of readiness' for India were received.

Regiment ordered to India.

'E' and 'G' troops were broken up, as the establishment was reduced from 8 to 6 troops, and a depôt troop was formed under the command of Captain Harpur.

Establishment altered to 6 troops and a depôt troop.

Being under orders for India, the establishment of the Fourteenth was fixed as follows from the 1st October 1875:—

Colonel,	1	
Lieutenant-Colonel,	1	
Major,	1	
Captains,	7	
Lieutenants and Sub-Lieutenants (the rank of 'Cornet' had been recently abolished),	14	
Adjutant,	1	
Riding-master,	1	
Quartermaster,	1	
Veterinary Surgeon,	1	
Total,	28	officers.
Non-commissioned officers and trumpeters,	52	
Rank and file,	465	
Total all ranks,	545	

Troop-horses, 456, of which 20 are for service at the cavalry depôt at Canterbury.

The actual strength of the regiment at this date was 608 non-commissioned officers and men, with 380 troop-horses. During the year 47 men deserted and 7 were given up as deserters from other corps.

One non-commissioned officer (Mr. J. Merrill) was promoted to the rank of Quartermaster, and he soon after exchanged to the 18th Hussars with Quartermaster G. F. Rumsey.

1876

On the 3rd January a party of 80 men, with the married families, under Captain Kentish and Sub-Lieutenant G. H. C. Hamilton[1] (recently promoted from the ranks of the 2nd Life Guards), proceeded by rail from Colchester to Portsmouth for embarkation in H.M.S. *Euphrates*.

On the 4th January the headquarters and service troops followed by the same route, and on the 5th January the regiment, under the command of Lieut.-Col. F. P. Campbell, sailed from Portsmouth for India in H.M.S. *Euphrates*.

On the 17th January the depôt, under Captain Harpur and Lieutenant Garrett, moved from Colchester to join the Cavalry Depôt at Canterbury, where the Commandant was Colonel Conyers Tower, C.B.

The strength of the depôt troop was 125 non-commissioned officers and men.

Roll of officers of the 14*th (King's) Hussars embarked in H.M.S. 'Euphrates' at Portsmouth on the* 4*th January* 1876, *for conveyance to Bombay.*

Lieutenant-Colonel—F. P. Campbell, commanding the regiment.
Major—William Arbuthnot.
Captains—J. H. Knox.
W. B. M'Taggart.
J. Dennis.
H. L'E. Malone.
Lieutenants—J. Kentish.
R. Garth.
A. J. English.
C. D. V. Tuthill.
J. S. Heron-Maxwell.
C. F. Lindsell.
G. C. Ricardo.
W. E. Stokes.
W. L. M'Call.
R. Owen.

[1] Afterwards Colonel G. H. C. Hamilton, commanding 14th Hussars.

Sub-Lieutenants—W. H. Burn.
A. C. King.
G. S. M. Palmes.
G. H. C. Hamilton.
Paymaster—Captain E. H. O'Dowd.
Lieutenant and Adjutant—The Honourable H. G. Gough.[1]
Veterinary Surgeon—F. F. Marshall.
Quartermaster—J. Merrill.
Riding-master—J. Harran.
Surgeon—T. M. Barrow, M.D.

Strength of non-commissioned officers and men, 425.

Officers' wives, 4; officers' children, 2.

Soldiers' wives, 57; soldiers' children, 100.

Sailed 5th January 1876; experienced pleasant weather throughout the whole voyage, and no casualty occurred.

On the 8th February the Fourteenth disembarked at Bombay and arrived by rail at Poona, where Major-General Lord Mark Kerr made an inspection. On the 14th February they arrived by rail at Bangalore, and remained stationed there for the next five years in the cavalry barracks. The route from Poona was *via* Sholapore, Karchore, and Arconum. Regiment arrives in India, February 1876.

Lieutenant-Colonel F. P. Campbell died at Bangalore on the 14th June of acute dysentery, after a week's illness, at the age of 39 years, when Major William Arbuthnot got command of the Fourteenth as Lieutenant-Colonel, and Captain J. H. Knox became Major. Lieutenant-Colonel Campbell was deeply regretted by officers and men alike. He was buried on the 15th June in the cemetery with full military honours, his funeral being largely attended by the residents of Bangalore. Lieutenant-Colonel—William Arbuthnot.

The strength of the service troops on the 1st December was 437 non-commissioned officers and men, with 393 troop-horses.

1877

On the 9th and 10th February, Major-General Elmhirst, C.B., Commanding the Mysore Division, inspected the regiment and

[1] Afterwards Colonel the Hon. G. H. Gough, C.B., commanding 14th Hussars.

was very pleased with all he saw. He said he considered the perfect state of the regiment could only result from the greatest attention to their respective duties on the part of every rank in it.

On the 8th September, His Excellency the Most Honourable Lord Lytton, G.C.S.I., Viceroy and Governor-General of India, honoured Lieutenant-Colonel Arbuthnot and the officers with his company at dinner in their mess-house. His Excellency was accompanied by his staff, among whom and the other guests were the following:—

Sir Alexander Arbuthnot, K.C.S.I.
The Honourable J. Bayley.
Mr. Thornton, C.S.I.
Lieutenant-Colonel Burne, C.S.I., Private Secretary.
Colonel Pomeroy Colley, C.B., Military Secretary.
Mr. Bernard.
Dr. Barnett.
Captain the Honourable G. P. Villiers, Grenadier Guards, A.D.C.
Captain Loch, A.D.C.
The Chief Commissioner of Mysore.

His Excellency paid a visit to the barracks, hospital, troop-bungalows, married-quarters, stables, etc., on the 9th September, and he again honoured the regiment with his presence, accompanied by his staff, and witnessed a theatrical entertainment given by the regimental club in the recreation room on another day, on which occasion he presented the famous comic singer of the regiment, Private Samson, with a seal as a memento.

On the 11th October the Fourteenth were inspected in the field by Lieutenant-General Sir Neville Chamberlain, G.C.B., G.C.S.I., Commander-in-Chief of the Madras Army. On the two previous days the Commander-in-Chief visited the stables and the barracks generally, and after his inspection made the following report: 'His Excellency considered the turn-out on all occasions, both as regards men and horses, most creditable.

'Reconnoitring duties, aided by the signallers of the regiment, were carried out very favourably. His Excellency directs that his commendation be expressed to Colonel Arbuthnot and his officers for the time and attention that have been given to this important practical part of light cavalry duties.'

During this year 2 men of the regiment died of cholera.

1878

On the 15th and 16th February, Lieutenant-General Elmhirst, C.B., made his annual inspection of the Fourteenth, after which he addressed Colonel Arbuthnot at the head of the regiment, and spoke in very flattering terms as to all he had seen, as well as of the good behaviour of all ranks since they came under his command.

The regiment remained stationed at Bangalore all this year.

Much regret was felt in the Fourteenth upon hearing of the death at Beverley, Yorkshire, on the 1st May, of Colonel P. S. Thompson, C.B., a former much-respected commanding officer.

On the 7th November the Fourteenth received orders to prepare for immediate active service in Afghanistan. The telegram came as follows:—'Clear the line: Hussars, 67th Regiment, 30th and 36th Native Infantry, and 2 companies Sappers for service.' Later intelligence informed that these regiments were to constitute part of a Madras Brigade which, with a Bombay Brigade, were to form a Reserve Division on the Lower Indus under the command of Major-General Primrose. Accordingly, all was got ready for service. Sword-blades were sharpened, scabbards blackened, belts and helmets browned. On the 24th November a telegram was received, 'Lower Indus Division modified for the present,' and on the 13th December came a telegram from the Government of

Ordered to Afghanistan.

Order for Afghanistan countermanded.

India which was almost a countermand, '14th Hussars not likely to be required to move yet awhile.' Some cholera broke out again this year, and one man died of sunstroke. Lieutenant W. E. Stokes died of acute mania at Madras on the 26th September.

On the 23rd and 24th December, in accordance with a special order from Headquarters of the Army applicable to corps warned for service, the Fourteenth were inspected by Colonel Hill Wallace, C.B., Royal Horse Artillery, Commanding at Bangalore. He was very pleased, and especially commended the non-commissioned officers who had been working as road sketchers, also the regimental signallers, and pointed out how very useful on service these acquirements would be.

1879

The whole of this year was spent at Bangalore.

Educational statistics of regiment.

At this period there were the following number of educational certificates in possession of the non-commissioned officers and men of the regiment :—

1st Class.	2nd Class.	3rd Class.	4th Class.	Total.
5	113	100	145	363

There were only 4 men who could neither read nor write.

On the 3rd May, Major-General W. Payne, C.B., commanding the Mysore Division, inspected the regiment and was very pleased.

1880

The whole of this year also was passed by the regiment at Bangalore. On the 21st and 22nd February, Major-General Payne, C.B., made his annual inspection.

There was a camp of exercise held at Begaum this year for eight days, where the Fourteenth took part in some interesting manœuvres.

The depôt troop still remained at Canterbury, where Colonel E. Napier was the Commandant.

Whilst the regiment was at Bangalore this year, Colonel

Arbuthnot selected 'The King of Prussia' as the regimental slow march (see p. 435). Regimental march.

The establishment of the regiment was fixed from the 1st July as follows:— Establishment, 1880.

8 Troops—27 Officers.
2 Warrant Officers[1] (1 Regimental Sergeant-Major and 1 Bandmaster).
41 Sergeants (the Trumpet-Major was now styled Sergeant-Trumpeter).
8 Farriers.
32 Corporals.
510 Privates.
400 Troop-horses.

1881

On the 13th February, 'orders of readiness' for active service in South Africa were received. Regiment ordered to South Africa, February 1881.

On the 14th February, Major-General Payne, C.B., made his inspection of the Fourteenth at Bangalore, and in his subsequent order he said: 'I bid farewell to the officers, non-commissioned officers and men, in the full assurance that they will maintain the bright reputation that has always distinguished the 14th King's Hussars.'

From the 18th to the 28th February the regiment left Bangalore for active service in the Transvaal, sailing from Bombay in three separate steamers of the British India Steamship Company's service, and landed at Durban from the 14th to the 26th March. Lands at Durban, March 1881.

One troop was left at Bangalore under Lieutenant Fraser, which formed a depôt, when the regiment proceeded on service.

'C' troop, under Captain R. Garth, left Bangalore on the 8th March, strength 100 men, 130 horses, and embarked in

[1] This was the first appointment of Warrant Officers, such as Regimental Sergeant-Major and Bandmaster.

transport ss. *Chufra*, 26th February, at Bombay, and landed at Durban on 14th March, then marched to Pietermaritzburg.

'A' and 'B' troops with headquarters under Lieutenant-Colonel Arbuthnot, strength 8 officers, 152 men, 179 horses, left Bangalore on the 22nd February, embarking at Bombay in the ss. *Booldana* (Captain Wood) on the 1st March, landed at Durban 18th to 20th March, and proceeded on the 21st to march to Pietermaritzburg. The third party under Major Knox, strength 10 officers, 167 men, 194 horses, left Bangalore the 28th February, embarked on the 11th March on the ss. *Hankow* at Bombay, and landed at Durban the 26th March, proceeding the next day towards Pietermaritzburg.

On arrival at Durban, information of the death of Major-General Sir G. Pomeroy Colley was received; also of an armistice of three days, and it was generally thought this meant an end of the Boer war. Owing to the surf at Durban, the Fourteenth disembarked in lighters. By the 11th April the whole regiment was concentrated under command of Lieutenant-Colonel Arbuthnot at Estcourt, where it remained till the 17th April, and then marched to Ladysmith. The regiment remained several months inactive at Ladysmith, during which there was not much to occupy the men, and several desertions took place into the Orange Free State; but by sending off mounted parties under an officer of the deserter's troop, this was gradually put a stop to, and most of the delinquents were brought back. One man (Private Kavanagh) got away on a stolen horse, and he subsequently claimed Queen's Jubilee pardon at Shorncliffe in 1887 and received it.

Lieutenant-Colonel—J. H. Knox.

On the 15th June, Major J. H. Knox became Lieutenant-Colonel, and succeeded Colonel William Arbuthnot in command of the regiment, Captain (Brevet-Major) F. S. Russell becoming Major in succession.

On the 1st July the new warrant was promulgated, by which Major F. S. Russell became second Lieutenant-Colonel, and the three senior Captains—Dennis, Kentish, and Hickman—became Majors, but still remained in command of troops.

During the stay at Ladysmith a stampede of the horses took place, in which several were lost, supposed to have been drowned in the Klip river; and there was a fire of the grass round camp one night, but the whole camp was quickly turned out—14th Hussars, 15th Hussars, and Welsh Regiment—and by their help the flames were got under, and only a few tents belonging to some wounded soldiers of the Royal Scots Fusiliers were damaged. Regimental Sergeant-Major Thompson died at Ladysmith, and Troop Sergeant-Major H. A. Pridgeon was promoted Regimental Sergeant-Major, with the rank of a warrant-officer, dated 1st July 1881. There were also several other deaths amongst the rank and file, and one sergeant (Rowley) was found dead at the bottom of the cliffs, several days after he was missed from camp, having probably fallen over in the dark.

Stampede and camp fire at Ladysmith.

On the 15th August, at Ladysmith, His Excellency Major-General Sir H. E. Wood, V.C., K.C.B., Governor and Commander-in-Chief of Natal, inspected the Fourteenth in marching order, and on the conclusion of the parade expressed himself pleased with what he had seen, and especially so with the condition of the horses and the riding of the men. The strength of the regiment was 24 officers, 484 non-commissioned officers and men, 415 troop-horses, and 43 officers' chargers.

One squadron ('D' and 'H' troops) under Major Kentish, numbering about 90 sabres, proceeded to Zululand to act as escort to His Excellency Major-General Sir H. E. Wood, on the 16th August. This squadron joined a squadron of the 6th Inniskilling Dragoons and a squadron of the 15th Hussars at Stale's Drift, Buffalo River, and proceeded thence, under the command of Colonel George Luck, C.B., 15th Hussars, to the Inhlazane Mountains in Zululand, returning to Ladysmith on the 12th September. One private was severely frost-bitten, and one troop-horse died on the march.

The regiment had entered Zululand in August, and it left Ladysmith the 5th November *via* Pinetown for the coast, having received orders to return to India. Several of the worst horses in the regiment were sold on the spot, the

Boers giving fabulous prices for them. Headquarters and right wing, under Lieutenant-Colonel Knox, embarked at Durban on the 13th November in the ss. *Calabria*, all hostilities with the Boers having ceased, and landed at Bombay on the 30th November, and were encamped there on the esplanade, marching on the 7th December for Secunderabad, where they arrived on the 21st January 1882, relieving the 12th Lancers.

Regiment arrived in India, November 1881.

The left wing, under Lieutenant-Colonel F. S. Russell, embarked at Durban on 9th December in ss. *Hankow*, landed in Bombay 31st December, and encamped on the Maidan, or esplanade, for a few days.

1882

The left wing marched from Bombay on 3rd January for Secunderabad, where it arrived on 21st February, the right wing having come there a month previously, as already related.

CASUALTIES IN NATAL.

		Officers.	Men.	Horses.
Increase.	Landed in Natal,	28	445	475
	Transferred to regiment,		15	
	Total increase,		15	
Decrease.	Died,		6	
	Destroyed,			8
	Cast and sold,			29
	Sent to England,	6	68	
	Deserted,		2	1
	Transferred,		5	56
	Discharged in Colony,		8	
	Invalids,		17	
	Total decrease,	6	106	94
	Landed at Bombay,	22	354	381

The following was communicated in a letter from the General Officer commanding at Natal, dated 19th November 1881, to the Quartermaster-General of the Forces, Horse Guards, London :—

'I have the honour to report for the information of H.R.H. the Field-Marshal Commanding-in-Chief, that the head-quarters of the 14th King's Hussars marched from Pinetown to Port Durban, a distance of thirteen miles, on the morning of the 14th November. The embarkation commenced at 7.30 A.M. same day, and all horses, men, and baggage had left the land by 3.15 P.M. No casualties occurred, and the Assistant Quartermaster-General reports that the regiment presented a most creditable appearance. It is the first time that a wing of a cavalry regiment complete with horses has been embarked from Durban, and any one acquainted with the bar and heavy surf at Durban will know the difficulties that embarking officers have to contend with, and the danger that is attendant thereon.'

Record embarkation of a wing of the 14th Hussars at Durban in less than eight hours' time.

On 28th February, at Secunderabad, Major-General Sir C. P. Keyes, K.C.B., commanding Hyderabad Subsidiary Force, inspected the Fourteenth. He expressed himself entirely satisfied with all he saw. He inspected again on the 19th September with an equally satisfactory result.

On 15th March, His Excellency the Commander-in-Chief in India, General Sir F. S. Roberts, Bart., V.C., G.C.B., C.I.E., inspected the regiment, and was satisfied with what he saw.

On 1st January 1882 the establishment of the Fourteenth was altered from Colonial to Indian, and fixed as follows :— Establishment.

Seven Troops (1 at Cavalry Depôt, Canterbury, with 20 of the troop-horses).

24 Officers.	7 Trumpeters.
2 Warrant Officers.	27 Corporals.
37 Sergeants.	461 Privates.
6 Farriers.	456 Troop-horses.

Colonel—C. W. Thompson.

On the 1st May, Major-General C. W. Thompson was appointed Colonel of the Fourteenth *vice* General J. Wilkie, deceased.

On the 10th June, Lieutenant-Colonel C. F. Morton exchanged into the Fourteenth from the Royal Dragoons with Lieutenant-Colonel Frank Shirley Russell, who went to the Royal Dragoons, and afterwards commanded that regiment.

On the 11th December field manœuvres commenced at Secunderabad, in which the Fourteenth took part.

Three men died of their wounds received on service.

1883

This year was passed at Secunderabad.

On the 20th January, Major-General Sir Charles Keyes, K.C.B., inspected the Fourteenth and was very pleased, and said he should report most favourably to His Excellency the Commander-in-Chief in India as to their efficient state.

From the 11th to 16th December the Fourteenth were present at the field manœuvres under His Excellency General Sir F. S. Roberts, Bart., V.C., G.C.B., C.I.E., Commander-in-Chief of the Madras army, and he expressed himself much pleased with the way the regiment acquitted itself, and particularly remarked on the good scouting and patrolling which it performed. The following is an extract from a letter written by His Excellency (Sir F. Roberts) to Colonel Keith Fraser, C.M.G. (A.A.G. at the Horse Guards):—

'Snowden, Ootacamund, 8*th August* 1884.

'The regiment with whom I have been most impressed is the 14th Hussars: their scouting is far above the average, and I attribute this in a great measure to the experience they gained in South Africa, which no doubt gave the men that self-reliance it is so difficult to acquire on the parade-ground. At some small manœuvres near Secunderabad in

1882,[1] it was a pleasure to see the way the 14th Hussars worked: wherever you moved you found an hussar scout watching you, and utterly declining to fight so long as his orders were to keep touch with the enemy.'

This year Major Kentish left the Fourteenth, and Captain Richard Garth succeeded him as Major.

On 17th December, Major-General Sir Charles Keyes, K.C.B., again inspected the Fourteenth, and said they were in a very serviceable condition, and he specially remarked on the conduct of the men since the Fourteenth came under his command, which he termed 'excellent.'

1884

The regiment remained at Secunderabad.

Four men died of fever during the year. The depôt remained at Canterbury, where the Commandant of the Cavalry Depôt was Colonel J. C. Le Quesne.

From the 22nd to the 29th November the Fourteenth took part in the camp of exercise held near Secunderabad, under His Excellency the Commander-in-Chief of the Madras army (Sir F. Roberts), and the latter expressed his complete satisfaction with their conduct and efficiency throughout the manœuvres.

1885

This year was passed in the same quarters at Secunderabad with the usual routine of Indian life.

On the 12th and 13th January, Major-General H. N. D. Prendergast, V.C., C.B., commanding the Hyderabad Subsidiary Force, inspected the regiment, and said he would report

[1] This reference of Lord Roberts's to the good scouting of the Fourteenth applied to the manœuvres held in 1882, apparently.

favourably on its general efficiency for the information of His Excellency the Commander-in-Chief.

Fatal lightning accident at Secunderabad, 28th May 1885.

This was a very hot season, and on the 28th May two promising young officers—Lieutenants G. H. Seymour and F. D. Bentley-Innes—were suddenly killed by lightning about 3.30 P.M. when playing billiards in the officers' mess. They had ordered their tandem to drive to a gymkhana, but as they saw a storm coming on they went back to the mess to play billiards. Suddenly a flash of lightning struck the mess-house. Just before it came young Seymour called to Potter, the mess butler, for a cheroot. The lightning struck him (Lieutenant Seymour) on the collar stud, ran down and tore the right seam of his trousers, and tore off the sole of his right boot. Lieutenant Bentley-Innes was struck on the poll of the neck. Neither of them spoke after they were struck, though Surgeon-Major Fraser did all in his power to restore animation. This sad event cast quite a gloom over the station, as the two young officers were much liked and respected by all. The various regiments in cantonment followed at their funeral next day. An anonymous donor used to send to the regimental sergeant-major of the regiment for some time afterwards each month on the day corresponding to the sad occurrence a cross of flowers, to be placed by their graves.

On the 15th June, Lieutenant-Colonel Knox became Brevet-Colonel, and Lieutenant-Colonel Morton obtained the same rank on the 1st July.

Lieutenant-Colonel—C. F. Morton.

Colonel J. H. Knox, commanding the regiment, died very suddenly of hepatitis (liver disease) on the 24th October; and Colonel C. F. Morton, from second Lieutenant-Colonel, succeeded to the command.

Major J. Dennis became second Lieutenant-Colonel on the 25th October; and Major J. Kentish from half-pay came back to the regiment.

Four men died of fever and two of liver disease during this year.

The depôt troop under Captain English and Lieutenant

Stoughton moved from Canterbury to Colchester on the 3rd December, and was attached to the provisional depôt there, under the command of Colonel E. A. Gore.

1886

This year the order came for the regiment to go home to England after eleven years' service in India and South Africa. Regiment ordered home.

Major-General B. L. Gordon, C.B., inspected at Secunderabad on February the 15th, 16th, and 17th.

Previous to the departure of the Fourteenth for England the following Divisional Order, dated Secunderabad, 22nd October 1886, was issued by Brigadier-General P. A. Carnegy, commanding at Secunderabad:—

'The 14th King's Hussars being about to leave for England, the Brigadier-General commanding, in bidding the regiment farewell, has much pleasure in placing on record the high opinion he entertains of this distinguished corps. During the five years that the Fourteenth have been in the command, the conduct of all ranks has been exceptionally good, and an example to the remainder of the garrison. On parade also, both officers and men have well upheld the character of the corps for smartness and dash. General Carnegy congratulates Colonel Morton in taking with him a body of non-commissioned officers and men who, he feels certain, will win from the home authorities the same approbation for their soldierly qualities as they have done from those in this country.'

On the 30th October the regiment, under Colonel C. F. Morton, embarked at Bombay, and sailed in H.M.S. *Serapis* for England, arriving 23rd November, and, landing at Portsmouth on the 25th November, proceeded thence by rail to Shorncliffe, there to be stationed. Regiment lands in England, 25th November 1886.

The voyage in H.M.S. *Serapis* was a very pleasant one, and the previous journey of the regiment (without horses) from Secunderabad to Bombay was accomplished by rail, halting at Wadi and Poona *en route*.

Roll of Officers embarked in H.M.S. 'Serapis.'

Lieutenant-Colonel—Brevet-Colonel C. F. Morton, commanding.
Majors—T. E. S. Hickman.
R. Garth.
Brevet Lieutenant-Colonel the Honourable G. H. Gough.
Captains—A. C. King.
C. E. S. Hemery.
Lieutenants—H. W. Mitchell.
H. Kirk.
A. B. Broadhurst.
L. A. Stransham.
L. J. Richardson.
J. P. Miller.
S. Robertson.
J. Murray.
Captain and Adjutant—G. H. C. Hamilton.
Major and Paymaster—H. J. Meares.
Quartermaster—F. Mugford.

Strength of non-commissioned officers and men embarked, 405; officers' wives, 3; children, 3; soldiers' wives, 24; children, 55.

The depôt troop under Captain English and Lieutenant Stoughton was moved from Colchester to Shorncliffe on the 31st May, arriving at the latter place on the 5th June, and continued being attached to Colonel Gore's Provisional Cavalry Depôt (which had been removed from Colchester) until the arrival of the service troops from India, when it joined headquarters at Shorncliffe, and the strength of the Fourteenth towards the end of the year was 536 non-commissioned officers and men, with 201 troop-horses transferred from the 7th Hussars. The whole of the regimental arms, accoutrements, saddlery, and equipment were now thoroughly inspected under the superintendence of Captain Hawes of

New arms and equipment issued at Shorncliffe.

the India Office, and new stores and equipment were supplied to replace all articles of obsclete pattern. A large quantity of new saddlery also was issued, and the regiment was armed with the new pattern solid-hilted sword and new Martini-Henry carbines, the non-commissioned officers receiving a new pattern revolver. Martini-Henry carbines issued.

On the 17th November, Lieutenant-Colonel J. Dennis exchanged to the 6th Dragoon Guards (Carabiniers); and Lieutenant-Colonel H. B. Hamilton from the Carabiniers joined the 14th Hussars as 2nd Lieutenant-Colonel.

Captain (Brevet Lieutenant-Colonel) the Honourable G. H. Gough became Major.

1887

The Fourteenth remained at Shorncliffe in huts the whole of this year. On the 13th April there was a brigade parade of all the troops in camp at Shorncliffe under Colonel Sir Baker Russell, K.C.B., K.C.M.G., A.D.C., commanding the troops, for the inspection of H.R.H. the Duke of Cambridge, Commanding-in-Chief, in which the Fourteenth took part; and after the review H.R.H. the Duke of Cambridge inspected all the remounts recently purchased by Colonel C. F. Morton for the regiment, to complete the establishment.

The Fourteenth were now reduced to the home establishment, and consisted of 8 troops, 24 officers, 2 warrant officers, 41 sergeants, 8 farriers, 8 trumpeters, 32 corporals, 378 privates, and 300 troop-horses. Establishment reduced, 1887.

Colonel Sir Baker Russell, K.C.B., K.C.M.G., A.D.C., commanding the troops, made his inspection on the 29th and 30th April.

On the 1st May the use of cruppers was discontinued. Cruppers abolished.

A wing of the Fourteenth took part in the Queen's Jubilee Review at Aldershot on the 9th July, being employed in At Queen's Jubilee Review, Aldershot.

keeping the ground, and afterwards returned by march route to Shorncliffe.

Lieutenant-Colonel—H. B. Hamilton.

On the 1st July, Colonel Morton was placed on half-pay, and Lieutenant-Colonel H. B. Hamilton was appointed to command the regiment. The second Lieutenant-Colonelcy was reduced from this date.

Firing team won Cambridge Shield at Wimbledon.

On the 23rd July, at Wimbledon, the regimental firing team, consisting of Troop Sergeant-Major Carr, Sergeant Upton, Private Fairbrass, and Private Upton, won the Royal Cambridge Challenge Shield, which is annually competed for by mounted teams from cavalry regiments in Great Britain and Ireland.

On the 10th and 11th August, Major-General Sir D. Drury Lowe, K.C.B., Inspector-General of Cavalry, made his annual inspection of the Fourteenth at Shorncliffe. Strength—23 officers, 2 warrant officers, 491 non-commissioned officers and men, 300 troop-horses, 16 officers' chargers.

On the 6th October the regiment took part in the field operations carried out by the troops composing the Dover and Shorncliffe garrisons, concluding with a march-past near the Royal Oak, on the Dover and Folkestone road, before H.R.H. the Duke of Cambridge, Commander-in-Chief.

Establishment altered.

The establishment of the regiment underwent several changes this year. On the 1st April the number of troop-horses was reduced from 300 to 250, and in consequence upwards of 40 troop-horses had to be transferred to the 19th Hussars in the autumn; but on the 1st October the establishment was again augmented to 300, and fresh remounts had consequently to be procured and trained afresh.

Captain Augustus John English became Major, and Captain Henry Wilmot Mitchell was appointed Adjutant.

The establishment was fixed from the 1st April 1887 at—

Colonel, ...	(1)
Lieutenant-Colonel, ...	1
Majors, ...	3

Captains,	6	
Lieutenants,	8	
Second Lieutenants,	3	
Adjutant,	1	
Riding-master,	1	
Quartermaster,	1	
Total,	24	officers.
Warrant officers,	2	
Sergeants,	49	
Rank and file,	410	
Total,	485	all ranks.

1888

On the 2nd April the Fourteenth took part in the Easter Monday volunteer review, combined with manœuvres of the regular troops of the South-Eastern District, between Folkestone and Dover; also a march-past at Dover before H.R.H. the Duke of Cambridge, Commander-in-Chief. Major-General Philip Smith, C.B., commanding the Home District, had the command of the regulars and volunteers on this occasion.

On the 3rd April the troops at Shorncliffe Camp under the command of Colonel Sir Baker Russell, K.C.B., K.C.M.G., A.D.C., paraded for a review held by Field-Marshal H.R.H. the Duke of Cambridge, Commander-in-Chief, in which the Fourteenth took part. Subsequently His Royal Highness saw the cavalry, consisting of the 14th Hussars and a wing of the 20th Hussars, exercised in brigade movements under Lieutenant-Colonel Graves, 20th Hussars, on which occasion His Royal Highness was pleased to express to Lieutenant-Colonel Hamilton, through the officer commanding the cavalry, how pleased he was at the appearance of the Fourteenth and the way in which it drilled.

His Royal Highness afterwards inspected the remounts

which had recently joined from the Army Remount Depôt, and subsequently, attended by his Staff, honoured the officers of the Fourteenth by lunching with them in their mess. Major-General Sir Redvers Buller, V.C., K.C.B., Quartermaster-General; Major-General Montgomery Moore, Commanding the South-Eastern District; Colonel Adolphus Stevens, A.D.C. to His Royal Highness; and Colonel Sir Baker Russell, K.C.B., K.C.M.G., A.D.C., commanding at Shorncliffe, were among those present.

On the 2nd May, Colonel Sir Baker Russell, K.C.B., K.C.M.G., A.D.C., commanding the troops at Shorncliffe, inspected the Fourteenth, and on the 24th May he saw them exercised in outpost duties between Shorncliffe and Hythe, and expressed on both occasions his high satisfaction with all he saw.

On the 7th May a slight reduction in the numbers of the establishment of the non-commissioned officers and men took effect.

At Aldershot. On the 4th June the Fourteenth marched to Aldershot, and encamped on Cove Common, near Farnborough, on the 9th June. A depôt was left at Shorncliffe in charge of Riding-master Odlum.

On the 11th June, Major-General Sir Drury Lowe, K.C.B., inspected the Fourteenth in 'marching order,' and expressed his full satisfaction at the appearance of the men and horses and with the general turn-out of the regiment.

On the 6th July, Captain C. E. S. Hemery died at Barnet of cerebral congestion. He was suddenly taken ill a few weeks previously, when driving the regimental coach at Red Hill. Lieutenant-Colonel Hamilton and the other officers from Aldershot attended his funeral at Barnet, where the family of the deceased officer resided.

On the 7th July the Fourteenth took part in a review at Aldershot, under Field-Marshal H.R.H. the Duke of Cambridge, Commander-in-Chief.

This was an unusually wet season, and both men and horses

were much exposed to the inclement weather in their camp, which was broken up towards the end of July; and after the regiment had taken part in the usual summer drills, it proceeded to Brighton and Hilsea to be quartered. Four troops, 'B,' 'D,' 'H,' and 'K,' under Major Garth, went to Hilsea; and 4, 'A,' 'C,' 'E,' and 'G,' with headquarters, under Lieutenant-Colonel Hamilton, arrived at Preston Barracks, Brighton, on the 25th July.

Wet season under canvas.

At Brighton and Hilsea.

On the 9th and 10th October, Major-General Sir D. Lowe, K.C.B., Inspector-General of Cavalry in Great Britain, inspected the Fourteenth at Brighton.

On the 13th October, 2 troops ('B' and 'D') under Major Garth joined headquarters at Brighton, leaving only 2 troops ('H' and 'K') under Captain King at Hilsea.

Major R. Garth exchanged into the 7th (Princess Royal's) Dragoon Guards, and Major M. A. Burke, an officer who had seen active service in Egypt with the 7th Dragoon Guards, came to the Fourteenth. Lieutenant Sir J. P. Miller, Bart., was appointed Adjutant, 1st September.

1889

The Regimental Almanack.

On 1st January the Regimental Almanack, which had been originally started in India several years ago by Schoolmaster Carnegie, but had not been continued since 1st January 1886, was thoroughly revised and corrected throughout by Lieutenant-Colonel H. B. Hamilton, and published for him as a copyright by Warrington and Co., Garrick Street, London, in an entirely new form, both on sheets of cardboard for the barrack-rooms, and also in book-form for officers and others. The commanding officer requested that a copy might always be kept hung up in each barrack-room, so that all ranks might have an opportunity of reading the exploits of their regiment since its first origin up to the present time; and he also expressed a hope that future commanding officers would continue to republish the Almanack annually, corrected up to date from the 'Regimental Digest of Services' kept in the orderly-room. The Almanack is so arranged as to present to view a short history of the Fourteenth, giving items of information for each day in the year. It also contains a roll of the officers, lists of former Colonels and Lieutenant-Colonels, with other details showing the present quarters and establishment, so as to interest officers and men in their regiment and to foster *esprit de corps*. Since 1889 it has been generally reprinted annually with up-to-date additions, and has become quite a regimental institution. A reprint of the Almanack for 1891 is inserted in this book.[1]

Uniform of the band.

During this year the dress of the bandsmen, which had latterly been of a regimental pattern, was assimilated to one uniform pattern in all hussar regiments by order of the Adjutant-General to the Forces.

Officers' Ball at Pavilion, Brighton, February 13th.

On 13th February the officers gave a grand full-dress ball at the Pavilion, Brighton, which was attended by upwards of four hundred guests.

On the 2nd and 3rd May, Major-General Montgomery Moore, commanding the South-Eastern District, made his annual inspection of the Fourteenth at Brighton. On the first day he saw a field-day of the regiment on the drill field in 'field-day order.' Afterwards the Major-General went round the barracks, and in the afternoon he had a general foot parade, and then examined all the regimental and troop books at the orderly-room. On the second day he saw 'G' troop at musketry and test firing on the rifle ranges.

The Major-General was good enough to express his great satisfaction with the excellent turn-out, and with the general set-up and smartness of officers and men, both on parade and in the field.

At Aldershot.

In June the regiment marched from Brighton and Hilsea to Aldershot for summer drills, and on the 28th June once more encamped on Cove Common, near Farnborough. A depôt was left at Preston Barracks, Brighton, under Captain Fraser.

On the 2nd July, at Aldershot, Major-General Sir Drury

[1] See page 431, etc.

Lowe, K.C.B., inspected the regiment in 'drill order' near its camp, and was satisfied with what he saw.

At Woolmer Forest.

In July the Fourteenth formed part of a cavalry column under Colonel E. Wood, C.B., which consisted of 16th Lancers under Lieutenant-Colonel Davison, 14th Hussars under Lieutenant-Colonel Hamilton, 19th Hussars under Colonel French,[1] with 'H' Battery Royal Horse Artillery, and proceeded to Woolmer Forest for outpost and reconnaissance duties, returning after a few days to their former encampment at Aldershot. The column was encamped at Woolmer from the 8th to the 13th July; each regiment had 2 squadrons made up to war strength.

On the 23rd July the Fourteenth took part in a review of Cavalry and Royal Horse Artillery, before Field-Marshal H.R.H. the Duke of Cambridge, Commander-in-Chief, in the Long Valley and Eelmore Plain, Aldershot.

Reviewed by the Emperor of Germany, 7th August 1889.

On the 7th August the regiment had the honour of taking part in the extensive sham fight and review of troops on the Fox Hills, near Aldershot, before H.I.M. William II., Emperor of Germany and King of Prussia, accompanied by the Head-quarters Staff and foreign representatives.

His Majesty was pleased to express his appreciation of the general appearance of the troops, as well as of the soldierlike manner in which the manœuvres were carried out by them, and he requested H.R.H. the Commander-in-Chief to convey this to the troops composing the Aldershot Division, and the other regiments, batteries, and battalions attached thereto who took part in the field-day at Aldershot, together with His Majesty's thanks to Lieutenant-General Sir H. E. Wood, V.C., K.C.B., K.C.M.G., and the officers and troops which were under his command on this occasion.

His Majesty was much struck with the efficiency in the field shown by the militia battalions and the volunteers.

Establishment.

This year the position of Regimental Paymaster-Sergeant was abolished, as well as that of Regimental Armourer-Sergeant. The former was transferred to the Army Pay

[1] Afterwards Major-General Sir J. D. P. French, K.C.B.

Department, and the latter became a member of the Corps of Armourer-Sergeants, but still remained attached to the regiment for duty.

At Brighton and Hilsea.

On the 9th August the Fourteenth left Aldershot for Brighton and Hilsea—6 troops and headquarters stationed at Brighton, under Lieutenant-Colonel Hamilton; 2 troops ('A' and 'B') at Hilsea, under Captains Kirk and R. M. Richardson.

On the 19th September the regiment was inspected at Brighton and Hilsea by Major-General Sir D. Drury Lowe, K.C.B., Inspector-General of Cavalry.

On the 16th December, Major (Brevet Lieutenant-Colonel) the Honourable G. H. Gough was promoted Brevet-Colonel.

1890

On the 15th January, second Lieutenant H. M. Tritton died from an accident in the hunting-field at Chailey, near Lewes. The untimely death of this popular and promising young officer was deeply regretted by all ranks, and the whole six troops of the Fourteenth mounted, in 'review order,' escorted his remains to a special service at St. Martin's Church, Brighton, and thence to the railway station for removal to Plympton, Devon, where they were interred in the family vault.

On the 17th March, Lieutenant-Colonel Hamilton received the brevet rank of Colonel. Colonel the Honourable G. H. Gough (senior major of the regiment), who had been serving on the staff since October 1887 as Deputy-Assistant Adjutant and Quartermaster-General in the Curragh District, rejoined the 14th Hussars for regimental duty on 1st January.

On 8th and 9th May, Major-General M. Moore, commanding South-Eastern District, inspected the Fourteenth at Brighton.

First day he saw a mounted parade in 'field-day order,' when the usual parade movements followed by field movements and some dismounted work were executed. In the afternoon foot parade in 'drill order,' and the Major-General went through the barracks and stables, including the regimental school and married quarters.

Second day he inspected the books, and also held a general kit and necessaries inspection of the whole regiment, going through each troop. Major-General Moore expressed his entire satisfaction with the appearance of the Fourteenth on both parades, as well as with the condition of the horses and the manner in which the parade and field movements were executed.

Queen's Birthday Parade held in Preston Park, 24th May 1890.

On the Queen's Birthday (24th May) the regiment marched in review order to Preston Park, where 1 squadron under Major Burke dismounted and fired a *feu de joie*, then re-mounted, and the 3 squadrons marched past at the walk, trot, and gallop, and performed the sword exercise and pursuing practice, much to the delight of several thousands of the inhabitants, the Mayor of Brighton having kindly granted the use of the park on the suggestion of Colonel Hamilton.

'Sainfoin' wins the Derby, 6th June 1890.

On the 6th June, 'Sainfoin,' a chestnut colt, owned by Captain and Adjutant Sir James P. Miller, Bart., won the Derby at Epsom. In 1884, Sir John Willoughby, Bart., an officer of the Royal Horse Guards, ran a dead heat for the Derby with his brown colt 'Harvester,' against Mr. J. Hammond's bay colt 'St. Gatien,' but it is believed there is no other instance of an officer of the army on full pay winning the Derby.

On the 14th July, Major-General Sir D. Drury Lowe, K.C.B., Inspector-General of Cavalry, inspected the Fourteenth in barracks at Brighton. He expressed to Colonel Hamilton his great satisfaction at the improvement he noticed in the riding of the non-commissioned officers and men since his last inspection; and after the foot parade he had the regiment formed up three sides of a square, and addressed it in complimentary terms.

Firing team wins Cambridge Shield at Bisley, second time.

On the 29th July the regimental firing team, consisting of Farrier-Sergeant Pollard, Sergeant Potter, Private Davis, and Private Upton, won the Royal Cambridge Challenge Shield at the Annual Army Rifle Meeting at Bisley, for the second time.

Tournaments held in Preston Park.

Whilst quartered at Brighton the Fourteenth gave three grand military tournaments, combined with pony races, at their annual athletic sports, in the Preston Park, which were witnessed by thousands of spectators. Captain and Adjutant Sir James Miller, Bart., was the honorary secretary and manager, and the results were eminently satisfactory, as a substantial balance, after paying all expenses, remained to the credit of the committee, which was handed over by the regiment to the Mayor of Brighton for distribution amongst the local charities. The tournaments took place on 26th September 1888, 5th September 1889, and 21st August 1890.

At Aldershot.

At the end of August the Fourteenth left Brighton and Hilsea, and were encamped at Aldershot on Jersey Brow, near Farnborough, on the 2nd September, the squadron from Hilsea having also arrived, preparatory to taking part in the great cavalry manœuvres about to take place in Berkshire, under Lieutenant-General Sir H. E. Wood, V.C., G.C.B., Director of Manœuvres.

Cavalry manœuvres in Berkshire.

The regiment was placed in the 1st Cavalry Brigade with the 8th Hussars, under Lieutenant-Colonel St. Quentin, the Brigadier being Major-General J. A. Le Quesne, and the Cavalry Division was commanded by Major-General Sir Baker Russell, K.C.B., K.C.M.G.

On the 8th September the brigade marched to Churm Camp.

On the 10th September, owing to scarcity of water-supply, the Fourteenth shifted its camp to Blewbury.

On the 13th the brigade was encamped at Uffington, where the Commander-in-Chief, accompanied by Colonel Lord Wantage, V.C., Lord-Lieutenant of Berkshire, and his staff, visited the encampment, and inspected the horse lines.

On the 16th September there was a grand review and march-past of the Cavalry Division on Chilton Downs, in presence of the Commander-in-Chief (H.R.H. the Duke of Cambridge), Lord Wantage, and a vast assemblage of spectators; and on 20th September, the manœuvres having come to an end, the regiments commenced to march back to Aldershot and their other destinations. The Fourteenth halted at Crookham Camp, and on the 22nd September arrived at their former camping-ground on Jersey Brow, Farnborough.

On 23rd September, Lieutenant-General Sir H. E. Wood, V.C., K.C.B., K.C.M.G., commanding the Aldershot Division, Major-General Sir Baker Russell, K.C.B., K.C.M.G., commanding the Cavalry Brigade, Aldershot, and Major-General J. C. Le Quesne, commanding 1st Brigade, Cavalry Manœuvres, inspected the regiment in 'watering-order'; and the Lieutenant-General commanding, as well as both the Major-Generals, complimented Colonel Hamilton on the smart appearance of the men and the excellent condition of the horses, after their recent hard work, which the Lieutenant-General considered was owing to the good care taken of them in camp, and the excellent regimental watering arrangements.

At Hounslow.

On 25th September the whole regiment marched to Hounslow, there to be quartered, detaching 'E' troop to Hampton Court under Captain Stanhope, and 'G' troop to Kensington Palace Barracks under Major Burke.

Musical ride at Royal Naval Exhibition in London.

At the Royal Naval Exhibition held this year in London, the Fourteenth supplied the men and horses for the musical ride, and during the continuance of the exhibition these men and horses were quartered in the Kensington Barracks.

On 8th October, Major-General Sir Drury Lowe, K.C.B., Inspector-General of Cavalry, made his annual inspection of the Fourteenth in the field. They paraded as strong as possible, in 'home marching-order,' on Hounslow Heath, when the usual parade and marching-past movements were executed; and subsequently the commanding officer exercised the regiment in a succession of field movements and manœuvres at a rapid pace. After this the Major-General saw some of the senior officers drill the regiment, and finished the inspection with some outpost work under the second-in-command.

At the conclusion, Sir Drury Lowe had the squadrons formed in close column, when he addressed Colonel Hamilton at the head of the regiment, the officers, warrant officers, non-commissioned officers and men present, expressing his approval of what he had seen, and particularly alluding to the manifest improvement he noticed throughout the regiment since his last inspection in the field a year ago.

The Inspector-General afterwards lunched with the officers, and this concluded his annual inspection, commenced at Brighton in July.

Establishment changes, 1890.

This year (from 1st April) a Sergeant-Master-Tailor was added to the establishment of the regiment, and Mr. Read, who for some time previously had most successfully carried out the duties of civilian master-tailor to the regiment, was enlisted and appointed Sergeant-Master-Tailor. An additional non-commissioned officer was also appointed to carry out the extra clerking work which had devolved on the orderly-room staff since the abolition of the post of Paymaster-Sergeant. The new clerk was given the title of orderly-room clerk, with rank of sergeant, and the former orderly-room clerk was styled for the future 'Orderly-room Sergeant.'

On 11th October, Major-General P. Smith, C.B., commanding Home District, made his inspection in barracks, and held a foot parade. He subsequently expressed himself highly pleased with the efficient appearance of the regiment, and the excellent state of everything he had seen within barracks and on parade.

Uniform.

A new pattern blue serge frock and service-cap were issued this year to cavalry regiments, for use at manœuvres and for drill purposes.

Officers' dress-call for mess introduced.

An officers' 'dress-call' for mess was introduced by Colonel H. B. Hamilton, and adopted by the regiment.

1891

'G' troop at Kensington, under Major Burke, was moved

into headquarters at Hounslow in the beginning of the year, and replaced by 'A' troop under Captain J. Murray.

On 18th April, Major-General P. Smith, C.B., commanding Home District, inspected the Fourteenth at Hounslow. He went through every portion of the barracks, and afterwards had a foot parade, at which the new clothing for 1891, just issued, was worn by the non-commissioned officers and men. The Major-General particularly remarked as to the excellent cut and fit of the tunics and overalls, and he praised the general smartness of the turn-out. Subsequently he addressed the regiment formed in quarter-column of squadrons, in very complimentary terms, and added these words: 'The regiment is in excellent order, and the marching and drill are very good.'

Regimental 'Standing Orders' introduced, 6th May 1891.

On the 6th May, Colonel H. B. Hamilton introduced a book of regimental 'Standing Orders' for the use of the regiment, none having previously existed, so far as was known at the time, and the want of such a book being very much felt and remarked on by inspecting generals on several occasions. There were no traces or mention of any previous code of standing orders preserved in the regiment, and none had existed in the memory of any one living who had previously served in the Fourteenth.[1] Extracts from the 'Standing Orders' were read aloud to the regiment on several successive parades in presence of the commanding officer at Hounslow, and on 31st May the 'Standing Orders' were printed and published by W. Mitchell and Co., of 10 Craig's Court, London, with the following preface:—

'As far as can be ascertained, these are the first 'Standing Orders' of the 14th (King's) Hussars which have ever existed, and they are now published as containing the most important regimental permanent orders on which the present system of the regiment is based. They do not pretend to be complete

[1] It has since been ascertained that from 1784 to 1792 an excellent code of regimental 'Standing Orders' did exist, but no traces of them are to be found. (See pp. 36, 37, and 50, 51.)

in any way; but it is hoped they will prove useful both to old and young soldiers. Every officer, troop sergeant-major, and sergeant of the regiment, is expected to keep a copy, both for his own guidance and to enable him to direct those under him in their duties.

HENRY B. HAMILTON, *Colonel*,
'Commanding 14th (King's) Hussars.

'HOUNSLOW BARRACKS, 31*st May* 1891.'

On 20th May, 4 troops under Major (Brevet-Colonel) the Honourable G. H. Gough marched to Aldershot to undergo their annual course of musketry training. They returned to Hounslow on 6th June, and on the 8th, 2 troops under Captain Kirk proceeded to Aldershot for the same purpose, returning to Hounslow on 24th June. The detachments at Kensington and Hampton Court had recently rejoined headquarters, their places being taken at both those stations by detachments of the Royal Horse Guards (Blue).

In June the Fourteenth took part in a cavalry brigade field-day on Wimbledon Common under Major-General P. Smith, C.B., commanding Home District, with the Household Cavalry, in which Colonel Charles Needham, 1st Life Guards, acted as Brigadier.

On 26th June, Major-General J. Keith Fraser, C.M.G., Inspector-General of Cavalry, made a minute inspection of the regiment in barracks, going through the troop-stables, troop-rooms, regimental institutes, workshops, etc. etc., and afterwards carefully examined the regimental and troop books, and the recently promulgated 'Standing Orders.' This completed the first day of his annual inspection of the regiment.

Lieutenant-Colonel—Hon. G. H. Gough.

On 1st July, Colonel H. B. Hamilton retired on half-pay after four years in command of the regiment, and Major the Honourable G. H. Gough (Brevet-Colonel) succeeded him. Captain A. C. King became Major in succession, and Major A. J. English became second in command.

4th July 1891.

Upon the occasion of the state visit of His Imperial

Majesty the German Emperor (William II.) to England, a detachment of the 14th (King's) Hussars (strength 10 officers, 160 men, and the band) marched from Hounslow to Windsor to keep the streets. The band paraded with their new drum-banners, presented to the regiment by Captain and Adjutant Sir James P. Miller, Bart., on which the device of the regimental badge, 'The Prussian Eagle,' is conspicuous.

Upon the occasion of the state visit of H.I.M. the German Emperor to the Guildhall, the regiment marched from Hounslow to London to keep the streets from Buckingham Palace and Pall Mall to the Haymarket. 10th July 1891.

Upon the occasion of the review of troops (24,000, principally volunteers) by the German Emperor on Wimbledon Common, the regiment marched from Hounslow to Wimbledon to keep the ground. On the same evening the regiment found a travelling escort of 14 non-commissioned officers and men, under command of Captain L. J. Richardson, to escort the Emperor and Empress from the Crystal Palace, at Sydenham, to Buckingham Palace. The Emperor expressed to the officer in command his very keen approbation with the manner the duty had been performed, and especially noticed the distinctive badge of the regiment, 'The Prussian Eagle.'[1] He subsequently sent his signed photograph to Captain L. J. Richardson. 11th July 1891.

On the 14th July the regiment proceeded by march route from Hounslow to Aldershot, under command of Colonel the Honourable G. H. Gough, for summer drills, and encamped at Bourley, being joined in brigade (2nd Cavalry Brigade) with the 2nd Life Guards and Royal Horse Guards (Blue), Major-General J. C. Le Quesne being the Brigadier of the 2nd Brigade, and Major-General Keith Fraser, C.M.G., Inspector-General of Cavalry, was in command of the Cavalry Division. At Aldershot.

On 16th July the Fourteenth, under Colonel Gough, took part in the review of the Aldershot Division by Her Majesty Queen Victoria, before whom they marched past, strength of Reviewed by Queen Victoria, 16th July 1891.

[1] This badge was granted in 1798. See p. 49.

regiment on parade being 17 officers, 219 non-commissioned officers and men.

On the 22nd July the regiment took part in a cavalry field-day on the Fox Hills, under H.R.H. the Duke of Cambridge, Commander-in-Chief, the reconnoitring squadrons towards the Fox Hills being furnished by the 14th Hussars. At the conclusion of the operations His Royal Highness's remarks were very complimentary to all.

Firing team wins Cambridge Challenge Shield for third time.

On the 24th July the regimental firing team, consisting of Farrier-Sergeant Pollard, Sergeant Potter, Private Davis, and Private Upton, once more, for the third time in five years, and the second time in succession, won the Royal Cambridge Challenge Shield at the Annual Army Rifle Meeting at Bisley.

From the 24th July to the 3rd August the regiment, as part of the 2nd Brigade, was encamped in Woolmer Forest.

At Leeds and Birmingham.

On the 4th August, headquarters, with 'A,' 'B,' 'C,' and 'D' troops, marched from Aldershot to Leeds by squadrons; 'H' and 'K' troops went to Birmingham, under Captains King and Robertson; 'E' and 'G' troops remained to go through their annual musketry practice under Major Burke, and subsequently marched to join the headquarters at Leeds on the 18th August, arriving on the 31st. The headquarters reached Leeds on the 19th, and the squadron for Birmingham arrived on the 12th.

The regiment was inspected in the field by Major-General J. K. Fraser, C.M.G., Inspector-General of Cavalry, on the 1st September, at Leeds. This concluded his annual inspection, commenced on the 26th June at Hounslow.

On the 18th September the regiment marched to York, and on the 19th September it was inspected, together with the Royal Dragoons, on Knavesmire, by Field-Marshal H.R.H. the Duke of Cambridge, Commander-in-Chief, returning to Leeds on the evening of the review. H.R.H. conveyed to the regiment the expression of his entire satisfaction with its turn-out and drill.

During this year a reduction (dated 1st April) in

the establishment of the regiment took place, which was fixed at—

Establishment reduced, 1891.

8 Troops.
24 Officers.
2 Warrant officers.
40 Sergeants.
8 Farriers.
8 Trumpeters.
33 Corporals.
4 Shoeing-smiths.
300 Privates.
58 Officers' chargers.
280 Troop-horses.

1892

The Fourteenth remained at Leeds and Birmingham till June.

Squadron organisation established, 1892.

The squadron organisation was introduced into the cavalry on the 1st April, and the 4 squadrons distinguished as 'A,' 'B,' 'C,' and 'D' squadrons.

'A' and 'B' troops became 'A' squadron.
'C' and 'D' troops became 'B' squadron.
'E' and 'G' troops became 'C' squadron.
'H' and 'K' troops became 'D' squadron.

On the 25th April, Major-General H. C. Wilkinson, C.B., commanding the North-Eastern District, inspected the regiment at Leeds, and expressed himself fully satisfied.

On the 4th June the regiment marched to York, and was encamped in the Infantry Barracks, and practised brigade drill on the Knavesmire with the Royal Dragoons from the 6th to the 10th June, under the superintendence of the Major-General commanding the North-Eastern District.

At Strensall Camp for drills.

On the 11th June, 3 squadrons ('A,' 'B,' and 'C') proceeded to Strensall Camp for summer drills. On the 30th June, 'D' squadron left Birmingham and reached Manchester on the 2nd July, and occupied quarters in Hulme Barracks. The depôt was left at Leeds till the 27th June, after the headquarters had proceeded to Strensall, when it proceeded to Manchester, the mounted portion reaching Hulme Barracks on the 3rd July.

On the 13th July, Major-General J. K. Fraser, C.M.G., Inspector-General of Cavalry, inspected the Fourteenth on Knavesmire, York.

Establishment augmented slightly, 1892.

On the 26th July the establishment was altered and augmented, to date from 1st April as follows :—

4 Squadrons.	8 Trumpeters.
24 Officers.	33 Corporals.
2 Warrant officers.	328 Privates.
40 Sergeants.	58 Officers' chargers.
8 Farriers.	280 Troop-horses.

At Manchester.

On 12th August the 3 squadrons from Strensall Camp arrived at Hulme Barracks, Manchester.

On 2nd September, as serious riots were apprehended at Winsford, a party of 166 non-commissioned officers and men, under Captains L. J. Richardson and Murray, proceeded dismounted by special train at 6.40 P.M. to the scene of disturbance. Their arrival put an end to the riot, and the party returned to headquarters on the 6th September. A telegram received from the Chief Constable of Cheshire on the 2nd September was the only warning given to the regiment. The Major-General commanding the North-Western District expressed his satisfaction at the promptitude shown and the good service rendered.

On the 30th September the regiment was inspected by Major-General Julian Hall, commanding the North-Western District. He saw a foot parade, and expressed his satisfaction.

Captain E. J. Tickell was appointed Adjutant.

1893

The Fourteenth remained this year in Hulme Barracks, Manchester.

On the 17th and 18th April, Major-General Julian Hall, commanding the North-Western District, made his annual inspection of the regiment.

First day, barracks generally and foot parade.

Second day, mounted parade on Manchester racecourse. He expressed his unqualified satisfaction with all he had seen.

On 4th and 5th May, Major-General J. K. Fraser, C.M.G., Inspector-General of Cavalry, inspected the regiment in riding-school, and held a foot parade.

On the 13th July, Major-General J. K. Fraser, C.M.G., Inspector-General of Cavalry, inspected the regiment in the field. After parade movements and field movements in Trafford Park, the Inspector-General witnessed 'B' squadron swimming their horses over a piece of water. The result of the inspection was thoroughly satisfactory.

On the 18th August a telegram came whilst the regimental sports were in progress, ordering 1 squadron, strength 100 men and horses, to proceed in aid of the civil power to Newport, Monmouthshire. Parts of 'A' and 'B' squadrons, under Captain Gilbert Hamilton, proceeded by train, arriving at Newport at 4.30 A.M. on the 19th, and were divided into 4 detachments at Crumlin, Pontypridd, Aberaman, and Bridgend. A large force of infantry was also employed. This decided action on the part of the authorities checkmated the rioters throughout the disturbed districts, who eventually went back to their work by the end of September.

Detachments went to Newport, etc., for riots.

On the 9th September this detachment returned to Manchester.

On the 12th September, Major-General William Arbuthnot, C.B., who formerly commanded the Fourteenth, died at Tooting, and was buried in London. Colonel the Honourable G. H. Gough and a party of officers and non-commissioned officers of the regiment proceeded there to attend his funeral, which took place on the 16th, in Brompton Cemetery.

On 16th September, 40 men and horses of 'D' squadron, under Captain Stacey and Lieutenant Brooksbank, went to Holywell, Flint, to be billeted there, owing to expected disturbances in the Bettisfield district.

Detachments sent to Flint for coal riots.

On the 22nd September this party proceeded to Mold, where the Buckley miners were creating disturbances, and rejoined headquarters on 17th October. The Chief Constable of the county wrote as to the conduct of the men—'I may say that their behaviour has won the respect of all classes.'

On 25th October, 35 men and horses of 'C' squadron went by rail to Wrexham to be billeted there, under Captain Gage and Lieutenant Henry, during the great coal strike. This party rejoined headquarters at Manchester, after the strike terminated, on 22nd November.

Martini-Metford carbines issued.

In November of this year Martini-Metford carbines and new swords, pattern 90, of English make, were issued to the regiment, and the Martini-Henry carbines, issued in 1886-87, were returned to ordnance stores.

Captain Gilbert H. Claude Hamilton became Major in the regiment *vice* M. A. Burke, retired.

1894

In March, Major-General J. K. Fraser, C.M.G., Inspector-General of Cavalry, made his first day's inspection in barracks, and saw the rides and a foot parade.

On 16th and 19th April, Major-General J. Hall, commanding North-Western District, inspected. On the first day the regiment carried out a reconnaissance against the 17th Lancers, quartered at Preston, the point of contact being Bolton-le-Moors-Hill. The greatest distance traversed was 32 miles; mean weight carried by troop-horses, 18 stone 8 lbs. On the second day of inspection, 19th April, the Major-General inspected books and barracks, and saw a foot parade, after which he addressed the regiment in flattering terms.

Strength:—24 officers; 1 warrant officer; 48 sergeants; 38 corporals; 8 trumpeters; 3 shoeing-smiths; 338 privates—total all ranks, 460. Number of officers' chargers, 58; troop-horses, 278.

On 21st May 1894, Her Majesty Queen Victoria came to Manchester to open the Ship Canal. The 14th Hussars and the 2nd King's Liverpool Regiment lined the route of the Royal procession through Manchester and Salford, a distance of 7½ miles. The Queen's escort, furnished by the Fourteenth, was commanded by Captain R. M. Richardson, accompanied by Lieutenants Hughes and Eley. For their services on this occasion the regiment received letters of approbation from the Chief Constables of Lancashire and Salford, and from the Mayor of Salford, who wrote: 'I was delighted to witness the smartness, good-temper, and discipline which your men displayed under rather trying circumstances along the route.' Opening of the Manchester Canal, 21st May 1894.

On 26th May, Colonel the Honourable G. H. Gough was awarded a Companionship of the Order of the Bath.

The Fourteenth remained at Manchester till June, when they proceeded to Ireland as follows:— Ordered to Ireland.

On 26th June, 'A' and 'D' squadrons proceeded by march route towards Liverpool, and embarked at Liverpool in ss. *Leitrim* for the North Wall, Dublin, whence they marched to Waterford and Cahir, arriving at Waterford on 2nd July, Cahir on 3rd July.

On 30th June, 'B' squadron and headquarters followed by the same route, and per ss. *Leitrim* to Dublin. The ss. *Leitrim* broke down on the voyage, and had to be towed into Dublin. Headquarters reached Cahir on 9th July. 'B' squadron reached Limerick on 9th July. At Cahir and out-quarters.

On 8th July, 'C' squadron proceeded by similar route to Liverpool, and sailed thence by ss. *Cavan* to Dublin, thence by march route to Fethard (half squadron), arriving 14th, and to Cahir (half squadron), arriving 16th July.

Two squadrons were at Cahir.
Half a squadron at Fethard.
Half a squadron at Waterford.
One squadron at Limerick.

On 4th August, Major-General J. K. Fraser, C.M.G., Inspector-General of Cavalry, had the second day of his annual inspection, commenced in March, and the detachment at Fethard came in to Cahir for the day of inspection.

On 16th August, Major-General J. Fryer, C.B., Commanding Cork District, made an inspection of the Fourteenth at Cahir, and subsequently he inspected the detachments at Waterford and Limerick. He went round the barracks in the morning, and had a marching-order mounted parade in the afternoon.

During August and September, Colonel the Honourable G. H. Gough, C.B., and Quartermaster Mugford were employed on temporary staff duties in the cavalry manœuvres in Berkshire, under the command of Major-General J. Keith Fraser, C.M.G., Inspector-General of Cavalry.

Establishment slightly augmented, 1894.

In October the establishment of the regiment was slightly augmented from 328 to 344 privates, and the number of troop-horses was increased at the same time from 280 to 300.

The actual strength on 31st December was 431 non-commissioned officers and men, with 298 troop-horses.

Changes in issue of army clothing introduced, 1894.

The whole system of army clothing underwent a change this year, and all articles when once issued became the property of the soldier. The old system under which all clothing belonged to the Government, and had to be returned or paid for, was finally abolished.

Captain H. M. Mitchell became Major *vice* A. C. King, retired.

1895

The Fourteenth passed the whole of this year at Cahir and out-quarters.

On 5th and 6th April, Major-General J. Fryer, C.B., commanding Cork District, made his annual inspection of the regiment at Cahir. He inspected the Fethard detachment on 3rd April, the detachment at Waterford on 8th April, and the squadron at Limerick on 9th April. In the morning he had a foot parade, and went round the barracks; in the afternoon he inspected the troops in the field in 'drill order.' On the 6th, at Cahir, he saw the riding-school work, gymnasium class, regimental and troop books, etc. The Major-General's remarks were very complimentary.

On 14th May, Field-Marshal the Right Honourable Viscount Wolseley, K.P., G.C.B., G.C.M.G., Commanding the Forces in Ireland, inspected the headquarters at Cahir. He saw the regiment mounted, and witnessed a march-past, then he made a careful tour round barracks, and afterwards lunched with the officers.

On 28th June, Major-General G. Luck, C.B., Inspector-General of Cavalry, inspected 'B' squadron at Limerick. On the 29th he inspected 'A,' 'C,' and 'D' squadrons in the field at Cahir. On 1st July he had squadron inspection in the field, each squadron separately, and every man mounted, then riding-school work; and in the afternoon he saw the squadrons separately, one in service marching order, one with stripped saddles, one without saddles; then the officers at stick practice; then all ranks armed with pistols at dismounted revolver practice.

Owing to rain on 2nd July the Inspector-General did not carry out his intended outpost and reconnaissance inspection.

The establishment was fixed as follows:— Establishment, 1895.

4 Squadrons.	8 Trumpeters.
24 Officers.	35 Corporals.
2 Warrant officers.	353 Privates.
39 Sergeants.	300 Troop-horses.
8 Farriers.	

1896

The Fourteenth remained at Cahir and out-quarters all this year.

On 1st and 2nd April, Major-General J. Fryer, C.B., Commanding Cork District, made his annual inspection at Waterford, Cahir, Fethard, and Limerick. He expressed his satisfaction with what he saw.

On 15th, 16th, and 17th April, Major-General G. Luck, C.B., Inspector-General of Cavalry, inspected the regiment at Limerick, Cahir, and Waterford, but only finished a portion of his annual inspection, completing it in September.

Lieutenant-Colonel—A. J. English.

On 1st July, Colonel the Honourable G. H. Gough, C.B., after five years in command, retired on half-pay, when Major A. J. English became Lieutenant-Colonel, and succeeded to the command of the Fourteenth.

At Curragh Camp for drills.

The Fourteenth went to Curragh Camp for the drill season on 1st July, and returned to previous quarters for the winter. Whilst at Curragh Camp they were encamped in Donelly's Hollow.

Lee-Metford magazine carbines issued, 1896.

On 3rd August, Lee-Metford magazine carbines, designated 'M.L.M.,' were issued to the regiment, and the Martini-Metford carbines were returned to ordnance stores.

On 6th August headquarters and 2 squadrons proceeded *en route* to Cahir, detaching parties to Fethard and Waterford. On 7th August, 1 squadron proceeded to the Kilkenny field manœuvres, under Captain Brown, from the Curragh Camp. On 11th August detachments of the regiment were sent to Carrick-on-Suir and Clogheen.

On 17th, 18th, and 19th August, Field-Marshal the Right Honourable Lord Roberts, G.C.B., G.C.S.I., C.I.E., Commander of the Forces in Ireland, inspected the regiment as follows—On 17th at Waterford; 18th at Fethard; 19th, headquarters at Cahir.

On 14th, 15th, and 16th September, the Inspector-General of Cavalry, Major-General G. Luck, C.B., inspected the regiment, beginning with the detachment at Waterford on 14th, and there was a reconnaissance carried out towards Cahir from Waterford on 15th. The strength of regiment was 20 officers, 2 warrant officers, 486 non-commissioned officers and men,—total all ranks, 508; officers' chargers, 54; troop-horses, 326. An augmentation in the establishment had taken place recently.

During the year 30 troop-horses were transferred from the

Fourteenth to the 3rd Dragoon Guards, but the effective strength of troop-horses was 324 at the end of the year.

On 3rd October, General C. W. Thompson, Colonel of the regiment, died at Braintree, Essex, and Lieutenant-General the Honourable C. W. Thesiger was appointed to succeed him. Colonel—Hon. C. W. Thesiger.

The following 'after order' was issued by Lieutenant-Colonel A. J. English, commanding the 14th Hussars, on 6th October, when he heard of the General's decease:—

'The Commanding Officer deeply regrets having to inform the regiment of the death of General Thompson, the Honorary Colonel of the regiment, which occurred on 3rd inst. General Thompson served with distinction in the regiment in India, and was present at the battles of Chillianwallah and Goojerat.

'As is well known, General Thompson had the greatest regard for the regiment. Out of respect for him the regiment will parade at 1.15 P.M. to-morrow, outside the stables, and the band will play the "Dead March in Saul."

'The band will not otherwise play for the next three days, and no trumpets will be sounded during that time.

'Officers will wear crape for 21 days.—By order,

'C. B. Tottenham, *Lieutenant*,
'*Acting Adjutant*, *14th Hussars.*
'A. J. E.'

On the 9th November, Colonel E. H. Holley, Commanding Royal Artillery, Cork District, and Colonel on the Staff, inspected the regiment in the absence on sick leave of Major-General J. Fryer, C.B. He saw the headquarters at Cahir on the 9th, and the detachment at Clogheen in the afternoon. He inspected the detachments at Carrick-on-Suir and Waterford on the 10th, and at Fethard on the 11th.

Major H. W. Mitchell retired from the regiment, and

Captains E. D. J. O'Brien and L. J. Richardson became Majors.

Captain R. C. Stephen was appointed Adjutant.

1897

On the 1st January, two squadrons and headquarters were at Cahir, a half squadron at Waterford, a half squadron at Carrick-on-Suir, a half squadron at Fethard, a half squadron at Clogheen.

In May, the half squadrons at Waterford and Carrick-on-Suir were withdrawn, one going to Cahir and the other to Fethard.

On 6th May, a squadron moved from Cahir and Clogheen (half from each) to Fethard, and half of 'C' squadron moved from Cahir to Clogheen. On 31st May, 'C' squadron moved from Cahir and Clogheen to Fethard, and 'A' squadron moved from Fethard to Cahir; at the same time half of 'B' squadron went from Cahir to Clogheen.

On 14th and 15th April, Major-General J. Fryer, C.B., Commanding Cork District, inspected the regiment at Cahir. He inspected at Clogheen on 16th; at Carrick-on-Suir, 17th; at Waterford, 22nd; and at Fethard on 23rd April.

The number of effectives (non-commissioned officers and men) was 492, and their religious denominations, nationalities, and educational statistics were as follows:—

Religious Denominations.

	Officers.	Warrant Officers.	Non-commissioned Officers and Men.
Church of England, . .	23	2	392
Presbyterians, . .			13
Wesleyans, . . .			23
Other Protestants, . .			5
Roman Catholics, . .			59
	23	2	492

NATIONALITIES.

	Officers.	Warrant Officers.	Non-commissioned Officers and Men.
English, ...	18	2	433
Scottish, ...	2		9
Irish, ...	3		50
	23	2	492

EDUCATIONAL ACQUIREMENTS.

Non-commissioned officers and men.

No. of men of inferior education, ...			8
No. of men who are better educated.	Holding certificates,	Third-class, .	64
		Second-class, .	99
		First-class, .	11
	Not holding certificates, .		310
			492

HORSES.

No. of each age.		*No. of each height.*	
Under 5 years, .	55	14½ hands and under,	0
5 to 8 years (inclusive),	148	15 ,, ,,	0
9 to 12 years (inclusive),	65	15½ ,, ,,	270
13 years and over,	56	16 ,, ,,	54
	324		324

Average age and service of the horses cast, 1896.

Age, 12½ years. Service, 8½ years.

Officers' sheepskins and shabracques discontinued.

On 20th March some new orders affecting the dress of officers and men were introduced by the Inspector-General of Cavalry, Major-General G. Luck, C.B. The use of sheepskins for all ranks and shabracques for officers was discontinued in the cavalry, the latter having already been discontinued for the ranks many years previously.

On 1st April the cavalry of the line was reorganised under

directions from the Inspector-General of Cavalry. Each regiment was to have 3 squadrons for service, and one reserve squadron. At the same time a new pattern saddle was issued to cavalry regiments, and the regimental establishment was augmented from 522 to 570 of all ranks, the number of troop-horses being raised from 325 to 343.

New cavalry saddles issued. Establishment increased, 1897.

Establishment of the 14th (King's) Hussars, 1897 :—

Officers :—	
Colonel,	1
Lieutenant-Colonel,	1
Majors,	4
Captains,	4
Lieutenants,	8
Second-Lieutenants,	3
Adjutant,	1
Riding-master,	1
Quartermaster,	1
Total,	24
Warrant Officers :—	
Regimental Sergeant-Major,	1
Bandmaster,	1
Total,	2
Non-commissioned Officers (Sergeants) :—	
Quartermaster-sergeant,	1
Farrier quartermaster-sergeant,	1
Sergeant-instructor in fencing,	1
Squadron sergeant-major rough rider,	1
Squadron sergeant-majors,	4
Squadron quartermaster-sergeants,	4
Orderly-room sergeant,	1
Orderly-room clerk,	1
Sergeant-trumpeter,	1

Armourer-sergeant (attached), . .	1
Saddler-sergeant, ...	1
Sergeant-cook, ...	1
Sergeant master-tailor, . . .	1
Sergeant-farriers, ...	4
Sergeants, ...	23
Total, . .	46

Rank and File :—

Trumpeters, ...	8
Corporals, ...	24
Corporal shoeing-smiths, . . .	3
Shoeing-smiths, ...	8
Saddlers, ...	4
Saddletree-maker, . . .	1
Privates, ...	460
Total, . .	508

Total of regiment, all ranks, . .	570
Horses (chargers and troopers), .	399

Establishment by Squadrons :—

	Officers.	Warrant Officers.	Non-commissioned Officers	Rank and File.	Officers' chargers.	Troop-horses. Riding	Troop-horses. Draught
'A' squadron, .	5		9	132	12	96	2
'B' squadron, .	5		9	132	12	96	2
'C' squadron, .	5		9	132	12	96	2
Reserve squadron,	3		8	111	7	38	2
'Headquarters,' .	5	2	11	1	13	9	
Totals, . .	23	2	46	508	56	335	8

N.B. (1) The reserve squadron in above table includes band, recruits, remounts, men in regimental employment. (2) 'Headquarters' includes the Lieutenant-Colonel, senior Major, Staff-Officers, Warrant Officers, etc.; staff-sergeants, etc. (3) The 'Colonel' is not included in the total of 23 officers; but in the total of 'all ranks' he is included.

Marched from Cahir to Newbridge.

In June the regiment marched from Cahir and out-quarters to cavalry barracks, Newbridge, detaching one squadron ('C') under Captain Brown to the Curragh Camp for vedette duty, where it arrived on 14th June.

'A,' 'B,' and 'D' squadrons and headquarters under Lieutenant-Colonel English reached Newbridge on the 16th and 17th, having commenced the march from Cahir on 11th June.

New pattern revolver-pistols issued ('Webley').

The sergeants were this year equipped with a new revolver-pistol, the 'Webley' (mark H.).

On 30th June, and again on 24th July, Major-General B. Combe, C.B., Commanding the Curragh District, inspected the regiment.

Regiment furnishes escorts for their Royal Highnesses the Duke and Duchess of York.

On the 14th August the regiment proceeded from Newbridge, and was encamped in Dublin under Lieutenant-Colonel English for a fortnight, during which time it supplied the escorts for their Royal Highnesses the Duke and Duchess of York, on the occasion of their visit to the Irish capital and its neighbourhood, returning to Newbridge on 28th August.

During their stay in Dublin, the Fourteenth, under Lieutenant-Colonel English, took part in the grand parade of troops held in the Phœnix Park under command of Field-Marshal the Right Honourable Lord Roberts, V.C., K.P., G.C.B., G.C.S.I., G.C.I.E., Commanding the Forces in Ireland, on which occasion Her Royal Highness the Duchess of York presented new colours to the 2nd Battalion of the Royal Fusiliers.

The strength of the 3 squadrons in Dublin was 18 officers, 344 men, and 275 horses.

Regiment entertained by 13th Hussars in Dublin.

During the stay of the Fourteenth in Dublin, their old friends the 13th Hussars, who were then quartered in the Marlborough Barracks, Phœnix Park, most hospitably entertained the whole regiment, and thus the friendship of both regiments, which was so marked in the old campaigning days of the war in the Peninsula, was most happily revived. Sic, VIRET IN ÆTERNUM.

On this occasion the officers of the Fourteenth used, as guests, those same mess-tables which their predecessors, about to embark for India, fifty-six years before (1841) had presented as a parting gift to the regiment which now entertained them.

On the 13th, 14th, and 15th September the Fourteenth were inspected by Major-General Sir George Luck, K.C.B., Inspector-General of Cavalry.

1898

On the 29th March, Major-General Sir G. Luck, K.C.B., Inspector-General of Cavalry, made his spring inspection of the regiment at Newbridge.

The number of carbines in possession of the regiment was 467, magazine Lee-Metfords; the number of Webley revolver-pistols, 62; the number of swords, cavalry pattern, 555.

On the 13th August, 1 squadron moved from Newbridge to Dublin, and returned from Dublin to Newbridge on the 31st August.

The total effectives on 1st December was as follows: —headquarters at Newbridge; 4 squadrons; 23 officers; 2 warrant officers; 589 non-commissioned officers and men,—total all ranks, 614. Horses: officers', 55; riding, 325; draught, 16. Soldiers' wives on the married roll (warrant officers, non-commissioned officers and men), 45. Children on the strength, males, 39; females, 54,—total 93. Strength of regiment, Dec. 1898.

1899

On the 22nd February, Lieutenant-Colonel A. J. English was placed on retired pay, when Major Gilbert H. C. Hamilton became Lieutenant-Colonel, and assumed command of the regiment. At the same time Major E. D. J. O'Brien became second in command, and Captain E. J. Tickell, recently Lieutenant-Colonel—G. H. C. Hamilton.

awarded a companionship of the 'Distinguished Service Order' for special services in West Africa, became a Major in the regiment.

Ordered to mobilise up to war strength.

Early in October the war in South Africa commenced, and the Fourteenth being now on the higher establishment of cavalry regiments on home service, were ordered on the 9th October to mobilise and to call up the reserves so as to be placed on a war strength. The higher establishment consisted of the following numbers :—

Officers,	23
Warrant Officers,	2
Sergeants,	45
Trumpeters,	8
Corporals,	30
Corporal shoeing-smiths,	3
Shoeing-smiths, saddlers, assistant saddlers, and saddle-tree maker,	16
Privates,	560
All ranks,	687
Horses,	465

Ordered to the Boer War in South Africa.

In regimental orders of the 18th October at Newbridge, the following appeared :—

'Departures—14th Hussars, "B" squadron on the 22nd inst. to embark in ss. *Jamaican*, and headquarters and "A" squadron on 23rd inst. in ss. *Siberian*, at Queenstown, for South Africa.'

Subsequently an order was received for the regiment to stand fast and await further orders. A few days later came a definite order cancelling the embarkation of the regiment, owing to a serious outbreak of influenza and 'pink-eye' amongst the horses. It was, however, intimated to the regiment that so soon as the horses were free from this epidemic, they would be moved to Aldershot, and in the event of more cavalry being required in South Africa, they would in all probability be sent out.

Inspection at Newbridge by Lord Roberts 26th October 1899.

On the 26th October, Field-Marshal the Right Honourable F. S. Lord Roberts, V.C., K.P., G.C.B., G.C.S.I., G.C.I.E., Commander-in-Chief in Ireland, inspected the regiment at Newbridge, in service marching-order, ready for active service, officers and men being dressed in khaki uniform, and at the conclusion of the review he spoke as follows:— 'Lieutenant-Colonel Hamilton, officers, non-commissioned officers and men of the 14th Hussars, I had some doubts whether I should be able to inspect you this morning, as there was a report that the epidemic of influenza amongst your horses might possibly prevent the regiment being sent on active service. Your commanding officer, however, informed me that the disease is evidently decreasing, and that there is a marked improvement in the condition of the sick horses to-day. I am very glad to hear this, for I should much regret anything occurring to prevent the 14th Hussars taking part in the important campaign now being carried on in South Africa. As the 14th Light Dragoons, the regiment had a splendid reputation. One only need read Napier's *History of the Peninsular War* to know how well it behaved during those trying years. Whether on escort duty, on reconnaissance work, or when engaged with the enemy, the discipline, intelligence, and bravery of the 14th Light Dragoons were proverbial. I went to India soon after the conclusion of the Punjaub war, and I used often to hear men who had served in it relate the good services performed by the 14th Light Dragoons at Ramnuggur, Chillianwallah, and Goojerat; and in later years, when I was on the staff of Lord Napier of Magdala, he delighted in telling me of the untiring way in which the 14th Light Dragoons worked day after day and night after night in the pursuit of the rebel Tantia Topee. No march was too long for them, no country too difficult. What this regiment did as the 14th Light Dragoons, I am confident it will do equally well as the 14th Hussars.

I congratulate you all most heartily on having been selected to form part of the force now being employed in South Africa, and I wish you all possible success. Should you return to Ireland at the close of the war, you may rest assured that you will receive a most cordial welcome. Good-bye, and God bless you.'

At Aldershot.

On the 15th November the wished-for order came, and the regiment, under Lieutenant-Colonel Gilbert Hamilton, left Newbridge next day, arriving on the 17th at the West Cavalry Barracks, Aldershot, having been conveyed by steamer and railway with all possible despatch.

Before leaving Newbridge, the horses of the Fourteenth not required for the depôt squadron at the Curragh Camp were handed over to the 21st Lancers, stationed at the Curragh. The Fourteenth, on arrival at Aldershot, were mounted by the Remount Establishment of the army with registered horses procured in large numbers from different hunts and other sources.

Regiment receives fresh horses.

The service squadrons ('A,' 'B,' 'C') and headquarters were constituted as follows:—

27 Officers.	446 Troop-horses.
1 Warrant Officer.	51 Chargers.
42 Sergeants.	1 Water-cart.
6 Trumpeters.	1 Maxim gun.
31 Corporals.	2 service small-arms ammunition wagons.
478 Privates.	

The names of the officers who proceeded to South Africa are as follows:—

Lieutenant-Colonel—G. H. C. Hamilton (in command).
Majors—E. D. J. O'Brien.
R. M. Richardson.
E. D. Brown.
Captains—A. M. B. Gage.
D. M. Miller.
R. G. Brooksbank.

Lieutenants—C. B. Tottenham.
W. Henry.
W. J. Lockett.
F. R. Lawrence, D.S.O., Adjutant.
H. J. Tilney (proceeded from India).
T. E. L. Hill-Whitson.
R. Blackett.
E. J. Jameson.
W. R. Campbell.
E. T. L. Wright.
Second-Lieutenants—J. B. Walker.
J. G. Browne.
Hon. H. Grosvenor.
B. M. Dawes.
C. H. B. Prescott-Westcar.
Quartermaster—F. Mugford (Honorary Captain).
Warrant Officer—Regimental Sergeant-Major A. Pridgeon.

Major E. J. Tickell, D.S.O., was employed in South Africa on 'special service,' away from the regiment, and Captain William Prevost joined the regiment in South Africa from sick leave in the spring of 1900, and was appointed Station Staff Officer, Lines of Communication, on 9th August in that year.

There were four attached officers, viz.—

Veterinary Captain J. G. O'Donel, A.V.D.
Captain C. Dalton, R.A. Medical Corps.
Captain P. R. Denny, 1st (King's) Dragoon Guards.
Lieutenant C. G. G. Hutchison, 21st Lancers.

Reserve squadron at Curragh Camp, November 1899.

A reserve squadron was left at the Curragh Camp to form a depôt whilst the regiment was on active service, where the regimental books and heavy baggage of the service squadrons were stored. This squadron was under the command of Captain J. Murray, and consisted of :—

Lieutenant—E. P. Philipson-Stow.

Second-Lieutenants—J. Harvey.
G. R. Scott.
J. F. Champion.
F. E. Lace.
Riding-Master—R. Odlum (Honorary Captain).
Warrant Officer—Bandmaster H. Hemsley.

428 Non-commissioned officers and men. 16 Soldiers' wives.
118 Horses. 47 Children.

'A and 'C' squadrons and headquarters embark, 13th December 1899.

On the 13th December, headquarters of the Fourteenth and two squadrons ('A' and 'C'), under Lieutenant-Colonel G. Hamilton, embarked in the hired transport ss. *Victorian*, and sailed from Southampton for Cape Town.

'B' squadron embarks, 21st December 1899.

The remaining squadron ('B'), under Major R. M. Richardson, left Aldershot on the 21st December 1899, and embarked at Southampton on board the hired transport ss. *Cestrian*. Major Richardson was taken ill at Southampton, and could not proceed with his men, Captain Brooksbank taking his place in command of the squadron; but Major Richardson was enabled to proceed subsequently by another ship, and rejoined his squadron in South Africa at De Aar on 22nd January 1900, and resumed command. The *Cestrian* experienced some very rough weather, and reached Cape Town on 10th January. This squadron disembarked, went to Maitland Camp, and remained there till 16th; thence it proceeded by rail to De Aar, arriving there 18th, and marched forward by road on the 29th, arriving at Orange River 2nd February, and at Zoutpans Drift 4th February.

1900

'B' squadron in Orange River Colony.

We will first recount the doings of the 'B' squadron:—On 9th February the Queen's chocolate was issued to all ranks, and the same day the march was continued to Ramah, on the Orange River, whence on 11th February the squadron started to take part in the relief of Kimberley as a portion of the Cavalry Division commanded by Major-General J. D. P. French.

On the 15th, the date of the relief of Kimberley, the 'B' squadron was engaged on the Modder River and outside Kimberley, and at Dronfield, north of Kimberley, on the 16th, when Squadron Quartermaster-Sergeant Ayres was severely wounded.

'B' squadron engaged at Kimberley, 15th and 16th February 1900.

On the 21st February this squadron, under Major R. M. Richardson, with Captain Brooksbank, Lieutenants Henry and Lockett, and Second-Lieutenants Walker and Dawes, was employed in the operations undertaken to surround General Cronje and his army of Boers at Paardeberg. Cronje surrendered on the 27th February, and the squadron crossed to the south of the river Modder on 6th March.

'B' squadron at Paardeberg, 21st to 27th February 1900.

On the 7th it was engaged with the enemy at Poplar Grove (Osfontein), and took part in a reconnaissance to Abraham's Kraal on the 9th; engaged on the 10th at Abraham's Kraal (Driefontein), and was present on 13th at the capitulation of Bloemfontein.

At Poplar Grove, Driefontein, and Bloemfontein, 7th to 13th March 1900.

Whilst at Bloemfontein on the 27th March, the 'B' squadron formed an escort for His Excellency Sir Alfred Milner,[1] G.C.M.G., K.C.B., Governor and Commander-in-Chief of the Cape of Good Hope and its Dependencies, and High Commissioner for South Africa, on the occasion of his visit to the capital of the Orange Free State (now the Orange River Colony) during its occupation by the British army under Field-Marshal Lord Roberts; and on the 30th March the whole squadron, officers and men, were present at the funeral of their late commanding officer, Colonel the Hon. G. H. Gough, C.B., who was employed on the staff, and whose death took place at Norval's Pont on the 28th March. The ceremony was also attended to the cemetery of Bloemfontein by Lord Roberts with all his generals and staff officers, the remains of the deceased officer having been previously conveyed from Norval's Pont, and met on arrival by a detachment of his old regiment under Lieutenant W. Henry. The other officers of the Fourteenth present on this melancholy occasion were Major R. M. Richardson, Captain R. G. Brooksbank.

'B' squadron attends funeral of Colonel Hon. G. H. Gough, C.B., at Bloemfontein, 30th March 1900.

[1] Afterwards Lord Milner.

Lieutenant W. J. Lockett, Second-Lieutenant J. B. Walker, and Second-Lieutenant B. M. Dawes.

'A' and 'C' squadrons and headquarters in Natal, 6th January 1900.

We must now return to the two squadrons ('A' and 'C') with headquarters of the Fourteenth, which reached Cape Town on 1st January in ss. *Victorian*, proceeded next day to Durban, and landed there on the 6th. From Durban they proceeded by train to Estcourt, and joined the army of Natal under General Sir Redvers H. Buller, V.C., G.C.B., K.C.M.G., which was operating on the Tugela for the relief of Ladysmith garrison besieged by the Boer army. The Fourteenth were posted to the 1st Natal Cavalry Brigade under command of Brigadier-General J. Burn-Murdoch, 1st (Royal) Dragoons, and in the same brigade were the 1st (Royal) Dragoons and the 13th Hussars.[1]

Engaged at Chieveley, 15th January 1900.

On 11th January 'C' squadron proceeded to Hodgson's Hill, 10 miles east of Estcourt, and returned on 13th. On the latter day 'A' squadron and headquarters marched from Estcourt to Frere. On the 15th half of 'A' squadron and headquarters proceeded to Chieveley to join Major-General Barton's Brigade, and took part in a demonstration. The mounted troops were commanded by Lieutenant-Colonel Gilbert Hamilton, 14th Hussars, and the machine gun, with 2 naval 12-pounders, opened fire at long ranges against the Boers, who were all entrenched and under cover in very strong positions, whence they kept up an occasional fire without doing much harm to our men.

Engaged at Hussar Hill, 23rd January 1900.

On the 23rd one squadron and headquarters, under Major E. D. J. O'Brien, took part in a reconnaissance in force commanded by Major-General G. Barton, C.B., which proceeded to Hussar Hill for the purpose of ascertaining what could be seen of the enemy's positions on Hlangwane Hill. The squadron of the Fourteenth was much pressed by the fire of the enemy, and Colonel Blagrove, 13th Hussars, who commanded the mounted force engaged on this occasion, compli-

[1] The Fourteenth were brigaded with both these regiments during the Peninsular War, with the 1st Royal Dragoons in 1810-11, and with the 13th Light Dragoons in 1813-14. See *ante*, pp. 71, 88, 135.

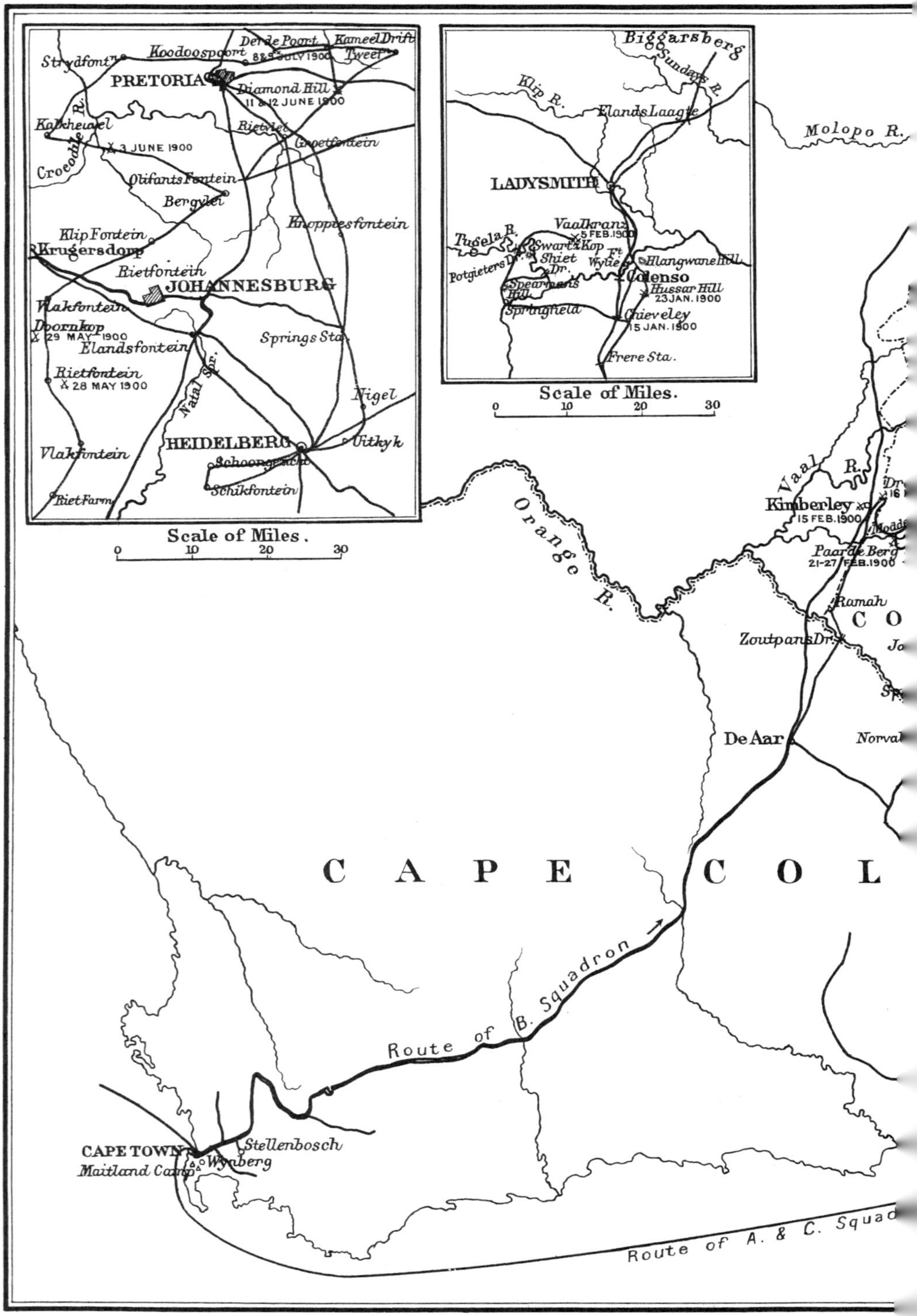

PRETORIA
JOHANNESBURG
Krugersdorp
HEIDELBERG
Scale of Miles.
LADYSMITH
Colenso
Scale of Miles.
Orange R.
Vaal R.
Kimberley
De Aar
CAPE COL
Route of B. Squadron
CAPE TOWN
Maitland Camp
Wynberg
Stellenbosch
Route of A. & C. Squad

London:

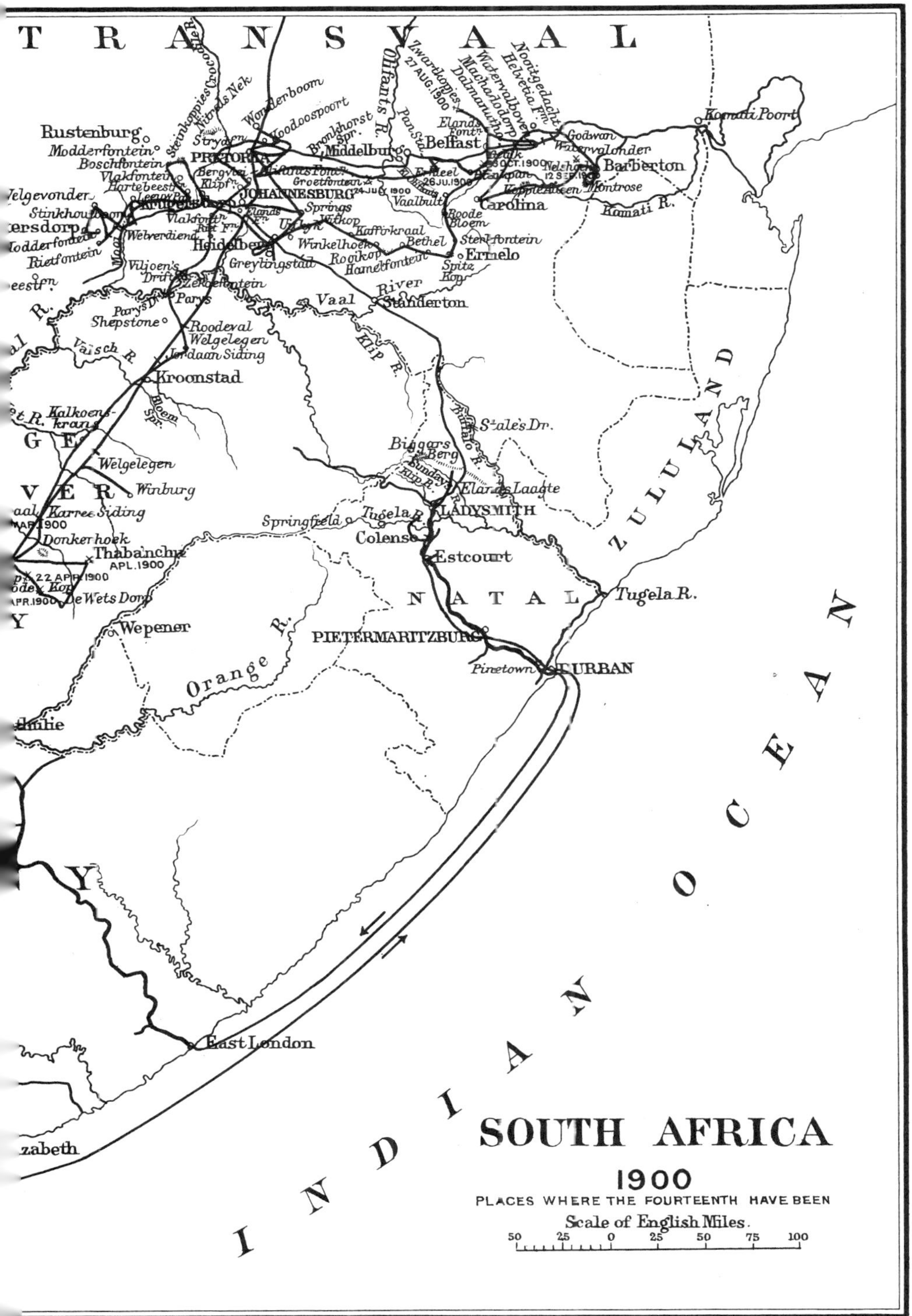

Stanford's Geog[l] Estab[t]

n & Co.

mented the officers and men for their excellent work and steady behaviour under fire. There were 25 casualties amongst the troops engaged, and Captain Dalton, Royal Army Medical Corps, attached to the Fourteenth, was severely wounded.

On 28th January the 2 squadrons and headquarters proceeded to Spearman's Hill *via* Springfield, and joined the mounted troops operating there under Brigadier-General the Earl of Dundonald, C.B.

At the action of Vaalkranz, 5th February 1900.

On 5th February the Fourteenth ('A' and 'C' squadrons and headquarters[1]), under Lieutenant-Colonel G. Hamilton, were present in the action of Vaalkranz, which was an attack on the position held by the Boers opposite Potgieters. Burn-Murdoch's Brigade was brought up to the front of Swartz Kop to threaten the enemy's right, but cavalry could do nothing owing to the nature of the ground, and the brigade remained all day in mass, frequently exposed to fire. The troops bivouacked on the ground they held, but at 3.30 A.M. next day the enemy's shot and shell falling amongst them necessitated a move to a less exposed position. It was not till the 8th February, however, that Sir Redvers Buller retired his army when he realised he was unable to force the position.

The Fourteenth were employed to cover the retirement of the troops, and they did not reach Springfield till 10th February.

On the 12th there was a long day's outpost work, and part of the regiment rode 40 miles in very hot weather, losing five horses, who died at once from exhaustion, and others subsequently. The Royal Dragoons were attacked and suffered several casualties, one officer wounded and several men killed and wounded. The Boers managed to elude us, and escaped across the Tugela at Shiet's Drift.

On 21st the 1st Cavalry Brigade marched to Chieveley, the Fourteenth acting as rearguard, and escorting a convoy to Pretorius Farm. They bivouacked *en route*, and joined the brigade next morning when on the march to a place east of Colenso, near the Tugela, where they remained till the 24th.

[1] The term 'headquarters' includes the Maxim gun belonging to the regiment.

Crossed the Tugela, 24th February 1900.

They were engaged on the 22nd and 23rd, and crossed the Tugela on the following day by pontoon bridge, forming up north of Fort Wylie. The Fourteenth lost one man and one horse drowned in the river during a patrol.

The ground over which the cavalry had to work in Natal was quite unsuited to that branch of the service, and owing to the fatigue of the long marches and want of water, there was an enormous loss in horseflesh incurred.

Relief of Ladysmith, 28th February 1900.

The relief of Ladysmith took place on the 28th February. On the 3rd March the relieving forces made a triumphal entry, the Fourteenth, under Lieutenant-Colonel Gilbert Hamilton, leading the procession, and passed through the town of Ladysmith under General Sir Redvers Buller, in the presence of its gallant defenders headed by their commander, Lieutenant-General Sir George S. White, V.C., G.C.B., G.C.S.I., G.C.I.E., after which they encamped beyond the town. It was a source of great surprise and disappointment to the cavalry that no pursuit was carried out when the Boers trekked.

On the 5th March the Fourteenth moved out to Elandslaagte, and reconnoitred the ground north of Sundays River.

On the 6th they reconnoitred the Biggarsberg, reporting that it was held by the Boers as an entrenched position.

'A' and 'C' squadrons and headquarters ordered to Orange River Colony, 6th March 1900.

Shortly afterwards orders were received for the 2 squadrons and headquarters to march to Durban, and after embarking there to sail to East London, thence to proceed by railway to Bethulie, and march onwards to Bloemfontein to join Lord Roberts's army there.

These squadrons embarked at Durban in ss. *Templemore* on 21st March, having previously received a draft of 115 men, under Second-Lieutenant Champion, and, arriving at East London on 22nd, proceeded by rail to Bethulie. Here they were supplied with a large number of remounts to complete the losses incurred on the voyage out and in Natal. The march onwards was commenced on 31st March, but as there was a long convoy of ox-wagons, as well as a large supply of army remounts, with other detachments of troops to be escorted to Bloemfontein, there were many delays; so that after passing Jagersfontein on 2nd April, and Bethany on 8th, Bloemfontein was not reached till the 11th, when the 2 squadrons and headquarters halted at Donkerhoek, 8 miles north of that city.

On the 14th, 'B' squadron, under Major R. M. Richardson, moved from Bloemfontein and joined the main portion of the Fourteenth at Donkerhoek, so that the whole regiment, under Lieutenant-Colonel Gilbert Hamilton, was now concentrated there as a portion of the Cavalry Division in the Orange River Colony commanded by Lieutenant-General French,[1] being posted to the 4th Cavalry Brigade under Major-General J. B. B. Dickson, C.B., in which were also the 7th (Princess Royal's) Dragoon Guards and the 8th (King's Royal Irish) Hussars.

Whilst at Bethulie the 2 squadrons had been almost entirely rehorsed by Argentine remounts and Cape ponies. The 'C' squadron, commanded by Major E. D. Brown, was mounted entirely on Cape ponies. The Argentines were neither fit for work nor acclimatised, as they had recently come off a long ocean voyage, and this may have accounted for the great losses subsequently incurred in horseflesh as soon as the regiment began to have hard work again. The 2 squadrons from Natal heard, on 1st April, at Springfontein, of the death of Colonel the Honourable G. H. Gough, C.B., who recently commanded the Fourteenth, and all ranks were deeply concerned at the loss of their former commanding officer, who was most highly esteemed and respected throughout the regiment.

A stampede unfortunately occurred in camp at Bloemfontein Stampede at Bloemfontein. on 19th April. The horses of the brigade were out grazing, some of them not hobbled, when suddenly the whole regiment of the 17th Lancers came galloping through the camp, practising an advance in line. In consequence of this the troop-horses of the 7th Dragoon Guards, 8th Hussars, and 14th Hussars were utterly scared, and galloped off in every direction. It took several days to collect them again, and some were never re-

[1] Major-General J. D. P. French had local rank of Lieutenant-General in South Africa whilst commanding the Cavalry Division.

covered by their own regiments. The Fourteenth lost about 53, and could only muster 350 horses for the relief of Wepener instead of 400, as would otherwise have been the case.

Relief of Wepener.

The Fourteenth took an active part in the operations undertaken in Orange River Colony for the relief of Wepener, towards the end of April.

On the 21st, the 3rd and 4th Cavalry Brigades were detailed to form part of the force under Lieutenant-General J. D. P. French, which proceeded from Bloemfontein with Major-General Pole-Carew's XIth Division, for the purpose of co-operating with the VIIIth Division commanded by Lieutenant-General Sir H. Rundle, K.C.B.

Engaged at Leeuw Kop and Roode Kop, 22nd to 24th April 1900.

These troops were continuously opposed, and encountered vigorous attacks made by the Boers with rifle-fire at long ranges. Engagements were fought on the 22nd, 23rd, and 24th April at Leeuw Kop and Roode Kop, and on the 25th the force came up with Sir H. Rundle's Division at Dewetsdorp, which place had just been evacuated by the enemy and occupied by us. The Fourteenth, in these engagements, lost Sergeant L. Cunningham,[1] mortally wounded at Leeuw Kop on 22nd; and on the 24th, at Roode Kop, Captain P. R. Denny (1st Dragoon Guards), attached to the 14th Hussars, was killed, and 9 men were wounded. Several of the latter died of their wounds shortly afterwards. Captain Denny was shot through the heart. He was deeply regretted by the whole regiment—a good officer, and possessed of a charming disposition which made him a great favourite. In an outpost affair near Thabanchu a few days later, Captain D. M. Miller of the Fourteenth was so severely wounded that the Boers, who had taken him prisoner, subsequently brought him into camp and handed him over to our army. Captain Miller had to be invalided and sent home to England as soon as he was able to travel. He afterwards recovered and was able to join the reserve squadron at Curragh Camp under Captain J. Murray nine months later.

Casualties of the Fourteenth.

For several days subsequently, at Thabanchu and the neigh-

[1] Sergeant Cunningham died the same night : a piece of shell penetrated his chest.

bourhood, the Fourteenth, as well as the rest of the cavalry, were incessantly engaged fighting with large bodies of the enemy who were hovering about. Having lost a considerable number of horses in these operations, as well as in the stampede at Donkerhoek, 180 fresh horses had to be procured from the Remount Depôt at Bloemfontein to complete deficiencies, and on the 7th May the Fourteenth advanced in brigade, under Major-General Dickson, as a portion of the Cavalry Division commanded by Lieutenant-General J. D. P. French which proceeded from Bloemfontein in the direction of Kroonstad, Johannesburg, and Pretoria. They made a wide detour to the left, and crossed the Vaal River at Parys. This was part of the great strategic movement of Lord Roberts's army on Pretoria. It was not till the 8th June that the Fourteenth reached Kameeldrift, to the north of Pretoria. The force which took part in this movement consisted of— Advance on Pretoria from Bloemfontein, 7th May 1900.

	No. of Men.
1 Cavalry Division, . . .	2,700
3 Divisions of Infantry, . .	18,800
1 Corps of Mounted Infantry, .	4,800
	26,300

With 132 guns.

The marches from Bloemfontein (Donkerhoek) were as follows:—

			Miles.
May	7.	To a place 4 miles north of Karree Siding, . .	20
,,	8.	Vet River, . . .	35
,,	9.	Kalkoens Krans, . .	25
,,	10.	Londerhaut, . . .	36
	11.	Valsch River Spruit, . .	25
			141
,,	12 to 19.	Jordaan Siding and Kroonstad.	
,,	20.	Kroom Bloem, . . .	12
,,	21.	Welgelegen, . . .	12

			Miles.
May	22.	Roodeval, . .	13
,,	23.	Shepstone, . . .	14
,,	24.	Viljoen's Drift (crossed Vaal River at Parys Drift), . .	20
,,	25.	Zeekoefontein, . . .	16
,,	26.	Riet Farm, . . .	8
,,	27.	Vlakfontein, . . .	20
,,	28.	Rietfontein, . . .	18
,,	29.	Vlakfontein, near Doornkop, .	10
,,	30.	Klipfontein, . . .	16
			159

On the 27th there was some skirmishing with the enemy.

Engaged at Rietfontein, 28th May 1900.

On the 28th the 4th Cavalry Brigade were much under fire. The Fourteenth had 2 men and 6 horses wounded. Major Brown, Captain Tottenham, and Lieutenant and Adjutant F. R. Lawrence, D.S.O., had their horses shot under them.

Engaged at Doornkop, 29th May 1900.

On the 29th, 2 squadrons of the 7th Dragoon Guards and 2 squadrons of the Fourteenth gallantly stormed and captured a kopje near Doornkop.

On 31st May, Johannesburg surrendered to Lord Roberts, and the Guards Division, under Major-General Pole-Carew, C.B., occupied it.

On 1st June the Fourteenth marched from Klipfontein to Bergvlei, 6 miles, and encamped there, 11 miles north of Johannesburg. On 3rd June the Cavalry Division marched onwards to Pretoria, making a wide detour to the left.

The marches were as follows from Klipfontein :—

			Miles.
June	1.	Bergvlei, ...	6
,,	3.	Kalkheuvel, . . .	30
,,	4.	Crocodile River, . .	10
,,	5.	Strydefontein, . . .	15
,,	6.	Koodoospoort (7 miles east of Pretoria), . . .	10
			71

On the 3rd June there was some fighting over difficult ground where the Boers made a stand in a defile. Three squadrons of the Fourteenth were engaged and successfully drove off the enemy. Engaged near Crocodile River, 3rd June 1900.

Pretoria was surrendered to Lord Roberts on the 4th June.

On 8th June the Fourteenth marched further north 8 miles to Kameeldrift.[1] There were only 13 officers of the Fourteenth who finished this long and arduous march to Pretoria, viz. :— Arrival at Pretoria, 8th June 1900.

Lieutenant-Colonel Hamilton
Major Brown.
Captain Tottenham.
Lieutenant Lockett.
Lieutenant Tilney.
Lieutenant Hill-Whitson.
Lieutenant Jameson.
Lieutenant Campbell.
Lieutenant Walker.
Lieutenant Browne.
Second-Lieutenant Honourable H. Grosvenor.
Lieutenant and Adjutant Lawrence, D.S.O.
Quartermaster Mugford (Hon. Captain).

During the advance on Pretoria, on the 12th of May, Captain Stuart Robertson, who had previously served many years in the Fourteenth, and who was in the Reserve of Officers, and belonged to the 3rd Battalion of the Black Watch (Royal Highlanders), came back and joined his old regiment as a duty officer. Unfortunately he was seized with illness a few days afterwards and taken to hospital at Kroonstad, where he succumbed to a severe attack of dysentery, and died on the 1st June, at the age of thirty-five years. His untimely loss was most deeply deplored by the whole regiment, as well as by his numerous friends amongst the old officers of the Fourteenth, with whom he was a great favourite. Death of Captain Stuart Robertson, 1st June 1900.

On the 11th and 12th June the Cavalry Division had considerable fighting at Diamond Hill, north-east of Pretoria, in which the Fourteenth took part, and had 2 men wounded. Considering that the men were exposed for 48 hours incessantly for the two days and nights to the enemy's fire from Engaged at Diamond Hill, 11th and 12th June 1900.

[1] Kameeldrift is about 14 miles north of Pretoria.

big guns, 'pom-poms,'[1] and rifles, the latter at 400 yards' distance, the casualties were very small. The Boers were all the time sheltered either by large rocks or by entrenchments. The cavalry acted dismounted, and were very skilfully handled. After this action the officers of the 4th Brigade were personally complimented by Lieutenant-General French for the manner in which the cavalry regiments had held their positions on both days, and he added, that had they been turned out of them it would have been no disgrace, as infantry could not have held them better.

On the 13th, after being a day at Tweefontein, where the troops bivouacked, they returned on the 14th to Kameeldrift, their former encampment. Here the Fourteenth remained for some time; they received a fresh supply of remounts, and their chief duties were patrols and outposts.

Affair at Derdepoort, 8th and 9th July 1900.

On the 8th and 9th July the Boers endeavoured to break through the British outposts at Derdepoort, where the Fourteenth had to ward off the attack. They were unsupported, as the 7th Dragoon Guards and 8th Hussars had been temporarily withdrawn to other positions for the whole chain was menaced. At Derdepoort the attack was successfully opposed and the enemy driven off. This was the occasion when the post at Nitral's Nek, held by a squadron of the Scots Greys, the Lincolns, and 2 guns of Royal Horse Artillery, was surprised by the enemy. One squadron of the Fourteenth was sent at the same time to support the 7th Dragoon Guards, who had some stiff fighting at Wonderboom, 4½ miles north of Pretoria, where they lost heavily, but cut their way out.

Great cavalry march commenced, 18th July 1900.

On the 18th July the Cavalry Division was at Olifantsfontein, and from this point a movement of British forces on a large scale commenced in an easterly direction.[2] The Boers' main laager of 5000 men was at Bronkhorst Spruit, and they retired skirmishing before our advance for several days.

[1] These are 'Vicars-Maxims.'

[2] The British force consisted of the Cavalry Division, Mounted Infantry Corps, Pole-Carew's Division, and Ian Hamilton's Division.

On the 24th, at Groetfontein, the cavalry came under long-range fire, but our Horse Artillery guns, being of too small calibre, could not reply with effect.[1] On the 25th a squadron of the Fourteenth, under Major Brown, when on outpost duty, at a drift over the Olifant's River, came under long-range fire of the enemy's Mausers, when the magazine Lee-Metford carbines of our cavalry were found quite unable to cope with the Boer rifles at long ranges. Groetfontein, 24th July 1900.

On the 26th the 4th Cavalry Brigade halted for several days at Erfdeel, on the Klein Olifant River, south of Middelburg. Here the Fourteenth and the Scots Greys were engaged, and the former regiment bivouacked on an exposed hill during a very wet night without any blankets or wood, and with only the food which was carried in the men's haversacks. Engaged at Erfdeel, July 26th, 1900.

On the 27th the brigade was at Pan Station on the Delagoa Bay Railway. The Fourteenth were now attached to the 1st Cavalry Brigade, and the 4th Cavalry Brigade was temporarily broken up, as Major-General Dickson had become Governor of Middelburg for a time.

By the 1st August the Cavalry Division had reached Blinkpan, 14 miles south of Belfast, where the Fourteenth were given a very long line of outposts to hold. The line was five miles in extent; it had to be held by 340 men and 250 horses, out of which 110 men were on duty (guard and outposts) every night. This part of the country was destitute of forage, as the Boers had cleared it nearly all away, and no hay was issued for the horses. General Sir Redvers Buller's troops were expected to come in touch with the Cavalry Division at this point, passing from Carolina towards Komati Poort. In consequence of the great scarceness of grass and forage, there was a considerable loss incurred in horse-flesh, as the marches were tedious and incessant.

On the 26th August the Cavalry Division moved from Geluk to Belfast, leaving General Sir Redvers Buller's troops, now in At Belfast, 26th August 1900.

[1] Major-General Dickson narrowly escaped a shell which fell close to the Fourteenth during this affair.

touch, at the former place, and joined Pole-Carew's (XIth) Division. On this day the Fourteenth were in the advance-guard with the Scots Greys and the Inniskilling Dragoons, and marched with an extended front. The enemy was soon discovered holding some rough ground, but was dislodged by our guns and carbine fire. After bivouacking 8 miles north of Belfast the cavalry proceeded on the 27th in the same order, the Fourteenth being in the advance-guard, and the enemy, as on the previous day, being driven in front of us. The bivouac was at Zwartkopjes. The Fourteenth were $1\frac{1}{2}$ miles on the left flank of the brigade, and the Boers attempted to drive them off a spur they held, but a troop, under Sergeant Dove, repulsed them, with a few of our men and horses wounded.[1]

Engaged at Zwartkopjes, 27th August 1900.

On the 28th the Cavalry Division marched eastwards and reached Elandsfontein. Here the Boers held a position, but after it had been shelled by our guns they retired. The country in this region is extremely hilly and most difficult for cavalry operations.

On the 29th the Cavalry Division advanced to within 4 miles of Watervalonder.

Engaged near Waterval-boven, 30th August 1900.

On the 30th the march was over a very rugged and precipitous region, and there was some hard work reconnoitring and occupying the hills near Watervalboven. The Fourteenth lost some horses shot. In the afternoon Major-General Pole-Carew's Division came up and relieved the Cavalry Division, when the latter bivouacked on the same ground as on the 29th, and on the next day (31st) marched to Machadodorp, leaving the XIth Division to proceed eastwards, where the country was quite impossible for cavalry to operate in.

At Machado-dorp, 31st August 1900.

The Cavalry Division moved from Machadodorp on 4th September, reached Carolina on the 7th, and marched towards Barberton on 9th. The distance from Carolina to Barberton is 70 miles. On the 9th there was some skirmishing, but the enemy retired.

[1] It was on the 26th and 27th of August that General Sir Redvers Buller's force was fighting about Dalmanutha.

On the 10th the advance-guard, under Lieutenant-Colonel Hamilton, was formed by the Fourteenth, 3 small squadrons of 6th Dragoon Guards (Carabiniers), and a pom-pom. The Cavalry Division bivouacked at Koppie Alleen on the 10th, and on the 11th marched towards a place on the Komati River, crossing the Buffels Spruit about 4 miles before coming to the river, and bivouacked on its banks.

Engaged at Nel's Hoek, near Komati River, 12th September 1900.

On the 12th there was considerable opposition encountered in the advance. A squadron of the Fourteenth, under Lieutenant Hill-Whitson, came across a party of the Boers on the right flank and drove them off. The remainder of the regiment had to dismount and lead their horses up a very rugged and steep hill, which they had been ordered to occupy by Brigadier-General Gordon, commanding the 1st Cavalry Brigade. Lieutenant the Honourable H. Grosvenor took a squadron to occupy a nek to the right, while Captain Tottenham, with his squadron, pushed on in advance and occupied a position on the higher hills beyond. This was effected without any loss. Lieutenant Grosvenor's squadron became hotly engaged, and his dismounted men did good work with their carbines, driving off the enemy and only losing one horse. Lieutenant Campbell had been ordered to push on as advance-guard to a convoy in front of the infantry (two battalions, commanded by Lieutenant-Colonel J. Spens, Shropshire Light Infantry); he came under a very hot fire from both flanks, but he took the kopje, the enemy retiring when he got within 300 yards of the position. Presently the infantry came up and occupied it. This was Nel's Hoek. Next day, 13th, the 1st Brigade, consisting of the Carabiniers, Scots Greys, Inniskillings, and Fourteenth made a dash for Barberton over Nel's Hoek, Lieutenant-General French leading. Captain Tottenham's squadron of the Fourteenth was sent forward from a hill, 7 miles from the town, where the cavalry was concentrated about 1.30 P.M., and brought back the Military Governor, Van der Post. At 5 P.M. the cavalry moved to a point 3 miles from Barberton, where they bivouacked, and the town was quickly evacuated by the Boers.

Dash for Barberton by cavalry under Lieut.-General French, 13th September 1900.

Barberton occupied, 15th September 1900.

Next day, the 14th September, the cavalry came close up to Barberton, and on the 15th the Union Jack was hoisted at the Court-house in presence of 100 men from each regiment.

Lieutenant the Honourable H. Grosvenor captures Boers, 17th September 1900.

On the 17th the Fourteenth marched on reconnaissance duty to Montrose Mine, and on the same day Lieutenant the Honourable H. Grosvenor, with a patrol of the Fourteenth, took prisoners a party of Boers hiding in a kloof. He first surrounded them by a chain of dismounted men armed with carbines, and when the Boers found shots coming from the direction in which they were trying to escape they very quickly surrendered.

On the 18th the Fourteenth returned, stopping at Ivy Mine, and reached Barberton on the 19th.

On the 25th a squadron of the Fourteenth, under Captain Lockett and Lieutenant the Honourable H. Grosvenor, went to join a reconnaissance-force under Major Scobell and a party of the Scots Greys towards Nel's Hoek.

On 2nd October marching orders came to proceed *via* Godwan River Station to Machadodorp, from the vicinity of Barberton. The Cavalry Division marched from Barberton 3rd October, reached Godwan Station, on Eland's River, on the 6th, Nooitgedacht Station on 7th (the scene of the recent detention of the British prisoners of war within wire kraals), and bivouacked same night at Watervalonder Station.

On the 8th the Division passed Helvetia Farm and reached Machadodorp, distant 77 miles from Barberton. At Machadodorp the Fourteenth received 100 remounts to replace casualties. Here the brigades were reorganised: Colonel B. T. Mahon, D.S.O., 12th Lancers, was given command of a brigade, consisting of the 8th Hussars, the 14th Hussars, and 'M' Battery Royal Horse Artillery; Major-General Dickson's Brigade was now made up by the 7th Dragoon Guards and Suffolk Regiment with some artillery, an ammunition column, and other details; Brigadier-General Gordon's (1st) Brigade, consisting of the 6th Dragoon Guards (Carabiniers),

2nd Dragoons (Scots Greys), 6th Dragoons (Inniskillings), and Royal Horse Artillery.

On the 12th, Mahon's Brigade, followed by the other brigades on 13th, marched for Heidelberg *via* Ermelo and Bethel. When near Geluk and Dalmanutha, about 14 miles from Machadodorp, Mahon's Brigade had an encounter with the enemy's patrols on its first day's march, but drove them off, and bivouacked there, on General Sir Redvers Buller's former camping-ground.

Very early next morning, the 13th October, Mahon's Brigade was shelled by the enemy, who had crept up through a ravine during the night, and had almost surrounded it. The Boers were about 1000 strong, Mahon's Brigade scarcely 600 all told. Without any delay Brigadier-General Mahon sent off the convoy, under escort of 2 small squadrons of the Fourteenth, towards Dalmanutha for safety, and then retired the 8th Hussars, the remainder of the Fourteenth holding the position so as to cover the retreat. Captain Tottenham's, the 'A' squadron, lined the trenches nearest the enemy, and was much exposed to fire. During the withdrawal of a picquet of the 8th Hussars which was nearly cut off, Lieutenant Wylam, 8th Hussars, and several men of his picquet were killed; Lieutenant Gilmore of the 8th Hussars and several men were wounded, and the adjutant of the 8th Hussars, Lieutenant Jones, was killed. The Fourteenth had 2 men and 5 horses killed, 9 men and 10 horses wounded, and Lieutenant Harvey had his horse shot under him. Major E. D. Brown, 14th Hussars, behaved most gallantly during this action: he brought a sergeant and a trumpeter out of action on his own horse one after the other, their horses having been shot, and himself remained behind under a very heavy fire to assist Lieutenant J. G. Browne to mount his horse, which was restive under the galling fire of the enemy and would not stand still. He was afterwards awarded the Victoria Cross for his conspicuous bravery on this occasion.[1] The 'M' Battery

Engaged at Geluk, 13th October 1900.

Casualties of 14th Hussars.

Major Brown, 14th Hussars, behaves gallantly.

[1] See Appendix C, p 619.

Royal Horse Artillery lost heavily, 1 officer killed, 1 officer wounded, and 11 men killed and wounded. The two pom-poms were of great service in helping to cover the retirement on this occasion. After a stiff fight the Boers gave up further pursuit, and the brigade was enabled to bivouac about 3½ miles back unmolested. The object of this attack was to seize the convoy for provisions and ammunition, of which the Boers were very short. The 1st and 4th Cavalry Brigades unfortunately were 10 miles distant at the time of this affair, and consequently unable to render any support to Mahon's Brigade.

The next marches of the brigades were as follows :—

			Miles.
October	15.	Vaalbult, . . .	18
,,	16.	Roode Bloem, . .	12
,,	17.	Sterkfontein, . .	12
,,	18.	Spitz Kop (a farm 3 miles from Ermelo), . .	10
,,	19.	Hamelfontein, . .	14
,,	20.	Bethel, . . .	16
			82

After leaving Ermelo on the 19th the rearguard of the Division was perpetually harassed by the enemy in small bodies, from a distance. Several farms where the 'white flag' treachery had been flagrantly perpetrated in this district were burned by the cavalry.

Lieutenant-General French's complimentary address at Bethel, 21st October 1900.

On the 21st, at Bethel, during a halt, the Lieutenant-General commanding the Cavalry Division ordered a parade of the Fourteenth at 11.30 A.M., as he wished to address the regiment. Lieutenant-General French spoke as follows :—

'Colonel Hamilton, officers, non-commissioned officers and men of the 14th Hussars,—I have been wishing for some time to address you as a regiment, and to thank you, each one individually, for your behaviour throughout this long and arduous campaign. The 14th Hussars have done so well at all times that it is difficult for me to mention any special occasions, but I can recall how well Colonel Hamilton led you, and what good work you did on the way to Dewetsdorp, also the following day at Roode Kop, when you maintained a difficult position under a heavy flank fire; and again the hot work at Thabanchu, when the regiment behaved so well. There are other occasions about which I have already spoken to your officers, viz. the march to Pretoria, when you did splendid work, the attack on Doorn Kop, south-west of Johannesburg, which contributed to the success of the operations; and later on, June the 11th and 12th, at the battle of Diamond Hill, where the regiment maintained an exposed position for forty-eight hours under a hot fire of artillery and musketry at close range. On this occasion most valuable service was rendered by checking the advance of the enemy, and enabling the 1st Brigade to work round his right, thus preventing the operations of our troops against the left flank of the enemy being interfered with.

'We have all fresh in our memories the action of yesterday week, the 13th of October, at Geluk. Brigadier-General Mahon reports to me the splendid and gallant behaviour of the 14th Hussars, and that owing to the steadiness of all ranks he was enabled to withdraw from a difficult position with fewer casualties than would otherwise have been the case.

'Throughout this long and trying campaign you have borne, without a murmur, short rations and exposure, hard work and harassing outpost duty; but the war is nearly over, and can only last a few weeks longer—perhaps for you only a few days, as we are going to Pretoria. It is unnecessary, in fact it would be an insult, for me to ask a regiment that has behaved throughout the campaign as you have to continue to its close the excellent conduct you have hitherto displayed.'

The march from Bethel to Heidelberg was as follows:—

		Miles.
October 22.	Rooikop, ...	12
,, 23.	Winkel Hoek, . . .	12
,, 24.	Kaffirkraal, . . .	12
,, 25.	Witkop, ...	12
,, 26.	Heidelberg, . . .	12
		60

All through these marches the Boers in small numbers hung about the flanks and rear of the cavalry, and took every opportunity they could of 'sniping.' On the 22nd, outside Bethel, the Fourteenth were engaged, but repulsed the attack aided by the guns, which shelled the enemy. On the 23rd the outposts were fired on, and on the 24th the rearguard was much pressed.

Major-General Dickson, C.B., after the Fourteenth were withdrawn from the 4th Cavalry Brigade, addressed the following complimentary letter, dated 29th October 1900, to Lieutenant-Colonel G. Hamilton :—

Complimentary letter from Major-General Dickson, C.B.

'My dear Colonel,—Had I been aware when your regiment left my brigade[1] that it was not returning to it, I should have had the pleasure of placing on record a farewell order stating how much I appreciated its services, and thanking all ranks for the ready assistance given on all occasions. The young officers commanding squadrons lead them with marked ability and success. The conduct of the men in the field, often engaged under a very heavy fire, was all that could be desired. Will you kindly inform all ranks how much I regret losing their services; and wishing you and the gallant Fourteenth prosperity and good fortune.—Believe me, yours very truly, (Signed) J. B. B. Dickson, *Major-General,*

'*Commanding 4th Cavalry Brigade.*

'*To* Lt.-Col. Hamilton,

'*Commanding 14th Hussars.*'

[1] The Fourteenth had belonged to the 4th Brigade until the reorganisation of brigades took place at Machadodorp, when they were posted to the brigade under Brigadier-General B. T. Mahon, D.S.O., 12th Lancers.

On 30th October the cavalry moved from Heidelberg to Springs, 20 miles. On 31st, Field-Marshal Lord Roberts, Commanding-in-Chief in South Africa, being *en route* by train from Pretoria to Johannesburg, came and inspected the Cavalry Division, which was formed up in line of masses, a battery of Royal Horse Artillery being with each of the three cavalry brigades. There was also present the half battalion of the Suffolk Regiment,[1] which had been attached to Major-General Dickson's Brigade on the march from Machadodorp.

At Pretoria, 3rd November, 1900.

On the 1st November the Cavalry Division marched to Rietfontein, 12 miles; 2nd, to Rietvlei, 10 miles; 3rd, to Pretoria, 12 miles. The Fourteenth encamped $2\frac{1}{2}$ miles west of Pretoria, this being the first occasion of their using tents since leaving Bloemfontein six months previously.

End of the great cavalry march.

This great cavalry march of French's Division now concluded has been classed as one of the notable achievements of the war. The difficulties of marching through an enemy's country with a small force and a huge convoy cannot be overestimated, and the cavalry as well as the horse artillery were under very serious disadvantages owing to the inferiority of the cavalry carbines to the enemy's rifles in point of range capacity, and the calibre of our horse artillery guns being unable to cope on equal terms with those brought against them by the Boers. The short rest at Pretoria was devoted to re-equipping the various regiments for future operations by which it was hoped to bring the war to a final conclusion.

Although the formal annexation to the British Empire of the Orange Free State took place on the 28th May, and although Lord Roberts hoisted the Union Jack at Pretoria on the 5th June, and published his proclamation annexing the Transvaal on the 1st September,[2] still the war is not yet

[1] Two companies of the Suffolk Regiment helped to drive off the enemy when the rearguard of the convoy was attacked *en route* to Heidelberg on 26th October.

[2] This proclamation had been issued on 4th July 1900.

brought to an end.[1] The Boers in considerable numbers, led by Generals De Wet, Botha, Delarey, and Viljoen, have for many months been carrying on a desultory guerilla warfare and dacoity of the worst type in various parts of the Transvaal, Orange River and Cape Colonies, thus postponing the final end of hostilities and the settlement of the country under British rule.

Whilst resting at Pretoria in November the Fourteenth refitted throughout. In the first place 300 remounts joined, and 212 dismounted men rejoined the regiment; secondly, clothing of all sorts was received, as well as new blankets, waterproof sheets, belts, arms, and a fresh supply of everything, including saddlery, required to complete existing deficiencies.

Lieut.-Colonel Gilbert Hamilton appointed Brigadier-General of Cavalry, 21st November 1900.

On the 21st November, Lieutenant-Colonel Gilbert H. C. Hamilton, 14th Hussars, was appointed Brigadier-General to command the 4th Cavalry Brigade, consisting of the 7th Dragoon Guards, 8th Hussars, 14th Hussars, and 'O' Battery Royal Horse Artillery.

March of the 4th Cavalry Brigade to Heidelberg, 22nd November 1900.

On the 22nd November the 4th Brigade started to march from Pretoria to Heidelberg *viâ* Springs, with orders to clear the country of marauding bands.

The brigade marched to Groetfontein, 16 miles. The 8th Hussars had been withdrawn from the 4th Brigade and sent to Brigadier-General Broadwood's Brigade to replace the Household Cavalry recently ordered home. The strength of the 4th Brigade was therefore only 920 horses exclusive of officers' chargers; the men were armed with rifles ranging to 2800 yards instead of carbines ranging only to 1200 yards as previously. On the 23rd the brigade marched to Knopiesfontein, 13 miles; on 24th to Springs, 13 miles. On the same day Lieutenant-General Lord Kitchener of Khartoum, G.C.B., K.C.M.G., General Officer Commanding-in-Chief the Forces, came to see the Brigadier, and returned to Johannesburg next day.

[1] December 1900.

Field-Marshal the Right Honourable Earl Roberts, V.C., K.G., K.P., G.C.B., G.C.S.I., G.C.I.E., being expected to pass down the line of railway to visit Colenso on his way home on the 29th or 30th November, the 4th Brigade was ordered to assist in guarding the line between Greylingstad and Heidelberg and to co-operate with Major-General Cooper.

On the 30th this brigade made a night-march in order to get round to the north of the commando of Boers under Hans Botha, who was reported to be occupying a strong position on the hills north-east of Heidelberg. Major-General Cooper was to attack him on the south, and Lieutenant-Colonel Arthur Colville's column to approach him on the east. The 4th Brigade bivouacked near Nigel, advanced early on 1st December to a point north-east of a range of hills supposed to be occupied by the Boers, and shelled the ridge at Uitkyk, which was occupied; but the Boers trekked to the east out of reach, and the most commanding hill of the whole range was then occupied by the troops of the brigade. Subsequently the entire range was taken possession of by the infantry of Lieutenant-Colonel Colville's column. The 4th Brigade captured 400 cattle and 6 mules and arrived in Heidelberg the same evening, having covered 10 miles' distance on the night of 30th November, and on the 1st December, 25 miles.

4th Brigade marches to intercept Hans Botha, 30th November 1900.

On 3rd December, 200 men, 1 pom-pom, and 10 wagons under Major Brown, 14th Hussars, cleared twelve farms in the vicinity, and brought in a large supply of cattle, sheep, mealies, and straw, after sending in the families to Heidelberg.

Employed in clearing farms, December 1900.

On 4th December the brigade covered 32 miles and cleared several farms with only one casualty, a man severely wounded, collecting 700 cattle, 2000 sheep, and 60 horses.

On the 7th the 4th Brigade left Heidelberg at 1 A.M., and arrived at Schoongezicht at 12 noon, co-operating with Lieutenant-Colonel Colville's movable column. At Nek, west of Schikfontein, a body of the enemy in a commanding position opposed further advance; but after our guns had shelled them out of it, the force cleared several farms in the vicinity, captured 60 cattle and 1000 sheep, and burned the Boers' laager. The 4th Brigade covered 30 miles that day, and marched on the 8th December 8 miles to Heidelberg, spreading out on a front of 8 miles so as to clear all the farms of forage and collect the cattle scattered on the hills. There were 21 farms cleared and 11 farms burned, the latter having been used either as bases of operations or for treachery. Total captures made by the 4th Brigade :—1500 sheep, 500 cattle, and 11 prisoners of war.

Captures made by 4th Cavalry Brigade, December 1900.

On 12th December the 4th Cavalry Brigade and Lieutenant-Colonel A. Colville's movable column, under Major-General Cooper, marched south from Heidelberg to clear out Boers who were occupying a tract of country 16 miles off. On the 13th the infantry attacked the enemy on the north-west flank, whilst the cavalry worked round to the south-east to cut them off: a large number of Boers escaped before it was possible to surround them owing to the difficult nature of the ground. The casualties were only 2 men wounded of the 7th Dragoon Guards. On the 14th the 4th Brigade moved 10 miles further to the north-east; on the 15th orders were received to return to Heidelberg and to proceed thence to Krugersdorp.

4th Cavalry Brigade marches to Krugersdorp, 16th December 1900.

On the 16th December the 4th Brigade marched from Heidelberg, and part of it reached Krugersdorp, 48 miles, on the 17th, being urgently required. News came that De Wet had broken through the circle surrounding him in the south-east corner of the Orange River Colony, and had got away to the north-east. The 4th Brigade had marched 22 miles on the 16th to Natal Spruit, when orders came for it to be entrained at Elandsfontein, 6 miles off, and to proceed to Cape Colony instead of to Krugersdorp, to act against De Wet's force which had escaped that way. Later on another order was received, that the brigade was to go to Krugersdorp by quickest route.

Early on 17th part of the brigade proceeded by march and part by rail to Krugersdorp, and late that night the 7th Dragoon Guards and a section of the Royal Horse Artillery were ordered to entrain next day for the south. This was effected on the morning of the 18th, and on the 19th a force marched from Krugersdorp under Lieutenant-General French to Steinkoppies, 15 miles, consisting of the 1st Cavalry Brigade under Brigadier-General Gordon, to which were added 2 squadrons of the 14th Hussars, 4 guns Royal Horse Artillery, and 2 pom-poms belonging to the 4th Cavalry Brigade, which was now broken up; the Scots Guards under Lieutenant-Colonel Pulteney, D.S.O.; an ox convoy and baggage, 500 men of the Dublin Fusiliers, 2 squadrons of the 14th Hussars, and a section of the Royal Horse Artillery, under Brigadier-General G. Hamilton. This was a relief force for Major-General Clements, who was opposed by 2000 Boers at Hekpoort, and the intention was to prevent the enemy escaping to the west. The 1st Brigade under Brigadier-General Gordon pushed on beyond Vlakfontein, and killed 49 of the enemy, but were not in time to prevent the escape of the main body, which got away to the west.

1st Cavalry Brigade and 14th Hussars march from Krugersdorp to relieve Major-General Clements's Brigade, 19th December 1900.

On 20th this force marched to Boschfontein, 14 miles, passing Major-General Clements, who was going westwards.

On 21st Brigadier-General Gordon was commanding the whole force, and Brigadier-General G. Hamilton, 14th Hussars, commanded the 1st Cavalry Brigade, with the whole of the Fourteenth and section of 'O' Battery Royal Horse Artillery. On reaching Vlakfontein, 10 miles, the Boers were found occupying two ridges to our right. The Scots Greys, Inniskillings, and 6th Dragoon Guards (Carabiniers) were in advance, with 2 guns and a pom-pom; the Fourteenth and other guns, close up, formed a second line. The enemy fled after the first line attacked, the shrapnel from our guns killing a few.

On 22nd the 1st Brigade and Fourteenth, with the rest of the force, moved to Hartebeestfontein, 3 miles, Lieutenant-General French and a squadron of the Inniskillings having previously gone to Krugersdorp. On the 23rd they moved to Stinkhoutboom, 25 miles.

On the 24th this force marched four miles to Welverdiend Station, making a reconnaissance in the afternoon to Mooi River and back. Lieutenant the Honourable H. Grosvenor, with a patrol of 24 men of the Fourteenth, was attacked by the enemy, numbering 70. He was outflanked, but cleverly avoided capture, losing 2 men and 11 horses.

Escape of Lieutenant the Honourable H. Grosvenor's patrol, 24th December 1900.

Brigadier-General Broadwood, 12th Lancers, arrived the same day with his brigade, consisting of 8th Hussars, 10th Hussars, and 12th Lancers, coming from Rustenburg. On 25th both brigades moved 10 miles north to Leeuw Pan, where there was no water; and on 26th to Rietfontein, near Vlakfontein. Beyers, with a commando of 1500 Boers, was now believed to be in the hills south of Welgevonden and west of Mooi River; Delarey, with 1400 Boers and 6 guns, was 12 miles to the north of Rietfontein. The British force was therefore halfway between these two bodies of the enemy. The intention was to attack Beyers from the north, while Brigadier-General Knox, now commanding Broadwood's Cavalry Brigade, attacked from the east; and a third column, under Colonel Kekewich, was to attack from the south. Lieutenant-General French was with Knox's Brigade. On the 27th the 1st Brigade moved south, the three columns converged on Beyers, but he escaped by the open door to the west. Delarey did not approach at all. The 1st Brigade bivouacked at Welgevonden, and marched towards Ventersdorp on the 28th, getting touch with Knox's Brigade, which was also marching on Ventersdorp, under Lieutenant-General French. The 1st Brigade was now ordered to march to Modderfontein, 14 miles distant.

Movements to cut off Beyers's commando fail, December 1900.

On the 29th December, Brigadier-General Gordon resumed command of the 1st Cavalry Brigade, and Brigadier-General G. Hamilton,[1] 14th Hussars, was appointed administrator and commandant at Ventersdorp, a town 12 miles distant from Modderfontein, and accordingly he proceeded there on

Brigadier-General G. Hamilton appointed commandant at Ventersdorp, 29th December 1900.

[1] Lieutenant-Colonel (local Brigadier-General) G. H. C. Hamilton was promoted Brevet-Colonel 29th November 1900.

the same day. The Fourteenth were now divided amongst three different brigades which had orders to assist in various operations about to take place for clearing the Southern Transvaal of the guerilla Boer commandoes still infesting the country: two squadrons went to the 1st Cavalry Brigade, one squadron to Major-General Clements's 12th Brigade, and the remaining squadron to Major-General Babington's Brigade. These brigades moved off during the last days of December, with the exception of Major-General Babington's, which did not leave Ventersdorp till 2nd January 1901. It is now time to close this Historical Record, which ends with the century. The war is not, however, yet finished, and we take leave of the gallant 14th (King's) Hussars, still well to the front on active service, and adding fresh lustre to their already high and distinguished reputation. It will remain for other, and it is to be hoped abler, hands to continue the narrative in the future.

14th Hussars divided amongst three separate brigades in Southern Transvaal, 31st December 1900.

REGIMENTAL ALMANACK OF THE 14TH (KING'S) HUSSARS—1891

'DOURO'

'FUENTES D'ONOR'

'VITTORIA'

'PENINSULA'

'CHILLIANWALLAH'

'PERSIA'

'TALAVERA'

'SALAMANCA'

'ORTHES'

'PUNJAUB'

'GOOJERAT'

'CENTRAL INDIA'

ROLL OF OFFICERS

Colonel.

Thompson, Lieut.-Gen. C. W. (Hon. Gen.), 1 May 1882.

Lieut.-Colonel.

Hamilton, H. B., 17 March '86. Com. Regt., 1 July '87. Bt.-Col., 17 March '90.

Majors.

Gough, Hon. G. H., 16 December 1885. (Brevet-Colonel, 16 December 1889.)
English, A. J., 17 August 1887.
Burke, M. A., 11 Dec. 1888. (28 July '86.)

Captains.

King, A. C., 12 July 1884.
Kirk, H., 1 July 1887.
Richardson, L. J., 24 March 1888.
Richardson, R. M., 25 July 1888.
Miller, Sir J. P., Bart. (Adjt.), 8 Sept. 1888.
Robertson, S., 18 December 1889.
Spencer-Stanhope, W., 11 June 1890.

Lieutenants.

Murray, J., 27 January 1886.
Tickell, E. J., 7 February 1885.
Miller, D. M., 8 August 1888.
Stacey, C., 8 August 1888.
Brooksbank, R. G., 5 December 1888.
Hamilton, J. D., 6 March 1889.
Hughes, H. B. L., 10 April 1889.
Eley, W. G., 12 March 1890.

Second-Lieutenants.

St. Maur, R. H., 23 October 1889.
Stephen, R. C., 2 April 1890.
Prevost, W., 8 November 1890.

Adjutant.

Miller, Sir J. P., Bart., Capt., 1 Sept. 1888.

Riding-Master.

Odlum, R. (Hon. Capt.), 9 June 1888.

Quartermaster.

Mugford, F. (Hon. Lieut.), 20 Sept. 1882.

Veterinary Surgeon (attached).

Spooner, W. B., 26 Nov. 1886. (Vet. surg. 1st class, ranking as capt., 26 Sept. '84.)

Supernumerary Officers.

In extra regimental appointments.

Capt. G. H. C. Hamilton, Adjt. 1st West York Yeomanry Cavalry, 22 Feb. '88.
Capt. H. W. Mitchell, Adjt. Middlesex Yeomanry Cavalry, 1 Sept. 1888.
Capt. A. B. Broadhurst, Adjt. Ayrshire Yeomanry Cavalry, 8 Sept. 1888.
Capt. E. D. Brown, Commandant Aux. Cav. School, Aldershot, 1 Jan. 1890.

Warrant Officers.

Haines, H., Regimental Sergeant-Major, 31 December 1889.
Hemsley, H., Bandmaster, 1 April 1887.

The Regiment returned from India 24th November 1886. *Stationed at Hounslow, Hampton Court, and Kensington since September* 1890.

Establishment :—24 officers, 423 men, 59 chargers, 281 troop-horses.

⁂ The lists of former Colonels and Lieutenant-Colonels are not reprinted in the Almanack, as they are contained on pages 462-64.

JANUARY

1	Th	1840 Straight swords issued. 1873 'D' Newbdge. to
2	Fr	1715 Orgnl. formation of regt. in E'gland [Curragh
3	Sa	1876 Detachment embarked on *Euphrates* for India
4	𝔖	1876 Hdqrs. embarked on *Euphrates* for India
5	M	1876 Sailed from Portsmouth for India
6	Tu	1867 Lieut.-Colonel P. S. Thompson to Colonel
7	W	1842 Inspected at Kirkee by Sir C. Napier
8	Th	1855 Received orders for Crimea at Meerut
9	Fr	1835 One troop Ennis to Galway
10	Sa	1816 Estbt. 8 troops of 4 S. 1 Tp. 1 Fr. 75 R. and F.
11	𝔖	1827 Dublin to Athlone and Ballinrobe
12	M	1835 One troop Longford to Castlebar
13	Tu	1849 Chillianwallah—Lieut. Cureton killed.
14	W	1858 'E' escorted siege-train from Indore to Sehore
15	Th	1835 One tr. Galway to Loughrea. '63 'D' Man-
16	Fr	1835 One tr. Loughrea to Gort [chester to Bury
17	Sa	1876 Depôt troop Colchester to Canterbury
18	𝔖	1876 At Port Said in H.M.S. *Euphrates*
19	M	1851 Left Lahore for Meerut
20	Tu	1876 At Suez. '83 Insptd. by M.-Gen. Sir C. Keyes
21	W	1882 Right wing at Secunderabad
22	Th	1845 Inspected at Kirkee by Major-Gen. MacNeil
23	Fr	
24	Sa	1855 Meerut to Kirkee
25	𝔖	
26	M	1860 Depôt from Maidstone to Newbridge
27	Tu	1815 A boat containing Lieut. Bridges and 37 men
28	W	1835 One troop Castlebar to Longford [captured
29	Th	1858 Major Scudamore at capture of Rathgur
30	Fr	
31	Sa	1858 Right wing at Barodia

FEBRUARY

1	𝔖	1872 Detachment Fermoy to Listowel
2	M	1859 To wear 'Persia' on cap-plates, etc.
3	Tu	1872 Detachment Ballincollig to Killarney
4	W	1872 Detachment Limerick to Listowel
5	Th	1842 L.-Corpl. J. Lewis found drowned at Kirkee
6	Fr	1876 Bombay signalled from *Euphrates*
7	Sa	
8	𝔖	1876 Disembarked at Bombay and arrd. at Poona
9	M	1829 Lapels for Light Cavalry abolished
10	Tu	1876 Poona to Sholapore
11	W	1860 Hdqrs. embarked for England under Colonel
12	Th	1876 Karchore to Arkonum [Scudamore
13	Fr	1889 Officers' Ball in Royal Pavilion, Brighton
14	Sa	1876 Ar. at Bangalore. '81 Ins. by M.-Gl. Payn, C.B.
15	𝔖	1848 Umballa to Ferozepore. '86 Ins. by M.-Genl.
16	M	1872 Detacht. Dingle to B'collig [Gordon, C.B.
17	Tu	
18	W	1881 'C' Bangalore to Bombay
19	Th	1866 'D,' 'E,' and 'C' Aldershot to Hounslow
20	Fr	1857 Kirkee to Persia [Standard
21	Sa	1849 Goojerat; Lt. Lloyd killed. Corp. Pain capd.
22	𝔖	1881 Hdqrs. 'A' and 'B' Bangalore for Bombay
23	M	
24	Tu	1866 Aldershot to Hounslow under Major Brown
25	W	1831 One troop Coventry to Kidderminster
26	Th	1881 'C' embarked at Bombay for Natal
27	Fr	1814 Orthes. 1846 Arrived at Agra
28	Sa	1881 'D,' 'H,' and 'K' Bangalore for Bombay

MARCH

1	𝔖	1881 Hdqrs. 'A' and 'B' embarked at Bombay
2	M	1832 One troop Clifton to Sudbury [for Natal
3	Tu	1858 Right wing at Muddenpore
4	W	1828 Athlone to Birmingham
5	Th	1811 Advance-guard in pursuit of Marshal Massena
6	Fr	1851 Lahore to Meerut
7	Sa	
8	𝔖	1814 Captain Townsend taken prisoner at Pau
9	M	
10	Tu	[Natal
11	W	1881 'D,' 'H,' and 'K' embarked at Bombay for
12	Th	1814 Captain Babington taken prisoner at Pau
13	Fr	1870 One squadron Athlone to Tuam
14	Sa	1810 Advcd. post of Gen. Hill's Corps at Aronches
15	𝔖	1820 Permitted to wear 'Talavera' on colours and
16	M	1846 Agra to Meerut [appointments
17	Tu	1890 Lieut.-Colonel H. B. Hamilton to Colonel
18	W	1881 Hdqrs. 'A' and 'B' disembarked at Natal
19	Th	,, ,, ,, ,, ,, ,,
20	Fr	1812 Badajos. '45 Two sq. returned from Kolapore
21	Sa	1858 Left wing at Jhansi, 70 miles in 2 days
22	𝔖	1828 Inspected at Dublin by Sir G. Murray
23	M	1815 Two squads. sailed from America for England
24	Tu	1800 Estab. to 90 rank and file per troop
25	W	1863 'H' Staleybridge to Manchester
26	Th	1857 'H' or Captain Prettejohn's at capture of
27	Fr	1863 'G' Ashton to Staleybridge [Mohamra
28	Sa	1871 'D' Fermoy to Clonmel
29	𝔖	1833 1 sq. em'kd. at Bristol for Dublin [14th L. D.
30	M	1872 1 troop Limerick to Galway. '47 Lt. Need to
31	Tu	1857 Hdqrs. at Bushire. 1848 Lt. Gall to Capt.

APRIL

1	W	1858 'A' or Capt. Need's distinguished at Betwa
2	Th	1833 H'low to Dublin. '88 Easter Vol. Rvw., Dover
3	Fr	1888 D. of Cambridge lunched with officers at Shorn-
4	Sa	1836 One tr. Dundalk to Carrickmacross [cliffe
5	𝔖	1830 Leeds to Brighton. 1858 Capture of Jhansi
6	M	1815 'Peninsula' on colours, etc. 1837 Glasgow to
7	Tu	1834 Dublin to Longford [Edinburgh
8	W	1834 One troop Dublin to Gort
9	Th	1841 Two troops Dorchester to Canterbury
10	Fr	1814 Toulouse. 1889 Establishment 281 horses
11	Sa	1835 Dress of Band same as Tptrs. 1881 Estcourt
12	𝔖	1832 Hdqrs. Stroud to Gloucester
13	M	1887 Reviewed by D. of Cambridge at Shorncliffe
14	Tu	1809 Advance-guard from Lisbon to Coimbra
15	W	1829 Overalls chd. from blue-grey to Oxford mix.
16	Th	1829 Lt.-Col. Townsend to command *vice* Lt.-Col-
17	Fr	1881 Estcourt to Ladysmith [Baker retired
18	Sa	1825 Hounslow to Cork. '39 B'ham. to Hounslow
19	𝔖	1839 Second party Leeds to Chichester
20	M	1825 Disembarked at Waterford
21	Tu	1855 Arrived at Kirkee from Meerut, 75 marches
22	W	1830 Recruiting suspd. '45 Col. Townsend died at
23	Th	1867 'A' and 'C' Hounslow to Newcastle [Cork
24	Fr	1839 Two troops Coventry to Hounslow
25	Sa	1858 'A,' 'C,' and 'E' under Col. Gall to Koonch
26	𝔖	1859 First party Gwalior to Bombay
27	M	1831 Three troops Coventry to Nuneaton
28	Tu	1855 Inspected at Kirkee by Brigdr.-Gen. Trydell
29	W	1887 Inspt. by Sir B. Russell, K.C.B., etc., at Sh'cliffe
30	Th	1830 Detacht. to Chichester. '41 Gen. Thompson
		[joined 14th Hussars as Lieut.

MAY

1	Fr	1882 Gen. Thompson app. Col. '87 Cruppers disc'd.
2	Sa	1889 Inspt. by M.-Gen. Moore. '88 Inspt. by Sir B.
3	𝕾	1809 Rvd. by Sir A. Wellesley at Coimbra [Russell
4	M	1839 Pistols for ranks discd. '67 Hdqrs. to Edinr.
5	Tu	1811 Fuentes d'Onor, Capt. Knipe wounded
6	W	1857 Persia to Kirkee. '67 'D' H'low to Hamilton
7	Th	1858 'C' took a gun at Koonch. '88 Estbt. 469
8	Fr	1890 Inspected by Maj.-Gen. Moore at Brighton
9	Sa	1829 Birmingham to Leeds. '68 'B' & 'D' to New-
10	𝕾	1809 Advcd. Picquets on march to Douro [bdge.
11	M	1848 Left wing Ferozepore to Lahore [of Douro
12	Tu	1809 Two sq. under Mj. Hervey distingd. at passage
13	W	1862 Dundalk to M'chester. '68 'A' & 'H' to New-
14	Th	1814 Rvd. at Bordeaux by Lord Dalhousie [bdge.
15	Fr	1871 'H' Waterford to Limerick, 'C' Carrick to
16	Sa	1815 Two sq. rejoined regt. at Hounslow [Bandon
17	𝕾	1833 Inspected at Dublin by Sir E. Blakeney
18	M	1868 'C,' 'G,' and 'K' Hamilton to Newbridge
19	Tu	1835 One troop Gort to Dundalk
20	W	1872 'D' Fermoy and 'G' Clogheen to Newbridge
21	Th	1868 Hdqrs. and 'E' to Newbridge
22	Fr	1858 Capture of Golowlee under Col. Scudamore
23	Sa	1859 Jhansi to Bombay. 1872 Hdqrs. B'collig to
24	𝕾	1834 Guidons in Lt. Dgs. discontd. [Newbridge
25	M	1874 'C' and 'D' Dublin to Aldershot
26	Tu	1813 Salamanca. 1836 Dundalk to Glasgow
27	W	1874 Hdqrs. 'E' and 'G' to Aldershot [lightng.
28	Th	1885 Lts. Bentley-Innes and Seymour killed by
29	Fr	1835 Inspt. at Dundalk by M.-Gen Macdonell
30	Sa	1874 Dismounted party Dublin to Aldershot
31	𝕾	

JUNE

1	M	1795 Clonmel to St. Domingo under Col. Carter
2	Tu	1838 One troop Carlisle to Bath
3	W	1860 Plymouth to Newbridge [command
4	Th	1811 Col. Hawker to Major-Gen.; Col. Hervey to
5	Fr	1859 Bombay to Kirkee [to Salamanca
6	Sa	1812 Advance-guard of Lord Lynedoch's Column
7	𝕾	1834 Inspected at Longford by Sir J. Buchan
8	M	1857 Left wing to Ahmednuggur
9	Tu	1888 Regt. encamped at Cove Common, Aldershot
10	W	1814 Bordeaux to Calais. '82 Lt.-Col. Morton to
11	Th	1819 Embarked at Dublin for H'nslow [14th Hrs.
12	Fr	1813 Burgos. 1832 Gloucester to Hounslow
13	Sa	1842 Insptd. at Kirkee by Sir T. McMahon, Bart.
14	𝕾	1841 Embarked for India. '76 Col. Campbell died
15	M	1881 Major J. H. Knox to Lt.-Col. [at Bangalore
16	Tu	1812 In front of Salamanca, 1 Sgt. and 1 Tr. kill'd.
17	W	[1858 Capture of Morar cantonments
18	Th	1838 Edinburgh to Birmingham
19	Fr	1830 Brighton. 1858 Capture of Gwalior
20	Sa	1832 Depôt, Coventry to Hounslow
21	𝕾	1813 Vittoria. 1858 'H' at Alipore
22	M	1860 'C' and 'E' at Gravesend for Newbridge
23	Tu	1833 Detachment Longford to Ballinasloe
24	W	1810 To Light Division under General Crauford
25	Th	1832 Four troops Hounslow to Kensington
26	Fr	1832 Four troops Kensington to Hounslow
27	Sa	1831 One squadron Coventry to Abergavenny
28	𝕾	1889 Regt. encamped Farnboro' Com., Aldershot
29	M	1830 Maj.-Gen. Sir E. Kerrison to be Colonel
30	Tu	1858 Lt. Leith received V.C. 1887 Col. Morton [to half pay

JULY

1	W	1887 Lt.-Col. H. B. Hamilton to command Regt.
2	Th	1812 Took Outposts on the Douro
3	Fr	1874 No. Camp to So. Cav. Bks., Aldershot
4	Sa	1813 Outpost duty in front of Maya, Pyrenees
5	𝕾	1872 'H' Newbridge to Curragh
6	M	1833 Hdqrs. Dublin to Newry
7	Tu	1888 Duke of Cambridge's review at Aldershot.
8	W	1889 Cavalry Column, Woolmer
9	Th	1887 Two sq. at Queen's Jubilee Review, Aldershot
10	Fr	1857 120 rank and file to Deccan Field Force
11	Sa	1810 Ciudad Rodrigo, 2 offs. and 11 men killed, 19
12	𝕾	1831 Birmingham to Glos'ter for riots [wounded
13	M	1875 Ald'shot to Coldingly. 89 Ald't from Woolmer
14	Tu	1890 Inspt. by Sir Drury Lowe, K.C.B., Brighton
15	W	1857 Hdqrs. to Deccan Field Force
16	Th	1812 Rearguard on the retreat from Rueda
17	Fr	1814 Disembarked at Dover & marched to London
18	Sa	1812 Castrillos, Maj. Brotherton wounded, 8 men
19	𝕾	1833 Two tr. Armagh to Dublin [& 20 hor. kill'd
20	M	1869 1st sq. D'blin to Crrgh. '74 Ald'shot to Wool-
21	Tu	1814 Rvd. at H'low by H.R.H. C.-in-C. [mer
22	W	1812 Salamanca, 4 men killed. '30 B'ton to London
23	Th	1812 Hanley's patrol. '87 Won Cambdge. Shield
24	Fr	1888 'B,' 'D,' 'H,' and 'K' detached at Hilsea
25	Sa	1810 To Torres Vedras. '88 Hdqrs. at Brighton
26	𝕾	1798 Rcd. title of 'The Duchess of York's Own'
27	M	1830 Title chngd. to King's Lt. Dns. by K. Wm. IV.
28	Tu	1809 Talavera, 5 officers wounded, 2 men and 30
29	W	1830 London to Birmingham [horses killed
30	Th	1866 At Buckingham Palace for Reform League
31	Fr	1819 Kersey overalls of dark or blue-grey [Riots

AUGUST

1	Sa	1797 America to Chelmsford. '13 Valley of Bastan
2	𝕾	1830 All Cavalry in red except R. H. Gds. (Blue)
3	M	1858 'C' and 'E' from Jhansi under Maj. Thompson
4	Tu	
5	W	1813 Outpost duty in front of Maya, Pyrenees
6	Th	1821 Estabt. to 6 tr. of 1 T.S.M., 2 S., 1 Fr. 1 Tr. &
7	Fr	1889 Empr. Germany's Revw. Ald't. [50 R. and F.
8	Sa	1865 Insptd. Ald'shot by Lord G. Paget. '89 'A'
9	𝕾	[and 'B' to Hilsea
10	M	1887 Insptd. by Sir Drury Lowe, K.C.B., at Shorn-
11	Tu	[cliffe
12	W	1889 Hdqrs. at Brighton
13	Th	1861 Lieut.-Col. Scudamore received Good Service
14	Fr	[Pension
15	Sa	1861 Prince of Wales dined with Regt. at New-
16	𝕾	1881 'D' and 'H' Ladysmith to Zululand [bridge
17	M	1861 Converted into Hussars
18	Tu	1849 Hdqrs. and right wing Ferozepore to Lahore
19	W	1831 Hdqrs. Tewkesbury to Gloucester
20	Th	1831 One troop Abergavenny to Llandilo
21	Fr	1871 'C' Bandon to Kinsale
22	Sa	1835 One troop Dundalk to Belfast
23	𝕾	1871 'C' Kinsale to Bandon
24	M	1861 Reviewed by Queen Victoria on the Curragh
25	Tu	1835 Detachment Maghera to Belfast
26	W	1839 Thanks of H.M. the Queen Dowager
27	Th	
28	Fr	1816 Establishment reduced to 440 rank and file
29	Sa	1831 One troop Dowlais to Abergavenny
30	𝕾	1856 Colonel Doherty, C.B., to England
31	M	1814 Two sq. sailed from Plymouth for G. of Mexico

SEPTEMBER

1	Tu	1831 One troop Llandilo to Abergavenny
2	W	1890 Regt. encamped at Cove Common, Aldershot
3	Th	
4	Fr	
5	Sa	[Bury
6	S	1841 Arrived at Bombay. 1862 'A' Manchester to
7	M	1799 Permitted to wear the Prussian Eagle
8	Tu	1846 Maj. J. W. King to 14th L.D. '90 In camp,
9	W	1890 Cavalry manœuvres began in Berks. [Churn
10	Th	1890 Regiment encamped at Blewberry
11	Fr	1826 Inspected at Dublin by Sir C. Grant
12	Sa	1861 'H' to Belturbet, 'G' to Belfast
13	S	1861 Newbridge to Dundalk. 1890 In camp,
14	M	[Uffington
15	Tu	
16	W	1890 Grand march-past, Chilton Downs, Berks.
17	Th	1873 'B' Newbridge to Dublin
18	Fr	1857 Captain R. H. Gall to Major [Brighton
19	Sa	1889 Inspected by Sir Drury Lowe, K.C.B., at
20	S	1877 The Viceroy visited regimental theatre at
21	M	1890 In camp at Crookham [Bangalore
22	Tu	1890 Encamped Ald'shot from cavalry manœuvres
23	W	1858 'C' and 'E' at Garotha under Mj. Thompson
24	Th	1816 Front peak of saddles cut down [Hounslow
25	Fr	1875 Insp. at Colchester by Dk. of Cambridge. '90
26	Sa	1831 Two tr. Gloucester to Abergavenny [H'slow
27	S	1811 Engaged with 4 regiments of enemy's lancers
28	M	1870 'B' Cahir to Fethard [at Espeja
29	Tu	1875 Insp. at Colchester by Sir T. W. McMahon, Bt.
30	W	1819 Lieut.-Col. Baker to command *vice* Sir F. [Hervey deceased

OCTOBER

1	Th	1841 Recd. 150 volunteers from 4th Lt. Dragoons
2	Fr	1841 Second party arrived at Bombay
3	Sa	1871 Insptd. at Ballincollig by Maj.-Gen. Wardlaw
4	S	1858 'B' and 'D' flying col. under Col. Scudamore
5	M	1844 Squadron to Kolapore on field service
6	Tu	1887 Duke of Cambridge's review near Dover
7	W	1874 Inspt. at Aldershot by Sir T. W. McMahon, Bt.
8	Th	1890 Inspt. at Hounslow by Sir Drury Lowe, K.C.B.
9	Fr	1888 Inspt. by Sir D. Lowe, K.C.B., at Brighton
10	Sa	1868 Inspt. at Newbridge by Brig.-General Little
11	S	1890 Inspt. by Maj.-Gen. P. Smith, C.B., Hounslow
12	M	1860 Inspt. at Newbridge by Major-General Parlby
13	Tu	1888 'B' and 'D' Hilsea to Brighton
14	W	
15	Th	
16	Fr	
17	Sa	1844 'A' squadron to Kolapore on field service
18	S	1869 Inspt. at Curragh by Brig.-Gen. Wardlaw
19	M	1868 Newbridge to Dublin
20	Tu	
21	W	1869 Half of second squadron Curragh to Longford
22	Th	1857 Left wing captured 3 guns and 3 standards
23	Fr	1839 Percuss. carbines issd. '69 Curragh to Long-
24	Sa	1885 Col. J. H. Knox died at Secunderabad [ford
25	S	1831 Two troops Gloucester to Sodbury
26	M	1831 Two troops Sodbury to Clifton
27	Tu	
28	W	1823 Lieut.-Gen. Sir J. Vandeleur to be Colonel
29	Th	1850 Inspt. at Lahore by Sir W. Gilbert [strong
30	Fr	1886 Regt. left Bombay in H.M.S. *Serapis*, 405
31	Sa	1831 One tr. under Mj. Beckwith to Bristol for riots

NOVEMBER

1	S	1831 Detachment Gloucester to Newport
2	M	1832 One troop Gloucester to Tewkesbury
3	Tu	1832 One troop Bristol to Gloucester
4	W	1832 Detachment Tewkesbury to Gloucester
5	Th	1881 Ladysmith for Durban
6	F	1832 One troop Abergavenny to Tredegar
7	Sa	1878 Received orders for service in Afghanistan
8	S	1832 One troop Tredegar to Abergavenny
9	M	1842 Draft of 1 sergt. and 99 men from England
10	Tu	1813 Entered France with Sir R. Hill's Division
11	W	1841 Lt.-Gl. Kerrison to Gl. '81 Pinetown for D'ban
12	Th	1841 Insptd. at Kirkee by Sir T. McMahon, Bart.
13	Fr	1881 Right wing embarked at Durban
14	Sa	1868 'C' and 'G' Dublin to Dundalk
15	S	1813 Outpost duty on the Nive
16	M	1812 Repulsed superior body of enemy's lancers
17	Tu	1886 Lieut.-Col. H. B. Hamilton to 14th Hussars
18	W	1837 Detachment Tewkesbury to Gloucester
19	Th	1835 Detachment Ballyshannon to Glentees
20	Fr	1838 One squadron Northampton to Coventry
21	Sa	1857 Lieut. Dew and picquet routed 300 natives
22	S	1848 R'nuggur; L.-Cl. Havelock & Cap. Fitzgerald
23	M	1857 Mundesor; Major Gall took 6 guns [killed
24	Tu	1878 Orders for Afghanistan cancelled
25	W	1886 Landed at Portsmouth, arrived at Shorncliffe
26	Th	1861 Major-General Beckwith to Lieut.-General
27	Fr	1845 Insptd. at Kirkee by Sir T. McMahon, Bart.
28	Sa	1868 'A' Dundalk to Dublin [Bt., to half-pay
29	S	1833 Recruiting susp. '64 Lt.-Col. Sir W. Russell,
30	M	1869 One squadron Nenagh to Newbridge

DECEMBER

1	Tu	1839 One squadron Coventry to Northleach
2	W	1830 Headquarters Birmingham to Oxford
3	Th	1830 One troop Kidderminster to Oxford
4	Fr	1830 One sq. Northleach to Burford [for Lisbon
5	Sa	1808 Embkd. at Falmouth under Lt.-Col. Hawker
6	S	1849 Insptd. at Meean Meer by the Gov.-General
7	M	1869 One squadron Newbridge to Athlone
8	Tu	
9	W	1813 Crossed the Nive, 2 men and horses drowned
10	Th	
11	Fr	1882 Field manœuvres at Secunderabad
12	Sa	1868 'B,' 'E,' 'H,' and 'K' Cork to Dublin
13	S	1813 Maj. Brotherton wounded and tkn. prisoner at
14	M	1810 With Light Division at Santarem [Hasparren
15	Tu	1845 Kirkee to Agra
16	W	1867 Detachment Glasgow to Hamilton [wounded
17	Th	1858 'A' and 'H' at Ranode. Major Prettejohn
18	Fr	1830 One troop Wellingborough to Northampton
19	Sa	1830 Hdqrs. Oxford to Coventry. '62 Bt. Lt.-Col.
20	S	1869 One sq. Limerick to Bal'robe [Need to half-pay
21	M	1830 One troop Northleach to Coventry
22	Tu	1835 One troop Dundalk to Monaghan
23	W	1808 Landed at Lisbon
24	Th	1878 Inspt. at Bangalore by Col. Wallace, R.H.A.
25	Fr	1818 Second assistant surgeon discontinued
26	Sa	1836 One troop Belturbet to Dundalk
27	S	
28	M	
29	Tu	
30	W	1815 Embkd. at Bristol for Ireland under Lt.-Col.
31	Th	1881 Left wing disembarked at Bombay [Baker

REGIMENTAL MUSIC

Regimental march, 'King of Prussia.'

The regiment has used 'The King of Prussia' as its regimental march for many years. It is a very fine march, and was introduced by the late Major-General William Arbuthnot, C.B., when he was Lieutenant-Colonel commanding the Fourteenth at Bangalore, in 1880.

Officers' dress-call for mess.

The officers' 'Dress-Call' for mess was introduced at Hounslow in 1890, by Colonel H. B. Hamilton, commanding the regiment, and previous to this there was no special call for this purpose in the 14th Hussars.

Regimental call.

The 14th Hussars' 'Regimental Call' is the one prescribed in the *Trumpet and Bugle-Sounds for the Army*, as follows :—

REGIMENTAL CALL

Mr. H. Hemsley, Bandmaster of the 14th Hussars, kindly arranged the regimental music for the piano, as follows :—

REGIMENTAL MARCH—'KING OF PRUSSIA'

Ped.
*
Ped.
*
8va
Pia.
Senza Ped.
loco.
ffor.
Animato.
Ped.

for. Marcato.
Ped.
Ped.
Ped.
Trio.
Pia. Con Grazia.
sf.
sf.
Pia.
Fine.
ffor. Energico.
D.C. Trio al Fine.

REGIMENTAL TROT—'UP, LIGHT LOO'

3
pia e leggerézza.
for.
pia e leggerézza.

REGIMENTAL GALOP—'RORY O'MORE'

CAVALRY BRIGADE TROT—'MONYMUSK'

CAVALRY BRIGADE GALOP—'BONNIE DUNDEE'

REGIMENTAL 'QUICK-MARCH'

for.
Pia.............
D.C.

OFFICERS' 'DRESS-CALL' FOR MESS

TABLE OF SERVICES AT HOME AND ABROAD

Years and Countries in which the Regiment has served since its Formation in July 1715 to July 1901.	Home.		Abroad. (Including the time occupied by sea voyages.)					
	Great Britain.	Ireland.	St. Domingo.	Portugal, Spain, and France.	India.	Persia.	South Africa.	Total Years and Months.
	Y. M.	Y. M.	Y. M.	Y. M.	Y. M.	Y. M.	Y. M.	Y. M.
1715-1717 (May)	1 10							1 10
1717-1742 (May)		25 0						25 0
1742-1747 (May)	5 0							5 0
1747-1795 (Feb.)		47 9						47 9
1795-1797 (Oct.)			2 8					2 8
1797-1808 (Dec.)	11 2							11 2
1808-1814 (July)				5 7				5 7
1814-1815 (Dec.)	1 5							1 5
1816-1819 (June)		3 6						3 6
1819-1825 (April)	5 10							5 10
1825-1828 (March)		2 11						2 11
1828-1833 (March)	5 0							5 0
1833-1836 (May)		3 2						3 2
1836-1841 (May)	5 0							5 0
1841-1857 (March)					15 10			15 10
1857 (May)						0 2		0 2
1857-1860 (June)					3 1			3 1
1860-1862 (May)		1 11						1 11
1862-1868 (May)	6 0							6 0
1868-1874 (May)		6 0						6 0
1874-1876 (Jan.)	1 8							1 8
1876-1881 (March)					5 2			5 2
1881 (Nov.)							0 8	0 8
1881-1886 (Nov.)					5 0			5 0
1886-1894 (July)	7 8							7 8
1894-1899 (Nov.)		5 4						5 4
1899 (Nov.-Dec.)	0 1							0 1
1899-1901 (July)							1 7	1 7
Total years and months in each country	50 8	95 7	2 8	5 7	29 1	0 2	2 3	186 0

	Years.	Months.
At Home,	146	3
Abroad,	39	9
1715-1901	186	0

Note.—Two squadrons (without horses) served near the Gulf of Mexico, at New Orleans, Mississippi River, North America, from August 1814 to May 1815.

CHANGES IN ESTABLISHMENT SINCE 1800

Date of Alteration.	Nature of Alteration.	Troops.	Sergeants.	Warrant Officers.	Trumpeters.	Farriers.	Corporals.	Privates.	Troop-horses.
Establishment on 25th December 1802,		8	28		8		24	456	436
ALTERATIONS SINCE ABOVE PERIOD.									
June 1804	Augmented,	10	44		10		40	760	854
Dec. 1804	Augmented,	10	54		10		50	950	1064
Dec. 1811	Augmented Schoolmaster-Sergeant, and reduced,	10	55		10		40	760	864
Dec. 1815	Reduced,	8	45		8	8	32	584	581
Aug. 1816	Reduced,	8	37		8	8	24	432	333
Aug. 1821	Reduced,	6	23		6	6	18	282	253
Oct. 1824	Hospital-Sergeant,	6	24		6	6	18	281	253
Jan. 1831	Orderly-Room Clerk and augmented horses,	6	25		6	6	18	280	271
April 1834	Reduced horses,	6	25		6	6	18	280	253
April 1838	Augmented horses,	6	25		6	6	18	280	271
Jan. 1841	Augmented,	9	55		12	8	40	627	701
April 1842	Hospital-Sergeant,	9	56		12	8	40	626	701
Dec. 1854	Saddler-Sergeant and augmented,	9	57		14	10	40	626	703
June 1857	Augmented,	10	59		14	10	40	626	703
April 1858	Schoolmaster-Sergeant reduced,	10	58		14	10	40	626	703
June 1860	Reduced,	8	40		9	8	32	537	428
Nov. 1861	Sergeant-Instructor of Musketry,	8	41		9	8	32	536	428
April 1862	Bandmaster,	8	42		9	8	32	535	428
April 1866	Sergeant-Cook and Sergeant-Instructor of Fencing,	8	44		9	8	32	533	428
April 1869	Reduced,	Sqdns. 4	43		9	4	24	443	344
April 1870	Reduced,	Troops 7	39		8	4	21	386	300
Feb. 1871	Augmented,	8	43		9	8	32	515	384
April 1872	Hospital-Sergeant reduced,	8	42		9	8	32	516	384
July 1881	Trumpet-Major to Sergeant-Trumpeter, and augmented,	8	41	2	8	8	32	510	400
Jan. 1882	Colonial to Indian establishment,	7	37	2	7	6	27	510	400
Oct. 1886	Home establishment. No Depôt,	8	41	2	8	8	32	378	300
April 1887	Horses reduced,	8	41	2	8	8	32	378	250
Oct. 1887	Horses increased,	8	41	2	8	8	32	378	300
April 1889	Armourer-Sergeant attached; Paymaster-Sergeant reduced.								
April 1890	Sergeant Master-Tailor and Orderly-Room Sergeant added.								
April 1891	Reduced,	8	40	2	8	8	33	300	270
April 1892	Augmented,	Sqdns. 4	40	2	8	8	33	328	280
April 1895	Augmented,	4	39	2	8	8	35	353	300
April 1897	Augmented,	4	42	2	8	4	24	460	343
Oct. 1899	Higher establishment, Offrs. 23	Sqdns. 4	45	2	8	4	30	560	465

N.B.—The rank of Troop Sergeant-Major was introduced in 1809.

CHANGES IN UNIFORM

GEORGE I.

14th Dragoons.

1715.—Uniform scarlet. Silver aiguillette.

GEORGE II.

1751.—Uniform scarlet; facings lemon colour. (See Warrant, p. 19.)

GEORGE III.

1768.—Uniform scarlet; facings lemon colour, without lapels. Silver lace as in 1751. Silver epaulette for officers and men instead of aiguillette. (See Warrant, p. 27.)

1773.—Officers wore embroidered epaulettes and button-holes.

14th Light Dragoons.

1776.—Cocked hats replaced by helmets. Half-lapels worn.

1784.—Uniform blue; facings lemon-yellow. (See Regulations, pp. 34, 35.)

1786.—Blue cloaks substituted for red ones. Black spatter-dashes worn.

1790.—Narrow silver lace worn on the shells and frocks of sergeants, in lieu of the silver chain which had been in use hitherto.

1796.—Officers' crimson silk sash to be tied on right side.

The 14th, or the Duchess of York's Own Regiment of Light Dragoons.

1798.—Prussian Eagle given by royal authority as regimental badge. Facings changed from lemon-yellow to orange, 1799. Queues ten inches long to be worn by cavalry.

1815.—Worsted webbed pantaloons and breeches discontinued in cavalry, and cloth overalls (grey speckled) to be worn with half-boots instead. A later order of 1815 says the overalls are to be blue-grey with white stripes. The men of the Fourteenth probably wore orange stripes about this date, and the officers silver lace stripes for full dress.

1816.—Overalls to be of kersey wove, and to be dark grey: adopted in 1819 by 14th Light Dragoons.

1820.—Blue-grey pantaloons authorised: adopted in 1821 by 14th Light Dragoons.

GEORGE IV.

1821.—Officers' clothing regulated by Royal Warrant, 25th April.

1823.—Light Dragoons to wear white and red horse-hair plumes instead of the feather as hitherto.

1824.—Brass shoulder-scales instead of worsted epaulettes.

1827.—Substitution of white worsted lace down the outer seam of the trousers, instead of stripes of cloth of the colour of regimental facings.

1829.—Half-lapels abolished. New shako adopted with Maltese cross ornament; the Prussian eagle was worn in centre of the cross until the King's crest was granted in 1832. Oxford mixture overalls introduced on the 15th April.

WILLIAM IV.

The 14th (King's) Light Dragoons.

1830.—Gold lace substituted for silver.

[1]1831.—Uniform scarlet instead of blue; facings blue instead of orange; overalls dark blue in place of light grey.

1832.—The King's crest granted.

1835.—Bandsmen clothed liked trumpeters.

[1] In August 1831 moustaches were abolished in the cavalry, excepting in the Life Guards, Royal Horse Guards, and regiments of hussars, and an order issued directing 'the hair of the non-commissioned officers and soldiers to be cut close to the sides and at the back of the head, instead of being worn in that bushy and unbecoming fashion adopted by some regiments.' (From Colonel R. S. Liddell's *Memoirs of the 10th Hussars*, p. 213.)

QUEEN VICTORIA.

1840.—Uniform blue; facings scarlet.

1841.—Cloaks made with sleeves.

1842.—Cloaks made without sleeves.

1849.—Yellow stripes on trousers adopted for light cavalry.

1855.—Light dragoons to wear jackets with five bars of gold lace. No sabretache.

1856.—Plain leather sabretache authorised for officers and sergeants of cavalry.

1857.—Shako covered with Paris velvet and leather top. Plume of hair five inches above top of chaco, gilt plume-socket, and corded ball with four upright rays of red and white colours.

1860.—Peaks to officers' caps discontinued, and gold stripes on their overalls adopted in lieu of cloth ones, with the addition of leather bootings. An embroidered silver eagle arm-badge approved for sergeants.

14th (King's) Hussars.

1861.—Regiment converted into Hussars on 17th August. Uniform blue; busby-bags yellow; plumes white. Scarlet sabretaches for officers.

1867.—A sterling silver Prussian Eagle arm-badge approved for sergeants, 14th Hussars; corporals to have German silver arm-badge.[1]

1871.—Knee-boots to be worn by cavalry instead of booted overalls as hitherto, and pantaloons introduced at same time. Officers of hussars to wear oval gold-gimp lace bosses on their boots, made detachable.

1888.—Sabretaches for non-commissioned officers and men discontinued.

1890.—Blue serge frocks and field-service caps introduced for manœuvres and drill purposes.

1893.—Officers to wear yellow cloth stripes, like the men, instead of gold lace stripes on their pantaloons, excepting in

[1] All staff-sergeants and sergeants of the Fourteenth wear above their chevrons as an arm-badge silver eagles, and all corporals wear German-silver eagles in the same manner.

levee dress; and to wear, in undress, sword-belts of white leather instead of gold lace. Head-ropes to be worn by officers' chargers, same as in the ranks, in lieu of steel chains.[1]

1897.—Shabracques discontinued for officers. They were abolished in the ranks about twenty-five years earlier. In the Fourteenth the approved pattern was the old light dragoon shabracque,[2] with rounded corners, fore and hind; lambskins (black) edged with yellow cloth. With the new pattern cavalry saddle, sheepskins are discontinued, but front and rear lambskins are worn. The officers' lambskins, in the Fourteenth, are edged with yellow cloth. The officers' chargers' throat ornament is white horse-hair, eighteen inches long, with brass ball and socket. The bridle-bosses have the Prussian Eagle encircled by a garter, within which is 'Fourteenth or The King's Hussars,' and surmounted by a crown.

1899.—Stable-jackets discontinued in the ranks; officers continue to wear them at mess only. Round forage-caps only used by the men for walking out, and by officers occasionally. The active service field-cap has been introduced generally into the cavalry. The pattern for hussars is scarlet, with gold piping for officers, and in the 14th Hussars an eagle badge is worn on the left side. Blue tartan frocks (blue serges) have been generally substituted for stable-jackets: with these the 14th Hussars, officers and men, wear the eagle badges on their collars, and steel shoulder-chains like other regiments. The non-commissioned officers and men, in addition, wear Arabic numerals ($\frac{14}{\text{H}}$) on their shoulders, fixed to the shoulder-chains. Khaki clothing has been introduced for service in the field. The 14th Hussars proceeded to South Africa wearing khaki. The officers wore eagle badges on their collars, and the men wore Arabic numerals on their shoulders ($\frac{14}{\text{H}}$). They all wore white helmets covered with khaki.

[1] Head-ropes had been worn on officers' chargers for several years previously, but only on manœuvres and for drill purposes.

[2] In the coloured plate (facing p. 380) of an officer in 'review order,' 1891, the officer's charger should carry a steel collar-chain instead of a head-rope, and in the coloured plate, 1894 (facing p. 388), the officer's charger should carry a head-rope instead of a collar-chain.

Uniform of Officers and Men.

Officers' Uniform in 14th Hussars.

Officers' cloaks and capes are lined with yellow. The mess waistcoat is scarlet cloth, with six bars of gold lace, and white shirt collars are worn at mess, without authority. No cockades are worn in their busbies by officers of the 14th Hussars, and this custom, which has existed many years, has been sanctioned in the *Dress Regulations for Officers of the Army*, published in 1900. The Prussian Eagle is worn on the officers' dress sabretache below the royal cypher and crown, with honorary distinctions below the eagle, as has been the custom for many years, and the eagle badge was similarly worn on the hind corners of the officers' shabracques. The sabretaches are lined with yellow leather, and the sword-knots are gold cord and acorn pattern, without any crimson, the same as those worn in the 13th Hussars. The busby-chains are lined with yellow, the leather lining being backed with velvet. The ostrich feather now worn in the busby of hussars generally is fifteen inches high from top of the busby to top of the feather.

The $\frac{1}{4}$-inch and $\frac{1}{8}$-inch stripe worn in the centre of the shoulder-belt, sabretache-lace, and sword slings respectively, in all hussar regiments, is simply of gold in the 14th Hussars, and in other regiments either buff, crimson, scarlet, or white.

The pouch worn in the Fourteenth is black leather, silver flap, and gilt metal ornaments. The shoulder-belt has the buckle, tip, and slide of silver, and silver engraved breast ornaments bearing Royal crest and Prussian eagle with chain and pickers are worn.

Regulations as to uniform of officers.

The whole of the uniform for officers is now clearly laid down in the *Dress Regulations for the Army*, published periodically. The latest edition is dated 1900.

Clothing regulations for the men.

The issue of clothing to the non-commissioned officers and men is laid down in the *Regulations for the Clothing of the Regular Army*, latest edition, 1898. Sealed patterns for each article are supplied annually (or as required) to regiments from the Royal Army Clothing Department, Pimlico, where the standard patterns are deposited when approved by the Sovereign.

GUIDONS OF THE REGIMENT AT VARIOUS PERIODS SINCE 1715

ORIGINALLY Light Dragoons had as a rule 3 guidons per regiment of 6 troops, but towards the end of the eighteenth century regiments mostly consisted of 8 troops, and some had as many as 10 troops, so that the number of standards or guidons was, as a rule, increased to an average of 4 per regiment; and in 1807 there were 5 per regiment when the establishment was at its highest, viz. 10 troops. The 1st Dragoon Guards with its 12 troops had 6 standards. Guidons.

A guidon measured 2 feet 10 inches long by 2 feet 4 inches deep. The royal warrant of September 1743 contained regulations for the standards and guidons of cavalry regiments, in which it is laid down that the guidons of Light Dragoons are to be of silk, the tassels and cords to be of crimson silk and gold mixed.

By royal warrant dated 1st July 1751, further instructions as to standards and guidons were issued. The Fourteenth were ordered to have their second and third guidons lemon-coloured, all 3 guidons to have the embroidery on them of silver, with the fringe of silver and red. They were ordered to bear on the second and third guidons the rank of the regiment thus: XIV. D. 1751.

Another royal warrant of December 1768 ordered for the first time as follows:

'The lance of the guidons (except those of the Light Dragoons) to be 9 feet long (spear and ferrule included). The flag to be 3 feet 5 inches to the end of the slit of the swallow-tail, and 2 feet 3 inches on the lance. Those of the Light Dragoons to be of smaller size.' 1768.

1777. In 1777 there is an entry in the War Office Miscellany Books of a charge of £13, 13s. 6d. made for one standard to complete deficiencies in the 12th Light Dragoons, 'as furnished to 14th Dragoons one pole, 11s. 4½d., one case, 10s.'

'The standard belts according to the richness, £7, 14s. 2d., though the 9th Dragoons paid £3, 17s. for one.'[1]

1822. In old times the standards or guidons of cavalry were carried by cornets, but in 1822, on the 30th November, an order was given for them to be carried by troop sergeant-majors.

1823. In 1823, on the 23rd January, new regulations for cavalry standards and guidons of Light Dragoons were issued. The size of the flag was ordered to be only 3 feet 2 inches to the end of the slit of the swallow-tail, and 1 foot 9 inches on the lance, having a fringe edging of crimson silk intermingled with silk of the colour of the regimental facings.

1832. In 1832 the Prussian eagle, which had since 1799 been the regimental badge, was authorised to be continued on the second and third corners of the guidons of the Fourteenth, at the same time that the King's crest (the Royal cypher within the garter) became the new regimental badge, in honour of the title of 'King's' conferred by His Majesty

1834. Guidons abolished in Light Dragoon regiments.

King William IV. On the 24th May 1834, the use of guidons in Light Dragoon regiments was finally discontinued by His Majesty's commands.

[1] *Standards and Colours of the British Army*, by S. M. Milne, Esqre. (1893).

LIST OF REGIMENTAL PLATE, PICTURES, ETC., BELONGING TO THE OFFICERS' MESS

PLATE

YEAR.	DESCRIPTION.	DONOR.
1851.	Claret-jug, . . .	Lieutenant Hickey.
1856.	Large jug, . . .	Lieutenant A. M'Gregor.
1861.	Equestrian centre-piece, .	Colonel C. Steuart, C.B.
1861.	Silver cup, . . .	Lieutenant Bradley.
1862.	Hoof snuff-box. 'Bob' joined 14th Light Dragoons 1832, died in 11th Hussars 1862, .	11th (P.A.O.) Hussars.
1862.	Claret-jug, . . .	Lieutenant Brudenell Bruce.
1862.	Claret-jug, . . .	Late Lieut. Franklyn's family.
1863.	Silver tray, . . .	Col. Sir Wm. Russell, Bart., C.B.
1863.	Bread-basket, . . .	Col. P. Scott Thompson, C.B.
1865.	Silver assiette, . . .	Lieut.-Col. P. S. Thompson.
1867.	Four silver dessert-dishes (with regimental costumes),	Capt. A. Clarke-Travers, Lieut. T. A. Henry, Lieut. Laing, Cornet Cobden, Capt. W. Arbuthnot, Capt. Bolton, Lieut. Hon. J. St. V. Saumarez.
1875.	Silver cigarette-box, . .	Detachment 6th Dragoon Guards (Carabiniers).
1890.	Silver cup, . . .	Major Hickman.
1890.	Silver tankard, . . .	Lieutenant H. St. Maur.
1891.	Silver tray, . . .	Lieutenant-Colonel Garth.
1892.	Cigar-cutter, . . .	12th Battery R.A.
1895.	Silver model of 'Sainfoin,' .	Captain Sir J. P. Miller, Bart.
1896.	Six old Irish silver baskets, .	Major Mitchell.
1896.	Twenty-four silver finger-bowls,	Captain Broadhurst.
	Silver cigarette-box, . .	Lieutenant Rowley.
	Gold snuff-box, . .	Lieutenant Watts.

YEAR.	DESCRIPTION.	DONOR.
	Two claret-jugs, . .	Lieut.-Colonel P. Campbell.
	Busby, etc., . . .	Captain Lynch Staunton.
	Silver cup, . . .	Baron de Grancy.
	Rifle cup, . . .	Won by regiment.
	Lloyd Lindsay Cup, . .	Won by regiment.
	Silver dessert-stand, . .	Lieutenant M'Taggart.
	Eagle cup, . . .	Lieutenant Dudley Ward.
	Silver tankard, . .	Lieutenant A. Brassey.
	Silver lamp, . . .	Lieutenant A. Brassey.
	Four statuettes of regiment, .	Colonel Wm. Arbuthnot.
	Four silver dessert-stands (oak pattern), . . .	Lieut. Mather, Lieut. Hon. F. Amherst, Capt. Digby Boycott.
	Ram's Head, . . .	Lieut. F. B. Chapman.
	Silver table bell, . .	Major R. M. Richardson.
	Large cigar-cutter, . .	Lieutenant-Colonel Russell.
	Palm vases and centre-piece, .	Captain Owen.
	Sugar vases, . . .	Captain Pemberton Campbell, Lieutenant Bradney.

PICTURES, ETC.

YEAR.	DESCRIPTION.	DONOR.
1887.	'Quatre Bras' (Thompson), .	Colonel H. B. Hamilton.
1887.	'Windsor Castle,' . .	Major A. F. English.
1891.	Weighing-chair, . .	Colonel H. B. Hamilton.
1894.	Dresses of Hussars, . .	Colonel H. B. Hamilton.
1894.	Dresses of Light Dragoons, .	Colonel H. B. Hamilton.
	Gilt clock, . . .	Lieutenant Merry.
	Silver clock, . . .	W. H. Dunne.
	Aneroid barometer, . .	Captain Kirk.
	'H.R.H. the Duke of Cambridge,' . . .	An old 'K.L.D.'
	'14th (King's) Hussars,' .	Captain Beaumont.
	'14th (King's) Hussars,' .	Captain Beaumont.
	'Day of Reckoning,' . .	Captain R. M. Richardson.
	'The Racecourse,' . .	Captain Sir J. P. Miller, Bart.
	'The Journey,' . .	Captain Sir J. P. Miller, Bart.
	'General Kellermann,' .	Captain H. B. L. Hughes.

SINCE 1891. DESCRIPTION.	DONOR.
'The First Steeplechase' (4),	Captain H. B. L. Hughes.
'Charge of the Cuirassiers,'	Colonel Hon. G. H. Gough.
Types of regiment,	Captain Stuart Robertson.
Portrait of Colonel J. Townsend, A.D.C.,	Captain Stuart Robertson.
'Patrol of 14th Light Dragoons in Peninsular War,'	Major F. B. Chapman.
Medals,	Captain Stuart Robertson.
Medals,	Colonel Hon. G. H. Gough.
Polo cup,	Captain C. Stacey.

REGIMENTAL RACE AND CHALLENGE CUPS

WITH LISTS OF WINNERS SINCE 1864

CONDITIONS.—*All Challenge Cups to be won three years in succession by the same officer before becoming his property.*

14TH (KING'S) HUSSARS STEEPLECHASE CHALLENGE CUP

'THE OLD REGIMENTAL CUP'

WINNERS

YEAR.	OWNER'S NAME.	HORSE'S NAME.
1864.	Captain Baskerville,	Dragoon.
1865.	Captain Baskerville,	Dragoon.
1866.		
1867.	Lieutenant S. T. Ashton,	Palermo.
1868.	Captain Hay Newton,	Viceroy.
1869.	Captain S. T. Ashton,	Merlon.
1870.	Colonel P. S. Thompson,	Conrad.
1871.	Lieutenant Lefroy,	Subeila.
1872.	Major Pemberton Campbell,	R. M.
1873.	Major Pemberton Campbell,	R. M.
1874.	Captain Knox,	Cruiskeen.

	YEAR.	OWNER'S NAME.	HORSE'S NAME.
	1875.	Major Pemberton Campbell, . . .	Dicky Dear.
	1888.	Lieutenant Sir J. P. Miller, Bart., . .	Tablet.
	1889.	Captain Kirk (rider, Lieutenant Tritton), .	Shelduck.
	1890.	Capt. Sir J. P. Miller, Bt. (rider, Lieut. Murray),	Battle Bell.
Cup won outright.	1891.	Colonel Honourable G. H. Gough (rider, Lieutenant Eley), . .	Hard Times.
Cup won outright.	1892.	Colonel Honourable G. H. Gough (rider, Lieutenant Eley), . .	Hard Times.
Cup won outright.	1893.	Colonel Honourable G. H. Gough (rider, Lieutenant Eley), . .	Hard Times.

CHALLENGE CUP PRESENTED BY CAPTAIN S. TUDOR ASHTON

WINNERS

YEAR.	OWNER'S NAME.	HORSE'S NAME.
1869.	Captain Henry, ...	Colleen.
1870.	Captain Lefroy, ...	Bayard.
1871.	Major Pemberton Campbell, . . .	Messenger.
1872.	Major Pemberton Campbell, . . .	R. M.
1873.	Captain Barrett, ...	Barmaid.
1874.	Lieutenant Merry, ...	Blarney.
1875.		
1876.		
1877.	Lieutenant Palmes, ...	The Widow.
1878.	Major Russell, ...	Carabinier.
1888.	Lieutenant Sir J. P. Miller, Bart., . .	Nugget.
1889.	Lieutenant Eley (rider, Lieutenant Stacey), .	Telegram.
1890.	Lieutenant Murray, ...	Ironsides.
1891.	Colonel Honourable G. H. Gough, . .	Hard Times.
1892.	Capt. R. M. Richardson (rider, Lieut. Brooksbank),	Scaleby.
1893.	Major English (rider, Captain Murray), . .	Richard.
1894.	Captain Stacey (rider, Lieutenant Tottenham), .	Bosbury.
1895.	Major Mitchell (rider, Lieutenant Eley), . .	Jack Tar.
1896.	Lieutenant Lockett, ...	Speculation.
1897.	Lieutenant Adam, ...	Erin's Queen.
1898.	Captain Brooksbank, ...	Triumph.
1899.	Lieut.-Col. G. C. Hamilton (rider, Lieut. Blackett),	Record.

CHALLENGE CUP PRESENTED BY CAPTAIN J. M. LEFROY

(Light Weight)

WINNERS

YEAR.	OWNER'S NAME.	HORSE'S NAME.
1872.	Captain Honourable F. Amherst,	Exchange.
1873.	Captain Laing,	Remus.
1874.	Captain Knox,	Cruiskeen.
1877.	Lieutenant Tuthill,	Lorna Doone.
1888.	Lieutenant Sir J. P. Miller, Bart.,	Tablet.
1889.	Lieutenant Tickell,	Caradoc.
1890.	Colonel Honourable G. H. Gough,	Trojan.
1891.	Lieutenant Brooksbank,	The Master.
1892.	Lieutenant Stephen,	Gipsy.
1893.	Colonel Honourable G. H. Gough,	Hard Times.
1894.	Colonel Honourable G. H. Gough,	No More.
1895.	Captain Brooksbank,	Triumph.
1896.	Lieutenant Adam,	The Emperor.
1897.	Lieutenant Adam,	May Queen.
1898.	Lieutenant Tottenham,	Whitehall.
1899.	Captain D. M. Miller,	Brunette.

CHALLENGE CUP PRESENTED BY COLONEL HONOURABLE G. H. GOUGH, C.B., TO REPLACE THE 'OLD REGIMENTAL CUP' WON BY HIM 1893

WINNERS OF THE 'GOUGH CUP'

YEAR.	OWNER'S NAME.	HORSE'S NAME.
1894.	Captain Murray,	Queensland.
1895.	Captain Murray (rider, Lieutenant Eley),	Queensland.
1896.	Col. Hon. G. H. Gough (rider, Lieut. Lockett),	Paraphine.
1897.	Captain Eley,	Sugar Loaf II.
1898.	Captain D. M. Miller,	The Buster.
1899.	Captain D. M. Miller,	The Buster.

SUBALTERNS' CHALLENGE CUP PRESENTED BY LIEUTENANT-COLONEL HONOURABLE G. H. GOUGH

WINNERS

YEAR.	OWNER'S NAME.	HORSE'S NAME.
1888.	Lieutenant Sir J. P. Miller, Bart.,	Darby.
1889.	Lieutenant Eley,	Ratcatcher.
1890.	Lieutenant Tickell,	Caradoc.
1891.	Lieutenant Stephen,	Gipsy.
1892.	Lieutenant Eley,	Joan.
1893.	Lieutenant Eley,	Joan.
1894.	Lieutenant Stephen (rider, Lieutenant Prevost),	Mistake.
Cup won outright. 1895.	Lieutenant Tottenham,	Handy Andy.
1896.	Lieutenant Tottenham,	Whitehall.
1897.	Lieutenant Tottenham,	Whitehall.

SUBALTERNS' CHALLENGE CUP PRESENTED BY COLONEL HONOURABLE G. H. GOUGH, C.B., ON HIS LEAVING THE REGIMENT, 1897

WINNERS

YEAR.	OWNER'S NAME.	HORSE'S NAME.
1898.	Lieut. H. J. Tilney (rider, Lieut. Tottenham),	Regulus.
1899.	Lieutenant Tottenham,	Lent Lily.

At the present time (1899) the Cups are competed for as follows:—

Ashton's Cup.—Heavy weight.—Point-to-point race.

Lefroy's Cup.—Light weight.—Point-to-point race.

Gough Cup.—Two and a half miles over a steeplechase course.

Subalterns' Cup.—Three miles over a steeplechase course.

There is also (annually) a Hunter's Race of £1 sweepstake, 21 sovereigns added.

In 1899 a new race, called 'The 14th Hussars Hunter's Plate,' was won by Captain Brooksbank's bay gelding 'Strathmore,' ridden by owner.

SOME RACES WON BY OFFICERS OF THE 14TH (KING'S) HUSSARS

1873

At Rugby, on the 4th March, the Grand Military Gold Cup, value 100 guineas, added to sweepstakes of £10 each and £100 added (24 subscribers), was won by Mr. John S. Heron Maxwell's (14th Hussars) 'Revirescat,' ridden by Mr. W. Hope Johnstone.

Betting—9 to 4 against 'Revirescat.' Won by 10 lengths. 16 runners.

At Rugby, on 5th March, the Military Weight-for-Age Stakes, won by Mr. J. S. Heron Maxwell's 'Revirescat,' ridden by Mr. W. Hope Johnstone.

Betting—3 to 4 on 'Revirescat.' Won by a distance. 'Revirescat' carried a stone penalty.

1890

The Derby, won by Captain and Adjutant Sir James P. Miller's, Bart., chestnut colt 'Sainfoin.'

1895

The Irish Point-to-Point Race (heavy weight), won by Colonel Honourable G. H. Gough's 'No More.'

1899

The Irish Army Point-to-Point Race (light weight), won by Lieutenant T. E. L. Whitson's brown gelding 'The Count.'

The Irish Military Steeplechase Plate, value £98. Ward Union Hunt Races, Fairy House, won by Captain D. M. Miller's bay gelding 'The Buster,' ridden by owner.

SUCCESSION OF COLONELS

	Lieutenant-General James Dormer, . .	July 22, 1715
	Lieutenant-General Clement Neville, .	April 9, 1720
	Lieutenant-General Archibald Hamilton, .	June 27, 1737
	Field-Marshal the Right Honourable James O'Hara, Lord Tyrawley, P.C., . .	July 24, 1749
	Lieutenant-General Louis Dejean, . .	Nov. 27, 1752
	Field-Marshal John Campbell (afterwards Marquis of Lorne and fifth Duke of Argyll), ...	April 5, 1757
	General Charles Fitzroy (afterwards Lord Southampton, ...	Sept. 11, 1765
	Lieutenant-General Daniel Webb, . .	Oct. 20, 1772
	General George Warde, . . .	Nov. 11, 1773
	General Sir Robert Sloper, K.B., . .	April 2, 1778
	General John W. Egerton (afterwards Earl of Bridgewater), ...	June 1, 1797
P.W.	General Sir John Ormsby Vandeleur, G.C.B.,	Oct. 20, 1823
P.W.	General Sir Edward Kerrison, Bart., G.C.H., K.C.B., ...	June 18, 1830
P.W.	General the Honourable Sir Henry Murray, K.C.B., ...	March 18, 1853
P.W.	Lieutenant-General William Beckwith, K.H.,	Nov. 12, 1860
	General Henry Richmond Jones, C.B., .	Feb. 24, 1871
	General John Wilkie, ...	May 15, 1873
	General Charles W. Thompson, . .	May 1, 1882
	Lieutenant-General Honourable Charles W. Thesiger, ...	Oct. 4, 1896

SUCCESSION OF LIEUTENANT-COLONELS

Name.	Date when Lieutenant-Colonel in the Regiment.	Date when Commanding the Regiment.	Brevet Rank, etc.
Henry Killigrew, . .	July 22, 1715		
Solomon Rapin, . .	1716-17		
William Boyle, . .	1717		
Beverly Newcomin, .	Dec. 1, 1720		
Henry Desgrangues, .	Nov. 12, 1731		
William Wright, . .	July 7, 1737		
James Norris, . .	July 14, 1749		
Thomas Erle, . .	Sept. 4, 1754	Sept. 4, 1754	Col., 1762; Major-Gen., 1770.
Sir J. Burgoyne, Bart.,	July 15, 1773	July 15, 1773	Col., Aug. 29, 1777; Major-Gen., 1782.
Grice Blakeney, . .	Nov. 19, 1781	Nov. 19, 1781	Col., 1790; Major-Gen., 1794; Lt.-Gen., 1802; Gen., 1809.
John Michel, . .	Dec. 5, 1799	April 29, 1802	Col., April 29, 1802; Major-Gen., 1809; Lt.-Gen., 1814; Gen., 1837.
Samuel Hawker, A.D.C.	June 12, 1800	1803	Col., 1808; Major-Gen., 1811; Lt.-Gen., 1831; Gen., 1848; K.C.H., 1831; G.C.H., 1836.
Neil Talbot, . .	Aug. 22, 1805		
Sir F. B. Hervey, Bart., C.B., K.H., A.D.C.,	Aug. 2, 1810	June 4, 1811	Col., June 4, 1814.
Charles M. Baker, .	Sept. 30, 1819	Sept. 30, 1819	
Jn. Townsend, A.D.C.,	April 16, 1829	April 16, 1829	Col., Nov. 23, 1841.
Wm. Havelock, K.H., .	April 30, 1841	April 23, 1845	
Edward Harvey, . .	April 23, 1845		Col. (H.-P.), 1854; Major-Gen., 1858.
J. W. King, C.B., .	April 25, 1848	Nov. 23, 1848	
Henry E. Doherty, C.B.,	Nov. 28, 1848	July 7, 1850	Col., Nov. 28, 1854; Major-Gen., 1863; Lt.-Gen., 1871; Gen., 1877.
Charles Steuart, C.B., .	July 7, 1850	Aug. 26, 1857	Col., Nov. 28, 1854; Major-Gen., Oct. 3, 1864; Lt.-Gen., May 6, 1873.
Charles P. Ainslie, .	Aug. 26, 1857		Col., Nov. 28, 1854; Major-Gen., 1862; Lt.-Gen., 1871; Gen., 1877.
Arth. Scudamore, C.B.,	June 19, 1861	June 19, 1861	Col., Nov. 9, 1862; Major-Gen., 1868.
Sir Wm. Russell, Bart., C.B.,	Oct. 8, 1861	Oct. 8, 1861	Col., July 20, 1863; Major-Gen., 1868.

SUCCESSION OF LIEUTENANT-COLONELS—*continued.*

NAME.	Date when Lieutenant-Colonel in the Regiment.	Date when Commanding the Regiment.	Brevet Rank, etc.
P. S. Thompson, C.B.,	Nov. 29, 1864	Nov. 29, 1864	Col., Jan. 6, 1867; Major-Gen., 1877.
Francis P. Campbell, .	April 29, 1875	April 29, 1875	
Wil. Arbuthnot, C.B., .	June 15, 1876	June 15, 1876	Col., 1881; Major-Gen., 1890.
John H. Knox, . .	June 15, 1881	June 15, 1881	Col., June 15, 1885.
F. S. Russell, C.M.G.,	July 1, 1881		Col., 1885; Major-Gen., 1897.
Charles F. Morton, .	June 10, 1882	Oct. 25, 1885	Col., July 1, 1885.
James Dennis, . .	Oct. 25, 1885		Col., Oct. 25, 1889.
Henry B. Hamilton, .	Nov. 17, 1886	July 1, 1887	Col., March 17, 1890.
Hon. G. H. Gough, C.B.,	July 1, 1891	July 1, 1891	Col., Dec. 16, 1889.
Augustus F. English, .	July 1, 1896	July 1, 1896	
Gilbert H. C. Hamilton,	Feb. 22, 1899	Feb. 22, 1899	Col., Nov. 29, 1900.

Note.—In 1799 there were two lieutenant-colonels, Blakeney and Michel.

In 1800 to 1802 there were three lieutenant-colonels, Blakeney, Michel, and Hawker.

In 1802, on reduction of establishment, Lieutenant-Colonel Hawker was placed on half-pay, but was brought back to the regiment to command it in 1803, *vice* Michel.

In 1805 to 1816 there were two lieutenant-colonels, .
- 1805 to 1810, Hawker and Talbot.
- 1810 to 1816, Hawker and Hervey.

In 1841 to 1860 there were two lieutenant-colonels, .
- 1841 to 1845, Townsend and Havelock.
- 1845 to 1848, Havelock and Harvey (King from April 1848 *vice* Harvey).
- 1848 to 1851, King and Doherty.
- 1851 to 1857, Doherty and Steuart.
- 1857 to 1860, Steuart and Ainslie.

In 1881 to 1887 there were two lieutenant-colonels, .
- 1881 to 1882, Knox and Russell.
- 1882 to 1885, Knox and Morton.
- 1885 to 1886, Morton and Dennis.
- 1886 to 1887, Morton and Hamilton (H. B).

SUCCESSION OF ADJUTANTS

Name.	Date of Appointment.	Rank on Appointment.	Remarks.
William Hamilton, . .	July 22, 1715	Cornet	He was succeeded by Thornborough.
G. Thornborough, . .	—— 1717		Reduced to 2s. a day, half-pay, June 24, 1717.
Nicholas Cowse, . .	May 18, 1743		Cornet, 1745; Adjt. second time, 1752.
Philip Mercier, . . .	Mar. 15, 1755	Qt.-master of a troop	Appointed Lieut., 2nd Bat. 23rd Foot, Dec. 7, 1756.
Jeremiah Hurley, . .	Dec. 7, 1756		
Arthur Ormsby, . . .	Nov. 27, 1766	Cornet	Previously Troop Quartermaster.
Thomas Paterson, . .	Nov. 27, 1771	Lieutenant	Previously Troop Quartermaster.
James Wilson, . . .	June 7, 1776		Born 1741, entered army 1759.
John Magee, . . .	Sept. 12, 1780		
James Latham, . . .	April 30, 1789	Lieutenant	
Lawrence Drought, . .	Jan. 6, 1796		Cornet, June 30, 1796.
George Humphrey, . .	Aug. 28, 1800	Lieutenant	
John Babington, . .	June 25, 1802	Cornet	
Benjamin Shotten, . .	Sept. 1, 1804	Cornet	Lieutenant, Mar. 20, 1806.
Charles M'Carthey, . .	May 5, 1814	Lieutenant	
Patrick Leary, . . .	Dec. 21, 1826	Cornet	Lieutenant, June 29, 1830.
William Clarke, . .	April 23, 1838	Cornet	Lieutenant, Jan. 4, 1841.
Richard P. Apthorp, . .	Nov. 25, 1842	Cornet	Lieutenant, Nov. 5, 1846.
Henry E. Reader, . .	Nov. 4, 1850	Lieutenant	
Harry T. Clements, . .	Dec. 1, 1854	Lieutenant	
James Giles, . . .	Sept. 9, 1859	Lieutenant	
Edward E. Digby Boycott,	June 10, 1862	Lieutenant	
F. J. S. Hay Newton, . .	Mar. 15, 1864	Lieutenant	
Joseph Harpur, . . .	May 16, 1865	Cornet	Lieutenant, July 20, 1866.
James R. Bray, . . .	Aug. 3, 1872	Lieutenant	
James Kentish, . . .	Mar. 15, 1873	Lieutenant	
Honourable H. G. Gough,	Oct. 23, 1875	Lieutenant	
C. D. V. Tuthill, . .	Aug. 27, 1879	Lieutenant	
G. H. C. Hamilton, . .	June 30, 1885	Lieutenant	Captain, July 24, 1885.
H. W. Mitchell, . . .	Sept. 1, 1887	Captain	
Sir James Percy Miller, Bt.,	Sept. 1, 1888	Lieutenant	Captain, Sept. 8, 1888.
E. J. Tickell, . . .	Sept. 1, 1892	Captain	Maj., Feb. 22, 1899, D.S.O.
R. C. Stephen, . . .	Sept. 1, 1896	Captain	
F. R. Lawrence, D.S.O., .	Nov. 18, 1899	Lieutenant	Captain, July 11, 1900.

MEMOIRS AND SERVICES OF THE COLONELS AND LIEUTENANT-COLONELS OF THE 14TH (KING'S) HUSSARS, A.D. 1715 TO A.D. 1900

COLONELS

LIEUTENANT-GENERAL JAMES DORMER[1]

APPOINTED COLONEL OF THE 14TH DRAGOONS, ALSO CALLED 'DORMER'S DRAGOONS,' JULY 22, 1715.

JAMES DORMER, sixth son of Robert Dormer, of Rousham and Dorton, Bucks, who died 1693, and of Anne Cottrell, his second wife, daughter of Sir Charles Cottrell, Master of the Ceremonies to Charles I., Charles II., and James I., and Ambassador at Brussels in 1663, was born on the 16th March 1679. He obtained his first commission as Lieutenant and Captain in the 1st Regiment of Foot Guards, 13th June 1700, and was wounded in that rank at the battle of Blenheim, 2nd August 1704,[2] where a brother officer of the same name, Lieutenant-Colonel Philip Dormer, was killed. James Dormer.

James Dormer early evinced strict attention to duty, zeal for the service, and personal bravery, which were recognised by the celebrated John, Duke of Marlborough, and in 1707 he was rewarded with the rank of Colonel in the army, being shortly after appointed Colonel of a newly raised Irish Regiment of Foot, with which he embarked for Spain in 1709, for service in the war of the Spanish succession. He distinguished himself at the battle of Saragossa in 1710, also in the advance to Madrid the same year, but he was surrounded when serving under General Stanhope in the mountains of Castile during the winter, and made a prisoner with his regiment at Brihuega. He appears to have been awarded £200 for his losses by pillage at Brihuega and at Bilbao on his way home on parole. He was exchanged, and on arriving in England was promoted to the rank of Brigadier-General. In 1712, after

[1] Lieutenant-General James Dormer was the founder and first colonel of the 14th Dragoons.

[2] New style, 13th August 1704.

James Dormer. the death of Lord Mohan in the notorious duel with the Duke of Hamilton, Dormer was appointed Colonel of Mohan's Regiment of Foot, which was disbanded the following year at the Peace of Utrecht (11th April 1713). In 1715, when an augmentation of the army took place, he was commissioned to raise a regiment of dragoons in the south of England, the present 'Fourteenth (King's) Hussars.' Dormer commanded a brigade during the Jacobite rising in Lancashire in 1715-16, and was engaged with the rebels at Preston, being wounded at the storming of the avenue leading to Lancaster. In 1720 he was transferred to the Colonelcy of the 6th Regiment of Foot, promoted to be Major-General in 1727, and to be Lieutenant-General in 1735. About 1727-28 he was Envoy-Extraordinary at Lisbon, where he had a dispute with Mr. Thomas Barnett, the British Consul.

In 1738 Dormer was rewarded with the Colonelcy of the 1st troop of Horse Grenadier Guards, which he retained till his death. In 1740 he was Governor of Hull. He died in 1741 at Crendon, Bucks. Lieutenant-General Dormer was a member of the Kit-Cat Club, and collected a fine library: he was an acquaintance of Swift. He must not be confused with Colonel Charles Dormer, who commanded Lord Essex's Dragoons, now 'The Queen's Own' 4th Hussars, and who fell at the head of that regiment at the battle of Almanza in 1707.

Lieutenant-General James Dormer died unmarried and bequeathed his Chersley estate to his cousin, Sir Clement Cottrell, Knight (afterwards Cottrell-Dormer), Master of the Ceremonies to George II. The portrait is from an oil painting in possession of his descendant, Captain Charles Cottrell-Dormer of Rousham Hall, Oxford, late of the 13th Hussars.

LIEUTENANT-GENERAL CLEMENT NEVILLE

APPOINTED COLONEL OF THE 14TH DRAGOONS, APRIL 9, 1720.

Clement Neville. Clement Neville entered the army at the Revolution in 1688, and served under King William III. in the Netherlands. He also served in the wars of Queen Anne; was promoted to the Lieutenant-Colonelcy of Munden's Regiment of Foot, with which he served in Spain, and signalised himself at the battle of Saragossa in 1710, but was made prisoner at Brihuega. He was shortly afterwards exchanged, and at the close of the campaign, on 15th November 1711, he was rewarded with the rank of Colonel in the army. At the Peace of Utrecht, which was on 11th April 1713, his regiment was disbanded, and in the summer of 1715 he was appointed the first Lieutenant-Colonel of the 13th Dragoons. In 1720 he was promoted to the Colonelcy of the 14th Dragoons, from which, on 27th June 1737, he was removed to the 8th Dragoons, and in 1739 he was promoted to the rank of Major-General. In 1740 he was appointed Colonel of the 6th Horse, now the 5th (Princess Charlotte of Wales's) Dragoon Guards, and in 1743 he was promoted to the rank of Lieutenant-General. He died in 1744. His photogravure is from an old mezzo-tint engraving by John Brooks from a painting by Hoar, representing him in Lieutenant-General's uniform, time of George II. Clement Neville.

LIEUTENANT-GENERAL ARCHIBALD HAMILTON

APPOINTED COLONEL OF THE 14TH DRAGOONS, JUNE 27, 1737.

Archibald Hamilton entered the army in November 1688, and at the Revolution he adhered to the Prince of Orange, afterwards King William III., under whose command he served in Flanders. In the reign of Queen Anne he served in Portugal and Spain, and his regiment (Mountjoy's Foot) was nearly annihilated at the battle of Almanza in 1707, where he was taken prisoner. This corps was subsequently incorporated into other regiments, and the officers sent home to recruit, and at the Peace of Utrecht, 11th April 1713, it was disbanded. In the summer of 1715 he was appointed Lieutenant-Colonel of the 11th Dragoons, and in May 1732 he was promoted to the Colonelcy of the 27th Foot, from which he was removed in 1737 to the Colonelcy of the 14th Dragoons. He became Major-General 1st January 1742, and Lieutenant-General about 1748. He died the following year. In St. Peter's Church, Dublin, there is a marble monument erected to Lieutenant-General Archibald Hamilton, bearing the family arms and crest borne by the Killyleagh and Abbotstown branches of the Irish house of Hamilton, the present representatives of which are Colonel Gawne W. Rowan Hamilton of Killyleagh and Lord HolmPatrick of Abbotstown.[1] The inscription records that Lieutenant-General Hamilton was at the siege of Derry in 1688, Archibald Hamilton.

[1] Colonel H. B. Hamilton (page 505) and Colonel G. H. C. Hamilton (page 509) both belong to the Abbotstown branch of the Irish house of Hamilton, and use the same arms and crest as are on this monument, but there is nothing to prove the lineage and descent of Lieutenant-General Archibald Hamilton in any pedigree of these families so far as can be ascertained; he may therefore have belonged to one of the other branches of the Irish Hamiltons, which are likewise descended from the Rev. Hans Hamilton, Vicar of Dunlop, the ancestor of the Earls of Clanbrassill (extinct).

Archibald Hamilton.

where he distinguished himself, and that he served abroad with reputation during all the wars of William III. and Anne, and died on the 15th July 1749.

FIELD-MARSHAL THE RIGHT HONOURABLE JAMES, LORD TYRAWLEY

APPOINTED COLONEL OF THE 14TH DRAGOONS, JULY 24, 1749.

The Rt. Hon. James, Lord Tyrawley.

The Honourable James O'Hara, Lord Kilmaine, and second Lord Tyrawley, was born in the year 1690. He was the only son of Sir Charles O'Hara, first Lord Tyrawley, of County Mayo, Ireland, and was appointed Lieutenant in the Royal Regiment of Fusiliers (commanded by his father) on the 15th March 1703. In 1706 he proceeded with his regiment to the relief of Barcelona. In the following year he served on the staff of the army in Spain, and was wounded at the battle of Almanza (SE. of Spain) on 25th April 1707, where, it is said, he was instrumental in saving the life of the British Commander, the Earl of Galway. Afterwards he served under the Duke of Marlborough, and was severely wounded in the wood of Tasniare, near Tournai, during the battle of Malplaquet, 11th September 1709. He served for several years at Minorca, and on 29th January 1713 obtained the Colonelcy of his regiment, the Royal Fusiliers (9th Regiment of Foot), in succession to his father. In 1722 Colonel O'Hara was awarded an Irish Peerage (Baron Kilmaine), and upon the decease of his father in 1724 he succeeded to the dignity of Baron Tyrawley. In 1727 he was appointed A.D.C. to His Majesty King George II. On 23rd November 1735 his lordship became Brigadier-General, and he was promoted Major-General on 2nd July 1739. In August 1739 he was removed from the Royal Fusiliers to the Colonelcy of the 5th Horse (now the 4th Royal Irish Dragoon Guards). In March 1743 Lord Tyrawley became Lieutenant-General, and in April of the same year he obtained the Colonelcy of the 2nd troop of Horse Grenadier Guards, from which, in 1745, he was removed to the 3rd troop of Life Guards, which gave him the privilege of taking the Court duty of Gold Stick.

In 1746, when King George II. had resolved to disband the 3rd and 4th troops of Life Guards, Lord Tyrawley was removed to the 10th Foot, and in 1749 he obtained the Colonelcy of the 14th Dragoons. On 27th November 1752 he was removed to the 3rd Dragoons, and in 1755 to the 2nd or Coldstream Regiment of Foot Guards. He became Governor of Portsmouth, 1st May 1759, was promoted General on 7th March 1761, and advanced to the rank of Field-Marshal, 10th June 1763. The Rt. Hon. James, Lord Tyrawley.

On the 28th January 1728 he was appointed Envoy-Extraordinary to the Court of Portugal, where he remained as Ambassador till 1741. He was very popular, and on his departure received from the King of Portugal fourteen bars of gold. He returned to England with three wives and fourteen children (Walpole's letters), and at once gained a reputation for wit at the expense of Lords Bath and Grantham and the House of Commons. From November 1743 to February 1745 he was Ambassador-Extraordinary at the Court of Russia, and on his return he received the command of the 3rd troop of Life Guards as already noted above.

In 1752 he returned as Ambassador to Portugal, and was also Governor of Minorca till 1756, when he was sent out on the Gibraltar expedition. On 14th December 1757 he was president of the court-martial on Sir John Mordaunt, having been relieved at Gibraltar on 16th April 1757. He was also president of the court-martial on Lord George Sackville in 1760. In 1762, when a Spanish invasion of Portugal was threatened, Tyrawley was appointed Plenipotentiary and General of the English forces, but was soon superseded as being too old, and he returned to England in 1763. He was sworn of King George III.'s Privy Council, 17th November 1762. Lord Tyrawley had a seat at Blackheath, and married Mary, daughter of Lieutenant-General Sir W. Stewart, second Viscount Mountjoy, but left no legitimate issue. He died at Twickenham, 13th July 1773, and was buried at Chelsea Hospital.

LIEUTENANT-GENERAL LOUIS DEJEAN

APPOINTED COLONEL OF THE 14TH DRAGOONS, NOVEMBER 27, 1752.

Louis Dejean served many years in the 1st troop of Horse Grenadier Guards, in which corps he rose to the rank of Lieutenant-Colonel, and in 1746 he was promoted to the Colonelcy of 37th Regiment of Foot, which was subsequently disbanded. On 16th October 1752 he was appointed to the Colonelcy of the 14th Dragoons, was promoted to the rank of Major-General in 1756, and removed on 5th April 1757 to the 3rd Irish Horse, now 6th Dragoon Guards (Carabiniers). In 1759 he was advanced to the rank of Lieutenant-General. Louis Dejean.

Louis Dejean.

He died at Dublin in 1764. His portrait is taken from an old mezzo-tint engraving kindly lent by S. M. Milne, Esq. The original painting was by Ph. Mercier, engraving by P. Faber.

FIELD-MARSHAL JOHN, DUKE OF ARGYLL

APPOINTED COLONEL OF THE 14TH DRAGOONS, APRIL 5, 1757.

John Campbell (Duke of Argyll).

John Campbell entered the army in the reign of King George II., served at the battle of Dettingen in Bavaria, fought on 26th June 1743, and in 1745 he was promoted to the Lieutenant-Colonelcy of the 54th Regiment, afterwards 43rd or Monmouthshire Light Infantry, with which corps he served a short period in the Netherlands. The rebellion breaking out in Scotland, he quitted Flanders, and in January 1746 he joined Lieutenant-General Hawley, with a thousand Argyllshire Highlanders, on the day of the unfortunate battle of Falkirk, 17th January 1746. He subsequently joined the Duke of Cumberland at Perth, and accompanied His Royal Highness to the North, and was present at Culloden, 16th April 1746. He was promoted to the rank of Colonel, and appointed Aide-de-Camp to His Majesty King George II. in November 1755. In the following month he was nominated Colonel of the 54th Regiment, then first embodied, from which in 1757 he was removed to the 14th Dragoons, and in 1759 he was promoted to the rank of Major-General and appointed Colonel of the Argyllshire Fencibles. In January 1761 he became Lieutenant-General. On the decease of his uncle Archibald, third Duke of Argyll, in 1761, his father, General John Campbell of the Scots Greys, succeeded to that title, and Lieutenant-General Campbell of the 14th Dragoons became Marquis of Lorne. In the following year he was appointed Commander-in-Chief in Scotland, and in 1765 he was removed to the Royal Regiment of Foot. He was again appointed Commander-in-Chief in Scotland in 1767, and in 1770 he became fifth Duke of Argyll. In 1778 he became General, was removed to the 3rd Foot Guards in 1782, and advanced to the rank of Field-Marshal in 1796. He became a great improver of land, was distinguished for his many social virtues, public and private, being highly honoured and respected in society, and he died lamented on 24th May 1806, aged eighty-two years. The photogravure of him, in General's uniform, is from a picture by Gainsborough, belonging to the late Duke of Argyll, recently in Argyll Lodge, Kensington, photographed by His Grace's kind permission in 1899.

GENERAL CHARLES FITZROY, LORD SOUTHAMPTON

APPOINTED COLONEL OF THE 14TH DRAGOONS, SEPTEMBER 11, 1765.

Charles Fitzroy (Lord Southampton).

Charles Fitzroy, son of Lord Augustus Fitzroy, grandson of Charles, second Duke of Grafton, and brother of Augustus Henry, Duke of Grafton, was born 25th June 1737. His mother was Elizabeth, daughter of Colonel William Cosby. He was appointed Ensign 1st Foot Guards, 1752; Lieutenant, 1756; promoted Captain, with rank of Lieutenant-Colonel, 1758; and appointed Colonel of the 119th Prince's Own Regiment of Foot in 1762. This regiment was disbanded in 1763. In 1765 he became Colonel of the 14th Dragoons, succeeding the Marquis of Lorne. He was present at the battle of Minden, 1st August 1759, as aide-de-camp to Prince Ferdinand of Brunswick, and carried the famous order for the advance of the cavalry, which Lord George Sackville (afterwards Sackville-Germain) neglected. He gave evidence at the court-martial which tried Sackville. In 1760 he was appointed groom of the bedchamber to the king, and resigned in 1762. He was engaged at the battle of Kirchderkern, 15th July 1761. On 20th October 1772 he was removed from the 14th Dragoons to the Colonelcy of the 3rd King's Own Dragoons. On 17th October 1780 he was raised to the Peerage as Baron Southampton, and on 27th December of the same year he became groom of the stole to the Prince of Wales. He moved the address to the throne in the House of Lords at the opening of Parliament in 1781, and made a speech on 18th February 1782 on Lord Carmarthen's motion protesting 'against the elevation to the Peerage of any person labouring under a heavy censure of a court-martial'—a motion aimed at Lord George Sackville-Germain, who had just been created Viscount Sackville of Drayton, denying that, as had been alleged or insinuated, the court-martial in question had been animated by a factious spirit. He also spoke, without definitely committing himself to either side, on the Regency Bill, 16th February 1789.

He became Colonel in 1762, Major-General in 1772, Lieutenant-General in 1777, and General on 25th October 1793. He married, on 27th July 1758, Anne, daughter of Sir Peter Warren, K.B., vice-admiral of the red, by whom he had issue nine sons and seven daughters. His eldest son, George Ferdinand, succeeded him. Lord Southampton was lord of the manor of Tottenham Court, Middlesex, and had his principal seat at Fitzroy Farm, near Highgate, the grounds of which he had laid out in the artificial style then in vogue. He died in 1797.

Charles Fitzroy (Lord Southampton).

His photogravure is from an old mezzo-tint engraving of a picture by Sir J. Reynolds, *circa* 1790, kindly lent by S. M. Milne, Esq.

LIEUTENANT-GENERAL DANIEL WEBB

APPOINTED COLONEL OF THE 14TH DRAGOONS, OCTOBER 20, 1772.

Daniel Webb.

Daniel Webb was born in 1703. He was an officer in the Eighth, now the Seventh, 'Princess Royal's' Dragoon Guards, for many years, at a period when the corps acquired a high reputation for discipline, efficiency, and valour, and was designated 'Ligonier's Horse.' He rose to the rank of Major in the regiment; commanded a squadron at the battle of Dettingen in 1743, where the regiment was highly distinguished under the eye of His Majesty King George II. He also commanded a squadron at the battle of Fontenoy in 1745. A few days after this battle he was promoted to the Lieutenant-Colonelcy in succession to Lieutenant-Colonel Francis Ligonier, who became Colonel of the 48th Foot. Lieutenant-Colonel Webb performed the duties of commanding officer of the Eighth Horse until November 1755, when he was rewarded with the Colonelcy of the 48th Foot; and, in 1759, he was promoted to the rank of Major-General. He served in Germany under Prince Ferdinand of Brunswick, and commanded a brigade of cavalry at the battle of Warbourg in 1760. In 1761 he was promoted to the rank of Lieutenant-General. In 1766 he was removed to the 8th Foot, and in 1772 to the 14th Dragoons, the command of which corps he retained until his decease in the following year (1773).

GENERAL GEORGE WARDE

APPOINTED COLONEL OF THE 14TH DRAGOONS, NOVEMBER 11, 1773.

George Warde.

George Warde was born in 1727, and for many years held a commission in the 11th Dragoons, becoming Major of that regiment in 1756. In 1758 he was promoted to the Lieutenant-Colonelcy of the 4th Dragoons, and he brought that corps into so excellent a state of discipline and efficiency, that he received the expression of the high approbation of His Majesty King George III. on several occasions, when His Majesty reviewed the regiment. He was promoted to the rank of Colonel in 1772. In the following year the king rewarded him with the Colonelcy of the 14th Dragoons; and four years later, in 1777, he was advanced to the rank of Major-General. In 1778 he was removed to the 1st Irish Horse, now the 4th (Royal Irish) Dragoon Guards, and he was promoted to the rank of Lieutenant-General in 1782. In 1792 he was appointed Commander-in-Chief in Ireland, and while in that country he devoted much of his time in bringing his regiment into a perfect condition for active service. He possessed sound ideas of what cavalry ought to be; he had an aversion to slow movements, and although nearly seventy years of age, he exercised his regiment five times a week, often leading it across the country, over hedge and ditch, to the astonishment of every one. In 1796 he was promoted to the rank of General. He was celebrated for philanthropy, and is represented by historians as 'a man of inviolable disinterested integrity, public and private; and the bestower of benefactions scarcely less secret than extensive.' He died in March 1803. George Warde.

GENERAL SIR ROBERT SLOPER, K.B.

APPOINTED COLONEL OF THE 14TH LIGHT DRAGOONS, APRIL 2, 1778.

Robert Sloper was appointed by King George II. to a commission in the 10th Dragoons, and in December 1755, at the augmentation of the army, His Majesty promoted him to the Majority of that regiment. In February 1759 he was promoted to the Lieutenant-Colonelcy of the 1st Dragoon Guards, and he held the command of that regiment during the remainder of the seven years' war in Germany, where he was repeatedly commended by Prince Ferdinand of Brunswick and other general officers under whose command he served. While under his orders, the 1st Dragoon Guards were preserved in a high state of discipline and efficiency. He was promoted to the rank of Major-General in 1771, and in 1778, King George III. rewarded him with the Colonelcy of the 14th Light Dragoons. He was promoted to the rank of Lieutenant-General on 30th November 1782, and to that of General on 3rd May 1796. In the following year he was removed to the 4th Dragoons. He was further rewarded, on 6th June 1788, with the dignity of a Knight Companion of the Most Honourable Order of the Bath, and was appointed Governor of Duncannon Fort. From 1784 to 1786 Lieutenant-General Sloper was Commander-in-Chief in India, and he died in the year 1802. Sir Robert Sloper.

GENERAL JOHN WILLIAM, EARL OF BRIDGEWATER

APPOINTED COLONEL OF THE 14TH LIGHT DRAGOONS, JUNE 1, 1797.

John William Egerton (Earl of Bridgewater).

John William Egerton (eldest son of the Reverend John Egerton, afterwards Lord Bishop of Durham) was appointed Cornet in the 7th Dragoons in January 1771; he became Captain in 1776, and was removed as Major, in 1779, into the 22nd Light Dragoons, from which he was removed to the 20th Light Dragoons in 1781. In 1782 he was promoted to the Lieutenant-Colonelcy of the 21st Light Dragoons, which corps was disbanded the following year in consequence of the termination of the American War. In 1790 he was appointed to the Lieutenant-Colonelcy of the 7th Light Dragoons, was promoted to the rank of Colonel in 1793, and to that of Major-General in 1795. He served some time on the staff in Ireland, and was removed to the eastern district of England in 1796. He appears to have been the general officer under whose directions the 14th Light Dragoons were dismounted at Clonmel in 1795, and embarked shortly afterwards at Waterford for Bristol, under orders for service in the West Indies. Then again we find him as Major-General on the staff at Chelmsford, when the Fourteenth, in very diminished numbers, arrived there from St. Domingo in 1797, in which year His Majesty conferred on him the Colonelcy of the regiment, and he was responsible for the remounting of the regiment and the refilling of its ranks on that occasion. On 29th April 1802 he was promoted to the rank of Lieutenant-General. On the decease of his cousin, Francis, third Duke of Bridgewater, in 1803, he succeeded to the title of Earl of Bridgewater, and in 1812 he was promoted to the rank of General. He retained the Colonelcy of the 14th Light Dragoons twenty-six years, and was particularly proud of the high reputation the regiment had acquired during the Peninsular War. He died in 1823.

The photogravure of him is from a portrait by H. Edridge, in the uniform of the 14th Light Dragoons, taken in 1805, with the Prussian Eagle conspicuous on the helmet. This drawing was kindly placed at the service of the author by the Right Honourable A. W. B., third Earl Brownlow, a descendant of Lord Bridgewater.

P.W. GENERAL SIR JOHN ORMSBY VANDELEUR, G.C.B.

APPOINTED COLONEL OF THE 14TH LIGHT DRAGOONS, OCTOBER 28, 1823.

John Ormsby Vandeleur was appointed Ensign in the 5th Regiment of Foot, 29th December 1781; Lieutenant in 67th Regiment, 21st July 1783; Lieutenant in 9th Regiment, 1788; Captain, 7th March 1792; removed to the 8th Light Dragoons, October 1792; Major, 1st March 1794; Lieutenant-Colonel, 1st January 1798; Colonel, 25th April 1808; Major-General, 4th June 1811; Lieutenant-General, 19th July 1821; General, 28th June 1838. From April 1794 to December 1795 he was employed in Holland, and present in the actions which took place under the Duke of York on 2nd and 18th May and on 1st June 1794-95, besides several minor affairs From August 1796 to October 1802 he was employed at the Cape, and from October 1802 to July 1806 he served in the East Indies. He served the campaigns there of 1803, 1804, and 1805, commanding the 8th Light Dragoons, and part of the time commanded a brigade of cavalry under General Lord Lake. He was present at the battles of Leswarree, Futtyghur, and Afzulghur, and at the sieges of Agra and Bhurtpore. At Leswarree, 1st November 1803, his brigade turned the enemy's left flank, charged and took 2000 prisoners, besides cutting many to pieces. In November 1804, at Futtyghur, the Mahratta chief Holkar was surprised and defeated, and on both these occasions Brigadier-General Vandeleur received the thanks of Lord Lake for his services. At Afzulghur, on 2nd March 1805, the Mahratta chief Ameer Khan was defeated, and Brigadier-General Vandeleur commanded the second line of the cavalry of the Bengal army under Major-General Smyth, consisting of the 8th Light Dragoons and 3rd and 5th Bengal Cavalry, which made a successful charge and defeated the enemy with great loss, Captain Dean's squadron of 8th Light Dragoons retaking the artillery which had been lost when the first line (24th and 25th Light Dragoons and a Bengal regiment of cavalry) was repulsed. In 1807 he commanded the 19th Light Dragoons. In 1811 General Vandeleur was given the command of a brigade as Major-General in the Peninsular War. He served with the Light Division, and was wounded at Ciudad Rodrigo, 19th January 1812, when leading the division to the breach at the time Major-General Crauford fell mortally wounded. He was with the Light Division at the battles of Salamanca and Vittoria in command of his brigade, which captured 300 prisoners a few days before the latter

Sir John Ormsby Vandeleur.

Sir John Ormsby Vandeleur.

battle, having intercepted and cut off a French brigade, and forced the remainder to disperse in the mountains.

In July 1813 he was given the command of a brigade of cavalry attached to the column under Lieutenant-General Lord Lynedoch, and afterwards under Lord Niddry, and participated in all the operations of that column. At the close of the Peninsular War he was directed by the Duke of Wellington to conduct one of the divisions of the British cavalry and artillery from Bordeaux to Calais, and in October 1814 was appointed to the staff of the army in Flanders, and served at the battle of Waterloo in command of the 4th Brigade of Cavalry, in which were the 11th, 12th, and 16th Light Dragoons. He afterwards commanded the whole of the British cavalry from the time that the Marquis of Anglesey was wounded till King Louis XVIII. entered Paris. Sir John Vandeleur received the Grand Cross of the Order of the Bath, a cross for Ciudad Rodrigo, and the battles of Salamanca, Vittoria, and Nive; was a knight of the second class of the Russian Order of St. Wladimir, and a commander of the Bavarian Order of Maximilian Joseph. On 12th January 1815 he was appointed Colonel of the 19th Light Dragoons, Colonel of the Fourteenth in 1823, and Colonel of the 16th Light Dragoons (Lancers) in 1830. He died in 1849. His portrait, in Lieutenant-General's uniform, is from an engraving by Villain, kindly given by Mrs. Frank E. Vandeleur, whose husband was a relative of Sir John's.

P.W. GEN. SIR EDWARD KERRISON, BART., G.C.H., K.C.B.

APPOINTED COLONEL OF THE 14TH LIGHT DRAGOONS, JUNE 18, 1830.

Sir Edward Kerrison, Bart.

Edward Kerrison was born in 1774. He was the only son of Matthias Kerrison, Esquire, of Hexne Hall, near Bungay, Norfolk. He became Cornet in the 6th Dragoons, 23rd June 1796; Lieutenant, 1st February 1798; Captain, 18th October 1798; Captain in the 7th Hussars, 1798; Major, 12th May 1803; Lieutenant-Colonel, 4th April 1805; Colonel, 4th June 1813; Major-General, 12th August 1819; Lieutenant-General, 10th January 1837; and General, 11th November 1851.

In 1799 Captain Kerrison, as he then was, served at the Helder in Holland, and was present at the battles of the 19th September, 2nd and 6th October. In 1808-9 he was in the campaign under Sir John Moore, and was present at Corunna, 16th January 1809. He commanded the 7th Hussars in the Peninsular, French, and Belgian campaigns of 1813, 1814, 1815, being present at the battles of Orthes, 27th February 1814; Toulouse, 10th April 1814; and Waterloo, 18th June 1815. He was also at the actions of Sauveterre and the passage of the Oleron in the Pyrenees. Sir Edward Kerrison, Bart.

In Spain, on the 25th December 1808, when engaged with the enemy on the plains of Leon, he was severely wounded, his arm being broken in two places, and subsequently at the battle of Waterloo he received a slight wound and his horse was shot under him. At the battle of Orthes, the charge headed by Major-General Lord Edward Somerset, in which Colonel Kerrison with the 7th Hussars took the chief part, was highly commended by the Duke of Wellington in his despatches. At Waterloo, after being wounded and his horse shot under him, he continued with his regiment, and took part in the occupation of Paris. On his return to England after the war, Kerrison was appointed a Knight Commander of the Bath. He was created a Baronet on the 27th July 1821. On 20th October 1813 he married Mary Martha, daughter of Alexander Ellice, Esquire, of Pittencrieff, Fifeshire, N.B. His only son, Edward Clarence Kerrison, was born in 1821, and became the second Baronet on the death of his father. His second daughter, Emily Harriet, became in 1834 the wife of Philip Henry, Earl Stanhope, the well-known historian: she died in 1873. From 1812 to 1818 Sir Edward Kerrison was M.P. for Shaftesbury in the Conservative interest; M.P. for Northampton, 1818 to 1824; M.P. for Eye, 1824 to 1852. He died at his London house in Great Stanhope Street on the 9th March 1853. He also had a residence at Wick, Sussex. Sir Edward Kerrison received the gold medal for Orthes, and the silver war medal with two clasps for Sahagun and Benevente, and numerous other distinctions. He was a Knight Commander of the Bath, and a Knight Grand Cross of the Order of Hanover.[1] The picture of Sir Edward Kerrison was taken when he was a colonel, in the uniform of the 7th Hussars, and is from an engraving by W. C. Edwards, published 1818, from a painting by M. A. Shee, Esq., R.A.

P.W. GENERAL THE HON. SIR HENRY MURRAY, K.C.B.

APPOINTED COLONEL OF THE 14TH (KING'S) LIGHT DRAGOONS, MARCH 18, 1853.

Henry Murray, fourth son of David, second Earl of Mansfield, by Louisa, daughter of Charles, ninth Baron Cathcart, and Countess of The Hon. Sir Henry Murray.

[1] The Order of the Guelphs.

The Hon. Sir Henry Murray.

Mansfield in her own right, was born 6th August 1784, entered the army as Cornet, 16th May 1800; became Lieutenant, 11th June 1801; Captain, 24th August 1802; Major, 26th March 1809; Lieutenant-Colonel, 2nd January 1812; Colonel, 22nd July 1830; Major-General 28th June 1838; Lieutenant-General, 11th November 1851; General, 6th February 1855. He served in Naples, Sicily, and Calabria in 1800-7. He was Aide-de-Camp to his uncle, Lord Cathcart, Commander of the Forces in Ireland in 1805; accompanied the expedition to Egypt in March 1807, and was present as an Aide-de-Camp at the attack on Alexandria, siege and storming of Rosetta, and on every other occasion when our troops were engaged. He served at Walcheren in 1809, including the siege and surrender of Flushing. He went with the 18th Hussars to the Peninsula in January 1813, and was present at the crossing of the Eslar. He commanded the Eighteenth at the action of Morales de Toro, in support of the 10th Hussars, but was injured in the knee on this occasion, and had to proceed to England. Served in the campaign of 1815, including the battle of Quatre Bras. He commanded the rear regiment of the column during the retreat of the following day, and at the battle of Waterloo he led the 18th Hussars in the brilliant charge of Sir Hussey Vivian's Brigade at the conclusion of the action. In 1842 Major-General Murray was appointed to the command at Limerick, and afterwards at Plymouth, where he remained till 1852, and was there during the riots of 1848. Sir Henry Murray was a Commissioner of Chelsea Hospital, and of the Royal Military College at Sandhurst.

His regimental career was a very varied one. He first became Cornet in the 16th (Queen's) Light Dragoons, then Lieutenant, 60th Regiment; then exchanged to 10th (Prince of Wales's) Light Dragoons; then became Captain-Lieutenant in the 20th Light Dragoons, and Captain when that rank (Captain-Lieutenant) was abolished in 1802. He was next appointed Major in the 26th (Cameronians) Regiment. Later he exchanged to the 18th Hussars, became its Lieutenant-Colonel in 1812, and commanded it for many years.

General Murray became Knight Commander of the Bath in 1860, and he died the same year on the 29th July. There is an interesting inscription to his memory on a monument in the Citadel Church at Devonport.

His photogravure, which represents him when quite young as an officer in the 18th Hussars, is from a pencil drawing (face coloured) by Cosway, belonging to Sir Henry's daughter, Miss Murray of Wimbledon Lodge, who kindly supplied it.

P.W. LIEUT.-GENERAL WILLIAM BECKWITH, K.H.,

APPOINTED COLONEL OF THE 14TH (KING'S) LIGHT DRAGOONS, NOVEMBER 12, 1860.

William Beckwith.

William Beckwith became Cornet, 16th Light Dragoons, 1813; Lieutenant, 1815; Captain, 1822; Major, 1828; Lieutenant-Colonel, 1833; Colonel, 1846; Major-General, 1854; Lieutenant-General, 26th November 1861; and General, 28th October 1869. He served in the Peninsular War with the 16th Light Dragoons from July 1813 to the end of the war in 1814, including the battles of the Nivelle, 9th, and the Nive, 10th December 1813, for which he had the war medal and two clasps. He was through the campaign of 1815, including the retreat on 17th June and the battle of Waterloo on 18th June 1815, and received the Waterloo medal. He served in the 14th Light Dragoons from 1817 to 1833, when he left on promotion to a half-pay Lieutenant-Colonelcy on 6th December 1833. In 1831 Major Beckwith, as he then was, greatly distinguished himself when in command of a squadron of the Fourteenth at the Bristol riots, and he was awarded the dignity of a Knight of the Hanoverian Order of the Guelphs in that year.

He died on 23rd February 1871.

GENERAL HENRY RICHMOND JONES, C.B.

APPOINTED COLONEL OF THE 14TH (KING'S) HUSSARS, FEBRUARY 24, 1871.

Henry Richmond Jones.

Henry Richmond Jones was appointed Cornet in the 6th Dragoon Guards (the Carabiniers) in 1825; Lieutenant, 1826; Captain, 1830; Brevet-Major, 1846; Major, 1850; Lieutenant-Colonel, 1851; Colonel, 28th November 1854; Major-General, 2nd April 1865; Lieutenant-General, 6th December 1873; and General, 1st October 1877. He commanded the Carabiniers in the Crimea from 14th August 1855, including the battle of the Tchernaya and the siege and fall of Sebastopol, for which he received the Crimean war medal with clasp, fifth class of the Medjidie, and Turkish medal. During the Indian Mutiny campaign of 1857-58 he commanded General Penny's column after that officer was killed by the first shot fired by the enemy in the action of Kirkrowlie, on 30th April 1858, when he captured a gun; was in command of a brigade of cavalry at the capture of Bareilly (Oude), 7th May 1858. He commanded the left column of Lord Clyde's force in the attack on Benhi Madhu's army at Dunderkéra, and the cavalry with Lord Clyde's troops in the Trans-gogra campaign, including the affairs at Magedia, Chandal, Baukee,

Henry Richmond Jones.

and the pursuit of the rebels to the Raptee. He received for these services the Indian Mutiny medal and the Companionship of the Order of the Bath.

General Jones served continuously for thirty-seven years in the Carabiniers (1825-62), became full Colonel of the Fourteenth in 1871, and of the Carabiniers on 15th May 1873. He held the latter appointment till his death in October 1881, having resided for many years at Leamington. His portrait is from a photograph kindly given by the General's only child, Mrs. Robert O. Milne of Leamington.

GENERAL JOHN WILKIE

APPOINTED COLONEL OF THE 14TH (KING'S) HUSSARS, MAY 15, 1873.

John Wilkie.

John Wilkie became Cornet, 11th May 1838; Lieutenant, 29th May 1840; Captain, 17th May 1844; Major, 27th February 1852; Lieutenant-Colonel, 10th February 1854; Colonel, 28th November 1854; Major-General, 6th March 1868; Lieutenant-General, 6th October 1876; and General, 21st December 1878.

He commanded the 10th Hussars in the Crimean campaign from 17th April 1855, including the capture of Tchorgaun, battle of the Tchernaya, siege and fall of Sebastopol, for which he received the medal with clasp, fifth class of the Medjidie, and Turkish medal. He died in 1882.

GENERAL CHARLES WILLIAM THOMPSON, K.S.F., J.P.

APPOINTED COLONEL OF THE 14TH (KING'S) HUSSARS, MAY 18, 1882.

Charles William Thompson.

Charles William Thompson, second son of General Thomas Perronet Thompson and his wife, Anne Elizabeth, daughter of the Rev. Thomas Barker of York, was born at Bombay, 21st November 1815, and received his first commission, 26th February 1836, as Ensign in the 81st Regiment. He served as Captain in the British Legion in Spain under Sir de Lacy Evans, previous to his entering the army, and was engaged at Arlaban on 16th, 17th, and 18th January 1835, at Hernani on 30th August 1835, and in the action of 5th May 1836, before San Sabastian, where he was severely wounded in the hip and hand. For these services he received the Spanish medal, and was made Knight, first class, of the Order of San Fernando. In 1841 he came as Lieutenant from the 81st Regiment to the 14th (King's) Light Dragoons, and became Captain in 1848. He was engaged with his regiment at Ramnuggur on 22nd November 1848, where his horse was wounded; also at Chillianwallah, 13th January 1849, where he commanded a squadron, and at Goojerat on 21st February 1849. He was present with the Fourteenth at the crossing of the Chenab, the Jhelum, and the Indus; at the surrender of the Sikh army at Rawul Pindhi, the capture of the bridge of boats at Attock, and the pursuit of the Afghans to Peshawur in April 1849. For this campaign he received the Punjaub war medal and two clasps, and on the 4th October 1880 he was awarded the good service pension. In 1849 he exchanged as Captain with Captain Pearson Scott Thompson into the 7th (Princess Royal's) Dragoon Guards, subsequently commanding this regiment for ten years, and returning with it from India in 1867, when he retired on half-pay as Colonel, living for a time at Dover, and afterwards settled at Wethersfield Place in Essex, where he was made a Justice of the Peace. He died there, 3rd October 1896, in his eightieth year, greatly and deservedly respected by all. He became Major, 1855; Lieutenant-Colonel, 1857; Colonel, 1862; Major-General, 1868; Lieutenant-General, 1880; Honorary General, 1881. In May 1882 he was appointed Colonel of the 14th (King's) Hussars, his old regiment, for which he had the deepest regard, as both he and his father had been engaged in it on active service,[1] and the Fourteenth held him in the highest respect. His last request was to be shrouded in his old military cloak, which, he was proud to relate, he had shared with a brother officer on the field of Goojerat when bivouacking during the night after the battle, and this wish was carried out. General Thompson was twice married, and left a numerous family. His second wife survived him. One son of the General's is a Captain in the 81st (Loyal North Lancashire) Regiment, and another is Major in the 7th Dragoon Guards, a detachment of which latter regiment carried the deceased General to his grave at Wethersfield.

Charles William Thompson.

LIEUT.-GEN. THE HON. CHARLES WEMYSS THESIGER

APPOINTED COLONEL OF 14TH (KING'S) HUSSARS, OCTOBER 4, 1896.

Charles Wemyss Thesiger, son of Frederick, the first Lord Chelmsford, Lord High Chancellor of England, was born 12th October 1831.

The Hon. Charles Wemyss Thesiger.

[1] General C. W. Thompson's father was Lieutenant in the 14th (Duchess of York's Own) Light Dragoons, 1812-14, and was present at the actions of Nivelle, Nive, Orthes, and Toulouse. He received the Peninsular War medal with four clasps, and was attached to the staff of Major-General (afterwards Sir Henry) Fane. He died a General in 1869, aged eighty-six years.

The Hon. Charles Wemyss Thesiger.

When just seventeen years of age, in 1848, he was given a cadetship in the then East India Company's service, and on 27th February 1849 was appointed Cornet in the 5th Madras Light Cavalry, with which regiment he served till 1853, when he was transferred to the 14th Light Dragoons. He served with the regiment in India both at Meerut and Kirkee, and on 17th June 1857 he went to the Inniskilling Dragoons. In 1858 he was appointed Aide-de-Camp to the Earl of Eglinton, Lord-Lieutenant in Ireland, but resigned that appointment the same year on his regiment being ordered to India. In 1860 he served in the China War as Aide-de-Camp to Brigadier-General Pattle (King's Dragoon Guards), commanding the cavalry brigade, and was present at the affairs of the 12th August and the 18th and 21st September 1860, at Taku Forts and Pekin, for which services he received the China War medal and two clasps. In 1867 he returned with his regiment (the Inniskilling Dragoons) to England, and commanded it as Lieutenant-Colonel from 1868 to 1878, when he retired on half-pay. On 3rd April 1878 he was appointed Inspecting Officer of Auxiliary Cavalry. In 1885, being promoted to the rank of Major-General, he was appointed to the command of the Curragh Brigade, and to the post of Inspector-General of Cavalry in Ireland. In 1890 he was advanced to the rank of Lieutenant-General, and awarded a pension of £100 a year for distinguished service. In 1892 he was appointed Colonel of the 5th (Royal Irish) Lancers, and in 1896 he was transferred to the 14th (King's) Hussars. He became Cornet, 1849; Lieutenant, 1853; Captain, 1858; Major, 1861; Lieutenant-Colonel, 1868; Colonel, 1873; Major-General, 1885; Lieutenant-General, 1890; retired, 1891.

LIEUTENANT-COLONELS OF THE 14TH (KING'S) HUSSARS

The first Lieutenant-Colonel of the regiment was

H. Killigrew.

HENRY KILLIGREW, appointed Lieutenant-Colonel and Captain of a troop in the newly raised regiment of Dormer's Dragoons on 22nd July 1715.

Sol. Rapin.

SOLOMON RAPIN became Lieutenant-Colonel in 1716 or 1717.

Wm. Boyle.

WILLIAM BOYLE became Lieutenant-Colonel in 1717.

B. Newcomin.

BEVERLY NEWCOMIN became Lieutenant-Colonel in 1720.

H. Desgrangues.

H. DESGRANGUES was Lieutenant-Colonel in 1731.

Wm. Wright.

WILLIAM WRIGHT was Lieutenant-Colonel, 7th February 1737. He was Lieutenant-Colonel of the regiment at Prestonpans, where

Sir John Cope's army was defeated by the rebels under the young Pretender on 21st September 1745, on which occasion Major Bowles of the 14th Dragoons greatly distinguished himself. Wm. Wright.

JAMES NORRIS became Lieutenant-Colonel in 1749. James Norris.

THOMAS ERLE was appointed Lieutenant-Colonel, 14th Dragoons, 4th September 1754; he remained at the head of the regiment as Lieutenant-Colonel till 15th July 1773, although he became a Major-General on 30th April 1770, and he was appointed Colonel of the 28th Foot, 13th July 1773. It is probable that he died in 1777, as his name does not appear in the Army List after that year. Thomas Erle.

MAJOR-GENERAL SIR JOHN BURGOYNE, BART.

APPOINTED LIEUTENANT-COLONEL OF THE 14TH DRAGOONS, JULY 15, 1773.

John Burgoyne was born in 1739, and was son of Sir Roger Burgoyne, Bart., of Sutton, Bedfordshire. He was cousin of Lieutenant-General the Right Honourable John Burgoyne, and entered the army at an early age. He served in the 7th Fusiliers, and was Major in the 52nd Regiment, becoming Lieutenant-Colonel of the 58th Foot in 1764, in Ireland, and afterwards of the 14th Dragoons in 1773. He became Colonel in 1777. Whilst commanding the regiment he is believed to have established the first code of 'Standing Orders' which the regiment ever possessed.[1] Sir John Burgoyne, Bart.

Sir J. Burgoyne commanded the regiment till 1781, when he was commissioned to raise the 23rd Light Dragoons for service in India, and having done so in Bedford, he proceeded with them to Madras, where he became Major-General in 1782, having previously, in 1780, succeeded his father in the baronetcy. It was during Sir John Burgoyne's command that the 14th Dragoons became 'Light Dragoons,' in 1776.

Sir John Burgoyne was under arrest in Madras for two years, having had a dispute with the East India Company. He was tried by court-martial, and honourably acquitted. He died in India in 1786. Sir John's son, Sir Montagu Burgoyne, who commanded the Scots Greys, brought an action against the Company at his father's death, and got very heavy damages. Sir John Burgoyne's portrait, in uniform of 23rd Light Dragoons, is from Romney's painting belonging to his great-grandson, Colonel Sir John Montagu Burgoyne, Bart., now of Sutton Park, Beds, who kindly allowed it to be photographed.

[1] See *ante*, pp. 36, 38, 50.

GENERAL GRICE BLAKENEY

APPOINTED LIEUTENANT-COLONEL, 14TH LIGHT DRAGOONS, NOVEMBER 19, 1781.

Grice Blakeney.

Grice Blakeney became Cornet in the 14th Dragoons in 1757-58; Lieutenant, 1761; Captain, 1767; Major, 1776; Brevet Lieutenant-Colonel, 1780; and succeeded Colonel Sir John Burgoyne, Bart., in the command of the regiment in 1781. He became Colonel, 1790; Major-General, 1794; Lieutenant-General, 1st January 1801; and General, 25th October 1809. He remained Lieutenant-Colonel of the Fourteenth several years after he became Major-General, until on the 29th April 1802 he was succeeded by Colonel Michel in the command.

It was during Colonel Blakeney's command that the regiment became (in 1798) the 'Duchess of York's Own Light Dragoons,' and up to 1790 the code of regimental 'Standing Orders' introduced by Colonel Sir John Burgoyne were adhered to, and very favourably noted by the Inspecting-Generals, especially by Lord Luttrell in 1785.

In 1799, at Canterbury, Major-General Wilford mentions that 'no established orders existed in the corps for its direction and guidance.'

Although General Grice Blakeney's name appears as the senior Lieutenant-Colonel of the Fourteenth in the Army Lists up to the year 1802, it is probable that after he became Major-General in October 1794, Major (Brevet Lieutenant-Colonel) Arthur Carter was practically the commanding officer till 1799, and from that date probably Lieutenant-Colonel Michel, who was the second Lieutenant-Colonel, took the command. During the campaign in St. Domingo (1795-97) the regiment out there was under command of Brevet Lieutenant-Colonel Arthur Carter, the Major. In his later years General Grice Blakeney was appointed Colonel of the 4th Royal Veteran Battalion, and remained so until 1814. He died in 1816.

GENERAL JOHN MICHEL

APPOINTED SECOND LIEUTENANT-COLONEL, 14TH (DUCHESS OF YORK'S OWN) LIGHT DRAGOONS, DECEMBER 5, 1799; COMMANDED THE REGIMENT, 1802-1803.

John Michel.

John Michel became Lieutenant-Colonel in the army, 24th August 1795; Colonel, 29th April 1802. He commanded the regiment, 29th April 1802-3, when he was placed on half-pay and succeeded by Lieutenant-Colonel Hawker. He entered the army 1781, and served at the siege of Fort St. Philip in Minorca, and was taken prisoner on its surrender in 1782. He was promoted Major-General, 25th October 1809; Lieutenant-General, 4th June 1814; and General, 10th January 1837. He died in 1844.

Walter L. Colls, Ph. Sc

Lieut-General Clement Neville,
Colonel of the 14th Dragoons,
1720-1737.

Walter L. Colls, Ph. Sc.

Lieutenant-General Louis Dejean,
Colonel of the 14th Dragoons.
1752-1757.

By permission of Mess[rs] Graves. Walter L. Colls, Ph. Sc.

General Charles Fitzroy, Lord Southampton,

Colonel of the 14th Dragoons,

1765-1772.

Walter L. Colls, Ph. Sc.

General John William Egerton, Earl of Bridgewater,
Colonel of the 14th (Duchess of York's Own) Light Dragoons,
1797-1823.

W. L. Colls. Ph. Sc.

General Sir Edward Kerrison, Baronet,
G.C.H. K.C.B.,
Colonel of the 14th (King's) Light Dragoons,
1830–1853.

Walter L. Colls, Ph. Sc.

General Henry Richmond Jones, C.B.,
Colonel of the 14th (King's) Hussars,
1871-1873.

Walter L. Colls. Ph. Sc.

Lieutenant General The Hon.ble C. W. Thesiger,

Colonel of the 14th (King's) Hussars.

1896.

Walter L. Colls, Ph.Sc.

General Sir Samuel Hawker, G.C.H.,
(Colonel of the 3rd Dragoon Guards)
Commanding 14th (Duchess of York's Own) Light Dragoons,
1803-1811.

Walter L. Colls, Ph. Sc.

Colonel Sir Felton Bathurst Hervey, Bart., C.B., K.H., A.D.C.,

Commanding 14th (Duchess of York's Own) Light Dragoons, 1811-1819,

(Sword of Honour presented by the Officers of the Fourteenth, 1812).

Walter L. Colls. Ph. Sc.

Lieutenant-Colonel Wm Havelock, K. H.,

Commanding 14th (King's) Light Dragoons,

1845-1848.

Walter L. Colls, Ph. Sc.

Colonel (afterwards Major General) A. Scudamore, C.B.,

Commanding 14th (King's) Hussars,

1861.

Walter L. Colls, Ph. Sc.

Colonel (afterwards General)
Henry E. Doherty, C.B.,
Commanding 14th (King's) Light Dragoons.
1850–1857.

Walter L. Colls, Ph. Sc.

Colonel (afterwards Major-General)
Sir William Russell, Baronet, C.B.,
Commanding 14th (King's) Hussars,
1861-1854.

Walter L. Colls. Ph. Sc.

Colonel (afterwards Major-General)
William Arbuthnot, C.B.,
Commanding 14th (King's) Hussars,
1876-1881.

Walter L. Colls. Ph. Sc.

Colonel The Hon. G. H. Gough, C.B.,

Commanding 14th (King's) Hussars.

1891-1896.

Walter L. Colls. Ph. Sc.

Lieutenant-Colonel G. H. C. Hamilton,

Commanding 14th (King's) Hussars.

1899.

GENERAL SIR SAMUEL HAWKER, G.C.H.

(*Colonel of the 3rd Dragoon Guards.*)

APPOINTED THIRD LIEUTENANT-COLONEL, 14TH LIGHT DRAGOONS, JUNE 12, 1800; COMMANDED THE REGIMENT, 1803-16.

Sir Samuel Hawker.

Samuel Hawker was born in 1763. He entered the army as a Cornet in the 16th Light Dragoons, 15th May 1779, and embarked with that regiment for the Netherlands, 24th April 1793, where he saw much active service. He commanded the picquet at the storming of the breach at Valenciennes, and was in temporary command of the regiment when senior Captain from the close of the year 1793. He led the 16th Light Dragoons, under the personal direction of H.R.H. the Duke of York, in a charge upon a nine-gun battery covered by cavalry, which was 'executed with great gallantry.'[1] This regiment under Captain Hawker was highly spoken of in the despatches of the Duke of York, commanding the troops. On 10th May 1794 the army was attacked at Tournay by the French, who were 30,000 strong. On this occasion the 16th Light Dragoons were led by Captain Hawker,[2] and charged a field battery of eight guns supported by cavalry and infantry, when Captain Hawker's horse was killed under him, and he was himself severely wounded in the chest. The 16th Light Dragoons returned to England in 1796, when Captain Hawker was personally thanked for his services by H.M. King George III., in the presence of the other members of the royal family.

He became Major, Sussex Fencible Cavalry, 6th April 1797. On 6th June 1799 he was appointed to the Lieutenant-Colonelcy of the Sussex Fencible Cavalry, and was removed to the 14th Light Dragoons as third Lieutenant-Colonel on 12th June 1800. On 25th April 1808 he was appointed one of the Aides-de-Camp to H.M. King George III. with the rank of Colonel in the army as a mark of His Majesty's approbation of his services, and of the efficient state of the 14th Light Dragoons under his command. In December 1808 he embarked for Portugal with his regiment, and was engaged in several actions with the French army, particularly at the memorable battle of Talavera (27th and 28th July), where the regiment distinguished itself, and was highly commended in the official despatches. Being severely wounded at Talavera, he returned shortly afterwards to England. He was promoted Major-General on 4th June 1811, and relinquished command of the regiment at that date. He was appointed to serve as a Major-General upon the staff of Great Britain on 25th November 1811, and was employed in

[1] *Historical Record of the 16th (Queen's) Light Dragoons (Lancers)*, by Richard Cannon, Esqre. (1842), pp. 37, 38.

[2] *Ibid.*

Sir Samuel Hawker.

command of the Eastern District until the 24th September 1814. He became Lieutenant-General, 19th July 1821, and was appointed Captain of Yarmouth Castle, in the Isle of Wight, on 22nd July 1829.

On 22nd April 1831 he was nominated by H.M. King William III. to the Colonelcy of the 3rd (Prince of Wales's) Dragoon Guards. In 1831 Lieutenant-General Hawker was nominated a Knight Commander of the Order of the Guelphs of Hanover (K.C.H.), and in 1836 he was advanced to the dignity of G.C.H. He became General on 28th June 1838, and died on 27th December following at the age of seventy-five, after a long and faithful service of nearly sixty years.

Sir Samuel Hawker's family are still settled at Binfield, Bracknell, Berkshire, where his eldest surviving son, Mr. F. A. Hawker, has the family residence; and Mr. Adolphus Hawker, late of the War Office, is another son. There were nine sons, all in their sovereign's service, either naval, military, or civil. The late Colonel Peter Hawker of Longparish, Hampshire, who was the author of *Instructions to Young Sportsmen*,[1] was a relative of the late Sir Samuel Hawker, and he served under him as a Captain in the 14th Light Dragoons in the Peninsular War. Captain P. Hawker also wrote a diary, published a few years ago, which contains some of his experiences with the old 14th Light Dragoons. Sir S. Hawker, in his later years, was a favourite at Court with H.M. King William IV., and on one occasion, when staying at Windsor Castle on a visit to the King, he had the good luck to hook a trout in the Thames, not far from the Castle, which weighed fourteen pounds. Fishing was one of the old General's favourite pastimes. When quite a young officer he commanded the escort of cavalry which escorted Lord George Gordon as a prisoner to the Tower in 1780. The photogravure of Sir Samuel is from a photograph of an oil painting by Lucas, taken in the uniform of the 3rd Dragoon Guards, of which he was Colonel 1831-38, kindly given by F. A. Hawker, Esq., his son.

LIEUTENANT-COLONEL NEIL TALBOT

APPOINTED SECOND LIEUTENANT-COLONEL, 14TH LIGHT DRAGOONS, AUGUST 22, 1805.

Neil Talbot.

Neil Talbot entered the army as an Ensign in the 27th Regiment, 25th June 1789; was promoted Lieutenant, 30th November 1791, Captain in the 118th Regiment, 10th July 1794. On 19th October 1796, Captain Talbot was removed to the 14th Light Dragoons. He was promoted to a Majority on 25th June 1802, and to a Lieutenant-

[1] *Instructions to Young Sportsmen in all that relates to Guns and Shooting*. 9th Edition. 1 vol., 8vo. (1844.)

Colonelcy on 22nd August 1805. In December 1808 he embarked with his regiment for the Peninsula. He was engaged in an affair with the enemy near Sexmiro on 11th July 1810, when a portion of the 14th Light Dragoons charged a square of French infantry 200 strong. The square withstood the charge and opened a terrific fire on the charging horsemen. The gallant Lieutenant-Colonel Talbot, a Quartermaster (M'Cormick), and eleven men of the regiment fell dead close up to the bayonets. The affair arose out of an attempt to cut off the French picquets on the Agueda, in front of Ciudad Rodrigo. Neil Talbot.

At the battle of Talavera, 28th July 1809, Lieutenant-Colonel Talbot had a horse killed under him, and from shortly after that battle, when Colonel Hawker was wounded and was obliged soon after to proceed to England, Lieutenant-Colonel Talbot practically commanded the regiment in the field until the day of his death, though Colonel Hawker was the actual Lieutenant-Colonel Commanding until his promotion to Major-General in 1811.

COLONEL SIR FELTON BATHURST HERVEY, BART., C.B., K.H., A.D.C.

APPOINTED SECOND LIEUTENANT-COLONEL, 14TH LIGHT DRAGOONS, AUGUST 2, 1810; COMMANDED THE REGIMENT, 1814-19.

Felton Bathurst Hervey entered the army as a Cornet in the 3rd Dragoon Guards on 6th May 1800, was promoted to a company of infantry on 9th July 1803, and removed to the 14th Light Dragoons, 28th July 1803. He was promoted Major on the 8th May 1806, and became second Lieutenant-Colonel of the regiment in succession to Lieutenant-Colonel Talbot. Colonel S. Hawker, who was the commanding officer of the Fourteenth at this time, had previously gone to England in consequence of his wound received at the battle of Talavera on 28th July 1809, and from that date to the day of his death, Lieutenant-Colonel Talbot had commanded the regiment in the field, and upon his lamented death, Lieutenant-Colonel Hervey assumed the command, and held it during the remaining four years of the war. It was under him that the Fourteenth became so famous as a light cavalry regiment, and was constantly employed on the outpost duties, while Lieutenant-Colonel Hervey gained a great reputation, having been highly distinguished by his soldier-like conduct on many notable occasions when he was present with the regiment. His chief actions were the following:—Passage of the Douro, 12th May 1809, where he lost an arm; Venta de Serra, 8th March 1811, where he made a suc- Sir Felton Bathurst Hervey, Bart.

Sir Felton Bathurst Hervey, Bart.

cessful charge with three squadrons of the 14th Light Dragoons against four squadrons of the 11th and 26th French Dragoons, and captured 14 men and 14 horses, losing only 2 men and 2 horses; Fuentes d'Onor, on 5th May 1811, where he had his horse killed under him and received a severe contusion; at Carpio on 25th September 1811, against the Lancers of Berg; after the capture of Badajoz on the night of the 10th April 1812, an enterprise against a large body of French cavalry; battle of Salamanca, 22nd July 1812; battle of Vittoria, 21st June 1813; battle of Orthes, 27th February 1814.

He became the only Lieutenant-Colonel and commanding officer of the Fourteenth on 4th June 1811, when Colonel Hawker was promoted to the rank of Major-General, and remained so until his death in 1819. On the 4th June 1814, Lieutenant-Colonel Hervey was appointed A.D.C. to the Prince Regent (afterwards George IV.) with the rank of Colonel in the army, and he was created a baronet for his services in 1818. At the battle of Waterloo he served on the personal staff of the Duke of Wellington, and was afterwards appointed Secretary to the Master-General of the Ordnance. His premature death on 24th September 1819 was deeply regretted by all ranks of the regiment and by his numerous friends. The following is a list of his medals, orders, crosses, etc.:—Crosses for the battles of Fuentes d'Onor, Salamanca, Vittoria, and Orthes; gold medals for the battles of Fuentes d'Onor, 1811, and Salamanca, 1812; medal for the battle of Waterloo, 1815; Companion of the Order of the Guelphs; Companion of the Order of the Bath; Russian Order of St. George of Wladimir; Austrian Order of Maria Theresa; Portuguese Order of Tower and Sword; Bavarian Order of Joseph Maximilian; Prussian Order of Merit; Knight of St. Henry of Saxony. The officers of the 14th Light Dragoons presented a sword to Sir Felton Hervey in 1813,[1] which is reproduced in the photogravure with his portrait, kindly supplied by Captain Sir F. Hervey Bathurst, Bart., Grenadier Guards.

LIEUTENANT-COLONEL CHARLES MASSEY BAKER

APPOINTED LIEUTENANT-COLONEL, 14TH LIGHT DRAGOONS, SEPTEMBER 30, 1819.

Charles Massey Baker.

Charles Massey Baker entered the army as an Ensign in the 27th Regiment on 30th March 1788; became Lieutenant, 30th November

[1] Sir Felton Hervey's picture at Somborne Park, Stockbridge, hangs with the sword below. The inscription attached to the latter is as follows:—'From the Officers of the XIVth Light Dragoons to Lieutenant-Colonel Hervey. MDCCCXIII.'

1791; and Captain, 22nd Dragoons, 31st August 1795. He was removed to the 14th (Duchess of York's Own) Light Dragoons on 3rd March 1804, and became Major, 30th January 1813. On the death of Colonel Sir F. B. Hervey, Bart., C.B., A.D.C., he was promoted Lieutenant-Colonel, and appointed to the command of the regiment 30th September 1819. He commanded the Fourteenth during the next ten years, and retired from the army, 15th April 1829. He was a Captain at the battle of Talavera on 28th July 1809, and his horse was killed under him in that action. Charles Massey Baker.

COLONEL JOHN TOWNSEND, A.D.C.

APPOINTED LIEUTENANT-COLONEL OF THE 14TH LIGHT DRAGOONS, APRIL 16, 1829.

John Townsend entered the army as a Cornet in the 14th Light Dragoons on 24th June 1805, was promoted to a Lieutenantcy, 27th February 1806, and to a troop on the 6th June 1811. He served in the Peninsula from December 1808 until taken prisoner near the city of Pau, in France, on 8th March 1814, including the different affairs of the 10th and 11th May, and in crossing the Douro on the 12th May 1809; battle of Talavera, July 1809; affair with the enemy's advanced posts on the 11th July 1810 in front of Ciudad Rodrigo, under the command of Lieutenant-Colonel Talbot, who was killed; passage of the Coa; skirmishes of the rearguard from Almeida to the lines of Torres Vedras in 1810; affairs during the enemy's retreat from Santarem to the frontiers of Spain, from 6th March to 4th April 1811; battle of Fuentes d'Onor, where he was wounded, 15th May 1811; affair with the enemy's lancers, 25th September 1811; siege of Badajoz, 1811; affairs with the enemy's cavalry at Usagre and Llerena in front of Salamanca, 10th April 1812, and near Castrillos, 18th July 1812; battle of Salamanca, 22nd July 1812; affair with the enemy's rearguard near Penaranda, 23rd July 1812; several skirmishes from Madrid to Ciudad Rodrigo, November 1812; and from May 26th, when near Salamanca, to the date of the battle of Vittoria on 21st June 1813, in which he took part; taking of a gun from the enemy near Pampeluna, and several engagements and skirmishes from the entry of the British army into France until the battle of Orthes on 27th February 1814. He embarked with the Fourteenth for America in October 1814, and was present at the attack on New Orleans on 8th January 1815. He was promoted to a Majority in the regiment John Townsend.

John Townsend.

on 13th September 1821, and to the Lieutenant-Colonelcy on the retirement of Lieutenant-Colonel Baker, 16th April 1829. He embarked in command of the regiment for India on the 24th May 1841, and was appointed Aide-de-Camp to Her Majesty Queen Victoria, with the rank of Colonel in the army, on the 23rd November 1841. He returned from India on leave of absence in the early part of the year 1845, and died at Castle Townsend, in Ireland, on the 22nd April 1845, after nearly forty years' service in the 14th Light Dragoons.

The photogravure of Colonel Townsend is from an oil portrait presented to the officers' mess by the late Captain Stuart Robertson, when serving in the Fourteenth.

LIEUTENANT-COLONEL WILLIAM HAVELOCK, K.H.

APPOINTED SECOND LIEUTENANT-COLONEL IN THE 14TH (KING'S) LIGHT DRAGOONS, APRIL 30, 1841; COMMANDED THE REGIMENT, 1845-48.

William Havelock.

William Havelock, son of H. Havelock, Esquire, formerly of Ingress Park, near Dartford, was born in 1795. He was educated at Charterhouse School, and then studied under a private tutor before entering the army, which he did on 12th July 1810, as Ensign in the 43rd Light Infantry. The celebrated Major-General Sir Henry Marshman Havelock, K.C.B., was his younger brother. At the age of fifteen he carried the colours of his regiment in the action of Coa. He was promoted Lieutenant on 12th May 1812, and became Captain in the 32nd Regiment on the 19th February 1818. He exchanged to the 4th Light Dragoons on the 19th July 1821, and was promoted Major on the 31st December 1830. Major Havelock was promoted from the 4th Light Dragoons to be second Lieutenant-Colonel on the augmentation of the 14th Light Dragoons to the Indian establishment, on the 30th April 1841, and succeeded to the command of the regiment on the death of Colonel Townsend, 23rd April 1845. His services in Europe were as follows:—In the Peninsular War from July 1810, till the end of the war in 1814, including the battles of Busaco, Sabugal, Salamanca, and Vittoria; passage of the Bidassoa; the battle of the Nivelle; the affair near Bayonne; and the battles of Orthes and Toulouse. He served also in the campaign of 1815, and was wounded at Quatre Bras. At the battle of Waterloo he was Aide-de-Camp to General Baron Alten, and received for his services the Cross of the Hanoverian Order. Subsequently he served at Corfu, and afterwards, on exchanging as Captain to the 4th Light Dragoons, proceeded with his regiment to Bombay, where he was Aide-de-Camp to Sir Charles Colville; then to Madras, where he was Military Secretary to Lord Elphinstone. After his promotion into the 14th Light Dragoons in 1841, he served in the various movements of Major-General Sir Charles Napier's force in the Bombay Contingent, and afterwards proceeded with the Fourteenth to the Punjaub, where on the 22nd November 1848 he fell mortally wounded, at the head of his regiment, when gallantly leading a charge against the Sikh forces of Shere Singh at Ramnuggur, on the banks of the Chenab. William Havelock.

Colonel Havelock married Miss Chaplin, related to the late Chief-Justice Tindal, and he left a large family. One of his sons, Colonel A. C. Havelock, was appointed Colonel in the Madras Staff Corps in 1885, and another is Sir Arthur E. Havelock, G.C.M.G., G.C.S.I., G.C.I.E., late Governor of Madras. Lieutenant-Colonel Havelock was a Knight of the Hanoverian Order of the Guelphs, and had received several war medals, to which no officer was better entitled for his thorough knowledge of his profession, as well as for his personal courage. Major-General Sir William Napier in his *History of the War in the Peninsula* mentions young Lieutenant Havelock, as he then was, of the 43rd Regiment, after the passage of the Bidassoa, at the second combat at Vera, in the Pyrenees, in October 1813. When on a critical occasion our allies, the Spaniards, were wavering under the heavy fire of two French regiments, posted behind a strong line of abattis, Lieutenant William Havelock, attached to Baron Alten's staff, was sent to ascertain how General Giron, who commanded the Andalusians, was progressing. He came up with them near the Puerto, between Vera and Sarre, and seeing the hesitation (Napier says), 'His fiery temper could not brook the check. Taking off his hat he called upon the Spaniards to follow him, and putting spurs to his horse, at one bound cleared the abattis and went headlong amongst the enemy. Then the soldiers, shouting for "El chico blanco"—"the fair boy," so they called him, for he was very young and had light hair—with one shock broke through the French, and this at the very moment when their centre was flying under the fire of Kempt's skirmishers from the Puerto de Vare."[1]

[1] *History of the War in the Peninsula*, vol. v. book xxii. chap. 4, p. 319.

William Havelock.

His picture is from a pencil-sketch copied from a water-colour portrait, in staff undress uniform, taken by Prince Soltikoff at Madras about the year 1842, when Havelock was Military Secretary to Lord Elphinstone, Governor of Madras, and was kindly supplied by his sons, Colonel A. C. Havelock, late Madras Staff Corps, and Sir Arthur Havelock, G.C.M.G., G.C.S.I., G.C.I.E., late Governor of Madras.

MAJOR-GENERAL EDWARD HARVEY

APPOINTED SECOND LIEUTENANT-COLONEL IN THE 14TH (KING'S) LIGHT DRAGOONS, APRIL 23, 1845.

Edward Harvey.

Edward Harvey entered the army as Cornet in the 4th Light Dragoons on the 24th March 1825; was promoted Lieutenant on the 4th May 1826, and Captain of infantry on the 12th October 1830. He exchanged to the 14th Light Dragoons, 27th December 1833, and was promoted Major on the 30th April 1841. On 31st December 1841 he received the brevet rank of Lieutenant-Colonel, and was appointed second Lieutenant-Colonel of the regiment on 23rd April 1845, on the decease of Colonel Townsend. He subsequently retired from the regiment in April 1848, and was succeeded as second Lieutenant-Colonel by Major J. W. King, who subsequently commanded the Fourteenth after Lieutenant-Colonel Havelock fell at Ramnuggur. Lieutenant-Colonel Harvey served at the investment of Kolapore, East Indies, in 1827; was employed on the staff in the operations against Ibrahim Pasha in Syria, 1840-41, and was present at the skirmish with the Egyptian troops near Askelon, 15th January 1841. He was placed on half-pay, 25th April 1848, as an unattached Lieutenant-Colonel, became Colonel, 20th June 1854, and Major-General, 26th October 1858. For his war services he received the brevet rank of Lieutenant-Colonel, the Syrian war medal, and a gold medal from the Sultan. After rejoining his regiment in India in 1842, subsequent to the Syrian campaign, he served with the wing of the 14th Light Dragoons, in 1844, at the bombardment and capture of the Mahratta fortresses of Panulla and Pownaghur, and was in command of the 3rd Brigade of the Field Force when, on the assault of Panulla by the 1st and 2nd Brigades, the fortress of Pownaghur was taken by the 3rd Brigade, for which he received in General Orders the thanks of the Bombay Government. He died in 1864.

LIEUTENANT-COLONEL JOHN WALLACE KING, C.B.

APPOINTED SECOND LIEUT.-COL., 14TH (KING'S) LIGHT DRAGOONS, APRIL 25, 1848; COMMANDED THE REGIMENT, 1848-50.

John Wallace King.

John Wallace King became Cornet, 24th March 1825; Lieutenant, 14th February 1828; Captain, 5th Dragoon Guards, 28th December 1832; Major, 1845; exchanged from 5th (Princess Charlotte of Wales's) Dragoon Guards into the 14th Light Dragoons with Major W. H. Archer, 8th September 1846; promoted second Lieutenant-Colonel, 14th Light Dragoons, 25th April 1848; and after the death of Lieutenant-Colonel Havelock at Ramnuggur he got command of the regiment. He commanded the regiment through the rest of the Sikh war, being present at the battles of Chillianwallah and Goojerat, and in the pursuit of the Sikhs across the Jhelum, and of the Afghans through the Khyber Pass. He died at Lahore in July 1850, when in command of the regiment. He received a C.B. for his services in the Punjaub, and was very favourably mentioned in Major-General Sir Joseph Thackwell's despatches after the battle of Goojerat.

GENERAL HENRY EDWARD DOHERTY, C.B.

APPOINTED SECOND LIEUT.-COLONEL IN THE 14TH (KING'S) LIGHT DRAGOONS, NOVEMBER 23; COMMANDED THE REGIMENT, 1850-57.

Henry Edward Doherty.

Henry Edward Doherty became Cornet, 31st December 1833; Lieutenant, 15th July 1836; Captain, 17th May 1839; Major, 22nd October 1847; Lieutenant-Colonel, 23rd November 1848; Colonel, 28th November 1854; Colonel on half-pay, 25th August 1857; Major-General, 8th February 1863; Lieutenant-General, 25th October 1871; General, 1st October 1877. He served with the 14th Light Dragoons throughout the Punjaub campaign of 1848-49, including the action of Ramnuggur, passage of the Chenab, battles of Chillianwallah and Goojerat, pursuit of the enemy across the Jhelum, and of the Afghans across the Indus through the Khyber Pass, for which he received the medal with two clasps. At Ramnuggur he brought the charging squadrons under Lieutenant-Colonel Havelock, who was killed, across the nullah and out of the enemy's entrenchments, and at the battle of Goojerat he commanded a mixed body of Her Majesty's Dragoons and Native Light Cavalry. He received a C.B. for his services. On the death of Colonel King, in July 1850, he succeeded to the command of

Henry Edward Doherty.

the regiment, which he retained till August 1857. He was placed on the retired list by the royal warrant of 25th June 1881, and he died in 1885.

His portrait was kindly supplied by his nephew, Major D. H. Doherty, late 3rd (King's Own) Hussars.

LIEUTENANT-GENERAL CHARLES STEUART, C.B.

APPOINTED SECOND LIEUTENANT-COLONEL, 14TH (KING'S) LIGHT DRAGOONS, JULY 7, 1850; COMMANDED THE REGIMENT, 1857-61.

Charles Steuart.

Charles Steuart became Cornet, 10th December 1825; Lieutenant, 5th February 1829; Captain, 9th November 1838; Major, 25th April 1848; Lieutenant-Colonel, 7th July 1850; Colonel, 28th November 1854; Major-General, 3rd October 1864; and Lieutenant-General, 6th May 1873. He served with the 14th Light Dragoons in the Punjaub campaign of 1848-49, including the battles of Chillianwallah, when he received a sabre-wound, and Goojerat, pursuit of the enemy across the Jhelum, and of the Afghans over the Indus through the Khyber Pass, for which he received the medal and clasps. He commanded a cavalry brigade in the Persian expedition of 1857, for which he received the medal, and C.B. on 1st January 1858; commanded the 2nd Brigade of the Central India Field Force under Sir Hugh Rose in 1858, and was present at the siege and capture of Rathgur, action of Barodia, relief of Saugor, and capture of Garrakota, forcing the Muddenpore Pass, siege and capture of Jhansi, and the battles of the Betwa and Koonch, for which he received the Indian Mutiny medal and clasp.

He succeeded to the command of the 14th Light Dragoons on 26th August 1857, and retired from the regiment when it was at Newbridge in June 1861. He was appointed Colonel of the 11th (Prince Albert's Own) Hussars, 19th January 1873, and died on the 22nd May 1873.

GENERAL CHARLES PHILIP DE AINSLIE

APPOINTED SECOND LIEUTENANT-COLONEL, 14TH (KING'S) LIGHT DRAGOONS, AUGUST 26, 1857.

Charles Philip de Ainslie.

Charles Philip de Ainslie, representative of the ancient family of De Ainslie of Dilphington, Roxburghshire, N.B., entered the service as Second-Lieutenant in the Rifle Brigade on 10th April 1825; became Lieutenant, 28th June 1826; Captain, 16th March 1830; Major, 14th October 1842; Lieutenant-Colonel, 22nd October 1847; Colonel, 28th

November 1854. He served in several cavalry regiments, in the 4th ('Queen's Own') Light Dragoons, in the 1st Royal Dragoons, 14th (King's) Light Dragoons, and 7th (Princess Royal's) Dragoon Guards. His first commission in the 14th Light Dragoons was that of Major, dated 3rd February 1843, when he came from the Royal Dragoons. He subsequently commanded the 7th (Princess Royal's) Dragoon Guards, and came back again to the 14th Light Dragoons in August 1857 as second Lieutenant-Colonel. He remained with the regiment till 28th August 1860, when the establishment was reduced to one lieutenant-colonel on returning from India, and he was placed on half-pay. He became Major-General, 1862; Lieutenant-General, 1871; General in 1877; and was appointed Colonel of the Royal Dragoons in March 1869. Charles Philip de Ainslie.

From June 1827 to May 1828, Lieutenant De Ainslie was Aide-de-Camp to Lieutenant-General Sir Thomas Bradford, K.C.B., Commander-in-Chief at Bombay.

From May 1840 to 1st April 1842, Captain De Ainslie was Aide-de-Camp to Major-General Lord Greenock, K.C.B., commanding the forces in Scotland. On the 25th August 1857, Colonel De Ainslie from half-pay came back to serve for the second time in the 14th (King's) Light Dragoons, and joining that corps in India as its second Lieutenant-Colonel, he commanded the cavalry at Kirkee, and afterwards as Brigadier-General at Jhansi. In May 1866, as Major-General, he commanded the troops in the Windward and Leeward Islands, with headquarters at Barbadoes, and afterwards at Jamaica.

He was the author of *Historical Record of the Royal Dragoons*, published by Chapman and Hall, 1887; and died at the Buckingham Palace Hotel, London, in 1890.

MAJOR-GENERAL ARTHUR SCUDAMORE, C.B.

APPOINTED LIEUTENANT-COLONEL, 14TH (KING'S) LIGHT DRAGOONS, JUNE 19, 1861.

Arthur Scudamore was appointed Cornet, 29th May 1835; Lieutenant, 18th February 1838; Captain, 22nd October 1847; Major, 1st June 1854; Lieutenant-Colonel, 20th July 1858; Colonel, 9th November 1862; and Major-General in 1875, with ante-date to 28th June 1868. He served with the 4th Light Dragoons throughout the campaign of 1839, under Lord Keane, in Afghanistan, including the siege and capture of Ghuznee, for which he received a medal; also throughout the Punjaub campaign of 1848-49 with the 14th Light Dragoons, Arthur Scudamore.

Arthur Scudamore.

including the action of Ramnuggur, where he was wounded (sabred in the face); passage of the Chenab, battle of Chillianwallah and battle of Goojerat, where he was dangerously wounded; and for these services he received the medal with two clasps. He commanded the regiment (in the absence of Colonel Steuart, C.B., who was commanding a brigade) during the campaign of 1858 in Central India, under Major-General Sir Hugh Rose, K.C.B., and was present at the siege and capture of Rathgur, action of Barodia, relief of Saugor, and capture of Garrakota. He commanded the detached force sent against Maltone[1] Pass, was present at the siege and capture of Jhansi, where he commanded the outposts during the investment; was present at the battle of Koonch, and at all the affairs during the advance on Calpee, including the action of Golowlee; also present at the action of Morar and recapture of Gwalior. He commanded a flying column for six months in the Gwalior and Jhansi districts, and for his distinguished services was three times mentioned in despatches, received the brevet rank of Lieutenant-Colonel, the C.B., and the Indian Mutiny medal, with clasp for Central India.[2] He was appointed to the command of the 14th Light Dragoons, 19th June 1861, and it was during his command that the regiment was converted into Hussars at Newbridge, on 17th August 1861. On the 8th October he exchanged to the command of the 7th Hussars with Lieutenant-Colonel Sir William Russell, Bart., C.B., who came to command the 14th Hussars.

On 1st April 1873, Colonel Scudamore was appointed to command the 34th Brigade Depôt at Exeter, and commanded a cavalry brigade in the Dartmoor Manœuvres held during the autumn months of 1873. He died in 1880.

COLONEL SIR WILLIAM RUSSELL, BART., C.B.

APPOINTED LIEUTENANT-COLONEL IN THE 14TH (KING'S) HUSSARS, OCTOBER 8, 1861.

Sir William Russell, Bart.

William Russell was born 5th April 1822. He was eldest son of Sir William Russell, Bart., M.D., of Charlton Park, Gloucestershire, who was created a Baronet in 1832. He entered the army as Cornet in the 7th (Queen's Own) Light Dragoons (Hussars) on 2nd July 1841, having succeeded his father as second Baronet in 1839. He served with his regiment in Canada and the East Indies, his commissions bearing

[1] Málthon (Malleson).

[2] Lieutenant-Colonel Scudamore was President of the court-martial which tried Tantia Topee at Goona and Sipri (Central India) in April 1859.

date as follows: Lieutenant, 27th February 1846; Captain, 16th April 1847; Major, 13th August 1857; Lieutenant-Colonel, 20th July 1858; Colonel, 20th July 1863, Major-General, 24th December 1868; retired with honorary rank of Lieutenant-General in 1881. He served on the Staff in the Crimea in 1855-56, and received the medal; served in the Indian campaign from February 1858 to March 1859 with the 7th Hussars; was present at the repulse of the enemy's attack on the Alumbagh, siege and capture of Lucknow, for which he received the brevet of Lieutenant-Colonel. He commanded the 7th Hussars at the affairs of Barree and Sirsee, action of Nawabgunge, occupation of Fyzabad, passage of the Goomtee at Sultanpore, throughout the Byswarra campaign, including the affairs of Pandoo Nuddee, Paleeghat, Hyderghur, and pursuit of Benhi Madho's force to the Goomtee; also in the Trans-Gogra campaign, including the affair near Churda and pursuit, taking the fort of Meejeedia, attack on Bankee with pursuit to the Raptee, advance into Nepaul and affair of Sitkaghat: for which services he was several times mentioned in despatches and received the C.B. and medal with clasp, having commanded a brigade of cavalry during some of the time in India. In October 1861 he exchanged with Lieutenant-Colonel Scudamore, C.B., into the 14th (King's) Hussars, and commanded the regiment from 1861 to 1864. He was Aide-de-Camp to the Lord-Lieutenant of Ireland, 1847 to 1852, and again in 1854; and Master of the Horse in Ireland, 1848-51. He was a Justice of the Peace for Gloucestershire and sat as M.P. for Dover (as a Liberal), 1857-59, and for Norwich, 1860-74. He died in 1892. Sir William Russell, Bart.

MAJOR-GENERAL PEARSON SCOTT THOMPSON, C.B.

APPOINTED LIEUTENANT-COLONEL, 14TH (KING'S) HUSSARS, NOVEMBER 29, 1864.

Pearson Scott Thompson entered the army as Cornet in the 7th Dragoon Guards, 5th August 1842. He became Lieutenant, 7th June 1844; Captain, 3rd September 1847; Major, 20th July 1858; Lieutenant-Colonel, 3rd July 1860; Colonel, 6th January 1867. He served against the emigrant Boers in 1845, beyond the Orange River in South Africa, and again throughout the Kaffir campaign of 1846-47 with the 7th Dragoon Guards. He exchanged to the 14th (King's) Light Dragoons in 1849 with Captain C. W. Thompson, who went to the 7th Dragoon Guards, and who was subsequently full Colonel of the 14th (King's) Hussars, 1882-96. Captain P. S. Thompson served with Pearson Scott Thompson.

Pearson Scott Thompson.

the 14th Hussars in the Central India Field Force in 1858 under Major-General Sir Hugh Rose, K.C.B., and was present at the siege and capture of Jhansi, 5th April 1858; at the capture of the fort of Lohari, 6th May 1858; action of Koonch, 7th May 1858; battle of Golowlee, 22nd May; capture of Calpee, 23rd May; capture of the Morar cantonments, 16th June; recapture of the town and fortress of Gwalior, 19th June 1858. He commanded a flying column in Bundelcund,[1] consisting of cavalry, artillery, and native infantry, 27th July 1858 to 2nd November 1858; commanded at the affair near the village of Garotha, captured three guns from the enemy, and commanded the cavalry at the affair of Malthone. For his services at Garotha he received the thanks of the Governor-General in Council at Bombay, 'for the dashing and successful attack made by the troops under his command,'—he was then a Captain. He was awarded the brevets of Major and Lieutenant-Colonel for his services in the field during the Indian Mutiny campaign.

On 16th June 1861 he became Major in the 14th Light Dragoons, and on 29th November 1864 was promoted to the Lieutenant-Colonelcy, succeeding Colonel Sir William Russell, Bart., C.B., in the command of the regiment. He retired on half-pay from the command of the regiment, 28th April 1875, having always received the highest credit from the authorities, more particularly for his excellent stable system and the management of his young horses. In 1869 he was awarded the Companionship of the Bath, in recognition of his distinguished services, and on 18th October 1876 he was appointed to the command of the Sunderland Brigade Depôt, which he held till his promotion to the rank of Major-General, 1st October 1877. He died at Beverley, Yorkshire, in 1878, deeply regretted, and he was always most highly esteemed by his brother officers and all who had served under him.

LIEUTENANT-COLONEL FRANCIS PEMBERTON CAMPBELL

APPOINTED LIEUTENANT-COLONEL, 14TH (KING'S) HUSSARS, APRIL 29, 1875.

Francis Pemberton Campbell.

Francis P. Campbell became Ensign, 3rd November 1854; Lieutenant, 9th March 1855; Captain, 14th January 1862; Major, 2nd December 1868; and Lieutenant-Colonel, 29th April 1875. He served

[1] Bundelkhand (Malleson).

with the 79th Highlanders in the Crimea from 12th July 1855, including the siege and fall of Sebastopol and assault of the Redan, on the 8th September, for which he received the medal with clasp and the Turkish medal. He also served with the 79th Highlanders in the Indian Mutiny campaign, including the siege and capture of Lucknow, action of Bareilly, Shahjehanpore, and Mohundeer. From September 1858 he served with the 1st Punjaub Cavalry, and was present at the capture of Rampore Kassia and pursuit, passage of the Gogra, affairs of Churda and attack on Bankee, with pursuit to the Raptee, advance into Nepaul, and affair at Sitkaghat. He also served with Colonel Sir Alfred H. Horsford's force on the frontier of Nepaul from January to June 1859, for which services he received another medal and clasp. Francis Pemberton Campbell.

He commanded the 14th Hussars when the regiment embarked for India in 1876; and on the 14th June 1876, after a short illness, he died of acute dysentery at Bangalore, at the early age of thirty-nine years, very deeply regretted.

MAJOR-GENERAL WILLIAM ARBUTHNOT, C.B.

APPOINTED LIEUTENANT-COLONEL, 14TH (KING'S) HUSSARS, JUNE 15, 1876.

William Arbuthnot, eldest son of the late Archibald F. Arbuthnot, Esq. (who was son of Sir William Arbuthnot, first Baronet), and the late Honourable Gertrude Sophia, daughter of the first Viscount Gough, was born 27th September 1838. He was educated at Eton, and entered the army as Ensign in the Rifle Brigade, 25th March 1856. He became Lieutenant, 3rd June 1859; exchanged into the 14th (King's) Hussars in 1861; became Captain, 29th November 1864; Brevet-Major, 15th August 1868; Major, 14th Hussars, 10th August 1870; Lieutenant-Colonel, 15th June 1876; Colonel, 15th June 1881; Major-General, 19th March 1890. William Arbuthnot.

He served with the Rifle Brigade in the Indian Mutiny campaign, 1857-58, including the operations on the Ramgunga, the siege and capture of Lucknow, and subsequent operations in Oude, for which he received the medal with clasp. He also served in the Abyssinian campaign, 1867-68, as Aide-de-Camp and Assistant Military Secretary to Major-General Lord Napier of Magdala, who commanded the expedition, and was present at the capture of Magdala, for which he was mentioned in despatches, and received the medal and the brevet rank of Major.

William Arbuthnot.

He served in the Transvaal campaign in the South African war of 1881, where he commanded the 14th (King's) Hussars. He was present in the Soudan Expedition, 1885, at Suakin, as Deputy Adjutant-General, for which services he received the medal with clasp, the bronze star, and the Companionship of the Bath.

Major-General Arbuthnot passed the final examination of the Staff College in 1863, and held the following Staff appointments:—A.D.C. to Lieutenant-General, Madras, 30th June 1858 to 27th February 1861; A.D.C. to Brigadier-General, Aldershot, 4th February 1862 to 30th September 1862; A.D.C. to Brigadier-General, Dublin and Curragh Division, 22nd March to 31st December 1864; A.D.C. to General Officer Commanding Abyssinian Expedition, 8th January 1868 to 17th April 1868; Assistant Military Secretary to General Officer Commanding Abyssinian Expedition, 18th April 1868 to 24th May 1868; Brigade-Major of Cavalry, Aldershot, 1st February 1869 to 15th August 1870; Garrison Instructor, Aldershot, 1st January 1873 to 20th January 1873; Deputy-Adjutant and Quartermaster-General, Malta, 1st April 1882 to October 1882; Brigadier-General, Egypt, 21st October 1882 to 24th December 1883; Deputy-Adjutant and Quartermaster-General, Malta, 25th December 1883 to 19th February 1885; Deputy-Adjutant and Quartermaster-General, Expeditionary Force, Suakin, 20th February 1885 to 13th June 1885; Assistant Adjutant-General, Headquarters of Army, 28th December 1885 to 1st April 1890.

He died near London, September 1893, and was buried in Brompton Cemetery. A detachment of the regiment was present at his funeral to do honour to his memory, as well as the Lieutenant-Colonel of the Fourteenth, Colonel the Hon. G. H. Gough, with others who had served under his command.

COLONEL JOHN HUNTER KNOX

APPOINTED LIEUTENANT-COLONEL, 14TH (KING'S) HUSSARS, JUNE 15, 1881.

John Hunter Knox.

John Hunter Knox was the son of Lieutenant-Colonel William Knox of Britas, near Thurles, Ireland, formerly of the 13th Light Dragoons. He entered the army as Cornet, 30th March 1858; became Lieutenant, 8th February 1861; Captain, 2nd October 1866; Major, 15th June 1876; Lieutenant-Colonel, 15th June 1881; Colonel, 15th June 1885. He served with the 14th Hussars in the Transvaal during the Boer campaign of 1881, and succeeded to the command of the regiment after Colonel Arbuthnot's retirement in June 1881. He died very suddenly when in command of the regiment at Secunderabad, India, on the 24th October 1885, of liver disease. John Hunter Knox.

MAJOR-GENERAL FRANCIS SHIRLEY RUSSELL, C.M.G.

APPOINTED SECOND LIEUTENANT-COLONEL, 14TH (KING'S) HUSSARS, JULY 1, 1881.

Francis Shirley Russell was born 13th December 1840, educated at Radley and Balliol College, Oxford, where he took his B.A. degree with honours in 1862. He entered the army as Cornet in the 14th Hussars, 6th February 1863; became Lieutenant, 23rd August 1864; Captain, 13th June 1868; Brevet-Major, 1st April 1874; Major, 14th Hussars, 15th June 1881; Lieutenant-Colonel, 1st July 1881; Colonel, 1st July 1885; Major-General, 20th January 1897; retired, 1898. He served in the Ashanti War from 17th December 1873, was attached to Wood's regiment, and commanded the post of Accrofooma on the lines of communication, for which services he received the brevet rank of Major and the medal with clasp. He served in the latter phase of the Zulu War of 1879 as Deputy-Assistant Adjutant-General, 2nd Division, and was present in the engagement at Ulundi on the 4th July 1879, for which he received the medal with clasp. He was present with the 14th (King's) Hussars in the Boer campaign in South Africa in the year 1881. Major-General Russell passed the final Staff College examination in 1873, and held the following staff appointments :— Francis Shirley Russell.

A.D.C. to General Officer Commanding the Forces in Ireland, 1st March 1869 to 30th June 1870.

Special Service, Ashanti Expedition, 4th December 1873 to 28th March 1874.

Instructor in Tactics, Royal Military College, 18th September 1875 to 27th December 1876.

Special Service, South Africa, 16th May 1879 to 20th October 1879.

Military Attaché, Berlin, 1887-90.

In 1881 he exchanged from the 14th Hussars to the Royal Dragoons with Lieutenant-Colonel C. F. Morton, and commanded the Royal Dragoons from 1885 to 1887.

In 1891 he was awarded the Companionship of the Order of St. Michael and St. George.

In February 1892, Colonel Russell was appointed to command the

Francis Shirley Russell.

Aberdeen Volunteer Brigade, and in 1895 he was returned in the Conservative interest as Member of Parliament for Cheltenham. He is a J.P. and Deputy-Lieutenant of the county of Aberdeen, where he has a country residence, 'Aden.' He is author of *Cavalry Field Duty*, *Russian Wars with Turkey*, *Memoir of the Earl of Peterborough*, and other works. He became Colonel of the 1st (Royal) Dragoons, 9th June 1900.

COLONEL CHARLES FALKINER MORTON

APPOINTED SECOND LIEUTENANT-COLONEL, 14TH (KING'S) HUSSARS, JUNE 10, 1882; COMMANDED THE REGIMENT, 1885-87.

Charles Falkiner Morton.

Charles Falkiner Morton, son of the late T. C. Morton, Esq., Barrister-at-Law, of Calcutta, was born 7th November 1843; educated at Eton, and entered the army as Cornet in the 1st Royal Dragoons, 18th February 1862; became Lieutenant, 21st July 1863; Captain, 12th June 1869; Major, 10th April 1881; Lieutenant-Colonel, 1st July 1881; Colonel, 1st July 1885. He served with the Royal Dragoons in Great Britain and Ireland, 1862-82, when he exchanged into the 14th (King's) Hussars with Lieutenant-Colonel F. S. Russell, and joined the regiment at Secunderabad, in India. Upon the death of Colonel Knox in 1885, Colonel Morton succeeded to the command of the regiment, which he retained till 30th June 1887, bringing the Fourteenth home from India in 1886, and retired on half-pay when quartered at Shorncliffe.

In July 1887, Colonel Morton commanded the wing of the Fourteenth which took part in the Queen's Jubilee Review at Aldershot on 9th July, his services being specially retained for a fortnight after 30th June to cover the Jubilee celebrations.

He was employed in the Army Remount Establishment as Assistant Inspector of Remounts in London from 19th October 1887 to 19th October 1892, and was placed on retired pay, 7th November 1900, after upwards of thirty-eight years' service in the cavalry.

COLONEL JAMES DENNIS

APPOINTED SECOND LIEUTENANT-COLONEL, 14TH (KING'S) HUSSARS, OCTOBER 25, 1885.

James Dennis.

James Dennis was born 13th August 1838; entered the army as Cornet in the 5th Dragoon Guards, 5th March 1861, and became

Lieutenant in the 5th Lancers, 12th October 1868, having been transferred to that regiment on 24th March 1863. He became Captain, 3rd Dragoon Guards, 10th June 1874, and exchanged to the 14th Hussars on 31st October 1874. He became Major, 1st July 1881; Lieutenant-Colonel, 25th October 1885; Colonel, 25th October 1889. He was Adjutant of the 5th (Royal Irish) Lancers from 25th June 1869 to 9th July 1871. He was present with the 14th Hussars in the Transvaal campaign of the South African War, 1881. On 17th November 1886 he exchanged into the 6th Dragoon Guards (Carabiniers) with Lieutenant-Colonel H. B. Hamilton (who came to the 14th Hussars), and commanded the Carabiniers from 1887 to 1891. He went on retired pay in 1891, after thirty years' service in the cavalry. James Dennis.

COLONEL HENRY BLACKBURNE HAMILTON

APPOINTED SECOND LIEUTENANT-COLONEL, 14TH (KING'S) HUSSARS, NOVEMBER 17, 1886; COMMANDED THE REGIMENT, 1887-91.

Henry Blackburne Hamilton, eldest son of the late Rev. Henry Hamilton, M.A., of 17 Devonshire Place, London, formerly Ensign in the 85th (Duke of York's) Light Infantry,[1] and afterwards Rector for many years of Thomastown, County Kildare, Ireland, and his wife, Frances Margaret, daughter of the late Ralph Peters, Esq., of Platbridge House, Wigan, Lancashire, was born 3rd July 1841; educated at Connaught House, Brighton, Elstree Hill School, and Eton; graduated B.A. at Christ Church, Oxford, in 1864, M.A., 1868; entered the army as Cornet in 6th Dragoon Guards (Carabiniers), 21st July 1864; became Lieutenant, 15th May 1866; Captain, 24th March 1869; Major, 1st July 1881; Lieutenant-Colonel, 17th March 1886; and Brevet-Colonel, 17th March 1890. He was Instructor of Musketry in the Carabiniers from 1866 to 1869, and served with them in the Afghan campaign of 1879-80, in the Khyber Division of the Cabul Field Force, commanded by Lieutenant-General R. O. Bright, C.B.[2] Henry Blackburne Hamilton.

In Afghanistan, in March 1880, Captain Hamilton, as he then was, commanded a detached troop of the Carabiniers with an expeditionary force despatched against the Mohmunds into the Chardeh Valley, under Lieutenant-Colonel G. C. Hodding;[3] also in June 1880, a detached

[1] Now the 2nd Battalion the King's (Shropshire) Light Infantry.

[2] Afterwards General Sir Robert O. Bright, G.C.B.

[3] Afterwards Brigadier-General G. C. Hodding, C.B.

Henry Blackburne Hamilton.

squadron (110 sabres) with the Lughman Valley expedition, commanded by Brigadier-General C. G. Arbuthnot, C.B.[1]

On 5th July 1880 he was present with his regiment in the successful attack on and destruction of the villages of Nargozi, Arab Khel, and Jokan by a mixed force commanded by Colonel Charles Acton of the 51st (King's Own) Light Infantry,[2] when he led the 1st squadron of the Carabiniers which opened the attack.[3] Received the Afghan War medal and was mentioned in despatches, being also recommended for a Brevet-Majority by his commanding officer (Colonel J. Fryer, C.B.). Exchanged in November 1886, as Lieutenant-Colonel, with Lieutenant-Colonel James Dennis, into the 14th (King's) Hussars, and was appointed to command that regiment 1st July 1887. After retaining the command for the usual period of four years, Colonel Hamilton went on half-pay, 1st July 1891, having been quartered with the Fourteenth at Shorncliffe, Brighton, Aldershot, and Hounslow.

During his command he revised and published annually the Regimental Almanack,[4] and he introduced and published a book of Regimental Standing Orders, none being then known in the regiment. The 'Standing Orders' introduced by Colonel Sir John Burgoyne, Bart., who commanded the Fourteenth from 1773 to 1781, had entirely disappeared, probably about the year 1792, and no traces of them remained.[5] In July 1896 Colonel Hamilton was placed on retired pay after thirty-two years' service in the cavalry.

COLONEL THE HONOURABLE G. H. GOUGH, C.B.

APPOINTED LIEUTENANT-COLONEL, 14TH (KING'S) HUSSARS, JULY 1, 1891.

The Hon. G. H. Gough.

The Honourable George Hugh Gough, son of the second Viscount Gough, was born 25th July 1852; educated at Eton and Trinity Hall, Cambridge, where he graduated B.A. in 1872; and entered the army as Cornet in the 14th (King's) Hussars, 28th October 1871. He became Sub-Lieutenant, 1st November 1871; Lieutenant, ante-dated, 28th October 1871; Adjutant, 23rd October 1875; Captain, 23rd July

1 Afterwards General Sir Charles G. Arbuthnot, G.C.B.

2 Now the 1st Battalion (The King's Own) Yorkshire Light Infantry.

3 See *Regimental Records of the 6th Dragoon Guards (Carabiniers)*, p. 66, by Captain Alexander Sprot. Chatham: Gale and Polden. 1888.

4 See *ante* p. 431, Regimental Almanack.

5 See *ante* p. 50.

1879; Brevet-Major, 18th November 1882; Major, 14th Hussars, 16th December 1885; Brevet Lieutenant-Colonel, 15th June 1885; Brevet-Colonel, 16th December 1889; Lieutenant-Colonel, 14th Hussars, 1st July 1891; and Substantive Colonel, 13th July 1898. The Hon. G. H. Gough.

He passed the final examination at the Staff College in 1883, and held the following staff appointments:—

A.D.C. to Lieutenant-General commanding 2nd Division Expeditionary Force in Egypt, 4th August 1882 to 2nd November 1882; D.A.A. and Q.M.-General, Curragh Camp, 1st July 1887 to 30th September 1889; Private Secretary to the Commander-in-Chief, January 1897 till October 1899, acting in the interval as Assistant Military Secretary at the Horse Guards from July to September 1898. Colonel Gough commanded the 14th (King's) Hussars from 1st July 1891 to June 1896, being stationed at Hounslow, Aldershot, Leeds, Manchester, and Cahir. He served in the South African War of 1881 in the Transvaal campaign as Aide-de-Camp to Brigadier-General Sir Thomas D. Baker, K.C.B., and afterwards to Major-General Sir H. E. Wood, V.C., G.C.M.G., K.C.B.; also in the Egyptian expedition of 1882 as Aide-de-Camp to Lieutenant-General Sir E. B. Hamley, K.C.B., K.C.M.G., commanding 2nd Division, and was at the battle of Tel-el-Kebir (horse killed), mentioned in despatches, received the medal with clasp, bronze star, brevet of Major, and fourth class of the Medjidie. In the Soudan expedition, 1884 and 1885, on the Nile, he was in command of the mounted infantry; was present at the action of Abu Klea (wounded), mentioned in despatches, and received two clasps and brevet of Lieutenant-Colonel. When H.M. Queen Victoria opened the Manchester Ship Canal on 21st May 1894, the 14th (King's) Hussars and the 2nd (King's) Liverpool Regiment lined the route of the royal procession through Manchester, on which occasion the whole of the troops, including the volunteers, were under the command of Colonel the Honourable G. H. Gough, and on the 26th of the same month he was awarded the Companionship of the Bath (Military Division).

When the South African War broke out, Colonel Gough was appointed Assistant Adjutant-General of the Cavalry Division in October 1899, and proceeded to Cape Colony, where he served till his much-lamented death, which occurred on the 28th March 1900 at Norval's Pont, to the inexpressible grief of his numerous friends and brother officers, with whom he was most popular. He was buried with full military honours on the 30th March, in the cemetery at Bloemfontein, a squadron of his old regiment being present at the ceremony. The *Daily News* correspondent at the seat of war states in

The Hon. G. H. Gough.

a wire published on 2nd April in the *Westminster Gazette*, that 'Lord Roberts and other officers holding high command followed the remains of Lord Wolseley's former private secretary to the grave. It was an impressive sight as the procession moved solemnly through the streets of Bloemfontein, the troops walking in a slow march, with arms reversed, and bands playing the "Dead March." The coffin was covered with the Union Jack and wreaths of flowers. At the cemetery the trumpeters sounded the "Last Post" over the grave of the soldier whom all comrades honour.'

The same journal also contains the following from a military correspondent in South Africa:—'Colonel the Honourable George Gough—George, as he was affectionately known to all his intimates—will be sincerely mourned by all who knew him. His friends were legion. Few officers were able so quickly to gain the affection and respect of all privileged to serve under him, and few officers will be so missed and so regretted. The last time I saw him was, I think, in 1895, when he won the open military point-to-point race in County Meath, over a big country, riding against all the best men of the Dublin garrison, and further handicapped by the fact that the country was absolutely strange to him. I shall never forget seeing him cantering home, bare-headed, some fifty lengths in front of the foremost of his pursuers, his face radiant with delight at his victory. I think that no officer in command of a regiment—and he was then in command of the 14th Hussars—has ever before or since won this race. When men come to command they are usually not so keen about racing over a country. Not so George Gough: no day was too long and no fence was too big for him. He was in every way the beau-ideal of a light cavalry officer.'

A former colleague writes to the *Army and Navy Gazette* of 7th April as follows:—'I cannot tell you how shocked we all are to hear the news of the death of Colonel the Honourable George H. Gough, C.B. A truer-hearted and more upright and honourable gentleman, a better soldier, and more courteous and industrious official, never served on the staff at headquarters. To know him was to feel affection for him; to be associated with him in office was to admire the abilities which were hidden under that quiet yet manly exterior. He was an officer who was capable of anything, for like most quiet men he was blessed with wonderful powers of judgment, whilst his knowledge of the duties and requirements of his own arm of the service was exceptional. In his death the army has indeed sustained a great loss, and throughout the War Office he will be sincerely mourned.'

LIEUTENANT-COLONEL A. J. ENGLISH

APPOINTED LIEUTENANT-COLONEL, 14TH (KING'S) HUSSARS, JULY 1, 1896.

Augustus John English was born 26th September 1851, and entered the army as a Sub-Lieutenant, 14th (King's) Hussars, 30th December 1871. He became Lieutenant, 30th December 1873; Captain, 16th January 1884; Major, 17th August 1887; and Lieutenant-Colonel, 1st July 1896. A. J. English.

He was extra Aide-de-Camp to the Governor of Madras, 12th June 1877 to 11th November 1878, and again from 11th January to 24th May 1881; and Aide-de-Camp from 2nd June 1879 to 22nd November 1880.

He retained the command of the Fourteenth till 21st February 1899, when he was placed on retired pay after twenty-seven years' service. During his command of the regiment he served at Cahir and Newbridge. In 1881 he was present with the Fourteenth in the Transvaal campaign in South Africa.

COLONEL GILBERT H. C. HAMILTON

APPOINTED LIEUTENANT-COLONEL, 14TH (KING'S) HUSSARS, FEBRUARY 22, 1899.

Gilbert Henry Claude Hamilton was born 30th July 1852, being eldest son of the late Lieutenant-General Henry Meade Hamilton, C.B., and Henrietta, daughter of the late Rev. Sir Erasmus Dixon Borrowes, Bart., of Barretstown Castle, County Kildare, Ireland. Gilbert H. C. Hamilton.

He was appointed Sub-Lieutenant, 14th Hussars, 1st September 1875; Lieutenant on same date; Captain, 24th July 1885; Major, 23rd August 1893; Lieutenant-Colonel, 22nd February 1899; Colonel, 29th November 1900; and he held the post of Adjutant, 14th Hussars, from 30th June 1885 till 31st August 1887. Colonel Hamilton was A.D.C. to Major-General, Madras, 28th September 1877 to 8th December 1888; A.D.C. to Brigadier-General, Afghan campaign, 11th January 1879 to 3rd June 1879, and from 11th January 1880 to August 1880. He was Adjutant to the Yorkshire Dragoons (Yeomanry), 22nd February 1888 to 17th April 1893, and he has commanded the 14th (King's) Hussars since 22nd February 1899.

The following account of Colonel Hamilton appeared recently in a number of the *Navy and Army Illustrated* newspaper:—

'Gilbert Hamilton holds a unique record, for he is the first

Gilbert H. C Hamilton.

cavalry officer in the army who has ever risen straight from the ranks to command a cavalry regiment, and he deserves the position he has won for himself, if ever a man did, for he has shown himself an exceptionally fine soldier and officer. He joined the 2nd Life Guards as a trooper at the age of seventeen and a half, and before he was twenty-three he had won his commission, being appointed to the regiment he now commands. His war services comprise the Afghan War of 1879, in which he was, first, aide-de-camp to Sir Sam. Browne, and afterwards galloper to the officer commanding the cavalry brigade at Cabul, and also the Boer War of 1881. In all military exercises Colonel Hamilton is *facile princeps.* He holds the record score for shooting with the Martini-Henry carbine, and won twice running the Inter-Regimental Revolver competition. So much for his shooting powers. As a swordsman he competed seven times at the Royal Military Tournament in the sword *versus* sword mounted competition, and for six years out of the seven was either first or second—in fact only on one occasion was he knocked out before reaching the final.'

Gilbert Hamilton comes of a soldier stock. His father, the late Lieutenant-General H. M. Hamilton, C.B., served in several campaigns. One of his brothers, Bruce Meade Hamilton, late of the East Yorkshire Regiment, is now Major-General Commanding 21st Brigade in the war in South Africa, and has already distinguished himself there in many engagements. He is a Staff College man. Hubert, also a Staff College man, is a Lieutenant-Colonel in the army, late Major in the Royal West Surrey Regiment, and gained the 'Distinguished Service Order' recently in the campaign in Egypt, when serving with the Egyptian army. He also is at the front, and is Military Secretary to Lieutenant-General Lord Kitchener of Khartoum, G.C.B., K.C.M.G., Commanding-in-Chief the Forces in South Africa; whilst the youngest, Keith, is serving in South Africa with his regiment the Oxfordshire Light Infantry, in which he is a Captain. Father and sons together have up to the present date earned twenty medals between them. Colonel Gilbert Hamilton was mentioned in despatches and received the medal for his services in the Afghan wars of 1878, 1879, and 1880, and he has been commanding the Fourteenth in the present campaign in South Africa until 21st November 1900, when he was appointed Brigadier-General to command the 4th Brigade of Cavalry in Lieutenant-General French's Cavalry Division. His portrait represents him wearing his active service dress (khaki) as he went to South Africa.

EXTRACTS FROM WAR OFFICE MS. ARMY LISTS AT THE PUBLIC RECORD OFFICE

1715

BRIGADIER DORMER'S REGIMENT OF DRAGOONS

Colonel	James Dormer	22nd July 1715
Lieutenant-Colonel	Henry Killigrew	22nd July 1715
Major	Solomon Rapin	22nd July 1715
Captains	Henry Pelham	22nd July 1715
	William Boyle	22nd July 1715
	Beverly Newcomin	22nd July 1715
Captain-Lieutenant	James Stevens	22nd July 1715
Lieutenants	Peter Davenport	22nd July 1715
	Cuthbert Smith	22nd July 1715
	Jonathan Pirke	22nd July 1715
	James Fleming	22nd July 1715
	Henry Lasalle	22nd July 1715
Cornets	Thomas Ellis	22nd July 1715
	Rigley Molyneux[1]	22nd July 1715
	Edward Strode	22nd July 1715
	Thomas Delahay	22nd July 1715
	Andrew Forrester	22nd July 1715
Surgeon	Caleb Harbett	22nd July 1715 (Ensign, 21st November 1710)
Chaplain	Peter Pickering	22nd July 1715
Cornet and Adjutant	William Hamilton	22nd July 1715

1730

COLONEL NEVILL'S REGIMENT OF DRAGOONS

Colonel	Clement Nevill	9th April 1720
Lieutenant-Colonel	Beverly Newcomin	1st December 1720
Major	Cuthbert Smith	12th February 1725
Captain	Richard Bowles	2nd May 1724
Captain-Lieutenant	Josias Patterson	13th September 1721
Lieutenants	John Watson	28th October 1715
	Thomas Ellis	5th April 1720
	William Hamilton	25th August 1720
	Rigley Molyneux	13th September 1721
	Nathaniel Forth	1st January 1725
Cornets	William Ross	3rd June 1717
	James Baillie	13th September 1701
	Peter Smith	14th April 1724 (Ensign, 4th November 1717)
	David Liesly	5th May 1724
	Henry Echlin	1st January 1725

[1] Or 'Molineux.'

1736

COLONEL NEVILL'S REGIMENT OF DRAGOONS

Colonel . .	Clement Nevill	*Lieutenants* . .	Nathaniel Forth
Lieut.-Colonel .	Henry Desgrangues		William Ross
Major . .	Cuthbert Smith		Francis Bayley
Captains . .	James Norris	*Cornets* . . .	Wroth Watson
	O'Brien M. Dilkes		James Baillie
	Richard Bowles		Henry Echlin
Capt.-Lieutenant	Josiah Paterson		John Maine
Lieutenants .	Alexander Knapton		Trevor Smith
	William Hamilton		

EXTRACT FROM AN OLD ARMY LIST

1740

FOURTEENTH, OR BRIGADIER HAMILTON'S REGIMENT OF DRAGOONS

(*Quartered in Ireland*)

RANK.	NAMES OF OFFICERS.	DATE OF PRESENT COMMISSION.	DATE OF FIRST COMMISSION.
Colonel . .	Archibald Hamilton .	27th June 1737	Cornet, 1688
Lieut.-Col. .	William Wright . .	7th July 1737	Cornet, 12th Aug. 1712
Major . .	Michael O'Brien Dilkes .	13th Jan. 1737	Cornet, 12th Aug. 1712
Captains .	R. Bowles	2nd May 1724	Cornet, 15th July 1719
	James Norris . . .	12th Dec. 1833	Cornet, 1st May 1705
	James Clark . . .	13th Jan. 1737	
Cap.-Lieut. .	Josias Paterson . .	13th Sep. 1721	Lieut., 21st Dec. 1708
Lieutenants .	Thomas Ellis . . .	5th April 1720	Cornet, 3rd July 1711
	William Hamilton . .	25th August 1720	Cornet, 22nd July 1715
	William Ross . . .	28th Feb. 1729	Cornet, 3rd June 1717
	Alexander Knapton .	30th August 1733	Ensign, 24th April 1706
	James Baillie . . .	29th June 1739	Cornet, 13th Sep. 1721
Cornets . .	Peter Smith . . .	14th April 1722	Ensign, 4th Nov. 1717
	John Maine . . .	1st May 1734	
	Trevor Smith . . .	2nd August 1734	
	Arthur Forth . . .	20th June 1739	
	George Nash . . .	22nd Nov. 1739	
	Henry Malone . .	22nd Nov. 1739	

EXTRACTS FROM WAR OFFICE MS. ARMY LISTS AT THE PUBLIC RECORD OFFICE

1745

Major-General Hamilton's Regiment of Dragoons

Colonel	Archibald Hamilton
Lieut.-Colonel	William Wright
Major	Cuthbert Smith
Captains	M. O'Brien Dilkes
	Richard Bowles
	James Norris
Capt.-Lieutenant	Josias Paterson
Lieutenants	Thomas Ellis
	William Hamilton
	Nathaniel Forth
	William Ross
	Alexander Knapton
Cornets	James Baillie
	Wroth Watson
	Peter Smith
	Henry Echlin
	John Maine
	Trevor Smith

1752

Lieutenant-General the Right Hon. Lord Tyrawley's Regiment of Dragoons

Colonel	Rt. Hon. Lord Tyrawley
Lieut.-Colonel	James Norris
Major	James Bailie
Captains	Marcus Norman
	John Pomeroy
	John Arabin
Capt.-Lieutenant	Peter Smith
Lieutenants	William Ross
	John Maine
	Trevor Smith
	Arthur Forth
	George Nash
Cornets	Henry Malone
	Andrew Jacob
	Nicholas Cowse
	James Bailie
	Lewellin Nash
	Thomas Goddard
Chaplain	Peter Vatass
Adjutant	Nicholas Cowse (18th May 1743; 3rd June 1752)
Surgeon	Thomas Irving (17th August 1747)

ALPHABETICAL LIST OF THE OFFICERS WHO SERVED IN THE 14TH DRAGOONS, A.D. 1715-54

(The names of the Officers of the Regiment at its formation, in 1715, are printed in italics.)

NAME.	RANK AND DATES SO FAR AS CAN BE ASCERTAINED.	REMARKS.
Arabin, John	Captain, serving in 1754	
Baillie, James, or Bailie or Baily	Lieut., 1739; Capt., 1746; Major, 19th June 1751	Cornet, 1721
Baillie, ——	Quartermaster, 1731	
Baily, James, or Bailie	Cornet, 1747; Lieut., 1754	
Bartellot, ——	Chaplain, December 1729	Was serving, 1741
Bayly, ——	Cornet, 1725	
Beauclerk, ——	Cornet, 1746	Was serving, 1748
Bentley, ——	Quartermaster, 1717 (or earlier)	Died 1718 or 1719
Bettesworth, ——	Cornet, 1717	Was serving, 1719
Blakeney, Grice	Cornet, 8th March 1747	Commanded regiment, 1781-99
Blennerhasset, D.	Quartermaster, 1754	
Bowles, Richard	Capt., 1724; Maj., *vice* Dilkes, 13th Mar. 1741-42	Wounded and greatly distinguished at battle of Prestonpans, 20th September 1745
Boyle, William	Captain, 22nd July 1715	Major, 1716, *vice* Rapin; Lieut.-Col., 1717, *vice* Rapin
Brett, Barnaby	Quartermaster, 1750	
Brown, Robert	Cornet, 20th June 1753; Lieutenant, 1755	Became Captain, 8th March 1757
Chenevix, Daniel	Captain, 4th April 1754	From Lieutenant, 3rd Horse
Clark, James	Capt., *vice* Dilkes, 1737	Was serving, 1748
Cowse, Nicholas	Cornet, 12th Sept. 1745	Twice Adjutant; retired on half-pay as Cornet of Horse, Dec. 1755
Davenport, Peter	Lieut., 22nd July 1715	
Dejean, Louis	Colonel, 27th Nov. 1752	Went in 1757 to 3rd Irish Horse, now 6th Dragoon Guards (Carabiniers)
Delahaye, Thomas	Cornet, 22nd July 1715	
Desgrangues, Hy.	Lieutenant-Colonel, 1731	Was in army up to 1752
Diggs, Thomas	Maj., 1717; Brev. Lieut.-Col. about June 1718	
Dilkes, M. O'Brien	Capt. 1723; Maj., *vice* Smith, 14th Jan. 1737-8	Was M.P. and got leave to attend Parliament, 1735-36; left regiment in 1741-42
Dormer, James	Capt. and Col., 22nd July 1715; Founder of the Regiment	Went to 6th Foot, 1720; died Lieutenant-General, 1742
Echlin, Henry	Cornet, 1st January 1725	Was serving up to 1739
Ellis, Thomas	Cornet, 22nd July 1715; Lieutenant, 1720	Was serving, 1740, as Lieutenant

LIST OF OFFICERS WHO SERVED IN THE 14TH DRAGOONS, 1715-54

NAME.	RANK AND DATES SO FAR AS CAN BE ASCERTAINED.	REMARKS.
Erle, Thomas	Lieut.-Col., 4th Sept. 1754	Left as Major-General, 1773
Fleming, James	Lieut., 22nd July 1715	Was serving, 1718
Ford, John	Lieut., 15th March 1755	Came from 4th Horse
Forrester, Andrew	Cornet (junior), 22nd July 1715	
Forth, Arthur	Cornet, 20th June 1739	Was serving, 1754
Forth, Nathaniel	Lieutenant, 1st Jan. 1725	First commission, 15th Sept. 1719; went to Neville's regiment, 1739
Hamilton, Wm.	Cornet, 22nd July 1715; Lieut., 25th Aug. 1720; Captain, 1746	Was employed in England, 1724, 1726, and 1742, to bring over recruit horses to the regiment in Ireland; was the first Adjutant of Dormer's Dragoons, July 22, 1715
Hamilton, ——	Major, 1725-27	
Hamilton, G.	Major, 1749-50	
Hamilton, Arch.	Colonel, 27th June 1737; Maj.-Gen., 1742	Died Lieut.-Gen., 8th July 1749
Hamilton, H.	Quartermaster, 1724	
Herbert,[1] ——	Surgeon, 1733	
Irving, Thomas	Surgeon, 17th Aug. 1747	
Jacob, ——	Surgeon, 1734	
James, Richard	Quartermaster, 1753	
Killigrew, Henry	Captain and Lieut.-Col., 22nd July 1715	Left regiment about 1717
Knapton, Alex.	Lieutenant, 1733; Lieutenant in army, 1707	Was serving as Lieutenant, 1740; first commission, 1706
Lasalle, Henry	Lieut., 22nd July 1715	Became Capt.-Lieut., 3rd June 1717
Leslie, ——	Cornet, 1733	
Lidserfe, ——	Quartermaster, 1740	
Maine, John	Cornet, 1734; Lieutenant, 1746; Capt.-Lieut., 1754	
Maine, ——	Quartermaster, 1746	Serving in 1748
Malone, Henry	Cornet, 1739	Was serving, 1740, as junior Cornet; died 1753
Medland, ——	Cornet, 1730	Serving, 1734
Molyneux, Rigley	Cornet, 22nd July 1715; Lieut., 1725 (or earlier)	Was serving, 1734
Monck, William	Cornet, 1742	
Morines, Peter	Capt., *vice* Boyle, 1717	
Nash, George	Cornet, 1739; Lieut., 1754	
Nash, Llewellin	Cornet, 1747; Lieut., 1757	
Neville,[2] Clement	Colonel, 9th April 1720	From Lieut.-Col., 13th Dragoons, he went to 8th Dragoons, 1737
Newcomin, Beverly	Captain, 22nd July 1715	Lieut.-Col., 1720; was serving in 1730 (or later)
Norman, ——	Captain, 1750	
Norris, ——	Captain, *vice* Pearce, 1733 or 1734; Major, 1746; Lieut.-Col., 1749	
Patterson, or Paterson, Josias	Lieut., 1717; Capt.-Lieut., 1731; Capt., *v.* Bowles, 1741-42 (13th March)	His first commission as Lieutenant was dated 21st December 1708

[1] This may have been 'Caleb Harbet,' who was the first surgeon appointed to the 14th Dragoons, July 22, 1715. The same names are often spelt in different ways in the old Army Lists.

[2] Or Nevill.

LIST OF OFFICERS WHO SERVED IN THE 14TH DRAGOONS, 1715-54

NAME.	RANK AND DATES SO FAR AS CAN BE ASCERTAINED.	REMARKS.
Pearce, Edward	Captain, 1724	Became Sir E. Pearce, Bart., about 1732; died about 1734
Pelham, Henry	Captain, 22nd July 1715	
Pepper, Thomas	Cornet, Jan. 1755; Lieutenant, 1756	
Pickering, Peter	Chaplain, 1717	Was serving, 1729
Pirke, Jonathan	Lieut., 22nd July 1715	
Pomeroy, John	Captain, 1750-52	
Rapin, Samuel	Captain and Major, 22nd July 1715; became Lieut.-Col., 1716 or 1717	Was wounded in the King's service, probably at Lancaster or Preston, though not mentioned by Cannon; got six months' leave for his wounds in 1716; left regiment, 1717
Ross, William	Cornet, 3rd June 1717; Lieutenant, 1729	Was serving as Lieut., 1740; first commission, Dec. 1716; retired, half-pay, in Tyrawley's regiment, 20th June 1753
Smith, Cuthbert	Lieut., 22nd July 1715; Captain about 1720; became Major, 1725	Was serving, 1730
Smith, John	Cornet, 12th March 1754	
Smith, Peter	Cornet, 14th April 1724; Lieutenant, 1741-42	First commission, 1717
Smith, Trevor	Cornet, 1734	Serving in 1740 as Cornet
Smith, ——	Quartermaster, 1738	Serving in 1740
Stevens, James	Capt.-Lieut., 22nd July 1715; Captain, 1717	
Strode, Edward, or *Stroud*	Cornet (senior), 22nd July 1715	Lieutenant, 3rd June 1717
Thornborough, G.	Was appointed Adjutant, 6th June 1717	He was reduced to 2s. a day half-pay, at age of twenty years
Tyrawley, James, Lord	Lieut.-Gen., appointed Colonel, 24th July 1749	From Colonel of the 10th Foot
Vatass, Peter	Chaplain, 24th Dec. 1745	Remained in the regiment till 1797, and appears to have been on leave the whole of his service. He was seventy-five years of age on retirement
Vernon, Edward	Lieut., 20th June 1753; Captain, 1756	
Waldegrave, Chas.	Quartermaster, 1725	He went with Lieut. Wm. Hamilton, in 1725, to Great Britain, to buy horses for the regiment in Ireland
Watson, Wroth	Cornet, 2nd Jan. 1722	Was serving, 1739; first commission, 1720
Wright, William	Lieutenant-Colonel, 7th July 1737	Was serving in 1746; commanded regiment at Prestonpans, 20th September 1745
Wrighton, ——	Quartermaster, 1740-41	

EXTRACTS FROM ARMY LISTS, A.D. 1754 TO 1901

14TH REGIMENT OF DRAGOONS

Stations (in 1754)—3. Phillipstown; 3. Tullamore. (In 1755)—1. Navan; 2. Ballyshannon; 3. Belturbet. (In 1756)—Ireland.

	1754	1755	1756
Colonel	L. Dejean	Lewis Dejean	Lewis Dejean
Lt.-Colonel	T. Erle	Thomas Erle	Thomas Erle
Major	James Bailie	James Bailie	Marcus Norman
Captains	Marcus Norman	Marcus Norman	John Arabin
	J. Arabin	John Arabin	Daniel Chenevix
	Daniel Chenevix	Daniel Chenevix	Edward Vernon
Capt.-Lieut.	John Mayne	John Mayne	John Mayne
Lieutenants	Trevor Smith	Trevor Smith	Trevor Smith
	Arthur Forth	Edward Vernon	Robert Brown
	George Nash	James Bailie	John Ford
	Edward Vernon	Robert Brown	Thomas Pepper
	James Bailie	John Ford	Arthur Molesworth
Cornets	Andrew Jacob	Andrew Jacob	Andrew Jacob
	Nicholas Cowse	Nicholas Crowse	Llewellin Nash
	Llewellin Nash	Llewellin Nash	Thomas Windh. Goddard
	T. Wind. Goddard	Thomas Windh. Goddard	John Smith
	R. Brown	John Smith	Robert Howard
	J. Smith	Thomas Pepper	James O'Brien
Chaplain	Peter Vatass	Peter Vatass	Peter Vatass
Surgeon	T. Irwin	Thomas Irving	Thomas Irving
Adjutant	Nicholas Cowse	Philip Mercier	Philip Mercier
Quartermasters	Hugh Hamilton	Hugh Hamilton	Hugh Hamilton
	D. Blennerhasset	Arthur Ormsby	Arthur Ormsby
	Barnaby Brett	Barnaby Brett	Barnaby Brett
	Joseph Laughee	Joseph Laughee	Joseph Laughee
	P. Mercier	Philip Mercier	Philip Mercier
	T. Paterson	Thomas Paterson	Thomas Paterson

14TH REGIMENT OF DRAGOONS

Station—Ireland. *Uniform*—Red ; facings, lemon yellow.

	1757	1758	1759
Colonel . .	Lewis Dejean	John Campbell	John Campbell
Lt.-Colonel .	Thomas Erle	Thomas Erle	Thomas Erle
Major . .	Marcus Norman	Marcus Norman	Marcus Norman
Captains . .	John Arabin	Daniel Chenevix	Robert Brown
	Daniel Chenevix	Robert Brown	John Ford
	Edward Vernon	John Ford	Thomas Pepper
Capt.-Lieut. .	John Mayne	John Mayne	John Mayne
Lieutenants .	Trevor Smith	Trevor Smith	Trevor Smith
	Robert Brown	Thomas Pepper	Arthur Molesworth
	John Ford	Arthur Molesworth	Llewellin Nash
	Thomas Pepper	Llewellin Nash	Robert Howard
	Arthur Molesworth	R. Howard	Thomas Windh. Goddard
Cornets . .	Andrew Jacob	Andrew Jacob	John Smith
	Llewellin Nash	Thomas Windh. Goddard	James O'Brien
	Thomas Windh. Goddard	John Smith	Grice Blakeney
	John Smith	James O'Brien	Philip Savage
	Robert Howard	Grice Blakeney	Henry Butler
	James O'Brien	Philip Savage	Thomas Patterson
Chaplain . .	Peter Vatass	Peter Vatass	Peter Vatass
Surgeon . .	Thomas Irving	Thomas Irving	Thomas Irving
Adjutant . .	Jeremiah Hurley	Jeremiah Hurley	Jeremiah Hurley
Quartermasters	Hugh Hamilton		
	Arthur Ormsby		
	Barnaby Brett		
	Joseph Laughee		
	Philip Mercier		
	Thomas Paterson		
Agent . .		Mr. Geo. Cockburn, Dublin	Mr. Geo. Cockburn, Dublin

14TH REGIMENT OF DRAGOONS

Station—Ireland. *Uniform*—Red ; facings, lemon.

	1760	1761	1762
Colonel . .	John Campbell, Maj.-Gen.	John Campbell, Maj.-Gen.	John, Marquis of Lorne, Maj.-Gen.
Lt.-Colonel .	Thomas Erle	Thomas Erle	Thomas Erle
Major . .	John Forde	John Forde	John Forde
Captains . .	Robert Brown	Thomas Pepper	Thomas Pepper
	Thomas Pepper	Arthur Molesworth	Arthur Molesworth
	Arthur Molesworth	Robert Howard	Robert Howard
Capt.-Lieut. .	John Mayne	John Mayne	John Mayne
Lieutenants .	Trevor Smith	Llewellin Nash	Llewellin Nash
	Llewellin Nash	Thos. W. Goddard	Thos. W. Goddard
	Robert Howard	James O'Brien	Grice Blakeney
	Thos. W. Goddard	Grice Blakeney	Philip Savage
	James O'Brien	Philip Savage	Henry Butler
Cornets . .	John Smith	John Smith	John Smith
	Grice Blakeney	Henry Butler	Thomas Patterson
	Philip Savage	Thomas Patterson	Richard O'Brien
	Henry Butler	Edward O'Brien	John Uniacke
	Thomas Patterson	John Uniacke	
	Edward O'Brien	John Head	
Chaplain . .	Peter Vatass	Peter Vatass	Peter Vatass
Adjutant . .	Jeremiah Hurley	Jeremiah Hurley	Jeremiah Hurley
Surgeon . .	Thomas Irving	Thomas Irving	Thomas Irving
Agent . .	Mr. George Cockburn, Dublin	Mr. George Cockburn, Dublin	Mr. George Cockburn, Dublin

	1763	1764	1765
Colonel . .	John, Marquis of Lorne, Maj.-Gen.	John, Marquis of Lorne, Maj.-Gen.	John, Marquis of Lorne, Maj.-Gen.
Lt.-Colonel .	Thomas Erle, Col.	Thomas Erle, Col.	Thomas Erle, Col.
Major . .	John Forde	Arthur Molesworth	Arthur Molesworth
Captains . .	Thomas Pepper	Thomas Pepper	Thomas Pepper
	Arthur Molesworth	Robert Howard	Robert Howard
	Robert Howard	Henry Butler	Henry Butler
Capt.-Lieut. .	John Mayne	John Mayne	John Mayne
Lieutenants .	Llewellin Nash	Llewellin Nash	Llewellin Nash
	Thos. W. Goddard	Thos. W. Goddard	Thos. W. Goddard
	Grice Blakeney	Grice Blakeney	Grice Blakeney
	Philip Savage	John Head	Glynne
	Henry Butler	Glynne	William Owen
Cornets . .	John Smith	Thomas Patterson	Thomas Patterson
	Thomas Patterson	William Owen	Robt. Thos. Hearn
	John Uniacke	Robt. Thos. Hearn	Matthew Jacob
	John Head	Matthew Jacob	Edward Smith
	James Butler	Edward Smith	Simon Pepper
	William Owen	Simon Pepper	Parsons
Chaplain . .	Peter Vatass	Peter Vatass	Peter Vatass
Adjutant . .	Jeremiah Hurley	Jeremiah Hurley	Jeremiah Hurley
Surgeon . .	Thomas Irving	Thomas Irving	Thomas Irving
Agent . .	Mr. George Cockburn, Dublin	Mr. George Cockburn, Dublin	Mr. George Cockburn, Dublin

14th Regiment of Dragoons

Station—Ireland. *Uniform*—Red ; facings, lemon.

	1766	1767	1768
Colonel . .	Charles Fitzroy	Charles Fitzroy	Charles Fitzroy
Lt.-Colonel .	Thomas Erle, Col.	Thomas Erle, Col.	Thomas Erle, Col.
Major . .	Arthur Molesworth	John Maine	John Maine
Captains . .	Thomas Pepper Robert Howard Henry Butler	Thomas Pepper Robert Howard Grice Blakeney	Thomas Pepper Robert Howard Grice Blakeney
Capt.-Lieut. .	John Mayne	John Mayne	John Mayne
Lieutenants .	Llewellin Nash Thos. W. Goddard Grice Blakeney Glynne William Owen	Thos. W. Goddard J. Conway Glynne William Owen Robt. Thos. Hearn Edward Smith	Thos. W. Goddard J. Conway Glynne William Owen Robt. Thos. Hearn Edward Smith
Cornets . .	Thomas Patterson Robt. Thos. Hearn Matthew Jacob Edward Smith Simon Pepper Parsons	Thomas Patterson Simon Pepper Wentworth Parsons Arthur Ormsby John Slack Oliver de Lancey	Thomas Patterson Simon Pepper Wentworth Parsons Arthur Ormsby John Slack Oliver de Lancey
Chaplain . .	Peter Vatass	Peter Vatass	Peter Vatass
Adjutant . .	Jeremiah Hurley	Arthur Ormsby	Arthur Ormsby
Surgeon . .	Thomas Irving	Thomas Irving	Thomas Irving
Agent . .	Mr. George Cock-burn, Dublin	Wm. Montgomery, Dublin	Wm. Montgomery, Dublin

Station—Ireland. *Uniform*—Red ; facings, lemon ; white, red, and green-striped lace.

	1769	1770	1771
Colonel . .	Charles Fitzroy	Charles Fitzroy	Charles Fitzroy
Lt.-Colonel .	Thomas Erle, Col.	Thomas Erle, Col.	Thos. Erle, Maj.-Gen.
Major . .	John Maine	John Maine	John Maine
Captains . .	Thomas Pepper Robert Howard Grice Blakeney	Thomas Pepper Robert Howard Grice Blakeney	Thomas Pepper Grice Blakeney Simon Pepper
Capt.-Lieut. .	John Mayne	Thos. W. Goddard	Thos. W. Goddard
Lieutenants .	Thos. W. Goddard William Owen Robt. Thos. Hearn Thomas Patterson Arthur Ormsby	William Owen Robt. Thos. Hearn Thomas Patterson Arthur Ormsby Simon Pepper	Thomas Patterson Arthur Ormsby Wentworth Parsons Oliver de Lancey Eyre Power French
Cornets . .	Simon Pepper Wentworth Parsons John Slack Oliver de Lancey Eyre Power French Josias Paterson	Wentworth Parsons John Slack Oliver de Lancey Eyre Power French Josias Paterson John Browne	John Slack Josias Paterson John Browne Otway Toler Benjamin Smith Hon. Joseph Dawson
Chaplain . .	Peter Vatass	Peter Vatass	Peter Vatass
Adjutant . .	Arthur Ormsby	Arthur Ormsby	Arthur Ormsby
Surgeon . .	Thomas Irving	Thomas Irving	Thomas Irving
Agent . .	Wm. Montgomery, Dublin	Wm. Montgomery, Dublin	Wm. Montgomery, Dublin

14TH REGIMENT OF DRAGOONS

Station—Ireland. *Uniform*—Red; facings, lemon; white, red, and green-striped lace.

	1772	1773	1774
Colonel . .	Charles Fitzroy	Daniel Webb, Lieut.-Gen.	George Warde
Lt.-Colonel .	Thomas Erle, Maj.-Gen.	Thomas Erle, Maj.-Gen.	John Burgoyne
Major . .	Thomas Pepper	Thomas Pepper	Thomas Pepper
Captains . .	Grice Blakeney	Grice Blakeney	Grice Blakeney
	Simon Pepper	Simon Pepper	Simon Pepper
	Wentworth Parsons	Wentworth Parsons	Wentworth Parsons
Capt.-Lieut. .	T. Wyndh. Goddard	T. Wyndh. Goddard	T. Wyndh. Goddard
Lieutenants .	Thomas Patterson	Thomas Patterson	Thomas Patterson
	Arthur Ormsby	Arthur Ormsby	Arthur Ormsby
	Oliver de Lancey	Oliver de Lancey	Eyre Power French
	Eyre Power French	Eyre Power French	Josias Paterson
	Josias Paterson	Josias Paterson	John Browne
Cornets . .	John Browne	John Browne	Otway Toler
	Otway Toler	Otway Toler	Benjamin Smith
	Benjamin Smith	Benjamin Smith	Hon. Joseph Dawson
	Hon. Joseph Dawson	Hon. Joseph Dawson	George Ross Lewin
	George Ross Lewin	George Ross Lewin	Samuel Faulkner
	Samuel Faulkner	Samuel Faulkner	Hon. J. Cunninghame
Chaplain . .	Peter Vatass	Peter Vatass	Peter Vatass
Adjutant . .	Thomas Patterson	Thomas Patterson	Thomas Patterson
Surgeon . .	Thomas Irving	Thomas Irving	Thomas Irving
Agent . .	Wm. Montgomery, Dublin	Wm. Montgomery, Dublin	Wm. Montgomery, Dublin

	1775	1776
Colonel	George Warde	George Warde
Lieut.-Colonel . .	John Burgoyne	John Burgoyne
Major	Thomas Pepper	Grice Blakeney
Captains . . .	Grice Blakeney	Simon Pepper
	Simon Pepper	Wentworth Parsons
	Wentworth Parsons	Henry Pigot
Captain-Lieutenant .	T. Wyndham Goddard	T. Wyndham Goddard
Lieutenants . . .	Thomas Patterson	Eyre Power French
	Eyre Power French	Josias Paterson
	Josias Paterson	John Browne
	John Browne	Otway Toler
	Otway Toler	Benjamin Smith
Cornets . . .	Benjamin Smith	George Ross Lewin
	George Ross Lewin	Samuel Faulkner
	Samuel Faulkner	Hon. J. Cunninghame
	Hon. J. Cunninghame	George Warde
	George Warde	John Blakeney
	John Blakeney	Chas. Stanley Monck
Chaplain . . .	Peter Vatass	Peter Vatass
Adjutant . . .	Thomas Paterson	Thomas Paterson
Surgeon . . .	George Borthwick	George Borthwick
Agent	Sir W. Montgomery, Bart., Dublin	Sir W. Montgomery, Bart., Dublin

14TH REGIMENT OF LIGHT DRAGOONS

Station—Ireland. *Uniform*—Red; facings, lemon; white, red, and green-striped lace.

	1777	1778	1779
Colonel . .	George Warde	G. Warde, Maj.-Gen.	R. Sloper, Maj.-Gen.
Lt.-Colonel .	John Burgoyne	John Burgoyne, Col.	John Burgoyne, Col.
Major . .	Grice Blakeney	Grice Blakeney	Grice Blakeney
Captains . .	Simon Pepper	Henry Pigot	Henry Pigot
	Henry Pigot	John Gordon	John Gordon
	John Gordon	Otway Toler	Otway Toler
Cap.-Lieut. and Captain .	Otway Toler	Hon. J. Cunninghame	Hon. J. Cunninghame
Lieutenants .	Josias Paterson	John Browne	John Browne
	John Browne	Samuel Faulkner	Samuel Faulkner
	Benjamin Smith	George Warde	Nich. Power French
	Samuel Faulkner	Nich. Power French	Chas. Stanley Monck
	Hon. J. Cunninghame	Chas. Stanley Monck	Robert Riddell
Cornets . .	George Ross Lewin	Hon. Ed. Warde	Henry Bambrick
	George Warde	Henry Bambrick	George Garward
	John Blakeney	George Harward	Richard Galway
	Chas. Stanley Monck	Richard Galway	Hugh H. Magan
	Hon. Ed. Warde	Hugh H. Magan	Francis Blasse
			James Dalrymple
Chaplain . .	Peter Vatass	Peter Vatass	Peter Vatass
Adjutant . .	James Wilson	James Wilson	James Wilson
Surgeon . .	George Borthwick	George Borthwick	George Borthwick
Agent . .	Sir W. Montgomery, Bart., Dublin	Sir W. Montgomery, Bart., Dublin	

	1780	1781	1782
Colonel . .	R. Sloper, Maj.-Gen.	R. Sloper, Maj.-Gen.	R. Sloper, Maj.-Gen.
Lt.-Colonel .	John Burgoyne, Col.	Sir J. Burgoyne, Bt., C.	Grice Blakeney
Major . .	Grice Blakeney	G. Blakeney, Lt.-Col.	Andrew Corbet
Captains . .	Henry Pigot	Otway Toler	Chas. Stanley Monck
	John Gordon	Chas. Stanley Monck	Sir J. Erskine, Bart.
	Otway Toler		Hugh H. Magan
Cap.-Lieut. and Captain .	Hon. J. Cunninghame	Hon. J. Cunninghame	Robert Compton
Lieutenants .	John Browne	Samuel Faulkner	George Dunbar
	Samuel Faulkner	Robert Compton	Francis Blasse
	Nich. Power French	Hugh H. Magan	Robert Wallace
	Robert Riddell	George Dunbar	John King
	Robert Compton	Francis Blasse	Richard Maginnis
Cornets . .	Henry Bambrick	F. C. Musgrave	Alexander Crawford
	Richard Galway	Robert Wallace	Joseph Keighley
	Hugh H. Magan	John King	John Gamble
	Francis Blasse	Alexander Crawford	James Leatham
	James Dalrymple	Joseph Kighley	John Hubert Moore
	F. C. Musgrave		
Chaplain . .	Peter Vatass	Peter Vatass	Peter Vatass
Adjutant . .	James Wilson	John Magee	John Magee
Surgeon . .	George Borthwick	George Borthwick	George Borthwick
Agents . .		Stephen and Robert Wybrants, Dublin	Stephen and Robert Wybrants, Dublin

14TH REGIMENT OF LIGHT DRAGOONS

Station (1783-84)—Ireland. *Uniform* (1783-84)—Red ; facings, lemon yellow ; white, red, and green-striped lace ; (from 1785) blue ; facings, lemon yellow.

	1783	1784	1785
Colonel	R. Sloper, Lt.-Gen.	R. Sloper, Lt.-Gen.	R. Sloper, Lt.-Gen.
Lt.-Colonel	Grice Blakeney	Grice Blakeney	Grice Blakeney
Major	Andrew Corbet	William Richardson	William Richardson
Captains	Chas. Stanley Monck	Chas. Stanley Monck	Sir Geo. Dunbar, Bt.
	Sir Jas. Erskine, Bt.	Hugh H. Magan	Hugh Tuite
	Hugh H. Magan	Sir Geo. Dunbar, Bt.	Lewis Saurin
Cap.-Lieut. and Captain	Robert Compton	Robert Compton	Robert Compton
Lieutenants	Sir Geo. Dunbar, Bt.	Robert Wallace	John King
	Robert Wallace	John King	Richard Maginnis
	John King	Richard Maginnis	Alexander Crawford
	Richard Maginnis	Alexander Crawford	James Leatham
	Alexander Crawford	James Leatham	Edward Dunn
Cornets	Joseph Keighley	Joseph Keighley	Joseph Keighley
	John Gamble	John Gamble	Avarell Lecky
	James Leatham	John Hubert Moore	Charles Norman
	John Hubert Moore	Avarell Lecky	Hon. Ab. Creighton
	Avarell Lecky	Charles Norman	William Connell
	Charles Norman		John Whaley
Chaplain	Peter Vatass	Peter Vatass	Peter Vatass
Adjutant	John Magee	John Magee	John Magee
Surgeon	George Borthwick	George Borthwick	George Borthwick
Agents	Stephen and Robert Wybrants, Dublin	Stephen and Robert Wybrants, Dublin	Stephen and Robert Wybrants, Dublin

	1786	1787	1788
Colonel	R. Sloper, Lt.-Gen.	R. Sloper, Lt.-Gen.	R. Sloper, Lt.-Gen.
Lt.-Colonel	Grice Blakeney	Grice Blakeney	Grice Blakeney
Major	William Richardson	Arthur Carter	Arthur Carter
Captains	Sir Geo. Dunbar, Bt.	Sir Geo. Dunbar, Bt.	Sir Geo. Dunbar, Bt.
	Hugh Tuite	Hugh Tuite	Hugh Tuite
	Lewis Saurin	George Montgomery	George Montgomery
Cap.-Lieut. and Captain	Robert Compton	Robert Compton	Robert Compton
Lieutenants	Richard Maginnis	James Leatham	James Leatham
	Alexander Crawford	John Enery	John Enery
	James Leatham	Charles Norman	Hon. Ab. Creighton
	Avarell Lecky	Hon. Ab. Creighton	David Mellefont
	John Enery	David Mellefont	John Thos. Buller
Cornets	Charles Norman	William Connell	William Connell
	Hon. Ab. Creighton	Dennis Bingham	Dennis Bingham
	William Connell	William Gore	William Gore
	John Whaley	Anthony Lynch	Anthony Lynch
	Denis Bingham	Rich. Thos. Nugent	Rich. Thos. Nugent
		Lorenzo Moore	Lorenzo Moore
Chaplain	Peter Vatass	Peter Vatass	Peter Vatass
Adjutant	John Magee	John Magee	John Magee
Surgeon	George Borthwick	George Borthwick	George Borthwick
Agents	Stephen and Robert Wybrants, Dublin	Stephen and Robert Wybrants, Dublin	Stephen and Robert Wybrants, Dublin

14TH REGIMENT OF LIGHT DRAGOONS

Uniform—Blue; facings, lemon yellow.

	1789	1790	1791
Colonel . .	Robert Sloper, K.B., Lieut.-Gen.	Robert Sloper, K.B., Lieut.-Gen.	Robert Sloper, K.B., Lieut.-Gen.
Lt.-Colonel .	Grice Blakeney	Grice Blakeney	Grice Blakeney, Col.
Major . .	Arthur Carter	Arthur Carter	Arthur Carter
Captains . .	Sir G. Dunbar, Bart. Hugh Tuite G. Montgomery, Bt.	Sir G. Dunbar, Bart. Hugh Tuite G. Montgomery, Bt.	Sir G. Dunbar, Bart. Hugh Tuite David Mellefont
Cap.-Lieut. and Captain .	Robert Compton	Patrick Miller	John Kindillon
Lieutenants .	James Leatham John Enery Hon. Ab. Creighton David Mellefont	James Leatham John Enery David Mellefont William Tomkins Samuel Stapleton	James Leatham William Tomkins Samuel Stapleton John Browne J. B. Wandesford
Cornets . .	William Connell William Gore Anthony Lynch Lorenzo Moore Henry Dod	William Connell Lorenzo Moore J. B. Wandesford Ed. Gerald Butler Geo. Allen Madden Rich. Isaac Starke	William Connell Lorenzo Moore Ed. Gerald Butler Geo. Allen Madden James Butler David Thos. Powell
Chaplain . .	Peter Vatass	Peter Vatass	Peter Vatass
Adjutant . .	John Magee	James Leatham	James Leatham
Surgeon . .	George Borthwick	George Borthwick	George Borthwick
Agents . .	Stephen and Robert Wybrants, Dublin	Stephen and Robert Wybrants, Dublin	Stephen and Robert Wybrants, Dublin

	1792	1793	1794
Colonel . .	Robert Sloper, K.B., Lieut.-Gen.	Robert Sloper, K.B., Lieut.-Gen.	Robert Sloper, K.B., Lieut.-Gen.
Lt.-Colonel .	Grice Blakeney, Col.	Grice Blakeney, Col.	Grice Blakeney, Col.
Major . .	Arthur Carter	Arthur Carter	Arthur Carter
Captains . .	Sir G. Dunbar, Bart. Hugh Tuite David Mellefont	Sir G. Dunbar, Bart. Hon.J.B.Wandesford Mungo Noble	Sir G. Dunbar Hon.J.B.Wandesford Mungo Noble
Cap.-Lieut. and Captain .	John Kindillon	John Kindillon	John Kindillon
Lieutenants .	James Leatham John Browne John B. Wandesford Stephen Mahon	James Leatham John Browne Thos. David Powell Hon. James Butler Robert Christmas	James Leatham John Browne Thos. David Powell Hon. James Butler Robert Christmas
Cornets . .	William Connell Ed. Gerald Butler James Butler David Thos. Powell	Dennis Pack William Cane John George Onge William Pigott	Dennis Pack John George Onge William Pigott Poitier
Chaplain . .	Peter Vatass	Peter Vatass	Peter Vatass
Adjutant . .	James Leatham	James Leatham	James Leatham
Surgeon . .	John Craven	John Craven	John Craven
Agents . .	Stephen and Robert Wybrants, Dublin	Stephen and Robert Wybrants, Dublin	Stephen and Robert Wybrants, Dublin

14TH REGIMENT OF LIGHT DRAGOONS

Uniform—Blue; facings, lemon yellow.

	1795	1796	1797	1798
Colonel .	Robt. Sloper, K.B., Lt.-Gen.	Robt. Sloper, K.B., Lt.-Gen.	Robt. Sloper, K.B., Gen.	John W. Egerton, Maj.-Gen.
Lt.-Colonel	Grice Blakeney, Maj.-Gen.	Grice Blakeney, Maj.-Gen.	Grice Blakeney, Maj.-Gen.	Grice Blakeney, Maj.-Gen.
Major .	Arthur Carter, Lt.-Col.	Arthur Carter, Lt.-Col. Sir G. Dunbar, Bt.	Arthur Carter, Lt.-Col. Sir G. Dunbar, Bt.	Arthur Carter, Lt.-Col. Sir G. Dunbar, Bt.
Captains .	Sir G. Dunbar, Bt., Maj., March 1794 Hon. J. B. Wandesford Hon. James Butler Hamilton Gorges Henry Browne G. H. Montgomery	John Kindillon Hon. James Butler Hamilton Gorges Henry Browne G. H. Montgomery James Kearnay	Hon. James Butler Hamilton Gorges Henry Browne G. H. Montgomery James Leatham William Nowlan	Hon. James Butler Hamilton Gorges Henry Browne William Nowlan Henry Shepherd Neil Talbot
Cap.-Lieut. and Cap.	John Kindillon	James Leatham	John George Onge	James Tyrrell
Lieuts. .	James Leatham John Browne Thomas D. Powell Robert Christmas John George Onge James Tyrrell	James Tyrrell John Craven Dennis Pack Poitier John Trump Potts Jas. Edward White John Hall	James Tyrrell John Craven Poitier John Trump Potts Jas. Edward White John Hall Pat. Nugent Savage John Lombard John Deane Edward Fitzgerald Edward Turton Oliver M'Causland	John Craven Poitier John Trump Potts Jas. Edward White Pat. Nugent Savage John Lombard Edward Turton Oliver M'Causland Charles S. Sloper John Chapman Nixon Brown W. F. Leatham Thos. Dean Pearce
Cornets .	Dennis Pack William Pigott Poitier Jas. Edward White John Sidwell Lodge John Craven	Edward Turton Oliver M'Causland W. F. Leatham H. Arnott Charles S. Sloper Large	W. F. Leatham H. Arnott Charles S. Sloper Large James Flanagan	James Flanagan Lawrence Drought Norman Newby T. Armstrong Peter Keogh Thomas Payter John Mahoney
Chaplain .	Peter Vatass	Peter Vatass	Peter Vatass	
Adjutant .	James Leatham	Lawrence Drought	Laurence Drought	Lawrence Drought
Surgeon .	John Craven	Hugh Arnot	Hugh Arnott	Ruxton
Assistant Surgeons				Samuel Newman Henry Killaly
Agents .	S. and R. Wybrants, Dublin	S. and R. Wybrants, Dublin	Messrs. Sloper and Allen, Gray's Inn	Messrs. Sloper and Allen, Gray's Inn

14TH (DUCHESS OF YORK'S OWN) REGIMENT OF LIGHT DRAGOONS

Uniform—Blue; facings, orange. *Badge*—Prussian Eagle.

	1799	1800	1801
Colonel	John Wm. Egerton, Maj.-Gen.	John Wm. Egerton, Maj.-Gen.	John Wm. Egerton, Maj.-Gen.
Lt.-Colonel	Grice Blakeney, Maj.-Gen.	Grice Blakeney, Maj.-Gen. John Michel	Grice Blakeney, Maj.-Gen. John Michel Samuel Hawker
Major	Arthur Carter, Col. Sir G. Dunbar, Bart., Lt.-Col.	T. O'Brien O'Loghlin Hon. James Butler	T. O'Brien O'Loghlin Hon. James Butler
Captains	Hon. James Butler Henry Browne William Nolan Neil Talbot Richard Pigot Thomas Smith	Henry Browne William Nolan Neil Talbot Richard Pigot Thomas Smith Pat. Nugent Savage	Henry Browne Neil Talbot Richard Pigot Thomas Smith William Savary Thos. Deane Pearce Frederick Villebois
Cap.-Lieut. and Captain	James Tyrrell	William O'Bryen	Pat. Nugent Savage
Lieutenants	Jas. Ed. White Pat. Nugent Savage John Lombard Edward Turton Oliver M'Causland Chas. Stokes Sloper John Chapman Nixon Brown Thos. Dean Pearce James Flanagan Ed. John Nugent D. John Mahoney Elrington Laing	John Lombard Edward Turton Oliver M'Causland Chas. Stokes Sloper John Chapman Nixon Brown Thos. Deane Pearce James Flanagan Ed. John Nugent D. John Mahoney Elrington Laing Edwin Saville James Lowry John Cox, Kt.	Edward Turton Oliver M'Causland Chas. Stokes Sloper John Chapman Nixon Brown Ed. John Nugent D. John Mahoney Edwin Saville James Lowry Matthew L. Bell Peter Keogh Samuel Dick S. Neville Bull George Humphrey
Cornets	Lawrence Drought Norman Newby Peter Keogh James Lowry William Cooper Edwin Saville	Lawrence Drought Peter Keogh George Collier Matthew Bell Samuel Dick	Cecil Bisshopp Hon. Chas. Butler T. Campbell Matthew Kemble Strethill Harrison
Paymaster		James Flanegan	James Flanegan
Adjutant	Lawrence Drought	Lawrence Drought	George Humphrey
Surgeon	John Kidston	John Kidston	John Kidston
Asst.-Surgeon	Henry Killaly	Robert Thompson	Robert Thompson
Vet.-Surgeon	Samuel Newman	Samuel Newman	
Agents	Mr. Collyer and Son, Lisle St., Leicester Square	Mr. Collyer and Son, Lisle St., Leicester Square	Mr. Collyer and Son, Lisle St., Leicester Square

14TH (DUCHESS OF YORK'S OWN) REGIMENT OF LIGHT DRAGOONS

Uniform—Blue; facings, orange. *Badge*—Prussian Eagle.

	1802	1803	1804
Colonel	John Wm. Egerton, Maj.-Gen.	John Wm. Egerton, Lt.-Gen.	John, Earl of Bridgewater, Lt.-Gen.
Lt.-Colonel	Grice Blakeney, Lt.-Gen. John Michel Samuel Hawker	John Michel, Col.	Samuel Hawker
Major	Henry Browne James Gambier	Henry Browne Neil Talbot	Henry Browne Neil Talbot
Captains	Neil Talbot Richard Pigot Thomas Smith Wm. Savary, Lt.-Col. Thos. Deane Pearse Frederick Villebois Chas. Auriol, Maj. Charles S. Sloper	Richard Pigot Thomas Smith William Savary Thos. Deane Pearse Samuel Dick	Richard Pigot, Maj. Thomas Smith Ed. Vincent Eyre Samuel Dick Hon. Chas. Butler Frederick Villebois John Chapman F. B. Hervey
Cap.-Lieut. and Captain	Ed. Vincent Eyre	Ed. Vincent Eyre	
Lieutenants	John Chapman Nixon Browne E. John Nugent D. John Mahoney James Lowry Matthew L. Bell Peter Keogh Samuel Dick George Humphrey Cecil Bishopp Hon. Chas. Butler	John Chapman Nixon Browne D. John Mahoney James Lowry Peter Keogh George Humphrey Cecil Bisshopp Hon. Chas. Butler Strethill Harrison	Nixon Browne D. John Mahoney Peter Keogh George Humphrey Cecil Bishopp Strethill Harrison Christ. Reddington Peter Hawker
Cornets	T. Campbell Matthew Kemble Strethill Harrison Christ. Reddington Ald. Richardson William Gaven	T. Campbell Christ. Reddington William Gaven Geo. Aug. Fenwick Charles Silvertop John Babington	T. Campbell William Gaven Charles Silvertop John Babington Valentine Knightly Robert Knipe T. Henry Skinner Daniel Capel
Paymaster	James Flanegan	James Flanegan	James Flanegan
Adjutant	George Humphrey	John Babington	J. Babington, Cornet
Surgeon	John Kidston	John Kidston	John Kidston
Asst.-Surgeons	Samuel Young Samuel Riviere	J. Hopkins Radford	George Garrett Charles Waite
Vet.-Surgeon	Robert Thompson	Robert Thompson	Robert Thompson
Agents	Mr. Collyer and Son, Park Place, St. James's	Mr. Collyer and Son, Park Place, St. James's	Mr. Collyer and Son, Park Place, St. James's

14th (Duchess of York's Own) Regiment of Light Dragoons

Uniform—Blue ; facings, orange. *Badge*—Prussian Eagle.

	1805	1806	1807
Colonel . .	John, Earl of Bridgewater, Lt.-Gen.	John, Earl of Bridgewater, Lt.-Gen.	John Wm., Earl of Bridgewater, Lt.-Gen.
Lt.-Colonels .	Samuel Hawker	Samuel Hawker Neil Talbot	Samuel Hawker Neil Talbot
Majors . .	Neil Talbot Richard Pigot	Richard Pigot Thomas Smith	Thomas Smith F. B. Hervey
Captains . .	Thomas Smith Ed. Vincent Eyre Hon. Charles Butler Frederick Villebois John Chapman F. B. Hervey D. John Mahony Chas. Massey Baker George Humphrey Peter Hawker	Edwd. Vincent Eyre Hon. Charles Butler Frederick Villebois John Chapman F. B. Hervey D. John Mahony Chas. Massey Baker George Humphrey Peter Hawker Peter Keogh	Edwd. Vincent Eyre Hon. Charles Butler John Chapman D. John Mahony Chas. Massey Baker George Humphrey Peter Hawker Peter Keogh Charles Silvertop Henry Morritt
Lieutenants .	Nixon Browne Peter Keogh Cecil Bishopp Strethill Harrison Christ. Reddington T. Campbell George R. Purdon Charles Silvertop John Babington Valentine Knightley	Cecil Bishopp Strethill Harrison T. Campbell Charles Silvertop John Babington Valentine Knightley Robert Knipe T. Henry Skinner Daniel Capel Joseph Dowson Charles Farquhar Thompson James Agnew Thomas Potter Milles James Humfrey George Stevenson Evelyn P. Dormer Hon. Henry Neville	Cecil Bishopp Strethill Harrison T. Campbell John Babington Valentine Knightley Robert Knipe Daniel Capel Joseph Dowson James Agnew Thos. Potter Milles James Humfrey George Stevenson Evelyn P. Dormer Hon. Henry Neville John Townsend Benjàmin Shotten Charles Eyre David Watts
Cornets . .	Robert Knipe T. Henry Skinner Daniel Capel Thos. Potter Milles James Humphrey George Stevenson Hon. J. Ger. de R. de Ginkel Frederick Holbrooke Hon. Henry Neville Benjamin Shotten	Benjamin Shotten John Townsend Charles Eyre David Watts Charles Read Augustus Atkins William Nourse Charles Knatchbull Houston Wallace John Gwynne	Charles Read William Nourse Charles Knatchbull Houston Wallace John Gwynne Lovell Ben. Badcock John Ponton Theophilus T. Ellis Thomas Smith William Wainman
Paymaster .	James Flanegan	James Flanegan	James Flanegan
Adjutant . .	Benjamin Shotten	Benjamin Shotten	Benjamin Shotten
Surgeon . .	John Kidston	John Kidston	John Kidston
Asst.-Surgeons	George Garrett	Widmer Matthew Spry	Widmer Matthew Spry
Vet.-Surgeon .	Robert Thompson	Robert Thompson	Robert Thompson
Agents . .	Mr. Collyer and Son, Park Pl., St. James's	Mr. Collyer and Son, Park Pl., St. James's	Mr. Collyer and Son, Park Pl., St. James's

14TH (DUCHESS OF YORK'S OWN) REGIMENT OF LIGHT DRAGOONS

Uniform—Blue; facings, orange. *Badge*—Prussian Eagle.

	1808	1809	1810
Colonel	John W., Earl of Bridgewater, Lt.-Gen.	John W., Earl of Bridgewater, Lt.-Gen.	John W., Earl of Bridgewater, Lt.-Gen.
Lt.-Colonels	Samuel Hawker	Samuel Hawker, Col., A.D.C.	Samuel Hawker, Col., A.D.C.
	Neil Talbot	Neil Talbot	Neil Talbot
Majors	F. B. Hervey	F. B. Hervey	F. B. Hervey
	Hon. Charles Butler	Hon. Charles Butler	Hon. Charles Butler
Captains	John Chapman	John Chapman	John Chapman
	D. John Mahony	D. John Mahony	D. John Mahony
	Charles M. Baker	Charles M. Baker	Charles M. Baker
	Peter Hawker	Peter Hawker	Peter Hawker
	Peter Keogh	Peter Keogh	Peter Keogh
	Valentine Knightley	Valentine Knightley	Valentine Knightley
	Hon. Henry Neville	Hon. Henry Neville	Hon. Henry Neville
	Thos. W. Brotherton	Thos. W. Brotherton	Thos. W. Brotherton
	Thomas P. Milles	Thomas P. Milles	Thomas P. Milles
	Patrick Stirling	Patrick Stirling	Patrick Stirling
Lieutenants	Cecil Bishopp	T. Campbell	T. Campbell
	T. Campbell	John Babington	John Babington
	John Babington	Robert Knipe	Robert Knipe
	Robert Knipe	Daniel Capel	Daniel Capel
	Daniel Capel	Joseph Dowson	Joseph Dowson
	Joseph Dowson	Evelyn P. Dormer	Evelyn P. Dormer
	Evelyn P. Dormer	John Townsend	John Townsend
	John Townsend	Benjamin Shotten	Benjamin Shotten
	Benjamin Shotten	Charles Reade	Charles Reade
	Charles Eyre	John Hill Forster	John Hill Forster
	David Watts	Houston Wallace	Houston Wallace
	Charles Reade	John Gwynne	John Gwynne
	John Hill Forster	Lovell B. Badcock	Lovell B. Badcock
	William Nourse	Theophilus T. Ellis	Theophilus T. Ellis
	Charles Knatchbull	William Wainman Phillips	William Wainman
	Houston Wallace	Thomas Smith	Thomas Smith
		Augustus Foster	Augustus Foster
		Samuel Hay	James Hay
		Thomas Clifton	Thomas Clifton
		William Jones	William Jones
			Theobald Shiel
Cornets	John Gwynne	Hon. A. Southwell	Hon. A. Southwell
	Lovell B. Badcock	William J. Pearson	William J. Pearson
	John Ponton	E. B. Impey	John Blackford
	Theophilus T. Ellis	I. W. E. Brydges	I. W. E. Brydges
	Thomas Smith		Hon. G. Fortescue
	William Wainman		Francis G. Keogh
	Augustus Foster		Edward Pellew
			Paul C. Parris
Paymaster	James Flanegan	James Flanegan	James Flanegan
Adjutant	Benjamin Shotten	Benjamin Shotten	Benjamin Shotten
Surgeons	John Kidston	John Ennis	John Ennis
Asst.-Surgeons	Widmer	Widmer	Widmer
	Matthew Spry	Matthew Spry	
		Wm. M'Gillivray	
Vet.-Surgeon	Robert Thompson	Robert Thompson	Robert Thompson
Agents	Mr. Collyer and Son, Park Pl., St. James's	Mr. Collyer and Son, Park Pl., St. James's	Mr. Collyer and Son, Park Pl., St. James's

14TH (DUCHESS OF YORK'S OWN) REGIMENT OF LIGHT DRAGOONS

Uniform—Blue ; facings, orange. *Badge*—Prussian Eagle.

	1811	1812	1813
Colonel	John Wm., Earl of Bridgewater, Lt.-Gen.	John W., Earl of Bridgewater, Lt.-Gen.	John W., Earl of Bridgewater, Lt.-Gen.
Lt.-Colonels	Samuel Hawker, Col., A.D.C.	Samuel Hawker, M.-Gen.	Samuel Hawker, M.-Gen.
	F. B. Hervey	F. B. Hervey	F. B. Hervey
Majors	Hon. Charles Butler	Hon. Charles Butler	Chas. Massey Baker
	John Chapman	John Chapman	Thos. W. Brotherton
Captains	Chas. Massey Baker	Chas. Massey Baker	Peter Hawker
	Peter Hawker	Peter Hawker	Thos. Potter Milles
	Thos. W. Brotherton	Thos. W. Brotherton	John Babington
	Thomas Potter Milles	Thos. Peter Milles	Daniel Capel
	John Babington	John Babington	Hon. Henry Percy
	Robert Knipe	Daniel Capel	Francis Anderson
	Daniel Capel	Hon. Henry Percy	Joseph Dowson
	Hon. Henry Percy	Francis Anderson	John Townsend
	Francis Anderson	Joseph Dowson	Lovell B. Badcock
	Joseph Dowson	John Townsend	William Wainman
Lieutenants	Evelyn P. Dormer	Evelyn P. Dormer	Evelyn P. Dormer
	John Townsend	Benjamin Shotten	Benjamin Shotten
	Benjamin Shotten	John Hill Forster	John Gwynne
	John Hill Forster	John Gwynne	Augustus Foster
	John Gwynne	Lovell B. Badcock	William Jones
	Lovell B. Badcock	Theophilus T. Ellis	Hon. Arth. Southwell
	Theophilus T. Ellis	William Wainman	John Blackford
	William Wainman	Augustus Foster	J. W. E. Brydges
	Augustus Foster	James Hay	Edward Pellew
	James Hay	Thomas Clifton	Francis Fowke
	Thomas Clifton	William Jones	Charles Ward
	William Jones	Theobald Shiel	Hon. Edward Cust
	Theobald Shiel	Hon. Arth. Southwell	Francis Hall
	Hon. Arth. Southwell	John Blackford	Archibald Douglas
	John Blackford	J. W. E. Brydges	James Clavering
	J. W. E. Brydges	Edward Pellew	William Rich. Ellis
	Hon. Geo. Fortescue	Francis Fowke	G. Newman Caswell
	Francis G. Keogh	Charles Ward	Henry White
	Edward Pellew	Hon. Edward Cust	John Gourlay
	Francis Fowke	Francis Hall	John Harvey Thursby
	Charles Ward	William Percy	T. P. Thompson
Cornets	Francis Hall	John Gourlay	Chas. G. Humphreys
	William Percy	James Clavering	Francis Wm. Taylor
	John Gourtay	William Rich. Ellis	William John Lyon
	James Clavering	G. Newman Caswell	Daniel James Webb
	Wm. Richard Ellis	John Harvey Thursby	
	G. Newman Caswell	Henry White	
	John H. Thursday	Chas. G. Humphreys	
		Francis Wm. Taylor	
		Alex. Mackintosh	
Paymaster	James Flanegan	James Flanegan	Samuel Rofe
Adjutant	Benjamin Shotten	Benjamin Shotten	Benjamin Shotten
Quartermaster	Jameson	Jameson	M'Carthy
Surgeon	James Taylor	Christopher Widmer	Arthur Richardson
Asst.-Surgeons	Widmer	Elijah Bush	Elijah Bush
	George Gardner		Thomas M. Perrott
Vet.-Surgeon	Robert Thompson	Robert Thompson	Robert Thompson
Agents	Mr. Collyer and Son, Park Pl., St. James's	Mr. Collyer and Son, Park Pl., St. James's	Mr. Collyer and Son, Park Pl., St. James's

14TH (DUCHESS OF YORK'S OWN) REGIMENT OF LIGHT DRAGOONS

Uniform—Blue; facings, orange; (from 1815) silver lace. *Badge*—Prussian Eagle. *Distinction* (1816)—'Peninsula.'

	1814	1815	1816
Colonel . .	John W., Earl of Bridgewater, Lt.-Gen.	John W., Earl of Bridgewater, Gen.	J. W., Earl of Bridgewater, Gen.
Lt.-Colonels .	S. Hawker, M.-Gen. F. B. Hervey	S. Hawker, M.-Gen. F. B. Hervey, Col.	S. Hawker, M.-Gen. F. B. Hervey, Col.
Majors . .	Chas. Massey Baker Thos. W. Brotherton	C. M. Baker, Lt.-Col. Thos. Wm. Brotherton, Lt.-Col.	C. M. Baker, Lt.-Col. Thomas W. Brotherton, Lt.-Col.
Captains . .	Thos. Potter Milles John Babington Daniel Capel Hon. Henry Percy Francis Anderson Joseph Dowson John Townsend Lovell B. Badcock William Wainman Augustus Foster	T. P. Milles, Major John Babington Daniel Capel Hon. H. Percy, Major Francis Anderson Joseph Dowson John Townsend Lovell B. Badcock William Wainman Augustus Foster	T. P. Milles, Major John Babington Daniel Capel Hon.H.Percy,L.-Col. Francis Anderson Joseph Dowson John Townsend Lovell B. Badcock William Wainman Augustus Foster
Lieutenants .	Evelyn P. Dormer Benjamin Shotten William Jones Hon. Arth. Southwell John Blackford J. W. E. Brydges Edward Pellew Francis Fowke Charles Ward Hon. Edward Cust Francis Hall Archibald Douglas James Clavering William Rich. Ellis John Gourlay John Harvey Thursby T. P. Thompson Chas. G. Humphreys Francis Wm. Taylor Wm. John Lyon James Bennett	Benjamin Shotten William Jones J. W. E. Brydges Francis Fowke Charles Ward Francis Hall Archibald Douglas James Clavering John Harvey Thursby Chas. G. Humphreys Francis Wm. Taylor James Bennett James Blake Charles M'Carthey George Hutchinson Hon. Charles Petre St. John Charlton Charles Wyndham Thos. Molyneux Stephen Simpson	Benjamin Shotten William Jones J. W. E. Brydges Francis Fowke Charles Ward Francis Hall James Clavering John Harvey Thursby C. Gard. Humphreys Francis Wm. Taylor James Bennett James Blake Charles M'Carthey George Hutchinson Hon. Charles Petre St. John Charlton Stephen Simpson G. Goodwin Barrett Robert C. Hammond William Wilton
Cornets . .	Daniel Jas. Webb Charles M'Carthey	Robt. C. Hammond David Brown William Wilton Constant E. Trent Francis O. Trent	Constant E. Trent Francis O. Trent John Thomas Brooks Peter Hammond Edward Lane Parry John St. Leger Harrison W. Sober
Paymaster .	Samuel Rofe	Samuel Rofe	Samuel Rofe
Adjutant . .	Benjamin Shotten	Charles M'Carthey	Charles M Carthey
Quartermaster	Henry Smith	Henry Smith	Henry Smith
Surgeon . .	Arthur Richardson	Arthur Richardson	Arthur Richardson
Asst.-Surgeons	Elijah Bush Daniel Owen Davis	Elijah Bush Daniel Owen Davis	Elijah Bush Daniel Owen Davis
Vet.-Surgeon .	Robert Thompson	Black	Alexander Black
Agents . .	Mr. Collyer and Son, Park Pl., St. James's	Mr. Collyer and Son, Park Pl., St. James's	Mr. Collyer and Son, Park Pl., St. James's

14TH (DUCHESS OF YORK'S OWN) REGIMENT OF LIGHT DRAGOONS

Uniform—Blue; facings, orange; silver lace. *Badge*—The Prussian Eagle.
Distinction—'Peninsula.'

	1817	1818	1819
Colonel	J. W., Earl of Bridgewater, Gen.	J. W., Earl of Bridgewater, Gen.	J. W., Earl of Bridgewater, Gen.
Lt.-Colonel	F. B. Hervey, Col.	F. B. Hervey, Col.	Sir Felt. B. Hervey, Bart., Col.
Majors	Chas. Massey Baker, Lt.-Col. Thos. Wm. Brotherton, Lt.-Col.	Chas. Massey Baker, Lt.-Col. Thos. Wm. Brotherton, Lt.-Col.	Chas. Massey Baker, Lt.-Col. Thos. Wm. Brotherton, Lt-Col.
Captains	Thos. P. Milles, Maj. John Babington Daniel Capel Hon. Henry Percy, Lt-Col. Francis Anderson Joseph Dowson John Townsend Lovell B. Badcock	Thos. Potter Milles John Babington Daniel Capel Hon. Henry Percy, Lt.-Col. Francis Anderson Joseph Dowson John Townsend Lovell B. Badcock	Thos. Potter Milles John Babington Daniel Capel Hon. Henry Percy, Lt.-Col. Francis Anderson Joseph Dowson John Townsend Lovell B. Badcock
Lieutenants	Benjamin Shotten William Jones Francis Fowke Charles Ward Francis Hall James Clavering John H. Thursby C. Gard. Humphreys C. M'Carthey, Adj. Hon. Charles Petre St. John Charlton	William Jones Francis Fowke Charles Ward James Clavering John H. Thursby C. M'Carthey, Adj. Hon. Charles Petre St. John Charlton Robt. C Hammond William Wilton William Beckwith	William Jones Francis Fowke Charles Ward James Clavering C. M'Carthey, Adj. Hon. Charles Petre St. John Charlton Robt. C. Hammond William Wilton William Beckwith
Cornets	Constant E. Trent Francis O. Trent Peter Hammond Edward Lane Parry John St. Leger Harrison W. Sober	Constant E. Trent Francis O. Trent Peter Hammond Edward Lane Perry John St. Leger Harrison W. Sober	Constant E. Trent Edward Lane Parry John St. Leger Harrison W. Sober Wm. T. Carruthers John Whitla Henry Gage
Paymaster	Samuel Rofe	Samuel Rofe	Samuel Rofe
Adjutant	Chas. M'Carthey, Lt.	Chas. M'Carthey, Lt.	Chas. M'Carthey, Lt.
Quartermaster	Henry Smith	Henry Smith	Henry Smith
Surgeon	Daniel O'Flaherty	Daniel O'Flaherty	Daniel O'Flaherty
Asst.-Surgeons	Elijah Bush Daniel Owen Davis	Elijah Bush Daniel Owen Davis	Elijah Bush Daniel Owen Davis
Vet.-Surgeon	Alexander Black	Alexander Black	Alexander Black
Agents	Mr. Collyer and Son, Park Pl., St. James's	Mr. Collyer and Son, Park Pl., St. James's	Mr. Collyer and Son, Park Pl., St. James's

14TH (DUCHESS OF YORK'S OWN) REGIMENT OF LIGHT DRAGOONS

Uniform—Blue; facings, orange; silver lace. *Badge*—Prussian Eagle. *Distinction*—'Peninsula.' *Additional distinctions from* 1821—'Talavera, 'Fuentes d'Onor,' 'Salamanca,' 'Vittoria,' 'Orthes.'

	1820	1821	1822
Colonel	J. W., Earl of Bridgewater, Gen.	J. W., Earl of Bridgewater, Gen.	J. W., Earl of Bridgewater, Gen.
Lt.-Colonel	Chas. Massey Baker	Chas. Massey Baker	Chas. Massey Baker
Major	Thos. Wm. Brotherton, Lieut.-Col. Thos. Potter Milles	Thos. Potter Milles Hon. Henry Percy, Lieut.-Col.	Thos. Potter Milles John Townsend
Captains	John Babington, Maj. Hon. Henry Percy, Lieut.-Col. Francis Anderson Joseph Dowson John Townsend Lovell B. Badcock, Major William Jones Francis Fowke	John Babington, Maj. John Townsend, Maj. Lovell B. Badcock William Jones Charles Ward Hon. Charles Petre Geo. Goodin Barrett St. John Charlton	John Babington, Maj. Lovell Ben. Badcock, Major William Jones Charles Ward Hon. Charles Petre Geo. Goodin Barrett
Lieutenants	Charles Ward Chas. M'Carthey, Adj. Hon. Charles Petre St. John Charlton Rob. Carr Hammond William Beckwith James Ormsby Edward Lane Parry John St. Leger	Chas. M'Carthey, Adj. William Beckwith Edward Lane Parry John St. Leger John Wm. Gage Edward H. Foster Harrison W. Sober Henry Gage Ed. Sherlock Gooch	Chas. M'Carthey, Adj. William Beckwith Edward Lane Parry John St. Leger John Wm. Gage Henry Gage Ed. Sherlock Gooch
Cornets	Harrison W. Sober Wm. T. Carruthers Henry Gage Ed. Sherlock Gooch J. Williams Willes Wm. James D'Urban Keith Alex. Jackson Thos. Rich. Baker	J. Williams Willes Wm. James D'Urban Keith Alex. Jackson Thos. Rich. Baker Christ. Musgrave John W. S. Smith T. D. E. Burroughs Horatio Ross	J. Williams Willes Wm. James D'Urban Sir Keith Alex. Jackson, Bart. Thos. Rich. Baker Christ. Musgrave John W. S. Smith
Paymaster	Samuel Rofe	Samuel Rofe	Samuel Rofe
Adjutant	Chas. M'Carthey, Lt.	Chas. M'Carthey, Lt.	Chas. M'Carthey, Lt.
Quartermaster	Henry Smith	John Houghton	John Houghton
Surgeon	Daniel O'Flaherty	Daniel O'Flaherty	Daniel O'Flaherty
Asst.-Surgeon	Elijah Bush	Elijah Bush	Elijah Bush
Vet.-Surgeon	Alexander Black	Alexander Black	Alexander Black
Agents	Mr. Collyer and Son, Park Place, St. James's	Mr. Collyer and Son, Park Place, St. James's	Mr. Collyer and Son, Park Place, St. James's

14TH (DUCHESS OF YORK'S OWN) REGIMENT OF LIGHT DRAGOONS

Uniform—Blue; facings, orange; silver lace. *Badge*—Prussian Eagle. *Distinctions*—'Peninsula,' 'Talavera,' 'Fuentes d'Onor,' 'Salamanca,' 'Vittoria,' 'Orthes.'

	1823	1824	1825
Colonel	J. W., Earl of Bridgewater, Gen.	Sir John Ormsby Vandeleur, K.C.B., Lieut.-Gen.	Sir John Ormsby Vandeleur, K.C.B., Lieut.-Gen.
Lt.-Colonel	Chas. Massey Baker	Chas. Massey Baker	Chas. Massey Baker
Majors	Thos. Potter Milles John Townsend	Thos. Potter Milles John Townsend	Thos. Potter Milles John Townsend
Captains	Lovell Ben. Badcock, Major William Jones Charles Ward Geo. Goodin Barrett William Beckwith Edward Lane Parry	Lovell Ben. Badcock, Major William Beckwith Edward Lane Parry Sir Thomas Ormsby, Bart. Hon. George Anson John Sam. St. Leger	William Beckwith Edward Lane Parry John Sam. St. Leger John Wm. Gage Henry Gage Ed. Sherlock Gooch
Lieutenants	Chas. M'Carthey, Adj. John St. Leger John Wm. Gage Henry Gage Ed. Sherlock Gooch J. Williams Willes Sir Keith A. Jackson, Bart.	Charles M'Carthey, Adjutant John Wm. Gage Henry Gage Ed. Sherlock Gooch J. Williams Willes Josh. Simmons Smith Ambrose Congreve	Charles M'Carthey, Adjutant Joshua S. Smith Ambrose Congreve Thos. Rich. Baker Christ. Musgrave J. Woolmore Sim. Smith
Cornets	Wm. James D'Urban Thos. Rich. Baker Christ. Musgrave John W. S. Smith Horatio Ross John Griffis, Riding-master Rich. Thos. Gilpin	Thos. Rich. Baker Christ. Musgrave John W. Simmons Smith John Griffis, Riding-master Rich. Thos. Gilpin Adam Gordon Duff George Rooke	Rich. Thos. Gilpin John Griffis, Riding-master Adam Gordon Duff George Rooke William Maxwell Charles Delmè Jas. Massey Dawson John Kennedy
Paymaster	Samuel Rofe	Samuel Rofe	Samuel Rofe
Adjutant	Charles M'Carthey	Chas. M'Carthey, Lt.	Chas. M'Carthey, Lt.
Quartermaster	John Houghton	John Houghton	John Houghton
Surgeon	Daniel O'Flaherty	Daniel O'Flaherty	Thomas Forster
Asst.-Surgeon	Elijah Bush	Elijah Bush	Elijah Bush
Vet.-Surgeon	Alexander Black	Alexander Black	Alexander Black
Agents	Mr. Collyer and Son, Park Place, St. James's	Messrs. Greenwood, Cox, and Hammersley	Messrs Greenwood, Cox, and Hammersley

14TH (DUCHESS OF YORK'S OWN) REGIMENT OF LIGHT DRAGOONS

Uniform—Blue; facings, orange; silver lace. *Badge*—Prussian Eagle.
Distinctions—'Talavera,' Fuentes d'Onor,' 'Salamanca,' 'Vittoria,' 'Orthes,' 'Peninsula.'

	1826	1827	1828
Colonel . .	Sir J. O. Vandeleur, K.C.B., Lt.-Gen.	Sir J. O. Vandeleur, K.C.B., Lt.-Col.	Sir J. O. Vandeleur, K.C.B., Lt.-Gen.
Lt.-Colonel .	Chas. Massey Baker	Chas. Massey Baker	Chas. Massey Baker
Majors . .	Thos. Potter Milles John Townsend	Thos. Potter Milles John Townsend	Thos. Potter Milles John Townsend
Captains . .	William Beckwith Edward Lane Parry John William Gage Henry Gage Edward S. Gooch John Methold	William Beckwith Edward Lane Parry John William Gage Edward S. Gooch John Methold Charles M'Carthey	William Beckwith Edward Lane Parry John William Gage John Methold Charles M'Carthey Joshua S. Smith
Lieutenants .	C. M'Carthey, Adj. Joshua S. Smith Ambrose Congreve Thos. Richard Baker J. Woolmore Sim. Smith Adam Gordon Duff George Rooke	Joshua S. Smith Ambrose Congreve J. Woolmore Sim. Smith John Griffis, Riding-master William Maxwell Charles Delmè Jas. Massey Dawson	Ambrose Congreve J. Woolmore Sim. Smith John Griffis, Riding-master William Maxwell Charles Delmè Jas. Massey Dawson John Kennedy
Cornets . .	John Griffis, Riding-master William Maxwell Charles Delmè Jas. Massey Dawson John Kennedy Charles Barton	John Kennedy Charles Barton Wm. B. Alexander Charles Abbott Edward Tenison Chas. John Griffiths Patrick Leary, Adj.	Charles Barton Charles Abbott Edward Tenison Chas. John Griffiths Patrick Leary, Adj. Henry Evans
Paymaster .	Samuel Rofe	Samuel Rofe	Samuel Rofe
Adjutant .	Chas. M'Carthey, Lt.	Patrick Leary, Cor.	Patrick Leary
Quartermaster .	John Houghton	John Houghton	John Houghton
Surgeon . .	Thomas Foster	Thomas Foster	Thomas Foster
Asst.-Surgeon .	Elijah Bush	Chas. Clarke Hughes, M.D.	Chas. Clarke Hughes, M.D.
Vet.-Surgeon .	Alexander Black	Alexander Black	Alexander Black
Agents . .	Messrs. Greenwood, Cox, and Hammersley	Messrs. Greenwood, Cox, and Hammersley	Messrs. Greenwood, Cox, and Hammersley

14TH (DUCHESS OF YORK'S OWN) REGIMENT OF LIGHT DRAGOONS

Uniform—Blue; facings, orange; silver lace. *Badge*—Prussian Eagle.
Distinctions—'Talavera,' 'Fuentes d'Onor,' 'Salamanca,' 'Vittoria,' 'Orthes,' 'Peninsula.'

	1829	1830
Colonel	Sir John Ormsby Vandeleur, K.C.B., Lt.-Gen.	Sir John Ormsby Vandeleur, K.C.B., Lt.-Gen.
Lieut.-Colonel . .	Charles Massey Baker	John Townsend
Majors	John Townsend	William Beckwith
	William Beckwith	Edward Lane Parry
Captains	Edward Lane Parry	John William Gage
	John William Gage	Joshua Simmonds Smith
	John Methold	J. Woolmore Sim. Smith
	Charles M'Carthey	Ambrose Congreve
	Joshua Simmons Smith	Charles Delmè
	J. Woolmore Sim. Smith	Christopher Musgrave
Lieutenants . . .	Ambrose Congreve	John Griffis, Riding-master
	John Griffis, Riding-master	William Maxwell
	William Maxwell	James Massey Dawson
	Charles Delmè	John Kennedy
	James Massey Dowson	Charles Barton
	John Kennedy	Charles Abbott
	Charles Barton	Edward Tenison
Cornets	Charles Abbott	Charles John Griffiths
	Edward Tenison	Patrick Leary, Adj.
	Charles John Griffiths	Edward Stanley Curwen
	Patrick Leary, Adj.	Henry Van Straubenzee
	Henry Evans	Henry Bowyer
	Edward Stanley Curwen	James Hodson
Paymaster . . .	Samuel Rofe	Samuel Rofe
Adjutant . . .	Patrick Leary	Patrick Leary
Quartermaster . .	John Houghton	John Houghton
Surgeon	Patrick Henry Lavens	Patrick Henry Lavens
Assistant-Surgeon . .	Joseph Huey	Joseph Huey
Veterinary-Surgeon .	Alexander Black	Alexander Black
Agents	Messrs. Greenwood, Cox, and Hammersley	Messrs. Greenwood, Cox, and Hammersley

14TH (THE KING'S) REGIMENT OF LIGHT DRAGOONS

Uniform (1831)—Blue; facings, orange; silver lace; (from 1832), scarlet; facings, blue; gold lace. *Badges*—Prussian Eagle, (and from 1832) the King's Crest. *Distinctions*—'Talavera,' 'Fuentes d'Onor,' Salamanca,' 'Vittoria,' 'Orthes,' 'Peninsula.'

	1831	1832	1833
Colonel	Sir Edward Kerrison, Bt., K.C.B., Maj-Gen.	Sir Edward Kerrison, Bt., G.C.H., Maj.-Gen.	Sir Edward Kerrison, Bt., G.C.H., Maj.-Gen.
Lt.-Colonel	John Townsend	John Townsend	John Townsend
Majors	William Beckwith	Wm Beckwith, K.H.	Wm. Beckwith, K.H.
	Edward Lane Parry	Edward Lane Parry	Edward Lane Parry
Captains	John Wm. Gage	John Wm. Gage	John Wm. Gage
	Joshua S. Smith	Joshua S. Smith	Joshua S. Smith
	J. Woolmore Sim. Smith	J. Woolmore Sim. Smith	J. Woolmore Sim. Smith
	Ambrose Congreve	Ambrose Congreve	Ambrose Congreve
	Charles Delmè	Charles Delmè	Charles Delmè
	Christ. Musgrave	Christ. Musgrave	James M. Dawson
Lieutenants	John Griffis, Riding-master	John Griffis, Riding-master	John Griffis, Riding-master
	Jas. Massey Dawson	James M. Dawson	Charles Barton
	Charles Barton	Charles Barton	Hon. Charles Abbott
	Hon. Charles Abbott	Hon. Charles Abbott	Edward Tenison
	Edward Tenison	Edward Tenison	Patrick Leary
	Chas. John Griffiths	Patrick Leary, Adj.	Francis H. Stephens
	Patrick Leary, Adj.	Edward S. Curwen	Charles Thornhill
	Edward S. Curwen	Henry Bowyer	James Phelps
Cornets	Henry V. Straubenzee	James Hodson	Edmund Royds
	Henry Bowyer	Robt. Alex. Lockhart	Arthur Surtees
	James Hodson	Charles Thornhill	Hon. B. C. Yelverton
	Robt. Alex. Lockhart	James Phelps	John Henderson
	Charles Thornhill	Edmund Royds	William H. Archer
Paymaster	Samuel Rofe	Samuel Rofe	Samuel Rofe
Adjutant	Patrick Leary	Patrick Leary	Patrick Leary
Quartermaster	John Houghton	John Houghton	John Houghton
Surgeon	Patrick H. Lavens	Patrick H. Lavens	Patrick H. Lavens
Asst.-Surgeon	Joseph Huey	Joseph Huey	Joseph Huey
Vet.-Surgeon	Alexander Black	Alexander Black	Alexander Black
Agents	Messrs. Greenwood, Cox and Co.	Messrs. Greenwood, Cox and Co.	Messrs. Greenwood, Cox and Co.

14TH (THE KING'S) REGIMENT OF LIGHT DRAGOONS

Uniform—Scarlet; facings, blue; gold lace. *Badges*—The King's Crest and Prussian Eagle.
Distinctions—'Talavera,' 'Fuentes d'Onor,' 'Salamanca,' 'Vittoria,' 'Orthes,' 'Peninsula.'

	1834	1835	1836
Colonel . .	Sir Edward Kerrison, Bt., G.C.H., Maj.-Gen.	Sir Edward Kerrison, Bt., G.C.H., Maj.-Gen.	Sir Edward Kerrison, Bt., G.C.H., Maj.-Gen.
Lt.-Colonel .	John Townsend	John Townsend	John Townsend
Major . .	Edward Lane Parry	Edward Lane Parry	J. Woolmore S. Smith
Captains . .	J. Woolmore Sim. Smith Charles Delmè James M. Dawson Charles Barton Edward Harvey Edward Tenison	J. Woolmore Sim. Smith Charles Delmè James M. Dawson Charles Barton Edward Harvey Edward Tenison	Jas. Massey Dawson Charles Barton Edward Harvey Edward Tenison Francis H. Stephens John Griffis
Lieutenants .	John Griffis, Riding-master Patrick Leary, Adj. Francis H. Stephens Charles Thornhill James Phelps Edmund Royds Hon. B. C. Yelverton John Henderson	John Griffis, Riding-master Patrick Leary, Adj. Francis H. Stephens Charles Thornhill James Phelps Edmund Royds Hon. B. C. Yelverton John Henderson	Patrick Leary, Adj. Charles Thornhill James Phelps Edmund Royds Hon. B. C. Yelverton John Henderson Wm. Henry Archer
Cornets . .	William H. Archer Charles Jones Lawrence C. Bayntum John B. Culpeper Henry E. Doherty	Wm. Henry Archer Charles Jones Lawrence C. Bayntum John B. Culpeper Henry Ed. Doherty	Lawrence C. Bayntum John B. Culpeper Henry Ed. Doherty C. Edmond Doherty Isaac Cornock
Paymaster .	Samuel Rofe	Samuel Rofe	Samuel Rofe
Adjutant . .	Patrick Leary	Patrick Leary	Patrick Leary
Quartermaster .	John Houghton	John Houghton	John Houghton
Surgeon . .	Patrick H. Lavens	Patrick H. Lavens	Patrick H. Lavens
Asst.-Surgeon .	Joseph Huey	Joseph Huey	Joseph Huey
Vet.-Surgeon .	Alexander Black	Alexander Black	Alexander Black
Agents . .	Messrs. Greenwood, Cox and Co.	Messrs. Greenwood, Cox and Co.	Messrs. Cox and Co.

14TH (THE KING'S) REGIMENT OF LIGHT DRAGOONS

Uniform—Scarlet; facings, blue; gold lace. *Badges*—The King's Crest and Prussian Eagle. *Distinctions*—'Talavera,' 'Fuentes d'Oncr,' 'Salamanca,' 'Vittoria,' 'Orthes,' 'Peninsula.' *Additional distinction in* 1838—'Douro.'

	1837	1838	1839
Colonel	Sir Edward Kerrison, Bt., G.C.H., Lt.-Gen.	Sir Edward Kerrison, Bt., G.C.H., Lt.-Gen.	Sir Edward Kerrison, Bt., G.C.H., Lt.-Gen.
Lt.-Colonel	John Townsend	John Townsend	John Townsend
Major	J. Woolmore S. Smith	J. Woolmore S. Smith	Charles Barton
Captains	Charles Barton Edward Harvey Francis H. Stephens Charles Thornhill James Phelps Edmund Royds	Charles Barton Edward Harvey Francis H. Stephens James Phelps Edmund Royds John Henderson	Edward Harvey Francis H. Stephens James Phelps John Henderson Wm. Henry Archer John B. Culpeper
Lieutenants	Patrick Leary, Adj. Hon. B. C. Yelverton John Henderson Wm. Henry Archer John B. Culpeper Henry Ed. Doherty C. Edmond Doherty	Patrick Leary, Adj. Wm. Henry Aicker John B. Culpeper Henry Ed. Doherty C. Edmond Doherty Geo. Main Fullerton Isaac Cornock	Henry Ed. Doherty C. Edmond Doherty Geo. Main Fullerton Isaac Cornock John Henry Tonge Chas. Tindal Griffis, Riding-master
Cornets	Isaac Cornock Chas. Tindal Griffis William Bodkin John Henry Tonge Andrew Robertson Archibald R. Miller	Chas. Tindal Griffis, Riding-master William Bodkin John Henry Tonge Andrew Robertson Arch. Robt. Miller William Faber	Andrew Robertson Arch. Robt. Miller William Faber Geo. K. M. Dawson William Clarke, Adj. Chas. H. J. Rich Henry R. Boucherett
Paymaster	Samuel Rofe	Samuel Rofe	Samuel Rofe
Adjutant	Patrick Leary	Patrick Leary	William Clarke
Quartermaster	John Houghton	William Clarke	Samuel Brodribb
Surgeon	Patrick H. Lavens	Patrick H. Lavens	Patrick H. Lavens
Asst.-Surgeon	Joseph Huey	Joseph Huey, M.D.	Jas. Wemyss Moffat
Vet.-Surgeon	Alexander Black	Alexander Black	James Geo. Philips
Agents	Messrs. Cox and Co.	Messrs. Cox and Co.	Messrs. Cox and Co.

14TH (THE KING'S) REGIMENT OF LIGHT DRAGOONS

Station—(in 1841) India. *Uniform*—Scarlet; facings, blue; gold lace. *Badges*—The King's Crest and Prussian Eagle. *Distinctions*—'Douro,' 'Talavera,' 'Fuentes d'Onor,' 'Salamanca,' 'Vittoria,' 'Orthes,' 'Peninsula.'

	1840	1841	1842
Colonel	Sir Edward Kerrison, Bt., G.C.H., Lt.-Gen.	Sir Edward Kerrison, Bt., G.C.H., K.C.B., Lt.-Gen.	Sir Edward Kerrison, Bt., G.C.H., K.C.B., Lt.-Gen.
Lt.-Colonels	John Townsend	John Townsend	John Townsend, Col. Wm. Havelock, K.H.
Major	Charles Barton	Charles Barton	Charles Barton Ed. Harvey, Lt.-Col.
Captains	Edward Harvey Francis H. Stephens John Henderson Wm. Henry Archer John B. Culpeper Henry Ed. Doherty	Edward Harvey Francis H. Stephens Wm. Henry Archer Henry Ed. Doherty George Weston Geo. Main Fullerton	Francis H. Stephens Wm. Henry Archer Henry Ed. Doherty George Weston Geo. Main Fullerton Charles Stewart Geo. Gardine Shaw Arch. Ed. Bromwich John Henry Tonge
Lieutenants	C. Edmond Doherty Geo. Main Fullerton Isaac Cornock John Henry Tonge Chas. Tindal Griffis, Riding-master Arch. Robt. Miller	John Henry Tonge Chas. Tindal Griffis, Riding-master Arch. Robt. Miller Geo. K. M. Dawson William Clarke, Adj. Robert H. S. Barry John H. Goddard	Chas. Tindal Griffis, Riding-master Geo. K. M. Dawson William Clarke, Adj. John H. Goddard James Boalth James Chambre Frederick Holder Arthur Scudamore Richard H. Gall John F. FitzGerald John A. Todd Rich. B. Prettejohn Henry F. Hodson Chas. W. Thompson Thomas W. Smith John Hamilton Gray Sydney L. Horton Charles Morant George E. Hillier
Cornets	William Faber Geo. K. M. Dawson William Clarke, Adj. Robert H. S. Barry John H. Goddard Wm. Warner Allen J. W. M. G. Hughes	Wm. Warner Allen J. W. M. G. Hughes Charles Morant Rd. H. S. Barry Joseph Cuff Barrett William Nettleship	Joseph Cuff Barrett William Nettleship Francis D. Gray Chas. Potts Rosser Wm. Edward Buller Richard P. Apthorp James Coster Robert J. Brown
Paymaster	Samuel Rofe	Samuel Rofe	Samuel Rofe
Adjutant	William Clarke	William Clarke	William Clarke
Quartermaster	Samuel Brodribb	Samuel Brodribb	Samuel Brodribb
Surgeon	Patrick H. Lavens	Patrick H. Lavens	Patrick H. Lavens
Asst.-Surgeons	Jas. Wemyss Moffat	Jas. Wemyss Moffat	J. Wemyss Moffat E. W. Stone, M.D.
Vet.-Surgeon	James Geo. Philips	James Geo. Philips	James Geo. Philips
Agents	Messrs. Cox and Co.	Messrs. Cox and Co.	Messrs. Cox and Co.

14TH (THE KING'S) REGIMENT OF LIGHT DRAGOONS

Station—India. *Uniform*—Blue; facings, scarlet. *Badges*—The King's Crest and Prussian Eagle. *Distinctions*—'Douro,' 'Talavera,' 'Fuentes d'Onor,' 'Salamanca,' 'Vittoria,' 'Orthes,' 'Peninsula.'

	1843	1844	1845
Colonel . .	Sir Edward Kerrison, Bt., G.C.H., K.C.B., Lt.-Gen.	Sir Edward Kerrison, Bt., G.C.H., K.C.B., Lt.-Gen.	Sir Edward Kerrison, Bt., G.C.H., K.C.B., Lt.-Gen.
Lt.-Colonels .	John Townsend, Col.	John Townsend, Col.	John Townsend, Col.
	Wm. Havelock, K.H.	Wm. Havelock, K.H.	Wm. Havelock, K.H.
Majors . .	Ed. Harvey, Lt.-Col.	Edward Harvey	Ed. Harvey, Lt.-Col.
	Francis H. Stephens	Charles P. Ainslie	Chas. Philip Ainslie
Captains . .	Wm. Henry Archer	Wm. Henry Archer	Wm. Henry Archer
	Henry Ed. Doherty	Henry Ed. Doherty	Henry Ed. Doherty
	George Weston	George Weston	George Weston
	Charles Stewart	Charles Steuart	Charles Steuart
	Geo. Gardine Shaw	Geo. Gardine Shaw	Geo. Gardine Shaw
	Arch. E. Bromwich	Arch. E. Bromwich	Arch. E. Bromwich
	John Henry Tonge	John Henry Tonge	John Henry Tonge
	Geo. K. M. Dawson	Geo. K. M. Dawson	Geo. K. M. Dawson
	William Clarke	William Clarke	William Clarke
Lieutenants .	Chas. Tindal Griffis, Riding-master	Chas. Tindal Griffis, Riding-master	Chas. Tindal Griffis, Riding-master
	John H. Goddard	John H. Goddard	John H. Goddard
	James Boalth	James Boalth	James Boalth
	James Chambre	James Chambre	James Chambre
	Frederick Holder	Frederick Holder	Frederick Holder
	Arthur Scudamore	Arthur Scudamore	Arthur Scudamore
	Richard H. Gall	Richard Herbert Gall	Rich. Herbert Gall
	John F. FitzGerald	John F. FitzGerald	John F. FitzGerald
	John A. Todd	John Augustus Todd	John Augustus Todd
	Rich. B. Prettejohn	Rich. B. Prettejohn	Rich. B. Prettejohn
	Henry F. Hodson	Henry Fred. Hodson	Henry Fred. Hodson
	Chas. W. Thompson	Chas. W. Thompson	Chas. W. Thompson
	Thomas W. Smith	Thomas Wm. Smith	Thomas Wm. Smith
	John H. Gray	John Hamilton Gray	Charles Morant
	Sydney L. Horton	Sydney L. Horton	George E. Hillier
	Charles Morant	Charles Morant	Joseph Cuffe Barrett
	George E. Hillier	George E. Hillier	William Nettleship
	Joseph Cuffe Barrett	Joseph Cuffe Barrett	Balcarres W. Ramsay
	William Nettleship	William Nettleship	Francis Delaval Gray
Cornets . .	Francis D. Gray	Francis Delaval Gray	Chas. Potts Rosser
	Charles P. Rosser	Chas. Potts Rosser	Wm. Edward Buller
	Wm. Edward Buller	Wm. Edward Buller	R. P. Apthorp, Adj.
	Rd. P. Apthorp	R. P. Apthorp, Adj.	James Coster
	James Coster	James Coster	R. Johnston Brown
	Robert J. Brown	R. Johnston Brown	William McMahon
	F. J. G. Whitehead	William McMahon	William Greenham
	William McMahon	William Greenham	Geo. Anthony Foster
Paymaster .	Samuel Rofe	Samuel Rofe	Samuel Rofe
Adjutant . .	Rd. P. Apthorp	Rich. P. Apthorp	Rich. P. Apthorp
Quartermaster .	Samuel Brodribb	Samuel Brodribb	Samuel Brodribb
Surgeon . .	Jas. Wemyss Moffat	Robt. Dunkin Smyth	Robt. Dunkin Smyth
Asst.-Surgeons	E. W. Stone, M.D.	Ed. W. Stone, M.D.	Ed. W. Stone, M.D.
	A. S. Thomson, M.D.	A. S. Thomson, M.D.	A. S. Thomson, M.D.
Vet.-Surgeon .	James Geo. Philips	James Geo. Philips	James Geo. Philips
Agents . .	Messrs. Cox and Co.	Messrs. Cox and Co.	Messrs. Cox and Co.

14TH (THE KING'S) REGIMENT OF LIGHT DRAGOONS

Station—India. *Uniform*—Blue; facings, scarlet. *Badges*—The King's Crest and Prussian Eagle. *Distinctions*—'Douro,' 'Talavera,' 'Fuentes d'Onor,' 'Salamanca,' 'Vittoria,' 'Orthes,' 'Peninsula.'

	1846	1847	1848
Colonel	Sir Edward Kerrison, Bt., G.C.H., K.C.B., Lt.-Gen.	Sir Edward Kerrison, Bt., G.C.H., K.C.B., Lt.-Gen.	Sir Edward Kerrison, Bt., G.C.H., K.C.B., Lt.-Gen.
Lt.-Colonels	Wm. Havelock, K.H. Edward Harvey	Wm. Havelock, K.H. Edward Harvey	Wm. Havelock, K.H. Edward Harvey
Majors	Chas. Philip Ainslie Wm. Henry Archer	Chas. Philip Ainslie John Wallace King	John Wallace King Henry Ed. Doherty
Captains	Henry Ed. Doherty George Weston Charles Steuart Geo. Gardine Shaw Arch. E. Bromwich John Henry Tonge Geo. K. M. Dawson William Clarke John H. Goddard	Henry Ed. Doherty George Weston Charles Steuart Geo. Gardine Shaw Arch. Ed. Bromwich John Henry Tonge Geo. K. M. Dawson William Clarke John H. Goddard	Charles Steuart Geo. Gardine Shaw Geo. K. M. Dawson William Clarke John H. Goddard William Wilmer Arthur Scudamore Wm. A. Gaussen Rich. Herbert Gall
Lieutenants	Chas. Tindal Griffis, Riding-master James Chambre Frederick Holder Arthur Scudamore Rich. Herbert Gall John F. FitzGerald John Augustus Todd Rich. B. Prettejohn Henry Fred. Hodson Chas. W. Thompson Thomas Wm. Smith Charles Morant George E. Hillier Joseph Cuffe Barrett William Nettleship B. D. W. Ramsay Francis Delaval Gray James Coster R. Johnston Brown	Chas. Tindal Griffis, Riding-master James Chambre Frederick Holder Arthur Scudamore Rich. Herbert Gall John F. FitzGerald John Augustus Todd Rich. B. Prettejohn Chas. W. Thompson Chas. Morant George E. Hillier Joseph Cuffe Barrett William Nettleship B. D. W. Ramsay James Coster R. Johnston Brown William McMahon William Greenham	James Chambre John F. FitzGerald John Augustus Todd Rich. B. Prettejohn Chas. W. Thompson William Nettleship R. Johnston Brown Rd. P. Apthorp, Adj. William McMahon Robt. T. Woodman Arthur Need William English Jas. W. Bennett John Theodore Ling Taylor L. Maynel Herbert Edwards Thomas Barrett Henry E. Reader Ambrose Lloyd
Cornets	Chas. Potts Rosser Wm. Edward Buller R. P. Apthorp, Adj. William McMahon William Greenham Geo. Anthony Foster Robt. T. Woodman William English	Wm. Edward Buller Rich. P. Apthorp Robt. T. Woodman William English Jas. W. Bennett John Cornes Herbert Edwards Thomas Barrett Ambrose Lloyd	Samuel K. Ibbetson Augustus J. Cureton William D. Boyd Hon. Rich. Walter Chetwynd William Spilling Wm. D'Urban Blyth Stanley O. Black John Dudgeon
Paymaster	Samuel Rofe	Samuel Rofe	Samuel Rofe
Adjutant	Rich. P. Apthorp	Rich. P. Apthorp	Rich. P. Apthorp
Quartermaster	Samuel Brodribb	John Palmer	
Surgeon	Robt. Dunkin Smyth	Robt. Dunkin Smyth	Archibald Stewart
Asst.-Surgeons	Ed. W. Stone, M.D. A. S. Thomson, M.D.	Ed. W. Stone, M.D. A. S. Thomson, M.D.	Robert B. Wigstrom Charles H. Fasson
Vet.-Surgeon	James Geo. Philips	James Geo. Philips	James Geo. Philips
Agents	Messrs. Cox and Co.	Messrs. Cox and Co.	Messrs. Cox and Co.

14TH (THE KING'S) REGIMENT OF LIGHT DRAGOONS

Station—India. *Uniform*—Blue; facings, scarlet. *Badges*—The King's Crest and Prussian Eagle. *Distinctions*— Douro,' 'Talavera,' 'Fuentes d'Onor,' Salamanca,' 'Vittoria,' 'Orthes,' 'Peninsula.'

	1849	1850	1851
Colonel	Sir Edward Kerrison, Bt., G.C.H., K.C.B., Lt.-Gen.	Sir Edward Kerrison, Bt., G.C.H., K.C.B., Lt.-Gen.	Sir Edward Kerrison, Bt., G.C.H., K.C.B., Lt.-Gen.
Lt.-Colonels	John Wallace King Henry Ed. Doherty	John W. King, C.B. H. Ed. Doherty, C.B.	H. Ed. Doherty, C.B.
Majors	Charles Steuart William Clarke	Charles Steuart William Clarke	Charles Steuart William Clarke
Captains	John H. Goddard William Wilmer Arthur Scudamore Wm. A. Gaussen Rich. Herbert Gall Thomas Garratt James Chambre John A. Todd Chas. W. Thompson	John H. Goddard William Wilmer Arthur Scudamore Wm. A. Gaussen Rich. Herbert Gall Thomas Garratt James Chambre John A. Todd P. S. Thompson	John H. Goddard William Wilmer Arthur Scudamore Wm. A. Gaussen Rich. Herbert Gall Thomas Garratt James Chambre John A. Todd P. S. Thompson
Lieutenants	Rich. B. Prettejohn William Nettleship Robert J. Brown Rd. P. Apthorp, Adj. William McMahon Robt. T. Woodman Arthur Need William English John T. Ling Taylor L. Mayne Herbert Edwards Thomas Barrett Henry E. Reader Samuel K. Ibbetson Wm. Dundas Boyd Hon. R. W. Chetwynd William Spilling	Rich. B. Prettejohn William Nettleship Robert J. Brown Rd. P. Apthorp, Adj. William McMahon Robt. T. Woodman Arthur Need William English John T. Ling Taylor L. Mayne Herbert Edwards Thomas Barrett Henry E. Reader Wm. Dundas Boyd Hon. R. W. Chetwynd William Spilling Wm. D'Urban Blyth Stanley O. Black John Dudgeon	Rich. B. Prettejohn Robert J. Brown Rd. P. Apthorp William McMahon Robt. T. Woodman Arthur Need William English John T. Ling Taylor L. Mayne Herbert Edwards Thomas Barrett Henry E. Reader Wm. Dundas Boyd Hon. R. W. Chetwynd William Spilling Wm. D'Urban Blyth John Dudgeon M. C. Smith, Riding-master Thos. Ed. Gordon Chas. Edwyn Wyatt
Cornets	Wm. D'Urban Blyth Stanley O. Black John Dudgeon Michael C. Smith Thos. Ed. Gordon John Holliday Fred. Vansittart Geo. A. Franklyn	Michael C. Smith Thos. Ed. Gordon John Holliday Fred. Vansittart Geo. A. Franklyn John P. Kennedy James Leith Egerton Ford	John Holliday Fred. Vansittart Geo. A. Franklyn John P. Kennedy James Leith Egerton Ford Henry T. Clements Frederick Coates
Paymaster	Samuel Rofe	Wm. Featherstonhaugh, Lieut.	Wm. Featherstonhaugh, Lieut.
Adjutant	Rich. P. Apthorp	Rich. P. Apthorp	H. Elmhirst Reader
Quartermaster	George Shenton	George Shenton	Thomas Bennett
Surgeon	Archibald Stewart	Archibald Stewart	Archibald Stewart
Asst.-Surgeons	Robt. B.-Wigstrom Chas. H. Fasson	Robt. B.-Wigstrom Chas. H. Fasson	Robt. B.-Wigstrom Chas. H. Fasson
Vet.-Surgeon		Alex. W. Caldwell	Alex. W. Caldwell
Agents	Messrs. Cox and Co.	Messrs. Cox and Co.	Messrs. Cox and Co.

14TH (THE KING'S) REGIMENT OF LIGHT DRAGOONS

Station—India. *Uniform*—Blue; facings, scarlet. *Badges*—The King's Crest and Prussian Eagle. *Distinctions*—'Douro,' 'Talavera,' 'Fuentes d'Onor,' 'Salamanca,' 'Vittoria,' 'Orthes,' 'Peninsula.' *Additional distinctions in* 1853—'Punjaub,' 'Chillianwallah,' 'Goojerat.'

	1852	1853	1854
Colonel	Sir Edward Kerrison, Bt., G.C.H., K.C.B., Gen.	Hon. Henry Murray, C.B., Lt.-Gen.	Hon. Henry Murray, C.B., Lt.-Gen.
Lt.-Colonels	H. Ed. Doherty, C.B. Charles Steuart	H. Ed. Doherty, C.B. Charles Steuart	H. Ed. Doherty, C.B. Charles Steuart, Col.
Majors	William Clarke John H. Goddard	William Clarke John H. Goddard	John H. Goddard William Wilmer
Captains	Wm. Wilmer, Major Arthur Scudamore Wm. A. Gaussen Rich. Herbert Gall Thomas Garratt James Chambre John A. Todd P. S. Thompson Rich. B. Prettejohn	Wm. Wilmer, Maj. Arthur Scudamore Wm. A. Gaussen Rich. Herbert Gall Thomas Garratt James Chambre John A. Todd P. S. Thompson Rich. B. Prettejohn Robert J. Brown	Arthur Scudamore Wm. A. Gaussen Rich. Herbert Gall James Chambre John A. Todd P. S. Thompson Rich. B. Prettejohn Robert J. Brown William McMahon Thomas Barrett
Lieutenants	Robert J. Brown Rich. P. Apthorp William McMahon Arthur Need William English John T. Ling Taylor L. Mayne Herbert Edwards Thomas Barrett H. E. Reader, Adj. Wm. Dundas Boyd Hon. R. W. Chetwynd William Spilling Wm. D'Urban Blyth John Dudgeon Thos. Ed. Gordon Chas. Edwyn Wyatt John Holliday, Riding-master Fred. Vansittart	Rich. P. Apthorp William McMahon Arthur Need William English John T. Ling Taylor L. Mayne Thomas Barrett H. E. Reader, Adj. Wm. Dundee Boyd Hon. R. W. Chetwynd William Spilling Wm. D'Urban Blyth John Dudgeon Thos. Ed. Gordon Chas. Edwyn Wyatt John Holliday, Riding-master Fred. Vansittart Thos. Price Gratrex John P. Kennedy	Arthur Need William English John T. Ling Taylor L. Mayne Hy. E. Reader, Adj. Wm. D'Urban Blyth John Dudgeon Thos. Ed. Gordon Chas. Edwyn Wyatt Fred. Vansittart James Leith Alex. E. M'Gregor Henry T. Clements Claud B. Whish W. H. T. C. Travers Standish R. Jackson Robert Chadwick Chas. W. Thesiger H. W. M. Hathway
Cornets	J. P. Kennedy, Ensign James Leith Egerton Ford Henry T. Clements Claudius B. Whish W. H. T. Clarke Standish R. Jackson George Ross	James Leith Egerton Ford Henry T. Clements Claudius B. Whish W. H. T. Clarke Standish R. Jackson George Ross	L. St. Patrick Gowan E. O. V. Haldane Lawrence Mackenzie Leonard Redmayne Geo. Meyrick Dew James Giles
Paymaster	W. Featherstonhaugh	W. Featherstonhaugh	W. Featherstonhaugh
Adjutant	Henry E. Reader	Henry E. Reader	Henry E. Reader
Quartermaster	Thomas Bennett	Thomas Bennett	Thomas Bennett
Surgeon	Archibald Stewart	Archibald Stewart	Archibald Stewart
Asst.-Surgeons	Robt. B.-Wigstrom Chas. H. Fasson	Robt. B.-Wigstrom Chas. H. Fasson	Robt. B.-Wigstrom Chas. H. Fasson
Vet.-Surgeon	Alex. W. Caldwell	William C. Lord	William C. Lord
Agents	Messrs. Cox and Co.	Messrs. Cox and Co.	Messrs. Cox and Co.

14TH (THE KING'S) REGIMENT OF LIGHT DRAGOONS

Station—India. *Uniform*—Blue; facings, scarlet. *Badges*—The King's Crest and Prussian Eagle. *Distinctions*—'Douro,' 'Talavera,' 'Fuentes d'Onor,' 'Salamanca,' 'Vittoria,' 'Orthes,' 'Peninsula,' 'Punjaub,' 'Chillianwallah,' 'Goojerat.'

	1855	1856	1857
Colonel	Hon. Henry Murray, C.B., Gen.	Hon. Henry Murray, C.B., Gen.	Hon. Henry Murray, C.B., Gen.
Lt.-Colonels	Henry E. Doherty, C.B. Charles Steuart, Col.	Henry E. Doherty, C.B., Col. C. Steuart, C.B., Col.	Henry Ed. Doherty, C.B., Col. C. Steuart, C.B., Col.
Majors	William Wilmer Arthur Scudamore	William Wilmer Arthur Scudamore	William Wilmer Arthur Scudamore
Captains	Wm. A. Gaussen Richard H. Gall John Augustus Todd Pearson S. Thompson Rich. B. Prettejohn Robert J. Brown William McMahon Thomas Barrett Arthur Need Henry E. Reader	Wm. A. Gaussen Richard H. Gall John Augustus Todd Pearson S. Thompson Rich. B. Prettejohn Robert J. Brown William McMahon Thomas Barrett Arthur Need Henry Brett	Wm. A. Gaussen Richard H. Gall John Augustus Tod Pearson S. Thompson R. B. Prettejohn R. Johnston Brown William McMahon Thomas Barrett Arthur Need
Lieutenants	William English John Theodore Ling Taylor L. Mayne Wm. D'Urban Blyth John Dudgeon Thomas Ed. Gordon Chas. Edwyn Wyatt Frederick Vansittart James Leith Alex. E. M'Gregor Henry T. Clements Claud B. Whish William Hy. Travers Clarke Travers Standish R. Jackson Robert Chadwick Chas. W. Thesiger H. W. M. Hathway L. St. Patrick Gowan E. O. V. Haldane	William English Wm. D'Urban Blyth John Dudgeon Thomas Ed. Gordon Chas. Edwyn Wyatt Frederick Vansittart James Leith Alex. E. M'Gregor H. T. Clements, Adj. Claud B. Whish William H. Travers Clarke Travers Standish R. Jackson Robert Chadwick Chas. W. Thesiger L. St. Patrick Gowan E. C. V. Haldane Lawrence Mackenzie Leonard Redmayne Geo. Meyrick Dew	William English Wm. D'Urban Blyth Thos. Ed. Gordon Chas. Edwyn Wyatt James Leith Alex. E. M'Gregor Hy. T. Clements Claud B. Whish Wm. Hy. Travers Clarke Travers Standish R. Jackson Robert Chadwick Chas. W. Thesiger L. St. Patrick Gowan Ed. O. V. Haldane Lawrence Mackenzie Leonard Redmayne George Meyrick Dew
Cornets	Lawrence Mackenzie Leonard Redmayne Geo. Meyrick Dew James Giles W. H. S. Beamish	James Giles Wm. Hy. Slingsby Beamish	James Giles Wm. H. S. Beamish Rich. P. Ridley Thomas Phillips
Paymaster	W. Featherstonhaugh	W. Featherstonhaugh	W. Featherstonhaugh
Adjutant	Henry T. Clements	Henry T. Clements	Henry T. Clements
Riding-master			Joseph Raiker
Quartermaster	Thomas Bennett	Thomas Bennett	Thomas Bennett
Surgeon	Archibald Stewart	Archibald Stewart	Archibald Stewart
Asst.-Surgeons	Chas. H. Fasson Thos. W. Fox, M.B.	Chas. H. Fasson Thos. W. Fox	Chas. H. Fasson Thos. W. Fox, M.D.
Vet.-Surgeon	William C. Lord	William C. Lord	William C. Lord
Agents	Messrs. Cox and Co.	Messrs. Cox and Co.	Messrs. Cox and Co.

14TH (THE KING'S) REGIMENT OF LIGHT DRAGOONS

Station—India; (1861) Dunkalk. *Uniform*—Blue; facings, scarlet. *Badges*—The King's Crest and Prussian Eagle. *Distinctions*—'Douro,' 'Talavera,' 'Fuentes d'Onor,' 'Salamanca,' 'Vittoria,' 'Orthes,' 'Peninsula,' 'Punjaub,' 'Chillianwallah,' 'Goojerat.' *Additional distinction in* 1860—'Persia.'

	1858	1859	1860	1861
Colonel	Hon. H. Murray, C.B., Gen.	Hon. H. Murray, C.B., Gen.	Hon. H. Murray, C.B., Gen.	William Beckwith, K.H., Maj.-Gen.
Lt.-Cols.	C. Steuart, C.B., Col. C. P. Ainslie, Col.	C. Steuart, C.B., Col. C. P. Ainslie, Col.	C. Steuart, C.B., Col. C. P. Ainslie, Col.	C. Steuart, C.B., Col.
Majors	Arthur Scudamore Rd. Herbert Gall, Lt.-Col.	Arthur Scudamore, Lt.-Col. Rd. Herbert Gall, C.B., Lt.-Col.	Arthur Scudamore, C.B., Lt.-Col. Rd. Herbert Gall, C.B., Lt.-Col.	Arthur Scudamore, C.B., Lt.-Col. Rd. Herbert Gall, C.B., Lt.-Col.
Captains	John A. Todd P. Scott Thompson R. B. Prettejohn Robert J. Brown William McMahon Thomas Barrett Arthur Need W. D'Urban Blyth Thos. E. Gordon Chas. Edwyn Wyatt	John A. Todd, Maj. P. S. Thompson, Maj. R. B. Prettejohn Robert J. Brown W. McMahon, Maj. Thomas Barrett Arthur Need, Maj. W. D'Urban Blyth Thos. E. Gordon Chas. Edwyn Wyatt	John A. Todd, Lt.-Col. P. S. Thompson, Maj. Rd. B. Prettejohn, Maj. Robert J. Brown William McMahon Arthur Need, Maj. W. D'Urban Blyth Thos. E. Gordon Chas. Edwyn Wyatt	John A. Todd, Lt.-Col. P. S. Thompson, Lt.-Col. R. B. Prettejohn, Lt.-Col. Robert J. Brown Arthur Need, Maj. W. D'Urban Blyth F. B. Chapman Wm. Hy. Travers Clarke Travers
Lieuts.	James Leith H. T. Clements Claud B. Whish Wm. H. Travers Clarke Travers S. Radley Jackson Robert Chadwick L. St. P. Gowan E. O. V. Haldane L. Mackenzie George M. Dew James Giles W. H. S. Beamish Richard P. Ridley	H. T. Clements, Adj. Claud B. Whish Wm. H. Travers Clarke Travers S. Radley Jackson Robert Chadwick L. St. P. Gowan E. O. V. Haldane L. Mackenzie George M. Dew James Giles W. H. S. Beamish Richard P. Ridley	Claud B. Whish Wm. H. Travers Clarke Travers Standish Radley Jackson Robert Chadwick Lawrence St. P. Gowan L. Mackenzie George Meyrick Dew Wm. Hy. Slingsby Beamish Robert Mather	Claud B. Whish Robert Chadwick Lawrence St. P. Gowan L. Mackenzie James Giles, Adj. Robert Mather Ed. Essex Digby Boycott Francis B. Eagle Henry Bradley
Cornets	Francis B. Eagle Henry Bradley Percy Dodgson Wm. A. Atcherley	Francis B. Eagle Henry Bradley Percy Dodgson Wm. A. Atcherley W. Sandys Brown Hon. F. Amherst Ed. W. Pritchard	Francis B. Eagle Henry Bradley Percy Dodgson Wm. A. Atcherley W. Sandys Browne Hon. F. Amherst Ed. W. Pritchard	Percy Dodgson Wm. A. Atcherley W. Sandys Browne Hon. F. Amherst Chas. A. P. Talbot F. J. S. H. Newton John A. Donnelly
Paymaster	Wm. Featherstonhaugh	Wm. Featherstonhaugh	Wm. Featherstonhaugh	Wm. Featherstonhaugh
Adjutant	Henry T. Clements	H. T. Clements, Lt.		James Giles
Rid.-mas.	Joseph Raiker	Joseph Raiker	Joseph Raiker	Joseph Raiker
Qr'master.	Thomas Bennett	Thomas Bennett	Thomas Bennett	Thomas Bennett
Surgeon	Archibald Stewart	E. J. Franklyn, M.D.	William Arden	William Arden
Ass.-Surg.	R. C. Lofthouse, M.D. Robt. B. Forsyth-Brown	R. C. Lofthouse, M.D. Robt. B. Forsyth-Brown	R. C. Lofthouse, M.D. Robt. B. Forsyth-Brown	R. C. Lofthouse, M.D.
Vet.-Surg.	Henry Dawson	Henry Dawson	Henry Dawson	Henry Dawson
Agents	Messrs. Cox and Co.	Messrs. Cox and Co.	Messrs. Cox and Co.	Messrs. Cox and Co.

14TH (THE KING'S) REGIMENT OF HUSSARS

Uniform—Blue ; facings, scarlet. *Badges*—The King's Crest and Prussian Eagle. *Distinctions*—'Douro,' 'Talavera,' 'Fuentes d'Onor,' 'Salamanca,' 'Vittoria,' 'Orthes,' 'Peninsula,' 'Punjaub,' 'Chillianwallah,' 'Goojerat,' 'Persia.' *Additional distinction in* 1864—'Central India.'

	1862	1863	1864
Stations	Manchester	Manchester	Aldershot
Colonel	Wm. Beckwith, K.H., Lt.-Gen.	Wm. Beckwith, K.H., Lt.-Gen.	Wm. Beckwith, K.H., Lt.-Gen.
Lt.-Colonel	Sir Wm. Russell, Bart., C.B.	Sir Wm. Russell, Bt., C.B.	Sir Wm. Russell, Bt., C.B., Col.
Majors	Rich. Herbert Gall, C.B., Lt.-Col. Pearson S. Thompson, Lt.-Col.	Rich. Herbert Gall, C.B., Lt.-Col. Pearson S. Thompson, Lt.-Col.	Pearson S. Thompson, Lt.-Col. R. Johnston Brown
Captains	John Augustus Todd, Lt.-Col. Rd. Buckley Prettejohn, Lt.-Col. R. Johnston Brown Arthur Need, Lt.-Col. F. Barclay Chapman Hy. Travers Clarke Travers Robert Chadwick John Baskerville	Rich. Buckley Prettejohn, Lt.-Col. R. Johnston Brown F. Barclay Chapman Wm. Hy. Travers Clarke Travers Robert Chadwick John Baskerville Lawrence Mackenzie Robert Mather	Rich. Buckley Prettejohn, Lt.-Col. F. Barclay Chapman Wm. Hy. Travers Clarke Travers Robert Chadwick John Baskerville Robert Mather Fra. P. Campbell Edward Essex Digby Boycott
Lieutenants	Lawrence Mackenzie James Giles, Adj. Robert Mather Ed. E. D. Boycott Henry Bradley William Arbuthnot William S. Browne Hon. Fred. Amherst John Hunter Knox	Ed. E. D. Boycott Henry Bradley William Arbuthnot Hon. Fred. Amherst John Hunter Knox Chas. A. P. Talbot F. J. S. H. Newton John Mather R. J. C. Thompson	Henry Bradley William Arbuthnot Hon. Fred. Amherst John Hunter Knox F. J. S. H. Newton John Mather R. J. C. Thompson John Bradney G. S. Lynch-Staunton
Cornets	Percy Dodgson Chas. A. P. Talbot F. J. S. H. Newton John Mather Rawdon John Cortland Thompson John Bradney Geo. Staunton Lynch-Staunton Jas. Colquhoun Revell Reade	Percy Dodgson John Bradney Geo. Staunton Lynch-Staunton Jas. Colquhoun Revell Reade Francis Shirley Russell Jas. Logan White John Geo. Glyn Shaw Thos. Kingston Baker	Percy Dodgson Francis Shirley Russell James Logan White Jas. George Glyn Shaw Thos. Kingston Baker James Crum Sam. Tudor Ashton Hon. John St. Vincent Saumarez
Paymaster	W. Featherstonhaugh	W. Featherstonhaugh	Wm. Featherstonhaugh, Lt. and Hon. Major
In. of Musketry	Lawrence Mackenzie	Lawrence Mackenzie	F. J. S. H. Newton, Lt.
Adjutant	James Giles	Ed. E. D. Boycott	Do. do.
Riding-master	Joseph Raiker	Joseph Raiker	Joseph Raiker
Quartermaster	Thomas Bennett	Thos. Howell Clark	Thos. Howell Clark
Surgeon	William Arden	William Arden	W. Arden, Surg.-Maj.
Asst.-Surgeon	R. C. Lofthouse, M.D.	R. C. Lofthouse, M.D.	R. C. Lofthouse, M.D.
Vet.-Surgeon	Henry Dawson	Henry Dawson	Henry Dawson
Agents	Messrs. Cox and Co.	Messrs. Cox and Co.	Messrs. Cox and Co.

14TH (THE KING'S) REGIMENT OF HUSSARS

Uniform—Blue; busby-bag, yellow; plume, white. *Badges*—The King's Crest and Prussian Eagle. *Distinctions*—'Douro,' 'Talavera,' 'Fuentes d'Onor,' 'Salamanca,' 'Vittoria,' 'Orthes,' 'Peninsula,' 'Punjaub,' 'Chillianwallah,' 'Goojerat,' 'Persia,' 'Central India.'

	1865	1866	1867
Stations	Aldershot	Hounslow	Edinburgh
Colonel	Wm. Beckwith, K.H., Lt.-Gen.	Wm. Beckwith, K.H., Lt.-Gen.	Wm. Beckwith, K.H., Lt.-Gen.
Lt.-Colonel	Pearson S. Thompson	Pearson S. Thompson	Pearson S. Thompson
Majors	R. Johnston Brown F. Barclay Chapman	R. Johnston Brown F. Barclay Chapman	R. Johnston Brown F. Barclay Chapman
Captains	Wm. Hy. Travers Clarke Travers Robert Chadwick John Baskerville Fra. P. Campbell Edward Essex Digby Boycott Henry Bradley William Arbuthnot Hon. Fred. Amherst	Wm. Hy. Travers Clarke Travers Robert Chadwick John Baskerville Fra. P. Campbell Edward Essex Digby Boycott William Arbuthnot Hon. Fred. Amherst Arthur Nassau Bolton	Robt. Chadwick John Baskerville Fra. P. Campbell Edward Essex Digby Boycott William Arbuthnot Hon. Fred. Amherst Arthur Nassau Bolton John Hunter Knox
Lieutenants	John Hunter Knox Fra. John Stuart Hay Newton John Mather Geo. Staunton Lynch-Staunton Francis S. Russell James Logan White Jas. Geo. Glyn Shaw Thos. Kington Baker Samuel Tudor Ashton	John Hunter Knox Fra. John Stuart Hay Newton John Mather Geo. Staunton Lynch-Staunton Francis S. Russell James Logan White Jas. Geo. Glyn Shaw Samuel Tudor Ashton Joseph Boulton	F. J. S. H. Newton John Mather G. S. Lynch-Staunton Francis S. Russell Samuel Tudor Ashton Joseph Boulton Hon. John St. V. Saumarez Joseph Harpur, Adjt. Thos. Allan Henry Geo. W. D. Kempson
Cornets	Hon. John St. Vincent Saumarez Joseph Harpur Thos. Allan Henry Jas. W. Robertson Geo. W. D. Kempson George Robt. Elwes Malcolm Alf. Laing Charles Kennett	Hon. John St. Vincent Saumarez Joseph Harpur, Adj. Thos. Allan Henry Geo. W. D. Kempson George Robt. Elwes Malcolm Alf. Laing Halstead S. Cobden James Poynter	George Robert Elves Malcolm A. Laing Halstead S. Cobden James Poynter Edmund Peel G. B. C. Yarborough Joseph L. Milligan
Paymaster	Wm. Featherstonhaugh, Lt. and Hon. Major	Wm. Featherstonhaugh, Lt. and Hon. Major	Wm. Featherstonhaugh, Lt. and Hon. Major
In. of Musketry	Francis S. Russell, Lt.	Francis S. Russell, Lt.	Fra. S. Russell, Lt.
Adjutant	F. J. S. Hay Newton, Lt.	J. Harpur, Cornet	Joseph Harpur, Lt.
Riding-master	Joseph Raiker	James Harran	James Harran
Quartermaster	Thos. Howell Clark	Thos. Howell Clark	Thos. Howell Clark
Surgeon	W. Arden, Surg.-Maj.	Ormsby Bowen Miller	Ormsby Bowen Miller
Asst.-Surgeon	Rich. Chapman Lofthouse, M.D.	Rich. Chapman Lofthouse, M.D.	Rich. Chapman Lofthouse, M.D.
Vet.-Surgeon	Henry Dawson	Henry Dawson	Henry Dawson
Agents	Messrs. Cox and Co.	Messrs. Cox and Co.	Messrs. Cox and Co.

14TH (THE KING'S) REGIMENT OF HUSSARS

Uniform—Blue; busby-bag, yellow; plume, white. *Badges*—The King's Crest and Prussian Eagle. *Distinctions*—'Douro,' 'Talavera,' 'Fuentes d'Onor,' 'Salamanca,' 'Vittoria,' 'Orthes,' 'Peninsula,' 'Punjaub,' 'Chillianwallah,' 'Goojerat,' 'Persia,' 'Central India.'

	1868	1869	1870
Stations	Dublin	Longford	Cahir
Colonel	Wm. Beckwith, K.H., Lt.-Gen.	Wm. Beckwith, K.H., Lt.-Gen.	Wm. Beckwith, K.H., Lt.-Gen.
Lt.-Colonel	Pearson S. Thompson, Col.	Pearson S. Thompson, Col.	Pearson S. Thompson, C.B., Col.
Major	Fred. B. Chapman Robert Chadwick	Fred. B. Chapman Fra. P. Campbell	Fred. B. Chapman Fra. P. Campbell
Captains	John Baskerville Fra. P. Campbell Edw. E. D. Boycott William Arbuthnot Hon. Fred. Amherst John Hunter Knox Fra. J. S. H. Newton John Mather	William Arbuthnot Hon. Fred. Amherst John Hunter Knox John Mather G. S. Lynch-Staunton Fra. Shirley Russell Samuel T. Ashton Joseph Boulton	William Arbuthnot Hon. Fred. Amherst John Hunter Knox John Mather Fra. Shirley Russell Samuel T. Ashton Hon. John St. V. Saumarez Geo. Robert Elwes
Lieutenants	G. S. Lynch-Staunton Francis S. Russell Samuel T. Ashton Joseph Boulton Hon. John St. V. Saumarez Joseph Harpur, Adj. Thos. Allan Henry Wm. John St. Aubyn George Robt. Elwes	Hon. John St. V. Saumarez Joseph Harpur, Adj. Thos. Allan Henry Geo. Robert Elwes Malcolm A. Laing Wm. H. F. Palmer Henry Bulkeley Edmund Peel C. B. C. Yarborough	Joseph Harpur, Adj. Malcolm A. Laing Wm. H. F. Palmer Henry Bulkeley Edmund Peel G. B. C. Yarborough Wm. S. Beaumont Albert Brassey Gerald E. M'Carthy Chas. Jas. M. Lefroy
Cornets	Malcolm A. Laing Halstead S. Cobden James Poynter Edmund Peel G. B. C. Yarborough Wm. S. Beaumont Albert Brassey	Wm. S. Beaumont Albert Brassey Chas. E. P. Chaplin Gerald E. M'Carthy Chas. Jas. M. Lefroy Wm. Bell M'Taggart Henry L'Est. Malone Jas. Reginald Bray	Wm. Bell M'Taggart Henry L'Est. Malone Jas. Reginald Bray W. H. D. Ward James Kentish
Paymaster	Wm. Featherstonhaugh, Lieut. and Hon. Major	Jn. Macartney, Hon. Captain	Jn. Macartney, Hon. Captain
Adjutant	Joseph Harpur, Lt.	Joseph Harpur, Lt.	Joseph Harpur, Lt.
Riding-master	James Harran	James Harran	James Harran
Quartermaster	Thos. Howell Clark	Thos. Howell Clark	Thos. Howell Clark
Surgeon	Ormsby B. Miller	Ormsby B. Miller	Ormsby B. Miller
Asst.-Surgeon	Edwin Wilson	William Hensman	William Hensman
Vet.-Surgeon	Henry Dawson	Henry Dawson	Henry Dawson
Agents	Messrs. Cox and Co.	Messrs. Cox and Co., and Sir E. Borough, Bt., Armit and Co., Dublin	Messrs. Cox and Co., and Sir E. Borough, Bt., Armit and Co., Dublin

14TH (THE KING'S) REGIMENT OF HUSSARS

Uniform—Blue; busby-bag, yellow; plume, white. *Badges*—The King's Crest and Prussian Eagle. *Distinctions*—'Douro,' 'Talavera,' 'Fuentes d'Onor,' 'Salamanca,' 'Vittoria,' 'Orthes,' 'Peninsula,' 'Punjaub,' 'Chillianwallah,' 'Goojerat,' 'Persia,' 'Central India.'

	1871	1872	1873
Stations .	Ballincollig	Newbridge	Dublin
Colonel . .	Wm. Beckwith, K.H., Gen.	Henry Rich. Jones, C.B., Maj.-Gen.	Henry Rich. Jones, C.B., Maj.-Gen.
Lt.-Colonel .	Pearson S. Thompson, C.B., Col.	Pearson S. Thompson, C.B., Col.	Pearson S. Thompson, C.B., Col.
Majors . .	Fra. P. Campbell William Arbuthnot	Fra. P. Campbell William Arbuthnot	Fra. P. Campbell William Arbuthnot
Captains . .	Hon. Fred. Amherst John Hunter Knox Fra. Shirley Russell Hon. John St. V. Saumarez Malcolm A. Laing Wm. H. F. Palmer Edmund Peel Wm. S. Beaumont	Hon. Fred. Amherst John Hunter Knox Fra. Shirley Russell Hon. John St. V. Saumarez Malcolm A. Laing Edmund Peel Wm. S. Beaumont Chas. Jas. M. Lefroy	Hon. Fred. Amherst John Hunter Knox Fra. Shirley Russell Hon. John St. V. Saumarez Malcolm A. Laing Joseph Harpur Alan C. Gardner Samuel Barrett
Lieutenants .	Joseph Harpur, Adj. G. B. C. Yarborough Albert Brassey Chas. Jas. M. Lefroy Wm. Bell M'Taggart Henry L'Est. Malone Jas. Reginald Bray Wm. H. D. Ward James Kentish	Joseph Harpur, Adj. Wm. Bell M'Taggart Henry L'Est. Malone Jas. Reginald Bray Wm. H. D. Ward James Kentish Thos. H. H. Garrett Arch. Wm. Merry Richard Garth	Wm. Bell M'Taggart Henry L'Est. Malone Jas. Reg. Bray, Adj. Wm. H. D. Ward James Kentish Thos. H. H. Garrett Arch. Wm. Merry Richard Garth
		Sub-Lieuts.	*Sub-Lieuts.*
Cornets . .	Thos. H. H. Garrett Wm. Ernest Hyder Arch. Wm. Merry	Llewellyn Wynn B. Sparrow Hon. H. G. Gough	Hon. H. G. Gough Aug. John English Hy. Rich. L. Howard Jn. S. H. Maxwell Chas. Fred. Lindsell
Paymaster .	Jn. Macartney, Hon. Captain	Jn. Macartney, Hon. Captain	Jn. Macartney, Hon. Captain
Adjutant . .	Joseph Harpur, Lt.	Joseph Harpur, Lt.	Jas. Reg. Bray, Lt.
Riding-master .	James Harran	James Harran	James Harran
Quartermaster .	Thos. Howell Clark	Thos. Howell Clark	Thos. Howell Clark
Surgeon . .	Ormsby B. Miller	Ormsby B. Miller	Ormsby B. Miller
Asst.-Surgeon .	Wm. Hensman	Wm. Hensman	Wm. Hensman
Vet.-Surgeon .	Fred. F. Marshall	Fred. F. Marshall	Fred. F. Marshall
Agents . .	Messrs. Cox and Co., and Sir E. Borough, Bt., Armit and Co., Dublin	Messrs. Cox and Co., and Sir E. Borough, Bt., Armit and Co., Dublin	Messrs. Cox and Co., and Sir E. Borough, Bt., Armit and Co., Dublin

14TH (THE KING'S) REGIMENT OF HUSSARS

Uniform—Blue; busby-bag, yellow; plume, white. *Badges*—The King's Crest and Prussian Eagle. *Distinctions*—'Douro,' 'Talavera,' 'Fuentes d'Onor,' 'Salamanca,' 'Vittoria,' 'Peninsula,' 'Punjaub,' 'Chillianwallah,' 'Goojerat,' 'Persia,' 'Central India.'

	1874	1875	1876
Station	Aldershot	Aldershot	India
Colonel	John Wilkie, Maj.-Gen.	John Wilkie, Maj.-Gen.	John Wilkie, Maj.-Gen.
Lt.-Colonel	Pearson S. Thompson, C.B., Col.	Pearson S. Thompson, C.B., Col.	Francis Pemberton Campbell, Lt.-Col.
Majors	Fra. P. Campbell William Arbuthnot	Fra. P. Campbell William Arbuthnot	William Arbuthnot
Captains	Hon. Fred. Amherst John Hunter Knox Fra. Shirley Russell Joseph Harpur Alan C. Gardner Samuel Barrett Algernon Edward S. Preston Wm. Bell M'Taggart	Hon. Fred. Amherst John Hunter Knox Fra Shirley Russell Joseph Harpur Alan C. Gardner Algernon Edward S. Preston Wm. Bell M'Taggart James Dennis	John Hunter Knox Fra. Shirley Russell Joseph Harpur Alan C. Gardner Wm. Bell M'Taggart James Dennis Henry L'Estrange Malone
Lieutenants	Henry L'Estrange Malone James Kentish, Adj. Thos. H. H. Garrett Archibald W. Merry Richard Garth	Henry L'Estrange Malone James Kentish, Adj. Thos. H. H. Garrett Archibald W. Merry Richard Garth Hon. Hugh George Gough Chas. Fred. Lindsell Augustus J. English Christ. D. V. Tuthill Arthur Abney Sandys John Shaw Heron-Maxwell	James Kentish Thos. H. H. Garrett Richard Garth Hon. Hugh George Gough, Adj. Augustus J. English Christ. D. V. Tuthil John Shaw Heron-Maxwell Chas. Fred. Lindsell Gerald C. Ricardo William Lockhart M'Call Wm. Ed. Stokes Richard Owen
Sub-Lieuts.	Hon. Hugh George Gough Augustus J. English John Shaw Heron-Maxwell Chas. Fred. Lindsell Fred. Gordon Blair Gerald C. Ricardo Wm. Henry Burn	Fred. Gordon Blair Gerald C. Ricardo Wm. Henry Burn	Wm. Henry Burn Ar. Chas. King Gilbert Hy. Claude Hamilton Guy St. Maur Palmes
Paymaster	Jn. Macartney, Hon. Capt.		Ed. Hy. O'Dowd
Adjutant	James Kentish	James Kentish, Lt.	Hon. Hugh George Gough, Lt.
Riding-master	James Harran	James Harran	James Harran
Quartermaster	Thos. Howell Clark		Jacob Merill
Surgeon	Ormsby Bowen Miller	Ormsby Bowen Miller	A. C. Gage, Surg.-Maj.
Vet.-Surgeon	Fred. F. Marshall	Fred. F. Marshall	Fred. F. Marshall
Agents	Messrs. Cox and Co., and Sir E. Borough, Bt., Armit and Co., Dublin	Messrs. Cox and Co., and Sir E. Borough, Bt., Armit and Co., Dublin	Messrs. Cox and Co.

14TH (THE KING'S) REGIMENT OF HUSSARS

Station—India. *Uniform*—Blue; busby-bag, yellow; plume, white. *Badges*—(1879) The Royal Crest within the Garter and Prussian Eagle. *Distinctions*—'Douro,' 'Talavera,' 'Fuentes d'Onor,' 'Salamanca,' 'Vittoria,' 'Orthes,' 'Peninsula,' 'Punjaub,' 'Chillianwallah,' 'Goojerat,' 'Persia,' 'Central India.'

	1877	1878	1879
Colonel . .	John Wilkie, Lt.-Gen.	John Wilkie, Lt.-Gen.	John Wilkie, Lt.-Gen.
Lt.-Colonel .	William Arbuthnot	William Arbuthnot	Wm. Arbuthnot
Major . .	John Hunter Knox	John Hunter Knox	John Hunter Knox
Captains . .	Fra. Shirley Russell Alan C. Gardner Wm. Bell M'Taggart James Dennis Henry L'Estrange Malone James Kentish Thos. H. Heaton Garrett Thos. Ed. Stopford Hickman Richard Garth	Fra. Shirley Russell Alan C. Gardner Wm. Bell M'Taggart James Dennis James Kentish Thos. Ed. Stopford Hickman Richard Garth Wm. Rose Wynter Wm. Noel Hill	Fra. Shirley Russell Alan C. Gardner Wm. Bell M'Taggart James Dennis James Kentish Thos. Ed. Stopford Hickman Richard Garth Wm. Rose Wynter Wm. Noel Hill
Lieutenants .	Hon. Hugh George Gough, Adj. Augustus Jn. English Christ. D. V. Tuthill John Shaw Heron-Maxwell Chas. Fred. Lindsell Gerald C. Ricardo Wm. Ed. Stokes William Lockhart M'Call Richard Owen Chas. Cecil Percy Stoughton	Hon. Hugh George Gough, Adj. Wm. Henry Burn Augustus Jn. English Christ. D. V. Tuthill John Shaw Heron-Maxwell Chas. Fred. Lindsell Gerald C. Ricardo Wm. Ed. Stokes Wm. L. M'Call Richard Owen Gilbert Hy. Claude Hamilton Guy St. Maur Palmes Chas. Cecil Percy Stoughton	Hon. Hugh George Gough, Adj. Wm. Henry Burn Augustus Jn. English Christ. D. V. Tuthill John Shaw Heron-Maxwell Gerald C. Ricardo Wm. L. M'Call Arthur Charles King Richard Owen Gilbert Hy. Claude Hamilton Guy St. Maur Palmes Chas. Cecil Percy Stoughton Arch. Allan Cammell
Sub-Lieuts. .	Wm. Henry Burn Arthur Charles King Gilbert Hy. Claude Hamilton Guy St. Maur Palmes Chas. Edwd. Skyring Hemery	Arthur Charles King Chas. Edwd. Skyring Hemery	Chas. Edwd. Skyring Hemery
Paymaster .	Ed. Hy. O'Dowd	Ed. Hy. O'Dowd, Hon. Capt.	Ed. Hy. O'Dowd, Hon. Capt.
Adjutant . .	Hon. Hugh George Gough, Lt.	Hon. Hugh George Gough, Lt.	Hon. Hugh George Gough, Lt.
Riding-master .	James Harran	James Harran	James Harran
Quartermaster .	Geo. Ferris Rumsey, Cornet	Geo. Ferris Rumsey	Geo. Ferris Rumsey
Surgeon . .	Arthur Charles Gage, Surg.-Major	Arthur Charles Gage, Surg.-Major	
Vet.-Surgeon .	Fred. F. Marshall	Wm. Albert Russell	Wm. Albert Russell
Agents . .	Messrs. Cox and Co.	Messrs. Cox and Co.	Messrs. Cox and Co.

14TH (THE KING'S) REGIMENT OF HUSSARS

Stations—(1880) India. (1881) India and South Africa. (1882) India. *Uniform*—Blue; busby-bag, yellow; plume, white. *Badges*—The Royal Crest within the Garter and Prussian Eagle. *Distinctions*—'Douro,' 'Talavera,' 'Fuentes d'Onor,' 'Salamanca,' 'Vittoria,' 'Orthes,' 'Peninsula,' 'Punjaub,' 'Chillianwallah,' 'Goojerat,' 'Persia,' 'Central India.'

	1880	1881	1882
Colonel	John Wilkie, Gen.	John Wilkie, Gen.	John Wilkie, Gen.
Lt.-Colonels	Wm. Arbuthnot	Wm. Arbuthnot	John Hunter Knox Fra. Shirley Russell
Majors	John Hunter Knox	John Hunter Knox	James Dennis James Kentish Thos. Ed. Stopford Hickman
Captains	Fra. Shirley Russell Alan C. Gardner Wm. Bell M'Taggart James Dennis James Kentish Thos. Ed. Stopford Hickman Richard Garth Wm. Rose Wynter Wm. Noel Hill Hon. Hugh George Gough	Fra. Shirley Russell Alan C. Gardner James Dennis James Kentish Thos. Ed. Stopford Hickman Richard Garth Wm. Rose Wynter Wm. Noel Hill Hon. Hugh George Gough	Richard Garth Wm. Rose Wynter Wm. Noel Hill Hon. Hugh George Gough
Lieutenants	Wm. Hy. Burn Aug. Jn. English Christ. D. V. Tuthill, Adj. Gerald C. Ricardo Wm. L. M'Call Arthur Chas. King Richard Owen Gilbert Hy. Claude Hamilton Guy St. Maur Palmes Chas. Ed. Skyring Hemery Chas. Cecil Percy Stoughton Arch. Allan Cammell	Wm. Hy. Burn Aug. Jn. English Christ. D. V. Tuthill, Adj. Gerald C. Ricardo Wm. L. M'Call Arthur Chas. King Richard Owen Gilbert Hy. Claude Hamilton Chas. Ed. Skyring Hemery Chas. C. P. Stoughton	Wm. Hy. Burn Aug. Jn. English Christ. D. V. Tuthill, Adj. Arthur Chas. King Richard Owen Gilbert Hy. Claude Hamilton Chas. Ed. Skyring Hemery Chas. C. P. Stoughton Henry W. Mitchell James Kemp Fraser Hugh Kirk Francis Jas. Norman
Sub-Lieuts.	Henry W. Mitchell	Henry W. Mitchell James Kemp Fraser Peter M. Staunton Hugh Kirk	
Paymaster	Ed. Hy. O'Dowd, Hon. Maj.	Ed. Hy. O'Dowd, Hon. Maj.	Ed. Hy. O'Dowd, Hon. Maj.
Adjutant	Christ. D. V. Tuthill, Lieut.	Christ. D. V. Tuthill, Lieut.	Christ. D. V. Tuthill, Lieut.
Riding-master	James Harran	James Harran	James Harran
Quartermaster	Geo. Ferris Rumsey	Geo. Ferris Rumsey	Geo. Ferris Rumsey
Vet.-Surgeon	Daniel C. Pallin	Daniel C. Pallin	
Agents	Messrs. Cox and Co.	Messrs. Cox and Co.	Messrs. Cox and Co.

14TH (THE KING'S) REGIMENT OF HUSSARS

Station—India. *Uniform*—Blue; busby-bag, yellow; plume, white. *Badges*—The Royal Crest within the Garter and Prussian Eagle. *Distinctions*—'Douro,' 'Talavera,' 'Fuentes d'Onor,' 'Salamanca,' 'Vittoria,' 'Orthes,' 'Peninsula,' 'Punjaub,' 'Chillianwallah,' 'Goojerat,' 'Persia,' 'Central India.'

	1883	1884	1885
Colonel	Chas. Wm. Thompson, Gen.	Chas. Wm. Thompson, Gen.	Chas. Wm. Thompson, Gen.
Lt.-Colonels	John Hunter Knox Chas. Falkiner Morton	John Hunter Knox Chas. Falkiner Morton	John Hunter Knox Chas. Falkiner Morton
Majors	James Dennis James Kentish T. E. Stopford Hickman	James Dennis T. E. Stopford Hickman Richard Garth	James Dennis T. E. Stopford Hickman Richard Garth
Captains	Richard Garth Wm. Rose Wynter Wm. Noel Hill Hon. Hugh George Gough	Wm. Rose Wynter Wm. Noel Hill Hon. Hugh George Gough Wm. Henry Burn	Wm. Rose Wynter Hon. Hugh George Gough Aug. John English Christ. D. V. Tuthill, Adj. Arthur Chas. King
Lieutenants	Wm. Henry Burn Aug. John English Christ. D. V. Tuthill, Adj. Arthur Chas. King Richard Owen Gilbert H. C. Hamilton Chas. E. Skyring Hemery Chas. C. P. Stoughton Henry W. Mitchell James Kemp Fraser Hugh Kirk Francis Jas. Norman Arthur Brooks Broadhurst Lawrence Anthony Stransham	Aug. John English Christ. D. V. Tuthill, Adj. Arthur Chas. King Gilbert H. C. Hamilton Chas. E. Skyring Hemery Chas. C. P. Stoughton Henry W. Mitchell James Kemp Fraser Hugh Kirk Francis Jas. Norman Arthur Brooks Broadhurst Lawrence Anthony Stransham	Gilbert H. C. Hamilton Chas. E. Skyring Hemery Chas. E. P. Stoughton Henry W. Mitchell James Kemp Fraser Hugh Kirk Francis Jas. Norman Arthur Brooks Broadhurst Lawrence Anthony Stransham Geo. Hosk Seymour Lionel Jas. Richardson F. Dunbar-Sinclair Bentley-Innes.
Paymaster	Ed. Hy. O'Dowd, Hon. Maj.	Ed. Hy. O'Dowd, Hon. Maj.	Edwin L. R. Thackwell, Capt.
Adjutant	Christ. D. V. Tuthill, Lieut.	Christ. D. V. Tuthill, Lieut.	Christ. D. V. Tuthill, Capt.
Riding-master	Thomas Graham	Thomas Graham	Thomas Graham
Quartermaster	Fred. Mugford	Fred. Mugford	Fred. Mugford
Agents	Messrs. Cox and Co.	Messrs. Cox and Co.	Messrs. Cox and Co.

14TH (THE KING'S) REGIMENT OF HUSSARS

Uniform—Blue; busby-bag, yellow; plume, white. *Badges*—The Royal Crest within the Garter and Prussian Eagle. *Distinctions*—'Douro,' 'Talavera,' 'Fuentes d'Onor,' 'Salamanca,' 'Vittoria,' 'Orthes,' 'Peninsula,' 'Punjaub,' 'Chillianwallah,' 'Goojerat,' 'Persia,' 'Central India.'

	1886	1887	1888
Stations	India	Shorncliffe	Shorncliffe
Colonel	Chas. Wm. Thompson, Gen.	Chas. Wm. Thompson, Gen.	Chas. Wm. Thompson, Gen.
Lt.-Colonels	Chas. Falkiner Morton James Dennis	C. Falkiner Morton, Col H. Blackburne Hamilton	Hy. B. Hamilton
Majors	Thomas E. Stopford Hickman James Kentish Richard Garth	Thomas E. Stopford Hickman James Kentish Richard Garth Hon. Geo. H. Gough, Brevet Lt.-Col.	Richard Garth Hon. Geo. H. Gough, Brevet Lt.-Col. Aug. John English
Captains	Hon. Geo. H. Gough Aug. John English Christ. D. V. Tuthill Arthur Chas. King Gilbert H. C. Hamilton, Adj.	Aug. John English Christ. D. V. Tuthill Arthur Chas. King Gilbert H. C. Hamilton, Adj. Charles E. Skyring Hemery Hy. Wilmot Mitchell	Arthur Chas. King Gilbert H. C. Hamilton Charles E. Skyring Hemery Hy. Wilmot Mitchell James Kemp Fraser Chas. C. P. Stoughton Hugh Kirk
Lieutenants	Charles E. Skyring Hemery Chas. C. P. Stoughton Hy. Wilmot Mitchell James Kemp Fraser Hugh Kirk Francis Jas. Norman Arthur Brooks Broadhurst Lawrence Anthony Stransham Lionel J. Richardson James Percy Miller Stuart Robertson Richard E. Galindo	Chas. C. P. Stoughton James Kemp Fraser Hugh Kirk Arthur Brooks Broadhurst Lawrence Anthony Stransham Lionel Jas. Richardson Jas. Percy Miller Stuart Robertson John Murray	Arthur B. Broadhurst Lionel Jas. Richardson Sir Jas. Percy Miller, Bart. Stuart Robertson John Murray
Second-Lieuts.			Denis M. Miller Cyril Stacey Rd. G. Brooksbank John Cheape Stewart John D. Hamilton Hy. B. L. Hughes
Paymaster	Edwin L. R. Thackwell, Capt.	Hy. John Meares, Hon. Major	Hy. John Meares, Hon. Major
Adjutant	Gilbert H. C. Hamilton, Capt.	Gilbert H. C. Hamilton, Capt.	H. W. Mitchell, Capt.
Riding-master	Thomas Graham	Rich. Odlum, Capt.	Rich. Odlum, Hon. Lt.
Quartermaster	Frederick Mugford	Frederick Mugford	F. Mugford, Hon. Lt.
Agents	Messrs. Cox and Co.	Messrs. Cox and Co.	Messrs. Cox and Co.

14TH (THE KING'S) REGIMENT OF HUSSARS

Uniform—Blue; busby-bag, yellow; plume, white. *Badges*—The Royal Crest within the Garter and Prussian Eagle. *Distinctions*—'Douro,' 'Talavera,' 'Fuentes d'Onor,' 'Salamanca,' 'Vittoria,' 'Orthes,' 'Peninsula,' 'Punjaub,' 'Chillianwallah,' 'Goojerat,' 'Persia,' 'Central India.'

	1889	1890	1891
Stations	Brighton	Brighton	Hounslow
Colonel	Chas. Wm. Thompson, Gen.	Chas. Wm. Thompson, Gen.	Chas. Wm. Thompson, Gen.
Lt.-Colonel	Hy. B. Hamilton	Hy. B. Hamilton	Hy. B. Hamilton, Col.
Majors	Hon. George Hugh Gough, Bt. Lt.-Col. Aug. John English Mich. Arthur Burke	Hon. George Hugh Gough Aug. John English Mich. Arthur Burke	Hon. George Hugh Gough, Col. Aug. John English Mich. Arthur Burke
Captains	Arthur Chas. King Gilbert Hy. C. Hamilton Hy. Wilmot Mitchell James Kemp Fraser, *Acting Paymaster* Hugh Kirk Arthur B. Broadhurst Percy Knight Hale Coke Lionel Jas. Richardson Sir Jas. Percy Miller, Bart.	Arthur Chas. King Gilbert Hy. C. Hamilton Hy. Wilmot Mitchell Hugh Kirk Arthur B. Broadhurst Lionel Jas. Richardson Robt. Mervyn Richardson Sir Jas. Percy Miller, Bart., Adj. Ed. Douglas Brown	Arthur Chas. King Gilbert Hy. C. Hamilton Hy. Wilmot Mitchell Hugh Kirk Arthur B. Broadhurst Lionel Jas. Richardson Robt. Mervyn Richardson Sir Jas. Percy Miller, Bart., Adj. Ed. Douglas Brown Stuart Robertson Walter S. Stanhope
Lieutenants	Stuart Robertson John Murray Ed. James Tickell Denis M. Miller Cyril Stacey Rich. Gylby Brooksbank	Stuart Robertson John Murray Ed. James Tickell Denis M. Miller Cyril Stacey Rich. Gylby Brooksbank John Douglas Hamilton Hy. Bodvel Lewis Hughes	John Murray Ed. James Tickell Denis M. Miller Cyril Stacey Rich. Gylby Brooksbank John Douglas Hamilton Hy. Bodvel Lewis Hughes Wm. Gardiner Eley
Second-Lieuts.	John Douglas Hamilton Hy. Bodvel Lewis Hughes John Arthur Rowley Wm. Gardiner Eley Hy. Maxwell Buller Tritton	Wm. Gardiner Eley Hy. Maxwell Buller Tritton Richard Harold St. Maur	Richard Harold St. Maur R. Campbell Stephen William Prevost
Paymaster	Jas. K. Fraser, Capt., *Acting Paymaster*		
Adjutant	H. W. Mitchell, Capt.	Sir Jas. Percy Miller, Bart., Capt.	Sir Jas. Percy Miller, Bart., Capt.
Riding-master	Richard Odlum, Hon. Lieut.	Richard Odlum, Hon. Lieut.	Richard Odlum, Hon. Capt.
Quartermaster	Fred. Mugford, Hon. Lieut.	Fred. Mugford, Hon. Lieut.	Fred. Mugford, Hon. Lieut.
Agents	Messrs. Cox and Co.	Messrs. Cox and Co.	Messrs. Cox and Co.

14TH (THE KING'S) REGIMENT OF HUSSARS

Uniform—Blue; busby-bag, yellow; plume, white. *Badges*—The Royal Crest within the Garter and Prussian Eagle. *Distinctions*—'Douro,' 'Talavera,' 'Fuentes d'Onor,' 'Salamanca,' 'Vittoria,' 'Orthes,' 'Peninsula,' 'Punjaub,' 'Chillianwallah,' 'Goojerat,' 'Persia,' 'Central India.'

	1892	1893	1894
Stations	Leeds	Manchester	Manchester
Colonel	Chas. Wm. Thompson, Gen.	Chas. Wm. Thompson	Chas. Wm. Thompson
Lt.-Colonel	Hon. George Hugh Gough, Col.	Hon. George Hugh Gough, Col.	Hon. George Hugh Gough, Col.
Majors	Aug. Jn. English Michael Ar. Burke Arthur Chas. King	Aug. Jn. English Michael Ar. Burke Arthur Chas. King	Aug. Jn. English Arthur Chas. King Gilbert Hy. Claude Hamilton
Captains	Gilbert Hy. Claude Hamilton Hy. Wilmot Mitchell Hugh Kirk Arthur B. Broadhurst Lionel Jas. Richardson Robt. Mervyn Richardson Sir Jas. Percy Miller, Bart., Adj. Ed. Douglas Brown Stuart Robertson Walter Spencer Stanhope John Murray	Gilbert Hy. Claude Hamilton Hy. Wilmot Mitchell Hugh Kirk Arthur B. Broadhurst Lionel Jas. Richardson Robt. Mervyn Richardson Ed. Douglas Brown John Murray Ed. Jas. Tickell, Adj. Ælla Molyneux Berkeley Gage	Hy. Wilmot Mitchell Ar. Brooks Broadhurst Lionel Jas. Richardson Ed. Douglas Brown John Murray Edward James Tickell, Adj. Ælla M. Berkeley Gage Cyril Stacey Denis Menezes Miller
Lieutenants	Edward Jas. Tickell Denis Menezes Miller Cyril Stacey Rich. Gylby Brooksbank Jn. Douglas Hamilton Hy. Bodvel Lewis Hughes Wm. Gardiner Eley R. Campbell Stephen	Denis Menezes Miller Cyril Stacey Rich. Gylby Brooksbank Hy. Bodvel Lewis Hughes Wm. Gardiner Eley R. Campbell Stephen William Prevost Chas. Bosvile Tottenham	Rich. Gylby Brooksbank Hy. Bodvel Lewis Hughes Wm. Gardiner Eley R. Campbell Stephen William Prevost Chas. Bosvile Tottenham William Henry J. M'Mahon Roberts
Second-Lieuts.	William Prevost Chas. Bosvile Tottenham Fra. Alex. Garden Stuart Barrington	William Henry J. M'Mahon Roberts	Wm. Jefferey Lockett Chas. Hilton Bindloss Chas. Claude Wallace
Adjutant	Sir Jas. Percy Miller, Bart., Capt.	Ed. Jas. Tickell, Capt.	Ed. Jas. Tickell, Capt.
Riding-master	Rich. Odlum, Hon. Capt.	Rich. Odlum, Hon. Capt.	Rich. Odlum, Hon. Capt.
Quartermaster	Fred. Mugford, Hon. Lt.	Fred. Mugford, Hon. Capt.	Fred. Mugford, Hon. Capt.
Agents	Messrs. Cox and Co.	Messrs. Cox and Co.	Messrs. Cox and Co.

14TH (THE KING'S) REGIMENT OF HUSSARS

Station—Cahir. *Uniform*—Blue; busby-bag, yellow; plume, white. *Badges*—The Royal Crest within the Garter and Prussian Eagle. *Distinctions*—'Douro,' 'Talavera,' 'Fuentes d'Onor,' 'Salamanca,' 'Vittoria,' 'Orthes,' 'Peninsula,' 'Punjaub,' 'Chillianwallah,' 'Goojerat,' 'Persia,' 'Central India.'

	1895	1896	1897
Colonel	Chas. Wm. Thompson, Gen.	Chas. Wm. Thompson, Gen.	Hon. Chas. Wemyss Thesiger, Lt.-Gen.
Lt.-Colonel	Hon. George Hugh Gough, Col.	Hon. George Hugh Gough, Col.	Aug. John English
Majors	Aug. Jn. English Gilbert Hy. Claude Hamilton Hy. Wilmot Mitchell	Aug. Jn. English Gilbert Hy. Claude Hamilton Hy. Wilmot Mitchell	Gilbert Hy. Claude Hamilton Edmond Donough J. O'Brien Lionel Jas. Richardson
Captains	Arthur Brooks Broadhurst Lionel James Richardson Robert Mervyn Richardson Ed. Douglas Brown John Murray Edward James Tickell, Adj. Ælla M. Berkeley Gage Cyril Stacey Denis Menezes Miller Rich. Gylby Brooksbank	Lionel James Richardson Robert Mervyn Richardson Ed. Douglas Brown Edward James Tickell, Adj. Ælla M. Berkeley Gage Denis Menezes Miller Rich. Gylby Brooksbank Hy. Bodvel Lewis Hughes	Robert Mervyn Richardson Edward Douglas Brown Edward James Tickell Ælla M. Berkeley Gage Denis Menezes Miller Rich. Gylby Brooksbank Hy. Bodvel Lewis Hughes Wm. Gardiner Eley R. Campbell Stephen, Adj.
Lieutenants	Hy. Bodvel Lewis Hughes Wm. Gardiner Eley R. Campbell Stephen William Prevost Chas. Bosvile Tottenham William Henry Wm. Jefferey Lockett Clement Geo. Montague Adam	Wm. Gardiner Eley R. Campbell Stephen William Prevost Chas. Bosvile Tottenham William Henry Wm. Jefferey Lockett Clement Geo. Montague Adam Freeling Ross Lawrence	William Prevost Chas. Bosvile Tottenham William Henry Wm. Jefferey Lockett Clement Geo. Montague Adam Freeling Ross Lawrence Harry Fred. Hamilton Hardy Harold James Tilney
Second-Lieuts.	Freeling Ross Lawrence Harry Fred. Hamilton Hardy	Harry F. Hamilton Hardy Harold James Tilney	Thos. Ernest Lynedoch Hill-Whitson C. Archibald Montgomery Howard
Adjutant	Edwd. James Tickell, Capt.	Edward James Tickell, Capt.	R. Campbell Stephen
Riding-master	Richard Odlum, Hon. Capt.	Richard Odlum, Hon. Capt.	Richard Odlum, Hon. Capt.
Quartermaster	Fred. Mugford, Hon. Capt.	Fred. Mugford, Hon. Capt.	Fred. Mugford, Hon. Capt.
Agents	Messrs. Cox and Co.	Messrs. Cox and Co.	Messrs. Cox and Co.

14TH (THE KING'S) REGIMENT OF HUSSARS

Uniform—Blue; busby-bag, yellow; plume white. *Badges*—The Royal Crest within the Garter and Prussian Eagle. *Distinctions*—'Douro,' 'Talavera,' 'Fuentes d'Onor,' 'Salamanca,' 'Vittoria,' 'Orthes,' 'Peninsula,' 'Punjaub,' 'Chillianwallah,' 'Goojerat,' 'Persia,' 'Central India.'

	1898	1899	1900	1901
Stations .	Newbridge	Newbridge	South Africa. Reserve Squadron—Curragh	South Africa. Reserve Squadron—Curragh
Colonel .	Hon. C. W. Thesiger, Lt.-Gen. Ⓡ	Hon. C. W. Thesiger, Lt.-Gen. Ⓡ	Hon. C. W. Thesiger, Lt.-Gen. Ⓡ	Hon. C. W. Thesiger, Lt.-Gen. Ⓡ
Lt.-Col. .	A. J. English	A. J. English	G. H. C. Hamilton	G. H. C. Hamilton, Col. (*Brig.-Gen., S. Africa*) (*s.*)
Majors .	G. H. C. Hamilton (*2nd in Command*) E. D. J. O'Brien L.J.Richardson(*y.*) R. M. Richardson	G. H. C. Hamilton (*2nd in Command*) E. D. J. O'Brien L. J. Richardson-Gardiner R. M. Richardson	E. D. J. O'Brien (*2nd in Command*) R. M. Richardson (*s.*) E. D. Brown E. J. Tickell, D.S.O. (*sp. ser., S. Africa*)	E. D. J. O'Brien (*2nd in Command*) R.M.Richardson(*s*) E. D. Brown, V.C. E. J. Tickel, D.S.O.
Captains .	E. D. Brown E. J. Tickell Æ. M. B. Gage J. Murray D. M. Miller R. G. Brooksbank W. G. Eley *R. C. Stephen*, Adj.	E. D. Brown *E. J. Tickell* Æ. M. B. Gage J. Murray D. M. Miller R. G. Brooksbank *W. G. Eley* (*y.*) *R. C. Stephen*, Adj.	Æ. M. B. Gage J. Murray D. M. Miller R. G. Brooksbank *W. G. Eley* (*y.*) *R. C. Stephen* (*y.*) W. Prevost C. B. Tottenham W. Henry *F. R. Lawrence*, D.S.O., Adj.	Æ. M. B. Gage J. Murray D. M. Miller R. G. Brooksbank *W. G. Eley* (*y.*) *R. C. Stephen* (*y.*) *W. Prevost* (*s.*) C. B. Tottenham W. Henry *F. R. Lawrence*, D.S.O., Adj. H. J. Tilney T.E.L.H.-Whitson
Lieuts. .	W. Prevost C. B. Tottenham W. Henry W. J. Lockett F. R. Lawrence H. F. H. Hardy H. J. Tilney T. E. L. Hill-Whitson	W. Prevost C. B. Tottenham W. Henry W. J. Lockett *F. R. Lawrence* H. F. H. Hardy H. J. Tilney T. E. L. Hill-Whitson R. Blackett	H. J. Tilney T.E.L.H.-Whitson R. Blackett E. J. Jameson E. P. Phil.-Stow W. R. Campbell E. T. L. Wright J. B. Walker J. G. Browne Hon. H. Grosvenor B. M. Dawes J. Harvey	R. Blackett E. J. Jameson E. P. Phil.-Stow W. R. Campbell E. T. L. Wright J. B. Walker J. G. Browne Hon. H. Grosvenor B. M. Dawes J. Harvey C. H. B. P.-Westcar G. R. Scott J. F. Champion
Sec.-Lieuts.	R. Blackett E. J. Jameson E. P. Philipson-Stow	E. J. Jameson E. P. Philipson-Stow	C. H. B. P.-Westcar G. R. Scott J. F. Champion F. E. Lace T. James C. Persse C. J. Stephen H. E. Joicey Hon. H. Robertson	F. E. Lace T. James C. J. Stephen H. E. Joicey J. C. Johnston A. L. Palmer V. H. Secker.
Adjutant .	R. C. Stephen, Captain	R. C. Stephen, Captain	F. R. Lawrence, D.S.O., Capt.	F. R. Lawrence, D.S.O., Capt.
Rid.-mas. .	R. Odlum, Hon. Captain	R. Odlum, Hon. Captain	R. Odlum, Hon. Captain	R. Odlum, Hon. Captain
Qu'rmaster	F. Mugford, Hon. Captain	F. Mugford, Hon. Captain	F. Mugford, Hon. Captain	A. F. Pridgeon, Hon. Lieut.
Agents .	Messrs. Cox and Co.	Messrs. Cox and Co.	Messrs. Cox and Co.	Messrs. Cox and Co.

LIST OF WARRANT OFFICERS OF THE 14TH (KING'S) HUSSARS

1881—1900

SERGEANT-MAJORS

H. A. Pridgeon, 1st July 1881.

H. Haines, 31st December 1889.

A. Smith, 1st January 1894.

A. F. Pridgeon, 25th January 1899.

BANDMASTERS

A. Sim, 1st July 1881.

R. Foster, 5th December 1883.

H. Hemsley, 1st April 1887 (B.M., 1st July 1881).

APPENDIX A

RAMNUGGUR AND CHILLIANWALLAH

RAMNUGGUR

EXTRACTS from the *Illustrated London News* of 27th January 1849.

Extract from the letter of an officer of the Bengal Horse Artillery, dated 1st December 1848:—

'There has been a skirmish at a place called Ramnuggur, on the river Chenab, about 40 miles north of Lahore. It was intended for a reconnaissance, but circumstances turned it into a cavalry fight, as I will endeavour to explain to you. Lord Gough, hearing that the enemy had drawn up in force on the opposite side of the river Chenab, determined to reconnoitre them and two fords across the river. For this purpose he detached two parties, the first consisting of the Ramnuggur.

3rd Light Dragoons, 8th Light Cavalry, Troop of Horse Artillery,

to examine one ford; the second party consisted of

14th (King's) Light Dragoons, Irregular Cavalry,
5th Light Cavalry, Troop of Horse Artillery,

to look at the other ford.

'The force left camp about 3 A.M., 22nd November, and arrived on the ground about 7 A.M.

'One of the parties was going along steadily as directed, when they came across the nullah, and seeing a large number of the enemy's cavalry beyond, Colonel Havelock ordered a charge, and, as the party were crossing the nullah, the enemy's guns on the opposite side opened on them at about 300 yards' range. The Sikhs had placed their guns in masked batteries, and, as you may suppose, the sudden discharge took our people by surprise; nevertheless they went on, seeing a large number of the enemy beyond the nullah. The ground was very heavy and sandy; a large portion of our cavalry got into a quicksand, and the horses, being somewhat exhausted by the march over the heavy ground, were not able to extricate themselves as soon as they might have done. The enemy's infantry were, in the meantime, behind large sand hillocks, and steadily firing into our men, who were also being fired upon by the large Sikh guns on the other side of the river.

'The Horse Artillery immediately unlimbered their guns and returned the fire as well as could be expected from six small guns. The retreat was then sounded, and in re-crossing the nullah one of our guns, as also two of our ammunition-wagons, stuck fast in the sand, the poles of the limber and the horses being shot; they had, however, previously spiked the gun so as to render it useless to the enemy. The Commander-in-Chief, hearing the firing, had ridden up to the spot. Meantime, Colonel Havelock was about to make another charge at the enemy with the 14th Light Dragoons, but the Commander-in-Chief sent an order by General Cureton, who commanded the Cavalry Division, to prevent the charge, and in conveying this order the General was shot through the heart. Colonel Havelock had also fallen. Colonel Alexander, 5th Cavalry, was shot through the right arm, and has since had it taken out of the socket. Several officers of the 14th Light Dragoons were killed and wounded, also several officers of the 5th Light Cavalry, whose names we have not heard. The gun and ammunition-wagons were abandoned by order of Lord Gough, as the fire from the enemy's guns was very severe. The guns on the opposite side of the river, and the portion of the enemy's force drawn up behind a quicksand, has been a regular trap into which we have fallen most wofully.'

Reported in the 'Illustrated London News.'

'By intelligence from India, dated up to 28th November, from the army of the Punjaub, Shere Singh with his army was occupying a position on the right bank of the Ravee,[1] whilst Lord Gough's camp was situated about twelve miles from Ramnuggur, on the left bank. According as our troops advanced under Lord Gough the enemy were reported in force at several places, but always evacuated them before they were occupied by us; they were at length, however, ascertained to have determined on making a stand at Ramnuggur, and General Cureton was directed to halt until more troops came up. Brigadier-General Campbell joined his camp on the 12th November, taking command as senior officer and bringing up another brigade.

'Considerable additions to the force of the camp were made during the five or six days following, and at length the Commander-in-Chief, having considered that the time had arrived for the commencement of operations, orders were issued late on the night of the 21st November, in the camp of Brigadier-General Campbell at Saharun, for the troops to parade on the following morning at three o'clock in front of the centre of the camp without sound of bugle, trumpet, or drum; and a strong detachment marched under the personal command of the Commander-in-Chief (who had come up from the

[1] This, evidently, should be Chenab—(Author).

headquarters' camp to superintend the proceedings of the morning) at the appointed time, and reached the left bank of the Chenab near Ramnuggur at an early hour, in the hope of surprising the detachment of the enemy who were known to be on the same side. It would appear, however, that these had early intimation of the movement of the British troops, and had all retired across except a few stragglers and a picquet which fled, leaving their tents pitched. The object of the movement on our side was, however, mainly to ascertain the real strength and position of the enemy, and this was in some degree effected on clearing Ramnuggur towards the left bank of the river.

'Then it became apparent that their camp was pitched along the right bank of the river, which was seen to be crowded by thousands of men, whose white and light-coloured garments strongly contrasted with the bodies of the British troops to which the eyes of our officers had been for some time past accustomed. In drawing out their fire to ascertain their real strength in guns, and pressing, perhaps, rather too zealously in pursuit of those of the fugitives who appeared to be within reach, while crossing at the ford with water to the waist, Lieutenant-Colonel Lane's troop of Horse Artillery (2nd of the 3rd Brigade) got into heavy sand before they were aware of it, and a leading horse having been shot, it was found when the order came to limber up and retire that one of the guns could not be moved. The Sikhs were not slow at perceiving this most inopportune embarrassment, and directed so hot a fire upon the spot that it was found necessary to abandon the gun, although a light field battery was brought up to cover it. But the fire of the enemy from guns that were sheltered, while ours were exposed in an open plain, was so heavy that nothing could be done beyond scattering a few of the nearest groups of the enemy and then retiring. The cavalry were ordered to do the same and take shelter from the enemy's round-shot behind a tope of trees to the left, where they remained quietly for a time, the infantry being during that interim drawn up on the right. Some of the enemy being subsequently discovered on the left bank of the river further down to the left, a large number of our cavalry, among whom were the 3rd and 14th Light Dragoons, moved down by order to dislodge them.'

The following letter from an officer, who was an eye-witness of the scene, graphically describes the interesting details of this brief but sanguinary combat which cost us so dearly:— Extracts of a letter from an eye-witness of the scene.

'CAMP RAMNUGGUR, *November* 25, 1848.

'On the morning of the 22nd inst., at 2 A.M., the orderly-sergeant of the "E" troop came into my tent at Deeda Singh camp and showed me the order-book with the following command:—"The troops will parade in marching order at a quarter before 3 A.M., without sound of trumpet or bugle, and form up on their respective alarm-posts." The morning was pitch dark when the order to move forward was given, and before we had advanced a mile, cavalry, camp followers, artillery, and infantry, were jumbled together. At length day broke, order was restored, and a report ran through the columns that Ramnuggur was in sight. Immediately after, our destination was confirmed, and the enemy was stated to be in position in our front. At 7 o'clock A.M. we reached Ramnuggur and saw the enemy and their camp in the distance. At half-past 7 o'clock the 3rd Light Dragoons, Holmes's Irregulars, and the Horse Artillery were pushed forward with Her Majesty's 61st Regiment in skirmishing order; and we, with the 5th Cavalry and some Native Infantry, were ordered to halt on the right of the city with the 2nd Europeans a little on the left rear. A few minutes later the enemy opened their fire. This continued about an hour, during which time we stood inactively admiring their shot and shells as they flew through or burst in the air. By about half-past 8 A.M. our guns had got into position and began replying to the enemy, when the order came for us to advance, which we did, and were halted about three-quarters of a mile from where the enemy's picquet had been. Their picquet-tents were still standing on an entrenched piece of ground, but their picquet had retired across the river. The round-shot now began to whistle near us, generally falling short, and the men began to get excited; when, about 9 o'clock A.M., the order came for us to move more to our left and get shelter from a tope of trees nearer to and more in front of the enemy's guns. Here we saw the 3rd Light Dragoons, who were more than a mile on our left, charging some Sikh cavalry, who, retiring before them, opened out and dispersed the moment they drew the Third within range of their batteries, which immediately opened on them. One man was killed and three wounded by the fire, while several horses suffered. A staff officer sent by Lord Gough ordered them to retire immediately out of fire, as the guns were on the other side of the river. While doing so a 9-lb. shot struck Captain Ouvry's[1] horse, passing clean through him, but the rider fortunately escaped. In a few moments after the horse was stripped by the Sikh cavalry, who pressed on their rear, trying again to draw them under fire. During all this time the round-shot was flying over and through the 14th Light Dragoons, and, strange to say, doing no harm. A little after 11 o'clock A.M. the enemy's cavalry came across the nullah that protected part of their front, and formed upon the left bank, to the right front of the Fourteenth, in great force, when the 14th Light Dragoons and 5th Light Cavalry received orders to charge them.

'Before saying anything of this charge, I must try to describe the Sikh position. The Sikhs to the number of 30,000 men occupied the right bank of the river Chenab, where they had a strong entrenched camp with several batteries erected. A little to the right front of Ramnuggur the river formed a bend; in it was an island containing a couple of acres of ground, and between that and the left bank the water was about 30 yards wide, with a precipitous fall from the left bank of from four to six feet before you got into its bed, which was in some parts four feet deep. This part is called the nullah, as the main branch of the river is on the right of the island. It is as nearly as possible in

[1] Afterwards Colonel H. A. Ouvry, C.B., 9th Lancers.

front of the centre of the Sikh position; on it were about 4000 men and a battery of six guns, while the approach to it was swept by a cross-fire from two batteries on the mainland. Knowing nothing of this position, and deeming the nullah the river, the Fourteenth, when ordered to charge, galloped on to the enemy's cavalry, who retired through the nullah on to the island, while the enemy's batteries opened their fire, and their infantry on the island poured in their volleys. Nothing daunted, Colonel Havelock cheering led on the first and second squadrons of the Fourteenth down to the bank, then into the nullah, crossed it at a gallop, sabred hundreds of the enemy under the most frightful shower of missiles from their guns and infantry. They then retired a short distance, formed up, were joined by the other squadrons, and the 5th Light Cavalry who had crossed a little higher up, and charged again. In this, the second charge, Colonel Havelock met his death, it was supposed, for he was not seen or heard of after General Cureton joined them with orders to retire, as, though the Fourteenth seemed so determined to destroy the enemy, they were utterly indifferent to their own loss. The Commander-in-Chief having cleared the left bank of the enemy, did not wish for more. While General Cureton was giving the order to retire, a matchlock ball struck him in the throat, and another in the forehead, and thus fell this glorious man, the finest cavalry officer of the day, at the head of that regiment in which as a private soldier, under the assumed name of "Roberts," he had commenced his career, and out of which he received his first commission. The 14th Light Dragoons then retired in order, formed up, and the roll was called, when 45 men were found missing and about 50 horses. Of the missing men, 14 were killed, and the remainder wounded. One of the killed, Sergeant Todd, had his head taken off by a round-shot. Colonel Havelock has not yet been found: the last seen of him was in the second charge, while he was crossing the nullah. His orderly states that both the Colonel and his horse fell wounded or killed, that he was hurried on, and he did not see him after. About twelve of the 5th Light Cavalry suffered, I am told, and one of the first round-shots fired at the charging regiments took off the arm of Colonel Alexander, who commanded the 5th Light Cavalry. Nothing could exceed the accuracy of the enemy's fire: their range was beautifully taken for certain points, showing that they must have discovered them previous to our advance; and our artillery officers say they never saw anything finer than the way their horse artillery were brought up to the edge of the river and formed up. No nation could exceed them in the rapidity of their fire. It is said that a Frenchman, late an officer in the Maharajah Runjeet Singh's service, and Aide-de-Camp to General Avitahile, named L'Enfant, commands them. No men could act more bravely than the Sikhs. They faced us the moment we came on them, firing all the time, and when we did come on them some opened out, and immediately after closed round us, while others threw themselves on their faces or turned their backs, protected by a shield from the stroke of the dragoon's sabre, and the moment that was given turned round, hamstrung the horse and shot the rider; while their individual acts of bravery were the admiration of all. Many stood before a charging squadron and singled out a man, after killing or wounding whom they themselves were cut down immediately; while many, before their blows could take effect, received the point of a sabre and fell in the act of making a cut.

'Amongst our officers, Captain Gall's personal courage was most conspicuous. He took single-handed one of the enemy's standards, but before he could get assistance he was knocked over and his right hand nearly severed from his body; some of his men, however, rushed to his rescue and saved him from receiving a mortal wound, though they could not again recover the standard which he had so hardly fought and suffered for.

'After Captain Gall was knocked over, a young cornet named D'Urban Blyth rode at the head of the troop, and while charging, saw Lieutenant McMahon fall wounded a little way off and a Sikh rushing forward to kill him. Cornet Blyth galloped forward, gave point, and sent his blade clean through the Sikh. On three different occasions afterwards he rushed out from his troop, and each time in single combat killed his man. Many private soldiers performed wonders. But I must not omit to mention that after Captain Fitzgerald fell, Captain Wilmer's troop was passing on their return, when they saw that he was alive. Captain Wilmer and four troopers dismounted and succeeded in bringing him in under the most frightful fire. All agree in one thing, however nobly the Fourteenth gained their laurels in the Peninsula, no charge they ever made could surpass this in gallantry, and yet no more than four or five of the officers and men had ever been under fire previously. It is impossible to say what loss the enemy sustained; but had the Fourteenth not been broken by jumping into the nullah, more than half of the regiment must have been destroyed, so severe was the fire, as all the shot, had they charged in close order, would have taken effect. I have heretofore omitted saying anything of Colonel King, as he had little to do beyond assisting to keep the men together and obey orders, until Colonel Havelock was killed. From the moment, however, his loss was known, Colonel King took up the command and ably did he carry out the duties that were entrusted to him. . . . To his watchful care the greater number of the wounded that were brought in owe their safety. Officers and men agree that his admirable conduct on that occasion proves him an able successor.

'Lord Gough visited the wounded yesterday, and expressed himself to each in the kindest terms about his injuries, and with the strongest praise of his brave conduct.

'As I conclude this we are erecting batteries and expecting to be joined by Brigadier General Wheeler's force, and two regiments of Eckford's brigade that remained behind at Lahore.

'The enemy are about being reinforced by Chuttur Singh, who has 40,000 men and about 60 guns with him, and report says they will try to turn our rear; but we only wish that, as the men are all in high spirits and determined to show them no quarter. The enemy's guns never cease firing, we are continually having their round-shot bowling into us, but doing little damage.'

Another writer observes:—'An attempt will doubtless be made to cast some blame upon the Commander-in-Chief for the result of these two affairs, but not justly. The facts are that General Cureton ordered in the first instance both movements, and if they had been carried out as he had wished, they would have been attended with the happiest results, but on both occasions mistakes in the execution, to a great degree unavoidable, marred the original plan. In both cases the ardour of our troops was too great. It was a rush who should get at the enemy first, but it must have been most gratifying to the Commander-in-Chief to witness the brilliant conduct of the regiments engaged and the intrepidity with which they were led by their officers.' Another account from the *Illustrated News*.

Casualties at the battle of Ramnuggur

Killed.

'Brigadier-General Cureton, shot through the heart; Colonel Havelock, 14th Light Dragoons, missing, but subsequently found killed; Captain Fitzgerald, 14th Light Dragoons, killed.

Wounded

'Lieutenant Hardinge, A.D.C., shot through the shoulder; Captain Scudamore, 14th Dragoons, sabred in the face; Captain Gall, 14th Dragoons, wounded in the hand; Lieutenant McMahon, 14th Dragoons, shot through the head; Lieutenant Chetwynd, spent ball in the side.

'14th Dragoons, 3 privates killed, 9 missing, 23 wounded, 5 contused, 25 horses wounded, 34 horses missing.

'3rd Light Dragoons, 5 privates wounded.

'5th Light Cavalry, Quartermaster-Sergeant killed by a round-shot, which first took off the arm of Colonel Alexander and then contused the foot of Lieutenant Reilly. Twelve privates killed, 15 privates wounded. Forty horses killed and wounded.

'8th Light Cavalry, Subadar-Major killed.

'12th Irregular Cavalry, Captain Holmes wounded.

'Horse Artillery, 1 private wounded, 2 Syces killed, 4 horses killed.

'Soon after this the Sikh army, which amounted to 30,000 with a park of heavy artillery, effected a retreat from their position and left the right bank of the Chenab on 3rd December under Shere Singh. They proceeded along the mountain chain in the direction of the river Jhelum after remaining for nearly a fortnight in the presence of our army, which, during the greater part of that period, was waiting the arrival of reinforcements. It is computed that in the various affairs which took place on the banks of the Chenab our killed and wounded have been about 400, while the loss of the enemy was 4000. By the latest advices received, dated 18th December, from Bombay, it is stated that Lord Gough had crossed the Chenab.'

[*End of Extracts from the 'Illustrated London News' of 27th January* 1849.]

Extracts from Official Despatches, etc.

General Lord Gough, G.C.B., Commander-in-Chief of the army in India, writes as follows:—

'Ramnuggur, *November* 23, 1848.

Extract from Lord Gough's despatch.

'Deeming it necessary to drive the rebel forces at this side the river across, and to capture any guns they might have had on the left bank, I directed Brigadier-General Campbell, with an infantry brigade, accompanied by the cavalry division, and three troops of Horse Artillery under Brigadier-General Cureton, to proceed during the night of the 21st from Saharun, four miles in front of my camp at Nonbulla, to effect this object. I joined the Brigadier at 3 A.M. to witness the operation.

'I witnessed with intense anxiety, but equally intense admiration, a charge made by Lieutenant-Colonel William Havelock at the head of the 14th King's Light Dragoons, who, I fear, misconceived the orders he received from the officer commanding the Cavalry Division, or, from the inequalities of the ground and the fearful dust occasioned by such a rapid movement, mistook the body he was instructed to charge, and moved upon and overwhelmed another much closer to the river, which exposed him to a cross-fire from the enemy's guns. I never witnessed so brilliant a charge, but I regret to say the loss was considerable, were it only in that of Brigadier-General Cureton, than whom a better or braver soldier never fell in his country's service. The brave leader of the 14th Light Dragoons, Lieutenant-Colonel Havelock, is missing. He charged into a gole of the enemy and has not since been seen, regretted by every soldier who witnessed his noble daring. The enemy suffered severely; numbers were precipitated into the river and drowned, and a standard was captured.

'The Goorchurras were more daring than I have before seen them, but the brilliant charges both of the 3rd and 14th Light Dragoons will have taught them a lesson they will not readily forget. This was a cavalry affair alone.'

Brigadier-General C. Campbell, C.B.,[1] Commanding 4th Division, writes:—

'Ramnuggur, *November* 27, 1848.

Extract from Brigadier-General C. Campbell's despatch.

'Captain Warner's and Lieutenant-Colonel Lane's troops of Horse Artillery were engaged. In withdrawing from the deep and heavy sand under the fire of the whole of the enemy's artillery, amounting to 28 guns posted on high

[1] Afterwards Field-Marshal Lord Clyde, G.C.B., K.C.S.I.

ground overhanging the river on the opposite bank, I regret to say that one gun and two ammunition-wagons of Colonel Lane's troop got embedded in the heavy sand and could not be recovered. The enemy observing this immediately crossed with great confidence the whole of his cavalry at numbers between 3000 and 4000: they clung to the banks of the river and kept under cover of the fire of their artillery on the opposite bank.

'This cavalry was charged on separate occasions by Her Majesty's 3rd and 14th Light Dragoons, and 5th and 8th regiments of Light Cavalry. His Lordship the Commander-in-Chief was an eye-witness of the brilliant conduct of these corps and of the intrepid manner in which they were led by their officers. The enemy were overthrown upon every occasion and fled for shelter to the river-side to be under the cover and protection of their artillery; but I regret to say these several defeats of the enemy's cavalry were not effected without much loss. Brigadier-General Cureton, commanding the cavalry of the army, was killed while leading a squadron of the 14th Light Dragoons to the support of the 5th Light Cavalry. I regret also to have to report that Lieutenant-Colonel Havelock, commanding Her Majesty's 14th Light Dragoons, is reported to be missing. He was last seen charging the enemy at the head of his noble regiment, and has not since been heard of.'

The following is an extract from the General Orders of the Commander-in-Chief in India, dated Camp Ramnuggur, 23rd November 1848:—

General Lord Gough's General Orders issued after the affair of Ramnuggur.

'The enemy were signally overthrown on every occasion, and only saved from utter annihilation by their flight to the cover of their guns on the opposite bank of the river. In the deaths of Brigadier-General C. R. Cureton, C.B., commanding Cavalry Division, and Lieutenant-Colonel Havelock, K.H., commanding 14th (King's) Light Dragoons, as well as Captain J. F. Fitzgerald of the same noble regiment, the service has sustained a loss which the Commander-in-Chief is sure the whole army will unite with him in lamenting.'

The following statements were given, almost *verbatim*, to the author in answer to his questions by surviving officers of the 14th Light Dragoons who were present at Ramnuggur.

CAPTAIN R. P. APTHORP, who was Lieutenant and Adjutant of the Fourteenth at Ramnuggur, says:—

'*February* 14, 1899.

'I recollect the whole circumstances of the doings of the regiment at Ramnuggur, as if it were only yesterday it occurred. I have the whole scene in my mind's eye now. There were three charges.[1] We changed our front twice, as these Sikh goles scampered off before we could get to them, in different directions.

'It was in the second change of front that Herbert Gall rushed out from the troop he was commanding to seize a standard of the enemy, and very nearly

[1] Captain Apthorp refers to the first advance of Havelock before he changed front the first time, and he calls that *advance* a *charge*. See *ante* p. 222 *n.* and p. 571.

had his right hand severed off. Some men of his troop rushed after him, but I, being near, stopped them, as it broke our line of advance. As it was, he never recovered the entire use of his hand again. It was a gallant act, but a very indiscreet one. It was Colonel Doherty who brought the regiment out of action. I never saw any artillery gun on our side of the nullah, and I do not think there were any. They played long bowls at us from the other side. What made Lord Gough so impetuous was, that these large goles of Sikh horsemen, who persisted in remaining on our side of the nullah, although our artillery had been peppering at them for half an hour, would not disperse, and merely seemed to get out of the way of the shells. They took care to keep a long distance off. The 3rd Light Dragoons were sent by Lord Gough to disperse them (before we were ordered to charge), and they sent them across the nullah, but they were too wary to follow them over the nullah, and then the enemy returned to our side again, and this so nettled Lord Gough that he sent Colonel Cureton to ascertain the reason of the 3rd Dragoons retreating, and the former was told of the dreadful nullah which separated us from the main body of the Sikh army; and when Lord Gough was told this by Cureton he pooh-poohed it, and told him to order the Fourteenth to advance and go across the nullah, and of course it was there we lost Colonel Havelock and Fitzgerald and about fourteen of our men, besides several wounded. On the other side of the nullah we came up to their guns and the whole force of the enemy—we were entirely disorganised by the confusion caused by jumping into the nullah and out of it, and which was lined by sharpshooters under the taking-off bank. It was, of course, helter-skelter afterwards until we emerged to our side of the nullah again, and it was some little time before the officers could restore anything like order as the Sikh horsemen followed us again. I, being well in the rear trying to halt our men, seeing these fellows coming on at us (the men generally knowing my voice well), I called out for skirmishers, and several men responded, and we covered our rear and kept the Sikh horsemen at bay, and then order was gradually restored. Colonel King was not in the charge at all. He was ordered to command the squadron which was left in support when we first advanced. He came up with this squadron as we were retreating, and it was a nucleus for us to form upon.

'I recollect poor Colonel Havelock telling me that he was going to charge the enemy with three squadrons, and that he ordered one squadron to be in reserve to support them, and ordered me to go and find Colonel King and tell him to take charge of the supporting squadron and move up slowly after us. I did so. In the meantime Havelock had moved off to the front with the attacking squadrons, and I had to gallop after them to catch them up. It was just then that Havelock was changing his front to the left as the Sikhs had scampered off in that direction. When we got up to them, they, knowing they were in a line for the easiest part for crossing the nullah, made direct for it, and we saw them scamper down the banks; therefore, Havelock had to change his front again to follow them over the nullah, so there must have been three distinct charges. There was no doubt that poor Havelock was killed at the nullah, as on recrossing, some of the men saw his charger (a grey horse) lying in the nullah, and one or two of the men were going to dismount and lead him back, but being near them, and on looking round I saw some of the Sikhs following us, I told the men there was no time to get hold of the Colonel's charger, but to mount and get to the rear as fast as they could. With regard to Cornet Blyth, he had not long joined us, and I recollect that after his sabring four or five of the enemy, and saving M[c]Mahon's life, I dismissed him at once from any further sword-drill.'

As to the question of whether there were two or three charges, Captain Apthorp says:—'I do not think it signifies much whether there were three charges or two. I considered Colonel Havelock's first advance was a charge, as he went direct for a large gole of the enemy's horsemen, but I was not with them, having been sent by Colonel Havelock to find Colonel King to direct him to take charge of the supporting squadron, and when I got up to the attacking squadrons Colonel Havelock was changing front to the left as the Sikh horsemen veered off in that direction. All I can recollect is that I saw the two charges, and we had to gallop a great pace to catch up the enemy, they having got well ahead of us while we were changing front. It was in this charge that poor Fitzgerald and *M*—— got surrounded by some of the straggling enemy (as we came up to them) slipping through our squadron intervals, and their horses being blown, got into the rear. Poor Fitzgerald must have ridden eighteen to twenty stone with all his accoutrements on (he always rode Cape chargers), and *M*—— was at that time no light-weight. You may rest quite satisfied that there was *no* charge by Colonel King's supporting squadron, as they met us as we returned over the nullah, and we re-formed our straggling squadrons on them, got our usual strength of skirmishers out, and then made an orderly retreat. The Sikhs then stopped from following us.

'With regard to the important point—that Havelock had orders through Cureton to charge over the nullah—I cannot corroborate it. What gave colour to it in my mind was, that Colonel Havelock (when he came up to me and directed me to find Colonel King, and give him directions to take charge of the supporting squadron) said, "I have just seen Cureton, and from what he says I expect to get immediate orders to charge to the front, as Lord Gough was displeased at the old 3rd Light Dragoons not pursuing the Sikhs further over the nullah."

'We all thought that this nullah was the course of the Chenab, and that there was a ford over it, until Colonel White told Cureton it was a dry nullah with very steep banks, and he did not consider it expedient to cross it with his regiment, as the Sikhs were in great force on the other side and with a lot of guns. I have no doubt Colonel Havelock thought this almost tantamount to an order, as he (Colonel Havelock) said Cureton told him that Lord Gough said he did not understand a dry nullah stopping cavalry. There is no doubt Lord Gough thought better of this, as when poor Colonel Cureton was killed by a musket-shot he was galloping to stop us from going over the nullah, Lord Gough having noticed Havelock forming up the squadrons for a final attack over the nullah. As for Sir Charles Gough, we know he was very biassed in his statements about the affair as well as Chillianwallah. It has always appeared to me that it was Sir Charles Gough's aim all the way through to screen Lord Gough; for although all must admit that the latter was as brave an officer as ever drew sword, yet he was totally void of discretion as a commander. We have only to look at his tactics at Chillianwallah, where the 24th Regiment was decimated on account of his persisting that they should advance straight to their front, and attack and capture the Sikh guns, although he had been told that the Sikh gunners had the exact range for grape-shot, where they could pepper and annihilate the poor Twenty-fourth, and this was the result; but nothing could check Lord Gough's ardour when he had made up his mind, and all his reply to some of his cautious generals was: "Tell them to take the Sikh guns with the 'cold steel,'" of course meaning their bayonets.'

VISCOUNT CHETWYND, who, as Lieutenant the Honourable R. W. Chetwynd, was present with the Fourteenth at Ramnuggur, has given the following facts in reply to inquiries:—

'The facts as to the squadrons of the Fourteenth at Ramnuggur are: Colonel Havelock went off with the 1st and 2nd squadrons in open column of troops, left in front, and made his first attack to his front, with the result that the 1st squadron, in which I was, got immediately into great confusion. The order for increasing the front I did not hear, and was afterwards told it was to form squadron on the move.

'The 4th squadron, led by Captain Scudamore, received some order from a staff officer, the result of which was that Scudamore, after an independent charge of his own, joined Havelock's second charge. Scudamore's wound, mentioned in the newspaper, was received in guarding his face from a cut.

'The 3rd squadron, led by Captain Wilmer, we met as we were retiring, a mob (we, not Wilmer's squadron). My recollection is that I saw Colonel King with the 3rd squadron, but would not state it without reserve. I have no recollection of ever hearing of this squadron charging, and do not believe they did. The story mentioned of Blyth having saved McMahon is exactly what I understood at the time. He also disposed of two or three more. The story of the sergeant's head being taken off by a round-shot is exactly as Blyth, who saw it happen, described it to me. Havelock's body was not recovered till after we had crossed the Chenab in pursuit of the retreating Sikhs. Havelock certainly survived his first attack and led off the three squadrons (1st, 2nd, and 4th) for his second. I saw him in front of the line. Doherty was the senior officer in the three squadrons retiring from Havelock's second attack. I remember hearing him call out for Havelock. These squadrons were quite broken up.

'The first attack (that of the two squadrons, 1st and 2nd) was, I imagine, pretty much a charge of the 2nd, the 1st following them in great confusion. I was afterwards told Havelock ordered increase of front from troop to squadron. I did not hear the order, and thought it was not repeated. It is possible Havelock did not give time for the increase of front, and that our left troop rushed ahead before we in the right troop could get into squadron. There was no interval between the first and second attack, which latter was made by the three squadrons, 1st, 2nd, and 4th, in line. The 1st, I think, outflanked the Sikh left.

'As to the Sikh guns, they, I believe, were on the other side of the river; anyhow I saw nothing of them. They opened at once when Havelock moved off with the right wing. Their shot flew over our heads in the rear troop. I saw one round-shot take a rear-rank man in the rear troop very soon after moving off. This troop lost four killed. Another of the four I saw hit by a matchlock, the same volley that hit me, apparently from an ambuscade on our right.

'As to the three charges Apthorp speaks of, there were three if you include Scudamore's with the 4th squadron on his way to join the right wing under Havelock. I do not see how Havelock could have charged twice to his front without the rear of the column getting up to him. When I emerged from the dust and confusion, the Colonel was in front of the three squadrons forming line. I can distinctly tell who led the 1st and 2nd squadrons at Ramnuggur: Doherty led the 1st and Goddard the 2nd; Wilmer and Scudamore led the 3rd and 4th both at Ramnuggur and Chillianwallah, and, I fully believe, in the order here given. At Chillianwallah, Goddard led the 1st and Thompson the 2nd; Goddard's troop ("H") changed squadrons with the "D," Garratt's.'

CHILLIANWALLAH

The mishap which occurred to Pope's Cavalry Brigade at the battle of Chillianwallah has by some been attributed to panic, and to this the very best troops in the world are sometimes subject. On this occasion, however, there appears to have been another very potent cause for failure in the faulty tactics displayed by the Brigadier himself. He appears to have utterly disregarded all recognised rules of cavalry leading by deploying the whole of his nine squadrons of cavalry in one single line without any supports whatever. Cavalry in attack requires due supports to follow up an advantage or retrieve a check; it also requires a reserve or point to rally on. The jungly and obstructive nature of the ground was wholly unsuited for such an extended front as that of nine squadrons of cavalry in one line; and to add to the difficulty and confusion of advancing in presence of the enemy in such formation, under such circumstances, the fire of his own Horse Artillery guns was masked by the Brigadier bringing his squadrons in front of them and overlapping them at a critical moment. Such flagrant mismanagement courted disaster and is quite enough to account for the sequel. Thackwell's narrative of the second Sikh War of 1848-49 gives many interesting particulars about the Fourteenth and the battle of Chillianwallah. It completely vindicates the character of Lieutenant-Colonel King, and proves that the 14th Light Dragoons were by no means so much to blame as has generally been believed.

'The day after the action, a court of inquiry into the conduct of the regiment was held by Major-General Sir Joseph Thackwell, with closed doors, and from what transpired, the result was most satisfactory to that much-abused but brave body of men.' [1]

Thackwell's *Narrative of the Second Sikh War* (p. 143, etc.) states that Brigadier Pope, who was a Lieutenant-Colonel of the Indian Native Cavalry, was quite unable to mount his horse without assistance at the time he was commanding a brigade of cavalry at the battle of Chillianwallah; also that it was asserted by some that the officer in command did give the order 'Threes About' for the purpose of placing the Horse Artillery in possession of a clear front, but if this was his object 'Threes Right' was the proper word of command, unless the Cavalry Brigade was parallel to the interval between Major-General Sir Walter Gilbert's Division and the Horse Artillery. The wound received by Brigadier Pope was a sword-cut on the head.

The following is the reference to the affair which appeared in Lord Gough's despatch, dated Camp Chillianwallah, 16th January 1849:—

Extract from Lord Gough's despatch, 16th January 1849.

'The brigade of cavalry under Brigadier Pope was not, I regret to say, so successful. Either by some order or misapprehension of an order they got into much confusion, hampered the fine brigade of Horse Artillery which, while getting into action against a body of the enemy's cavalry that was coming down upon them, had their horses separated from their guns by the false movements of our cavalry, and, notwithstanding the heroic conduct of the gunners, four of those guns were disabled to an extent which rendered their withdrawal at the moment impossible. The moment the artillery was extricated and the cavalry re-formed, a few rounds put to flight the enemy that had occasioned this confusion. With this exception the conduct of the troops generally was most exemplary.'

The late General C. W. Thompson and Viscount Chetwynd have written an article in the *Journal* of the Royal United Service Institution, published in October 1895, which was a reply to an article in the same journal which was written by General Sir Charles Gough, V.C., G.C.B., and published in March 1895, and as the former article gives the true and authentic account of what happened at Chillianwallah, it is reproduced *verbatim* in these pages, so that all interested in the regiment may know exactly what these two officers saw with their own eyes on the occasion in question.

[1] Thackwell's *Narrative of the Second Sikh War of* 1848-49 (1851), p. 142.

EXTRACT from the *Journal* of the Royal United Service Institution, October 1895, Vol. xxxix., No. 212, pp. 1021-26.

By GENERAL C. W. THOMPSON, Colonel, 14th (King's) Hussars, and VISCOUNT CHETWYND, late Lieutenant, 14th Light Dragoons.

Sir Charles Gough's account of the battle of Chillianwallah, in the last March number of the R.U.S.I. *Journal*, has stirred the memories of some survivors of the 14th Light Dragoons, who are anxious that the whole truth should be known of the strange mishap which befell Pope's cavalry brigade on that occasion.

After standing dismounted for some time in column during the afternoon of the 13th January 1849, listening to the heavy firing on our left, but unmolested by the enemy, the brigade was ordered to mount and deploy, which it did deliberately, two squadrons of the 9th Lancers under Major (afterwards Sir) Hope Grant on the extreme right, then three squadrons of Native cavalry in the centre, with four squadrons of the 14th Light Dragoons on the extreme left of the brigade—nine squadrons in all—standing as above stated and not chequered by wings in the manner depicted at p. 241 of Sir Charles Gough's account. As commanding the 2nd squadron of the Fourteenth (the 7th from the right of the general line of the brigade), I had a good view to front and flanks, and can attest that, to the best of my belief and recollection, the whole of the Native cavalry were on our right, forming the centre of the brigade line. The Fourteenth were on the left of the brigade from the first, and remained so throughout the day. Having previously drawn swords, the brigade was now ordered to advance at a trot, without a skirmisher or 'scout' in front, or a man in support or reserve in rear, through broken, jungly ground, where some of the enemy's horsemen were seen to loiter, watching our movements. Brigadier Pope himself led the line in front of the Native cavalry, forming the centre by which we had been ordered to dress and regulate our pace, when insensibly its 'trot' dwindled to a 'walk,' and then came to a dead halt at the sight of a few Sikh horsemen peering over the bushes. Of course the flanks of the brigade had to do the same, being guided by the fluctuations of the centre which were not always clearly visible in the thick jungle, but were conformed to more by sound than by sight. I then saw Colonel King, commanding the 14th Light Dragoons, gallop to the Brigadier in front, energetically pointing with his sword towards the enemy's position and evidently urging an attack, which the other seemed unable to make up his mind to order. The Sikhs seeing the hesitation, a handful of their horsemen, some forty or fifty in a lump, charged boldly into the thick of the Native cavalry, who instantly turned with the cry 'threes about,' and disappeared for the rest of the day—at least I saw none of them.

This word of command, uttered authoritatively, was unfortunately repeated by the remaining squadrons in succession, but was no sooner found to be a mistake (as it might have been at a field-day), than the 'halt' and 'rally' were sounded amid redoubled shouts of 'halt!' from the officers, and the European Lancers and Dragoons were found in an open space like a ploughed field in the jungle facing to the front, where Lord Gough and staff shortly after rode by and were received with 'carried swords.' Why the order was not immediately given to advance and recover the two abandoned guns I never knew; but the Brigadier had been badly wounded in the retreat (not in the advance, as stated by Sir Charles Gough), the men were naturally disappointed by the unexpected failure, and perhaps it was prudent not to attempt too much at the time. Among the sights and sounds of the rallying troops, which have never faded from my recollection for the last forty-six years, I was much struck by the speech of a dragoon who, reining up his horse in line with the others, exclaimed: 'Ah, poor old Billy Havelock, if you had been here this would not have happened,' referring not to his immediate commanding officer—who had done all that a good soldier could do under the circumstances—but to the general handling of the brigade, which every one could see was pitiable in the extreme. Those who remember *El chico blanco*, 'the fair boy' of Napier's *History of the Peninsular War*,[1] and witnessed the gallantry with which he 'rode into the jaws of death' at the head of his regiment at Ramnuggur a few weeks before, will appreciate the force of the dragoon's homely remark. Colonel William Havelock, K.H., was the elder brother of Sir Henry Havelock, the hero of Lucknow, whose statue stands in Trafalgar Square, and at the time of his death in November 1848, was much the more distinguished of the two.

Of the Honourable East India Company's (not Royal) Horse Artillery attached to the brigade I remember little, and do not think that they fired a shot during the advance. Following uselessly in rear, it was rumoured at the time that when the artillery officers complained to the Brigadier that he was masking their guns in such a way as to prevent their opening fire, nothing was done to rectify this essentially false position—not even the simple expedient of dragging the guns into the squadron intervals, where they would have been comparatively safe, if unable to fire. In rear they were and in rear they remained until the line turned, when they turned with it, adding much to the confusion by blocking the way and some of them sticking fast and upsetting among the bushes, where they were captured by the enemy.

As illustrations of the scrambling nature of the *mêlée*, where so much was left to individual action and so little to superior command, I may mention that during one of our short rallies, followed by the enemy, Lieutenant Augustus John Cureton, a gallant youth of eighteen years of age, son of a gallant father, Brigadier-General Cureton, killed at Ramnuggur a few weeks previously, was seen to turn back and ride alone into the jungle, from which his horse shortly returned riderless; and I saw Cureton's body brought into camp a fortnight later, recognisable only by the sleeve of a regimental jacket on one arm.

[1] Vol. v. p. 139.

In the course of the fight in the jungle several hand-to-hand encounters took place, not all to the advantage of the enemy. Major Steuart, of the Fourteenth, overtook a Sikh horseman belabouring an artillery officer (Captain Huish, I believe), and blew him out of the saddle *en passant* by a snap pistol-shot in the breast. Being immediately attacked by another Sikh, sword in hand, the major had not time to return his pistol (a long single-barrelled old-fashioned 'horse pistol'), but guarded with the barrel, from which the native's sabre glanced off, inflicting a slight cut inside the right arm, and Steuart came into camp bleeding profusely, but not seriously. His antagonist fortunately did not renew the attack, but rode away into the jungle.

One of our troop-sergeant-majors seeing a *ghorchurra* (Sikh horseman) conveniently in front gave him a prod in the back, where the point of the sword became so firmly fixed that the exertions of neither party, pulling different ways, could separate them until the dragoon's sword-knot broke and the Sikh rode off with the sword sticking in his back, apparently little the worse. Probably the presence of chain armour under a cotton-quilted jacket or *mirzai*, as generally worn by natives in cold weather, might account for this singular occurence.

In his *Reflections* on Chillianwallah, Sir Charles Gough is mistaken in stating (p. 245) that 'the charge [of Pope's brigade] was badly delivered, and, instead of increasing the pace, the line was brought almost to a trot at the moment of collision.' There was, in fact, no charge and no collision except the partial one above related, and on this point I can speak decidedly from my position as leader of the second squadron of the Fourteenth enabling me to take an uninterrupted view of the front as far as the centre, by which we were ordered to dress.

Had the 'charge' or even the 'gallop' been sounded all would have been well, for the men were in high spirits, and on drawing swords and trotting they fully expected the charge to follow (though there was hardly any enemy visible in front among the bushes of the broken jungle), when the gradual decrease of pace and sudden halt in the centre struck the first vague note of suspicion that something was wrong somewhere, though no one knew *what* it was nor *where*. And upon this the cry of 'threes about' arose from the Native cavalry, was passed down the line from squadron to squadron, and the catastrophe took place. The experience of the two squadrons of the 9th Lancers on the extreme right appears to have been much the same, as related by their commander, Major (afterwards Sir) Hope Grant, in his official report of the 15th of January, given in his *Life*, vol. i. pp. 136-138.

'The 9th Lancers,' he writes, 'were dressing upon the 6th [Native] Light Cavalry, I think. . . . There were some few of the enemy now seen in our front, but nothing in the force to stop any body of Europeans. . . . The two squadrons were going along with the line steadily, and no hesitation was evinced; on the contrary, the flank-men were engaged with some of the enemy, and doing their duty, when the whole line checked and went about from the left, and my squadrons, certainly without a word from me, turned round too.'

It will be observed that Major Grant states above that 'the whole line checked and *went about from the left*,' i.e. *his* left, which doubtless was the case. My experience as second squadron leader of the Fourteenth when in line, was that the movement or 'wave' of retreat, together with the apparently authoritative words of command, came from the right, which would show that the disorder originated at some intermediate point between the two British regiments, viz. at or near the centre of the brigade line, held by the Native cavalry, as really was the case. Sir Charles Gough labours to disprove this by placing the Fourteenth at or near the centre (where they never were), and, speaking of the whole brigade as though the troops composing it were all of the same stamp, concealing the individuality of the Native cavalry under the generic title of 'British.' 'Now occurred what, happily, is a rare event in the annals of British cavalry,' he says—as if the Native cavalry of that day had any right to be considered British beyond the fact of being armed, mounted, and paid by the East India Company; or as if they were not on this occasion the authors of the mischief.

At that time, forty years ago, as Sir Charles will perhaps remember, there was no love lost between the Queen's and Company's services, and the mishap of Pope's brigade (himself a Company's officer) was hailed with something not unlike satisfaction by the local troops, as involving the reputation of two regiments of Peninsular fame, and cloaking the short-comings of their own favourite Native cavalry.

In corroboration of the above account, I am happy to possess the following letter from the Serrefile of the 2nd squadron, the Hon. R. W. (now Viscount) Chetwynd, whose reminiscences in great measure confirm and supplement my own. With one witness in front and another in rear of the line, as he and I were then placed, at different points of view, yet at no great distance from each other, it is hardly possible to suppose that anything of importance could escape our observation. May the truth of our evidence tend to clear up the mystery of this 'inexplicable' defeat and place the saddle of responsibility 'on the right horse'!

C. W. THOMPSON, *General,*
Colonel of the 14th (King's) Hussars.

July 5, 1895.

MY DEAR THOMPSON,—Having been the Serrefile of the squadron of the 14th Light Dragoons, which you led at Chillianwallah, I should like to state to you, as the present full colonel of the regiment, my impression of the account by Sir Charles Gough of Pope's brigade (including the Fourteenth), published in the March number of the *Journal* of the Royal United Service Institution.

Sir Charles begins by saying that the Brigadier 'was to blame for his manner of handling his cavalry. Without consideration he ordered the nine squadrons under his immediate command to advance to the attack in one long line, without support or reserve, thereby preventing the guns from opening fire.' So far Sir Charles is, I believe, perfectly correct, but now begin his mistakes. The first I shall mention is, as to the wounding of the Brigadier. I believe it occurred in the retreat, and that he was still leading the line in person when it turned; in which case the Brigadier's being wounded would have no part in breaking down the advance, as Sir Charles suggests it had. I now come to another mistake, of greater importance. Sir Charles says: 'the charge was badly delivered, and instead of increasing the pace, the line was brought almost to a trot at the moment of collision,' clearly implying that the order to gallop had been given and acted on.

A complete misstatement from beginning to end, as regards the Fourteenth, for they received no order whatever to gallop, and consequently continued at the trot. As for delivering a charge, or any collision, I saw nothing of the kind. There was only one increase of pace in the Fourteenth from the walk to the trot. The reception of the order for that by your squadron was to me a fine and impressive sight.

As to the going about, Sir Charles speaks of some 'wholly inexplicable' cause. This 'inexplicable' cause was, in your squadron, exactly what the Duke of Wellington stated it to have been in the House of Lords: 'a word of command from some unauthorised person.' I heard the word and obeyed it, as did the men in front of me, and so we began trotting back again. Sir Charles further describes the going about as commencing in 'the centre regiment' and 'about the centre of the brigade,' having previously placed the Fourteenth between two wings of Native cavalry. He is, I believe, right in saying that the going about commenced in the centre, but wrong in placing the Fourteenth there, they being, I believe, on the left of the brigade. This has its importance, but not equal to that of what follows.

Proceeding to the retreat, Sir Charles refers to it twice, in one place speaking of a 'portion of the brigade,' in the other of the whole. He means, I think, the same thing in both places, viz the nine squadrons forming the line led by the Brigadier, a part of the brigade being detached to cover the flank. These troops Sir Charles describes as 'breaking into a reckless stampede, galloping to the rear, and riding right down upon the ten guns . . . upsetting and disabling them.' Now, any one deriving his information from this description would certainly understand that these troops, including the Fourteenth, turned, went off at a gallop, and rode straight into the guns, upsetting and disabling them. Very different from what I saw in your squadron.

The squadron came about as already described, there being, as far as I could see just previously, only scattered horsemen in their front. Presently, as we were trotting to the rear, I heard a counter-order, which checked us, but was not obeyed; in my opinion, as I will explain, from the want of something to halt upon. I shortly saw ahead two of the ten guns Sir Charles describes as being ridden over and upset. The sight of them at once steadied us, because it supplied what was wanting—a common halting-point. There was every appearance of a halt upon the guns, when, as we were approaching with our attention fixed on them, off they started, with a fatal effect upon us. But this is not riding over guns and upsetting them; on the contrary, they upset us. As regards these two guns then, Sir Charles's description is, beyond a doubt, very unjust to the Fourteenth, even if—which is quite possible—they afterwards came to grief.

Further than that, it is, I think, not unreasonable to look upon what I saw in your squadron as some indication of the *morale* of the other three at the same time. They may or may not have had assistance, such as we had in the counter-order and seeing the guns in time. Some such assistance was wanted; as the Duke of Wellington pithily expressed it, 'a movement in retreat is not a movement in advance.'

In my opinion, founded on the incident of the guns, it was the fact of the Fourteenth being unsupported that made the going about fatal, and occasioned the loss of the guns and artillerymen. Supporting troops in the place of those guns would have stopped the mischief at once.

Those unsupported guns I take to have been in great danger, in any event, from the moment the unsupported cavalry advanced in front of them, and to have been the victims of bad generalship, as the Fourteenth themselves were.

I object then to this narrative of Sir Charles's, as unjust to the Fourteenth, from its misstatements of fact, both as to the advance and the retreat, and also from its general character—a short dry statement that the brigade was badly commanded, without a word to connect the results with this cause. Positively, the word '*support*' only occurs once in Sir Charles's narrative; the matter of leadership is then put aside, and the alleged results are attributed to the cavalry alone, as though the guns and the leadership had no part in them.

Whether such an account from a general officer of Sir Charles Gough's services should remain uncorrected is a question I leave to you, as the full Colonel of the Fourteenth. If you determine to communicate with the Editor of the *Journal* on the subject, you are welcome to send him this letter if you think fit.—Yours sincerely,

CHETWYND,
Late Lieutenant, 14*th Light Dragoons.*

TO GENERAL THOMPSON,
Colonel of the 14*th Hussars.*

Monument at Maidstone.

Some two and a half years after these events took place the Officers of the 14th (King's) Light Dragoons caused a monument to be erected at Maidstone to perpetuate the memory of their fallen comrades. Maidstone was the place where the depôt of the regiment had been stationed during the years the Fourteenth were in India, hence many of the officers and men were well known to the inhabitants of the town and its neighbourhood. Mr. R. Westmacott, junior, was the sculptor who designed and carried out the memorial. The monument stands eight feet high and four feet wide. It was placed in the All Saints' Collegiate Church at Maidstone in June 1851, and bears the following inscription :—

Sacred to the memory of
LIEUTENANT-COLONEL WILLIAM HAVELOCK, K.H.
He served in Portugal, Spain, and France, at Quatre Bras, where he was wounded, and at Waterloo. He fell at the head of his regiment, charging the Sikhs, at Ramnuggur, on the Chenab, on the 22nd November 1848, aged 56 years.

CAPTAIN JOHN FOSTER FITZGERALD.
He died on the 26th November 1848, of wounds received in action at Ramnuggur, aged 28 years.

LIEUTENANT AUGUSTUS JOHN CURETON.
Killed at the battle of Chillianwallah on the 13th of January 1849, aged 18 years.

LIEUTENANT AMBROSE LLOYD.
Killed at the battle of Goojerat on the 21st of February 1849, aged 26 years.

SERGEANT JOHN HARWOOD,
CORPORAL WILLIAM PARKER TODD,
and Privates
JOHN ALDERTON, WILLIAM ALPINE, RICHARD BAGG, WILLIAM BRAZEUR, CHARLES FOX, JOHN HATTON, RICHARD HUNGERFORD, BENJAMIN JENNINGS, JAMES RAINES, CHARLES TUTTELL, JOHN WARD, GEORGE WILLIAMS, killed on the 22nd November 1848, and GEORGE ATKINS, DAVID EVANS, GEORGE TOOKEY, killed on the 13th January 1849.

The Officers of the 14th (King's) Light Dragoons
erect this monument to
their Comrades
Who fell in the Campaign of the Punjaub.

'*Be thou faithful unto death.*'
Rev. II. 10.

APPENDIX B

Despatches of Major-General Sir Hugh Rose, K.C.B., etc.

EXTRACTS FROM DESPATCHES of MAJOR-GENERAL SIR HUGH ROSE, K.C.B., BRIGADIER-GENERAL SIR R. NAPIER, K.C.B., and other Officers, relative to JHANSI, KOONCH, MUNDESOR, RATHGUR, GARRAKOTA, BETWA, LOHARI, CALPEE, GWALIOR, MORAR, JOWRA-ALIPORE, and RANODE.

JHANSI DESPATCHES

From MAJOR-GENERAL SIR HUGH ROSE, K.C.B., Commanding Central India Field Force, to the CHIEF OF THE STAFF.

CAMP MOTE, *April* 30, 1858.

Extracts from Jhansi despatches, dated Camp Mote, 30th April 1858.

SIR,—I have the honour to report for the information of His Excellency the Commander-in-Chief the operations of my force against the fortress and fortified city of Jhansi, on the 20th ultimo. The 2nd Brigade, under my command, arrived at Limra, one day's march from Jhansi. My 1st Brigade had not yet joined me from Chanderi. The same day I sent Brigadier C. Steuart, with the cavalry and artillery, as follows, to invest Jhansi, viz.—

Six guns Horse Artillery.
14th Light Dragoons, 325 rank and file.
3rd Light Cavalry, 140 ,, ,, ,,
Hyderabad Contingent, 476 sabres.

I arrived the 21st ultimo with the remainder of my brigade before Jhansi. The picquets of the cavalry sent on the day before had sabred about 100 armed men, Bundeelas, endeavouring to enter Jhansi, having been summoned by the Ranee to defend it.

Flying camps of cavalry, under Majors Scudamore and Gall, 14th Light Dragoons.

I established seven flying camps of cavalry as an investing force round Jhansi, giving to Major Scudamore half a troop of Horse Artillery, and later to Major Gall two 9-pounders. These camps detached to the front outposts and vedettes, which watched and prevented all issue from the city day and night. Each camp, on any attempt being made to force its line, was to call on the other for help. I gave directions also that the roads from the city should be obstructed by trenches and abattis.

I had made arrangements on the 30th March for storming, but the general action of the Betwa on 1st April with the so-called 'army of the Peishwa,' which advanced across the Betwa to relieve it, caused the assault to be deferred.

On the 2nd instant I issued a Division Order for the assault of the defences of the city wall, of which a copy with a plan of attack was furnished to the officers in command. I have the honour to enclose copies of reports from Brigadier Stuart, commanding my 1st Brigade, and Brigadier C. Steuart, commanding my 2nd Brigade, of the operations of their respective columns against Jhansi. Whilst engaged in the town I received a report from the officer commanding one of the Hyderabad Cavalry flying camps that a large body of the enemy, flying from the town, had tried to force his picquet; that a few had succeeded, but that the main body, from 350 to 500 strong, had been driven back, and had occupied a high and rocky hill to the west of the fort; that he had surrounded the hill with cavalry till reinforcements were sent. I immediately ordered out from the camps of the two brigades the available troops of all arms against the hill. The enclosed report from Major Gall, 14th Light Dragoons, shows how satisfactorily these rebels were disposed of. Lieutenant Park was killed whilst gallantly leading on a party of the 24th Bombay Native Infantry along the ridge of the hill. The Ranee's father, Mammo Sahib, was amongst the rebels. He was wounded on the hill and captured some days afterwards, and hanged at the Tokim Bagh. The next day Brigadier Stuart and myself occupied the rest of the city by a combined movement united by Major Gall, who spiritedly scaled the bastion iron gate from his flying camp and captured the gun that was there and threw it down the ramparts. The following morning a wounded Mahratta ectanca of the Ranee was sent into me from Captain Abbott's flying camp. He stated that the Ranee, accompanied by 310 Velaitees and 25 sowars, fled that night from the fort; that after leaving it they had been headed back by one of the picquets, when the party and the Ranee separated, she herself taking to the right with a few sowars in the direction of the intended flight to Bandin. The observatory also telegraphed: 'Enemy escaping to the north-east.' I immediately sent off strong detachments of Her Majesty's 14th Light Dragoons, 3rd Light Cavalry, and Hyderabad Cavalry, to pursue, with guns to support them, as it was said Tantia Topee had sent a force to meet her. I also sent Brigadier Steuart with cavalry to watch the fords of the Betwa.

1st Brigade under Brigadier Stuart; 2nd Brigade under Brigadier C. Steuart, 14th Light Dragoons.

Jhansi captured, 4th April 1858.

Pursuit of the Ranee by 14th Light Dragoons and Native Cavalry.

In sight of Bandin, 21 miles from Jhansi, the cavalry came in view of the irregular horse sent to meet the Ranee, which separated, probably with a view to mislead her pursuers as to her real course. Lieutenant Dowker, Hyderabad Cavalry, was sent by Captain Forbes through the town of Bandin, whilst he, with the 3rd Light Cavalry and 14th Light Dragoons, passed it by the left. In the town, Lieutenant Dowker saw traces of the Ranee's hasty flight, and her tent, in which was an unfinished breakfast. On the other side of the town he came up with and cut up 40 of the enemy, consisting of Rohillas and Bengalee irregular cavalry. Lieutenant Dowker was gaining fast on the Ranee, who with four attendants was seen escaping on a grey horse, when he was dismounted by a severe wound and obliged to give up the pursuit. The Ranee's flight was the signal for a general retreat. Early in the morning I caused the outskirts of the city to be scoured with cavalry and infantry. It will give some idea of the destruction of insurgents which ensued when a party of the 14th Light Dragoons alone killed 200 in one patrol. The rebels, who were chiefly Velaitees and Pathans, generally sold their lives as dearly as they could, fighting to the last with their usual dexterity and firmness.

Two hundred rebels killed by a party of 14th Light Dragoons.

I beg leave to bring to the favourable notice of the Commander-in-Chief the conduct of the troops under my command in the siege, investment, and capture of Jhansi. They had to contend against an enemy more than double their numbers behind formidable fortifications, who defended themselves afterwards from house to house in a spacious city, often under the fire of the fort, then later in the suburbs, and in very difficult ground outside the walls. The investing cavalry force were, day and night, for seventeen days on arduous duty, the men not taking their clothes off, the horses saddled and bridled up at night. The nature of the defence and the strictness of the investment gave rise to continual and fierce combats, for the rebels, having no hope, sought to sell their lives as dearly as possible. But the discipline and gallant spirit of the troops enabled them to overcome difficulties and opposition of every sort, to take the fortified city of Jhansi by storming, subduing the strongest fortress in Central India, and killing 5000 of its rebel garrison. According to the first reports which I received, 3000 rebels were killed, but those received since the withdrawal of the seven flying camps make the loss of the enemy amount to above 5000 killed. Native accounts received by Brigadier Wheeler at Saugor make the loss of rebels to amount to more than 5000. I beg to recommend to His Excellency for gallant and good service in investing the fortress and city of Jhansi, Major Scudamore, 14th Light Dragoons, the senior officer in command of flying camps; Major Gall, H.M.'s 14th Light Dragoons; Major Forbes, C.B., commanding 3rd Cavalry; Captain Abbott and Lieutenant Dowker, Hyderabad Cavalry. I beg leave to state the obligations I am under to the following officers for the services which they have rendered me during the siege operations and capture of Jhansi:—

Five thousand rebels killed at Jhansi.

Officers of 14th Light Dragoons specially mentioned for gallant conduct and good service.

Brigadier Stuart, commanding 1st Brigade.
Brigadier C. Steuart, C.B., commanding 2nd Brigade.
Major Scudamore, commanding H.M.'s 14th Light Dragoons, &c.
Captain Todd, Major of Brigade, &c.[1]

I have, &c.

(Signed) HUGH ROSE, *Major-General,*
Commanding Central India Field Force.

[1] Other names, not connected with the Fourteenth, are omitted above.

No. 236.

From BRIGADIER C. STEUART, C.B., 14th Light Dragoons, commanding 2nd Brigade C.I.F.Force, to the ASSISTANT ADJUTANT-GENERAL, C.I.F.Force.

SIR,—In obedience to orders received through you, the brigade under my command moved in two columns on the morning of the 3rd April to the assault of the town of Jhansi.

. . . .

Captain Todd, Brigade-Major, and Captain Leckie, Deputy-Assistant Quartermaster-General of the 2nd Brigade, on this as on every previous opportunity have afforded me every assistance, etc. etc.—I have, etc.,

(Signed) C. STEUART, *Commanding 2nd Brigade C.I.F. Force.*

Total return of ordnance captured in the town of Jhansi on the 3rd April 1858 by the force under the command of Major-General Sir Hugh Rose, K.C.B. :—10 brass guns; 1 brass howitzer; 15 iron guns. Return of ordnance captured at Jhansi, 3rd April 1858.

Captured in the fort of Jhansi on 5th April 1858:—8 iron guns; 1 brass gun.

(Signed) THOS. S. HAGGARD, *Lieutenant,*
Commissary of Ordnance, C.I.F. Force.

KOONCH DESPATCHES

From MAJOR-GENERAL SIR HUGH ROSE, K.C.B., to GENERAL SIR WILLIAM MANSFIELD, K.C.B., Chief of the Staff.

CAMP GOLOWLEE, *May* 24, 1858.

SIR,—I have the honour to report to you for the information of His Excellency the Commander-in-Chief that the approach of Brigadier Smith's Brigade from Rajputana to Goona having secured Jhansi from attack by Kotah and Bundelcund rebels, I recalled Lieutenant-Colonel Lowth, commanding 86th Regiment, whom I had detached with a column to watch the road from Jhansi to Goona, and I marched with the 1st Brigade of my force from Jhansi on the 25th ultimo on Calpee. I left at Jhansi for its garrison the force detailed below, forming part of the 2nd Brigade, viz. :— Extract from Koonch despatches, Camp Golowlee, May 24, 1858.

Headquarter wing, 3rd Bombay European Regiment.
Eight companies 24th Bombay Native Infantry.
Hyderabad Cavalry, 100 sabres.
Three guns Bhoopal Artillery.
Half company Sappers and Miners.

I left there also Brigadier C. Steuart of the 14th Light Dragoons, with the remainder of his brigade, with orders to bring up to me the 71st Regiment and two troops of the 3rd Light Cavalry. I joined Major Gall's force at Pooch, 16 miles from Koonch, on 1st May. I had the honour to report on the 17th inst. the movements of this officer's movable column, as well as those of Major Orr's field force. I received information from Sir Robert Hamilton and Major Gall, whom I had detached along the road from Jhansi to Calpee with a flying column to watch the enemy and obtain information of their movements, that the Sepoy garrison of Calpee of all arms, reinforced by 500 Velaitees under the Ranee of Jhansi, cavalry from Kotah, and guns and troops from disaffected Rajahs—the whole under the command of Tantia Topee—had occupied Koonch, and thrown up entrenchments, which they had armed, to defend the roads leading to the town from Jhansi, and that they were determined to make a vigorous opposition at Koonch to my advance against Calpee. All the accounts agreed that the rebels were strong in cavalry, consisting of mutineers from Bengal, regular and irregular regiments. Koonch is an open town, but is difficult to attack because it is surrounded by woods, gardens, and temples with high walls round them, every one of which is a defence.

Koonch.

My left, the 1st Brigade, was resting with its left flank on the village of Nagapoore; my centre, the 2nd Brigade, under Brigadier C. Steuart, was in the village of Choman; my right, Major Orr's force, in front of the village of Ormree. I gave the orders that as soon as the three columns had taken up the position which I have mentioned they were to advance against the town and endeavour to effect a lodgement in it. When we came within sight of Koonch we perceived vedettes and strong picquets of the enemy's cavalry outside the wood. They conformed to our flank movement, and posted themselves nearly opposite to Nagapoore. A few rounds of shrapnel from Captain Lightfoot's guns emptied some of their saddles, and they disappeared into the wood.

The rebel infantry now showed in force behind a long wall to our front and in the wood to the left of it. I had marched the 1st Brigade a distance of 14 miles from Lohari that morning for the purpose of surprising the enemy by the flank movement, and not giving them time to alter their plan of attack. I ordered the men's dinners to be cooked for them, to rest and refresh them, and meantime battered the wall with the two 18-pounders and the 8-inch howitzer.

Lieutenant-Colonel Gall reconnoitres the enemy at Koonch.

Lieutenant-Colonel Gall galloped gallantly into the wood to reconnoitre the enemy. Although he was within half musket range of them they did not fire at him, because the shelling from our Horse Artillery had caused confusion in their ranks. He ascertained that the infantry to the left had retreated further into the wood, having in their rear a large body of cavalry; that the siege-guns had driven the enemy from the cover of the wall, but that some way in rear of it was posted a large body of infantry with elephants. I determined to drive the enemy out of the wood, gardens, and temples surrounding Koonch, and then to storm the town and a dilapidated mud fort on a rising ground, a strong position which was opposite to the right of the 1st Brigade.

I effected the operation by throwing the left wing of Her Majesty's 86th Regiment, under Major Stuart, and the whole of the 25th Bengal Native Infantry, under Lieutenant-Colonel Robertson, into skirmishing order, their flanks supported by the half-troop Horse Artillery and a troop of Her Majesty's 14th Light Dragoons, and Captain Ommaney's battery and two troops of Her Majesty's 14th Light Dragoons. I left Captain Woolcombe's battery, one troop 14th Light Dragoons, and the right wing 86th Regiment in a second line in reserve, under command of Lieutenant-Colonel Lowth. The rapidity and precision with which this formation was simultaneously made must have surprised the sepoys. The first line advanced, notwithstanding the artillery and musketry fire, through the whole north part of the town and took the fort. The troop 14th Light Dragoons made a circuit to their left, took all the obstacles to their front, and then brought their left shoulders forward. Just as the 86th Regiment and myself with the 25th were about to enter the town, Brigadier C. Steuart, 2nd Brigade, observed that a large number of the rebel infantry, strongly posted in cultivated ground, threatened the right of the line of attack of his brigade. He moved up Captain Field's battery and Captains Thompson's and Gordon's troops of 14th Light Dragoons and one troop of 3rd Regiment Hyderabad Cavalry to dislodge them. The enemy held the position obstinately, and it was not till a portion of the infantry of the 2nd Brigade moved down on them from another direction that they retreated, when Captain Gordon, whom I beg to recommend to His Excellency for his conduct on this occasion, with his troop and the cavalry above mentioned, charged and broke the mass, cutting up several of them, but topes of trees favoured the escape of the remainder. The 2nd Brigade, under Brigadier Steuart, owing to some misconception on his part, did not effect a lodgment in the town, but moving round to the south of it, his artillery and cavalry joined in the pursuit.

Battle of Koonch, 7th May 1858.

Four troops 14th Light Dragoons engaged in 1st Brigade.

Movements of 2nd Brigade.

Two troops 14th Light Dragoons charge the enemy.

Captain Gordon specially mentioned.

The cavalry of both brigades and Major Orr's force (except a party left to watch the Jaloun road and my rear), one troop of Horse Artillery, Captain Field's guns and the four guns No. 13 Light Field Battery, went in pursuit.

Cavalry of 1st and 2nd Brigades sent in pursuit.

The pursuit was commenced by Captain William McMahon's squadron and Captain Blyth's troop 14th Light Dragoons charging, the first the right and the latter the left of the enemy's skirmishers. A piece of very heavy ploughed land caused a check in the pace, under a heavy fire, of Captain McMahon's squadron; but the heavy ground was not broad, the squadron got through it, Captain McMahon leading the way, and cut to pieces the enemy, who fought fiercely to the last. Captain McMahon received three sabre-wounds, but he continued the pursuit to the last. I beg to recommend him for his gallant conduct and unvarying zeal and attention to his duties. On the centre the Horse Artillery opened a hot fire, and the cavalry charged the skirmishers. The enemy now threw back the extreme right of their skirmishers so as to enfilade our line of pursuit. I directed Captain Prettejohn to form line to the

Captains McMahon, Prettejohn, and Blyth specially brought to notice for gallant conduct.

left, charge and cut off the enfilading skirmishers, which he did effectually. This officer, the horses of his own troop being knocked up, placed himself with well-timed zeal at the head of a troop with fresh horses which was without an officer, and continued the pursuit with them to the end. I beg to submit his name to the favourable consideration of His Excellency, as well as the names of Captain Blyth, 14th Light Dragoons, and Captain Abbott, commanding 3rd Cavalry Hyderabad Contingent, who each very gallantly charged and captured a gun from the retreating enemy under a heavy fire. In the course of the pursuit (up to seven miles from Koonch) more guns and ammunition were captured by the cavalry.

The sun, fatigue, and scarcity of water told on my artillery and cavalry, a great part of whom were Europeans, and had been marching and engaged for 16 hours, so I marched them back at sunset to Koonch.

The enemy must have lost about 500 or 600 men in this action and pursuit.

Nine guns, etc., captured.

Nine guns and quantities of good English ammunition and stores furnished to the Gwalior Contingent were captured. Tantia Topee had disappeared at Koonch as rapidly as he had done at the Betwa, leaving to its fate at the most critical moment the force which he had called into existence under the pompous title of 'The Army of the Peishwa.'

Brigadier C. Steuart, C.B., commanding 2nd Brigade, mentions that his staff, Captain Todd, Major of Brigade, and Captain Leckie, Deputy-Assistant Quartermaster-General, afforded him every assistance. Enclosed are returns of killed and wounded, and of the guns and ordnance stores captured in the action.—I have, etc. (Signed) HUGH ROSE, *Major-General,*

Commanding C.I. Field Force.

Casualties of 14th Light Dragoons at Koonch.

ABSTRACT OF CASUALTIES (14TH LIGHT DRAGOONS).

Left wing.—5 men wounded.

Right wing.—4 men killed; 1 man died of his wounds.

Wounded.

Captain William McMahon, wounded severely (sword-cut in right hand and leg); 12 men wounded.

Casualties owing to sunstroke.

Horses of right wing.—3 killed; 6 wounded; 4 missing.

Captain Need, Lieutenant Travers, and 16 men, struck down by the sun, of whom 2 men died.

MUNDESOR DESPATCH

No. 10.

No. 47 *of* 1858.

'London Gazette,' 11*th March* 1858.

Extracts from Mundesor despatch.

The Right Honourable the Governor in Council has very great satisfaction in publishing for the information of the army the following report made to His

Excellency the Commander-in-Chief, by Brigadier Stuart, commanding the Malwa Field Force (now 1st Brigade Nerbudda Field Force), of his successful operations against the insurgents assembled near Mundesor during four days, from the 21st to the 24th November last:—

No. 11.
No. 201 *of* 1857.

EXTRACTS from the REPORT of BRIGADIER STUART, commanding the Malwa Field Force, and published in the *London Gazette*, dated 11th March 1858.

'The rebel enemy at Mundesor, hearing of our approach, had posted picquets entirely covering the country over which we were advancing, and observing our picquets thrown out, they mustered in some force outside the walls of the town, and appeared inclined to attack. I, however, contented myself with reinforcing the picquets, and leaving the whole charge of the front to Major Robertson, 25th Regiment Bombay Native Infantry. The field officer of the day returned to camp about 3 o'clock P.M. I received intimation from him that the enemy were advancing in force, and threatened both our flanks and centre at the same time. I accordingly moved out to meet them. They advanced steadily, with banners flying, and appeared in great numbers. On approaching our right front they were most gallantly charged by Lieutenant Dew, Her Majesty's 14th Light Dragoons, who with some of his men occupied that ground as a picquet. Captain Orr, commanding 3rd Regiment Cavalry Hyderabad Contingent, supported Lieutenant Dew, and the enemy were driven back with great loss, and before our guns, which had quickly moved up, could open upon them. The attack on our centre was repulsed by a few rounds of our artillery, whilst that on our left was successfully met by the Field Force under Major Orr. The enemy having been thus driven back at all points, were pursued for some distance—in fact until they nearly reached the walls of the town. Just previous to the camp being marked out intelligence came that Heera Sing's baggage had just left the village of Goraria, on the Neemuch road, and a party of cavalry, about 300, were observed in a north-westerly direction; so Her Majesty's 14th Light Dragoons, left wing, under Major Gall, and the 1st and 4th Regiments Cavalry Hyderabad Contingent, galloped off in pursuit, the 3rd Cavalry Hyderabad Contingent remaining as a reserve. They caught up the enemy about two miles south of Peeplia, and after cutting up about 200 of them, returned to camp.

Lieutenant Dew, 14th Light Dragoons, charges with his picquet.

Major Gall with left wing 14th Light Dragoons and Hyderabad Cavalry, charge a body of the enemy and kill about 200.

On the 23rd November my line advanced, covered by skirmishers. The enemy's infantry, with banners flying (many of them green), moved down to meet us through the intermediate fields of high jowarree, and their guns opened fire. I immediately halted my line and replied to the fire with Captains

Another engagement with the rebels at Mundesor.

Hungerford and Woolcombe's batteries, at a range of about 900 yards. After a few rounds I again advanced the line and permitted Captain Hungerford to move his half-battery to a position on our right front, from which he could enfilade the enemy. After an advance of about 300 yards our line was again halted, and firing resumed, that from both batteries being very effective. A most gallant charge was then made on the enemy's guns by the escort of Her Majesty's 14th Light Dragoons attached to Captain Hungerford, led by Lieutenant Martin, who found, however, that the position was still very strongly held by the enemy's infantry, and was compelled to retire, he himself being very severely wounded. Captain Hungerford's half-battery was again advanced to within 100 yards, and after a round or two of grape, the guns were at once charged and captured, the enemy flying in great numbers into the village to the right. I learned that during the afternoon, when we were hotly engaged in the front, a strong body of the enemy from Mundesor attacked our rear, and endeavoured to carry off the siege-train, baggage, etc. They were, however, most gallantly repulsed on every occasion. In one of these attacks I regret to say that Lieutenant Redmayne, Her Majesty's 14th Light Dragoons, was killed while most gallantly leading his men against the enemy. Notwithstanding the many attempts made by the enemy to press upon and harass our rear, it gives me great satisfaction to be able to state that not a particle of baggage was lost nor a follower injured. On this occasion Lieutenant Leith, commanding a squadron 14th Light Dragoons, appears to have done good service. A perusal of the report made by Captain Gall, commanding left wing Her Majesty's 14th Light Dragoons, will convey to His Excellency the Commander-in-Chief some idea of the good service performed by all under his command. I most fully concur in Captain Gall's report,[1] and beg to recommend to the consideration of His Excellency all the officers and men mentioned by him. Of Captain Gall himself I must in justice add, that a more able, zealous, and hard-working officer I never met with, nor one more worthy of distinction.

Lieutenant Martin, with artillery escort 14th Light Dragoons, charged enemy's guns, and was severely wounded.

Enemy attacked our rear; Lieutenant Redmayne charged and was killed.

Lieutenant Leith specially mentioned.

Captain Gall specially mentioned.

No. 12.

Casualties of 14th Light Dragoons at Mundesor, 21st to 24th Nov. 1857.

Nominal roll of officers and men of the 14th Light Dragoons killed and wounded in the engagements with the insurgents before and in the vicinity of Mundesor from 21st to 24th November 1857.

Killed.—Lieutenant Redmayne.

Wounded.—Lieutenant James Leith.
Lieutenant L. Gowan.
Lieutenant C. Martin.
Regimental Sergeant-Major T. Clark.
Troop Sergeant-Major S. Whittaker.
Privates Cooper, Harris, O'Neill, Remmington, and Buchanan.

[1] This Report is not available. *See* Despatches.

RATHGUR DESPATCH

EXTRACT from the REPORTS of MAJOR-GENERAL SIR HUGH ROSE, K.C.B., commanding Central India Field Force, and published in the *London Gazette*, dated 11th May 1858, relating to Rathgur.

No. 2.

I moved in the order of march which I always adopted when near the rebels, as a precaution against their system of surprises, that is, a line of flankers of H.M.'s 14th Light Dragoons on each side of the road 50 yards in front of the leading file of the advance-guards, which, with a file of irregulars, has charge of the guide; another line of irregular cavalry, 150 yards in echelon, in front of the outward flanks of the Fourteenth, and, should thick jungle border the road, a company of infantry, in extended order on each side of it, to support the flanks of the men of the 14th Light Dragoons and the advance-guard. By this means all dangerous ground is searched, surprises are almost impossible, and spies lying concealed at a great distance from the road are frequently seized. I followed with four guns of the Horse Artillery and a troop of H.M.'s 14th Light Dragoons in support under Lieutenant-Colonel Turnbull, ordering the rest of my force to follow, with the exception of Captain Hare's infantry and guns, which remained at the ford to prevent the rear being cut off. I was not at all sure that my camps with the siege artillery and numerous stores, left with a small force at Rathgur under Brigadier Steuart, might not be attacked during my absence, as they had been before. I therefore halted in the village only for a short time in order to rest the troops, who had been on duty for at least five days, and marched back the same night to Rathgur, as they had been marching or engaged 15 hours. The enemy's loss was severe—they themselves state it to be from 400 to 500, which is not surprising, as they were exposed to a well-directed fire for a length of time. Amant Sing, their ablest military leader, as well as a nephew of Fazil Mahomed Khan, were killed, and the Rajah of Banpore was wounded. The 14th Light Dragoons had one horse killed.

Extracts from Rathgur despatches.

GARRAKOTA DESPATCHES (*Extracts*)

No. 6.

CAMP KOOMEERI, *March* 9, 1858.

'. . . However, I ought to add that even if the Paunch Ghat had been occupied, it would have been quite impossible to have invested completely Garrakota with a force of my numbers, which, strictly speaking, was only sufficiently strong to guard its camp and work the field and siege artillery. My force was more than usually weak, as, in order to protect Saugor, in my absence I had left there a troop of H.M.'s 14th Light Dragoons and two companies of the 24th Regiment Bombay Native Infantry. I had also sent the 31st and 42nd Bengal Native Infantry to Koray to guard Saugor from the possibility of an attack from the

Captain Hare with Captain Need and a troop of 14th Light Dragoons cut up 70 to 100 rebels at the river Beas.

north. I have observed that nothing alarms the rebels more than a move to cut off their retreat. I sent off immediately Captain Hare with a half-troop of Horse Artillery, a troop of H.M.'s 14th Light Dragoons (under Captain Need), and a troop of the Hyderabad Contingent Cavalry in pursuit of the rebels, by a route which, from information I had received, I thought would enable them to cut into their line of retreat. They had, as I supposed, made a round to the south, and then turned northwards towards Shaghur. Captain Hare came up with the rear of the rebels as they had just crossed the river Beas, at the village of Beas.[1] The river was not practicable for guns. Captain Hare therefore sent the troop of H.M.'s 14th Light Dragoons, under Captain Need, and the Hyderabad Cavalry across the river after the enemy. They cut up 70 or 100 of them, of whom the greater part were mutineers of the 52nd and other native regiments. They had all abundance of ammunition and copper caps. Captain Hare speaks very highly of the conduct on this occasion of Captain Need and his troop. He describes Captain Need as a 'good and dashing officer,' and adds, 'he killed with his own hands five of the rebels, of whom three were sepoys, and pursued the rebels with his gallant troop till dark.' I beg to recommend Captain Need to your Excellency, as well as Captain Hare for his intelligence and activity in pursuing and coming up with the enemy.

Captain Need specially brought to notice for gallant conduct at Garrakota.

EXTRACTS from the *London Gazette*, published 31st May 1858.

No. 23.

The ADJUTANT-GENERAL of the Army to the SECRETARY to Government Secret Department, Bombay.

ADJUTANT-GENERAL'S OFFICE,
HEADQUARTERS, MAHABLESHWAR,
10*th April* 1858.

No. 140.

SIR,—In continuation of the letter from this department (No. 2701) under date of the 29th ultimo, transmitting a despatch from Major-General Sir Hugh Rose, K.C.B., commanding Central India Field Force, dated Koomeeri, 9th idem, I am desired by the Commander-in-Chief to request you will bring to the notice of the Right Honourable the Governor in Council the favourable mention made by the Major-General of the under-mentioned officers in the affair with the insurgents, which resulted in the evacuation of the strong fortress of Garrakota:—Lieutenant-Colonel Liddell, commanding 3rd European Regiment; Captain Lightfoot, commanding No. 18 Light Field Battery; Captain Hare, Hyderabad Contingent; Captain Need, H.M.'s 14th Light Dragoons, etc. etc.

His lordship in Council will doubtless observe that Major-General Sir Hugh Rose, K.C.B., has particularly noted the conduct of Captains Hare and Need and Lieutenant Sturt, and His Excellency therefore solicits that the Right

[1] Or Biás (Malleson).

Honourable the Governor in Council will accord to them, as also to the other officers, the approbation they have so well deserved.—I have, etc.

(Signed) EDWARD GREEN, *Colonel*,
Adjutant-General of the Army.

BETWA DESPATCH

Extracts from Betwa despatch.

(Published in the *London Gazette*, 10th August 1858.)

No. 1.

MAJOR-GENERAL SIR HUGH ROSE, K.C.B., commanding Central India Field Force, to the ADJUTANT-GENERAL, Bombay Army.

CAMP POOCH, *April* 30, 1858.

Action of the Betwa, 1st April 1858.

SIR,—I have the honour to report to you, for the information of His Excellency the Commander-in-Chief in India, that on the 1st April the forces under my orders fought a general action with the so-called army of the Peishwa, which attempted to relieve Jhansi while I was besieging it, and gained a complete victory over it, pursuing two miles beyond the river Betwa, taking 18 guns, of which one was an 18-pounder, one an 8-inch mortar, two 12-pounders, and two English 9-pounders, and killing upwards of 1500 rebels. For some time past Sir Robert Hamilton had given me information that Tantia Topee, a relative and the agent of Nana Sahib, had been collecting and organising a large body of troops in the neighbourhood of Mhow and Nowgong, in Bundelcund,[1] which was called 'the Army of the Peishwa,' and had displayed the standard of that abolished authority. After the fall of Chanderi[2] this army was reinforced by the numerous rebel troops, sepoys from Calpee and Bundeelas, who had besieged and taken it. Towards the end of last month I received constantly reports that the force, estimated at 20,000 or 25,000 men, with 20 or 30 guns, was advancing against me. On the 30th ultimo, Sir Robert Hamilton informed me that its main body had arrived at Burra Saugor, about three miles from the Betwa, would cross that river during the night and attack me next morning. The enemy crossed the Rajpore, the upper ford, in great numbers on the 30th March, preceded by an advance-guard of Velaitees, and took up after sunset a position in order of battle opposite the rear of my 2nd Brigade.

Eighteen guns captured.

The details below show how weak I was compared with the enemy:—

Artillery—3 siege guns; 16 light field guns.
14th Light Dragoons, 243 rank and file.
Hyderabad Cavalry, 207 sabres.
86th Regiment, 208 rank and file.
3rd Bombay European Regiment, 226 rank and file.
24th Bombay Native Infantry, 298 ,, ,,
25th Bombay Native Infantry, 400 ,, ,,

Two hundred and forty-three men of 14th Light Dragoons engaged at the Betwa.

[1] 'Bandalkand' (Malleson). [2] 'Chandairee' in original despatch.

I sent Major Orr with a party of his cavalry along the road to the Betwa to watch the enemy's movements.

I drew up my force across the road from the Betwa, half a mile from my camp. On the right flank of my first line, the 2nd Brigade, I placed Lieutenant Clark's Hyderabad Horse, a troop 14th Light Dragoons, and four guns Horse Artillery; in the centre, detachments of the 24th Bombay Native Infantry and 3rd Europeans, three heavy guns and detachments Hyderabad Infantry; on the left flank, Captain Lightfoot's battery and two troops 14th Light Dragoons.

The second line, my 1st Brigade, was in contiguous columns at quarter distance, consisting of a weak troop 14th Light Dragoons on the right, and Hyderabad Cavalry on the left flank, in the centre Her Majesty's 86th Regiment, Captain Woolcombe's battery of 6-pounders, and Captain Ommaney's battery of 9-pounders, with detachments of the 25th Bombay Native Infantry.

I threw out picquets and lines of vedettes of the 14th Light Dragoons and Hyderabad Cavalry well to my front and flanks. The Velaitees' outposts called out during the night that they were very numerous, that we were very few, that in the morning they would finish us off! My force was not in position till long after dark. Both ourselves and the enemy slept on our arms opposite each other.

Hearing that the enemy were crossing in large numbers at the lower, the Kolwar ford, with the object of turning my left flank and forcing their way along the Burragong road, through Major Scudamore's flying camp into Jhansi, I detached Brigadier Stuart, in the middle of the night, with the 1st Brigade along the Burragong road, about eight miles to the village of Burragong, close to the river Betwa, where he could oppose and outflank the enemy, who had crossed by the ford above the village. The departure of the 1st Brigade left me without a second line. I was therefore obliged to withdraw the detachments of the 24th Native Infantry from the first line, and make a second line of them. The enemy, before daybreak, covered by a cloud of skirmishers, advanced against me. My outposts retired steadily, closing to each flank. Shortly after the enemy opened a very heavy artillery, musket, and matchlock fire on my line from the whole of his front, to which my batteries answered steadily. The enemy were in rear of a rising ground. I ordered my first line of infantry to lie down, the troop of Horse Artillery to take ground diagonally to the right and enfilade the enemy's left flank. In this movement a round-shot broke the wheel of a Horse Artillery gun. Captain Lightfoot took up an advanced position to his left front, which made the fire of his battery much more efficacious. Whilst the enemy were suffering from the fire of the troop and battery I directed Captain Prettejohn, 14th Light Dragoons, to charge with his troop, supported by Captain McMahon, 14th Light Dragoons, with his troop, on the enemy's right flank, and I charged myself on their left with Captain Need's troop, 14th Light Dragoons, supported by a strong troop of

Hyderabad Cavalry. Both attacks succeeded, throwing the whole of the enemy's first line into confusion and forcing them to retire.

Captain Prettejohn's, Captain McMahon's, and Captain Need's troops of the 14th Light Dragoons charge successfully.

I beg to do justice to Captain Need's troop: they charged with steady gallantry on the enemy's left, which was composed of the best rebel troops, the Velaitees and sepoys, who, throwing themselves back on the right and resting the flanks of their new line four or five deep on the rocky knolls, received the charge with a heavy fire of musketry. We broke through the dense line, which flung itself amongst the rocks, and bringing our right shoulders forward took the front line in reverse and routed it. I believe I may say that what Captain Need's troop did on this occasion was equal to breaking a square of infantry, and the result was most successful, because the charge turned the enemy's position and decided in a great measure the fate of the day.

Captain Need specially mentioned; Lieutenant Leith recommended for V.C.

I have the honour to recommend to His Excellency's favourable consideration Captain Need and his devoted troop, and Lieutenant Leith, who saved Captain Need's life, for which I have ventured to recommend him for the 'Victoria Cross.'

Captain Prettejohn's squadron charges enemy.

The enemy's right gave way before the squadron of the 14th Light Dragoons, under Captain Prettejohn, reached them; he pursued and cut up several of them. I moved forward the whole of the artillery and cavalry in pursuit. Severe combats occurred between the pursuing cavalry and the fugitives, who fought with desperation.

Two troops of 14th Light Dragoons pursue across the Betwa.

I ordered two troops of the 14th Light Dragoons and the Hyderabad Cavalry across the Betwa. The enemy kept up a heavy fire on us as we crossed the ford and ascended the steep road leading up the opposite bank. The 14th Light Dragoons and Hyderabad Cavalry gallantly surmounted all opposition and sabred the rebels who still held their ground.

Eighteen guns and 2 standards captured.

About a mile and a half from the Betwa the 12-pounder, being the eighteenth and last gun of the rebel army, was captured. Two standards also were captured.

Horses and men being completely exhausted by incessant marching and fighting during the last forty-eight hours, and being now nine miles from Jhansi, I marched the troops back to camp.

Fifteen hundred of the enemy killed.

I beg leave to bring to the favourable notice of the Commander-in-Chief the conduct of the force under my command, which, without relaxing in the least the arduous siege and investment of a very strong fort and fortified city, garrisoned by 10,000 desperate men, fought, with the few numbers left in camp, a grand action with a relieving army, beat and pursued them nine miles, killing 1500 of them, and taking from them artillery, stores, and ammunition. The officers whom circumstances called prominently into action, and who, profiting by the opportunity, did valuable service, were Brigadier Stuart and the officers whom he mentions—Lieutenant-Colonel Turnbull, Bombay Horse Artillery; Captain Lightfoot, Bombay Artillery; Captain Need, 14th Light Dragoons; Lieutenant Leith, 14th Light Dragoons, etc. etc. Sergeant Gardiner, 14th Light Dragoons, attacked and killed a cavalry soldier as well as two armed

Sergeant Gardiner, 14th Light Dragoons, brought to notice.

Captain Prettejohn mentioned.

men on foot. His gallant conduct at Dhar had been previously honourably mentioned. The conduct of the men of the 14th Light Dragoons was so uniformly good that their commanding officer finds it difficult to bring any particular case of good conduct to notice. I am much indebted to the following officers for their zeal and assistance to me during the action, to Major Orr, commanding Hyderabad Contingent Field Force; Captain Prettejohn, commanding 14th Light Dragoons; Captain Hare, commanding Regiment Hyderabad Force, and Lieutenant Haggard, Commissary of Ordnance in command of the siege-train; as also to my Staff—Captain Macdonald, Assistant Quartermaster-General; Captain Wood, Assistant Adjutant-General; Captain Rose, Rifle Brigade, my Aide-de-Camp, and Lieutenant Lyster, 72nd Bengal Native Infantry, my Interpreter.—I have, etc.

(Signed) HUGH ROSE, *Major-General,*
Commanding Central India Field Force.

EXTRACTS from Return of Killed and Wounded of the Central India Field Force during the engagement with the enemy on the 1st April 1858 on the Betwa.

No. 5.

HER MAJESTY'S 14TH LIGHT DRAGOONS.

Casualties of 14th Light Dragoons on 1st April 1858 at the battle of the Betwa.

Killed—5 men; 11 horses.
Wounded—24 men; 16 horses.

Names of the killed and wounded:—

Regimental Sergeant-Major Thomas Clark, .	slightly wounded.
Sergeant John Myers, . . .	,, ,,
Private Thomas Ransom, . . .	killed.
. Walter Roberts, . . .	severely wounded.
,, Cornelius Gray, . . .	,, ,,
,, Leonard, ...	slightly wounded.
Lance-Sergeant William Roxby, . .	killed.
Private R. Barker, ...	,,
,, J. Leigh, ...	,,
,, W. Watkin, ...	,,
Sergeant Thomas Bowen and 9 privates, .	slightly wounded.
,, Wm. Parkins and 5 privates, .	severely wounded.
Private J. Waite, ...	dangerously wounded.
,, J. Byott, ...	,, ,,
,, Richard Baker, . . .	,, ,,

Total casualties in force.

Total casualties of the force engaged—15 killed; 66 wounded.

Return of Horses killed and wounded of the Central India Field Force on 1st April 1858.

Casualties among horses.

1st troop Horse Artillery, . .	2 killed.
Her Majesty's 14th Light Dragoons, .	11 killed; 16 wounded.
Total of the forces engaged, .	13 killed; 16 wounded.

(Signed) A. H. Wood, *Captain*,
Assistant Adjutant-General, Central India Field Force.

LOHARI DESPATCHES

EXTRACTS from the *London Gazette*, 31st August 1858.

Extracts from Lohari despatches.

No. 1.

General Orders by the Governor-General of India, Military Department.

No. 198 *of* 1858.

ALLAHABAD, *June* 5, 1858.

Reference to the Report of Major Gall, 14th Light Dragoons, of his capture of fort of Lohari.

The Right Honourable the Governor-General is pleased to direct the publication of the following despatch from the Deputy Adjutant-General of the Army, No. 373A, dated 31st May 1858, forwarding one from Major-General Sir Hugh Rose, K.C.B., enclosing a report from Major R. H. Gall[1] of Her Majesty's 14th Light Dragoons, of his capture of the fort of Lohari on the 2nd May.

The Governor-General desires to express the entire approval of the gallant conduct of the officers and men engaged under Major Gall in this affair, and his cordial appreciation of the merits of Major Gall himself, both in the present affair and on all occasions on which he has been employed.

No. 2.

The DEPUTY ADJUTANT-GENERAL of the Army, to the SECRETARY to the Government of India.

No. 373A.

HEADQUARTERS CAMP, FUTTEHGURH,
May 31, 1858.

SIR,—By desire of the Commander-in-Chief, I have the honour to transmit for the information of the Right Honourable the Governor-General a copy of a despatch[2] from Major-General Sir Hugh Rose, K.C.B., dated 17th instant, enclosing a report from Major R. H. Gall of H.M.'s 14th Light Dragoons, of his capture of the fort of Lohari on the 2nd idem.

[1] Report not available. *See* Despatches.
[2] Published in *London Gazette*, July 28, 1858, p. 3543.

His Excellency heartily concurs in the praise bestowed by the Major-General upon Major Gall and the officers and men engaged in this gallant affair.—I have, etc. (Signed) H. W. NORMAN, *Major*,
Deputy Adjutant-General of the Army.

BETWA DESPATCH (*Additional*)

EXTRACT from the *London Gazette*, published 24th December 1858.

VICTORIA CROSS.

Lieutenant (now Major) Leith, 14th Light Dragoons, gazetted to V.C., 24th December 1858.

14th Light Dragoons, now of the 6th Dragoons.	Lieutenant, now Brevet-Major, James Leith.

For conspicuous bravery at Betwa on the 1st April 1858, in having charged alone and rescued Captain Need of the same regiment when surrounded by a large number of rebel infantry.

Date of Act of Bravery, 1st April 1858.

Despatch from Major-General Sir Hugh Henry Rose, G.C.B., dated 28th April 1858.

CALPEE DESPATCHES

From MAJOR-GENERAL SIR HUGH ROSE, K.C.B., Commanding F.D.A. and Field Forces, to MAJOR-GENERAL SIR WILLIAM MANSFIELD, K.C.B., Chief of the Staff of the Army in India.

GWALIOR, *22nd June* 1858.

Extracts from Calpee despatches, dated Gwalior, 22nd June 1858. Difficulties to which the troops were exposed on march from Jhansi to Calpee, May 1858.

SIR,—In reporting to you for the information of the Commander-in-Chief in India my operations against Calpee, it is my duty, in justice to the unvarying devotion and discipline of the troops under my command, to state the new and very serious difficulties which beset them after leaving Jhansi. They had to contend not only against the rebel army, fighting as usual with all the advantages on their side of very superior numbers and knowledge of the ground, but they had to encounter also a new antagonist, a Bengal sun at its maximum of heat. This formidable ally of the rebel cause was more dangerous than the rebels themselves: its summer blaze made havoc amongst troops, especially Europeans, who, already exhausted by months of over fatigue, and want of sleep by continued night watchings and night marches, were often exposed to its rays, manœuvring or fighting, as at Koonch, from sunrise to sunset. At Koonch the thermometer was 115°; before Calpee, 118° in the shade; and on the march to Gwalior it burst in an officer's tent at 130°.

Whilst my force suffered so much from sunstroke, they were deprived in a great measure of its antidote—water. Between Jhansi and Calpee we found no streams; all was well-water. The wells, which are neither numerous nor abundant, being of extraordinary depth as we approached the Jumna, which increased the difficulties of obtaining water. Forage also was very scarce. The information which I had collected on the road, and a reconnaissance made by Lieutenant-Colonel Gall, H.M.'s 14th Light Dragoons, with his usual skill, confirmed all I had the honour to state in my report of the action at Koonch as to the enemy's elaborate lines of defences for the protection of Calpee, on the main road from Koonch to that fortress.

Scarcity of water.

Scarcity of forage.

Lieutenant-Colonel Gall's reconnaissance of Koonch and Calpee.

The Jumna is fordable at Golowlee, which stands in the nullahs running down to the Jumna just outside the dangerous labyrinth of ravines which surround Calpee.

On the 14th May I marched with the 1st Brigade and Major Orr's force for Golowlee, which I reached with no other opposition than an attack on the baggage by the rebel cavalry concealed in a ravine. They were put to flight by a troop of the 14th Light Dragoons, which, in anticipation of an ambuscade, I had sent to reinforce the rearguard. In this march we crossed the high road from Jubbulpore to Calpee. On my arrival at Golowlee I sent two of the Hyderabad Cavalry across the Jumna to Lieutenant-Colonel Maxwell, commanding a column of the Bengal army, who was on the left bank, and from whom I was to receive a supply of ammunition for the siege of Calpee, to make good the large amount which my force had expended in the sieges of Chandhairee (or Chanderi) and Jhansi. My force had now marched from Bombay to the Jumna, and had effected a union with the Bengal army, the immediate result of which was a combined operation of Bengal and Bombay troops against Calpee. Having heard at Golowlee that Major Forbes, commanding rearguard of the 2nd Brigade, leaving Etawa[1] was hard pressed, and hearing a heavy cannonade in his direction, I marched with the following troops to his assistance:—

Rebels attack baggage at Golowlee, 14th May 1858.

Repulsed by 14th Light Dragoons.

Communication established with Bengal army on left bank of the Jumna.

Half troop Bombay Horse Artillery.
One troop 14th Light Dragoons.
One troop Hyderabad Cavalry.
Three guns No. 4 Light Field Battery.
38th and 25th Regiments Bombay Native Infantry.

The enemy were pressing forwards. I immediately gave orders to the troops who were retiring from the village of Muttra, to reoccupy and hold it at any price; ordering up in their support, at a trot, the half-troop Horse Artillery, half of No. 4 Light Field Battery, a troop of the 14th Light Dragoons, and the 3rd Hyderabad Cavalry, with some Native Infantry. Brigadier Stuart, commanding the 1st Brigade, at Golowlee, and Captain Hare, commanding at Tehree, met the attack on Golowlee with vigour. Out of 36 men of the 14th Light Dragoons forming part of our forage escort, 17 were brought back to camp in dhoolies after only two hours' exposure to sun. When I speak of springing to their arms, I ought to make special mention of H.M.'s 14th Light

Affair at Golowlee.

Severity of the sun.

[1] Otherwise Etora.

14th Light Dragoons specially named.

Dragoons, for the admirable order and celerity with which their in-lying and out-lying picquets mounted on the frequent occasions when I turned them out, on alarms or sudden attacks of the enemy. Their vedettes and patrols also were always watchful and intelligent. My first and most important instructions were to take Calpee. On the morning of the 22nd May (1858) I made the following disposition of my troops to resist the expected attack:—The picquets on right front of the 86th Regiment were reinforced by the remainder of the 86th Regiment in skirmishing order. In support were three guns of No. 4 Light Field Battery, one troop 3rd Bombay Light Cavalry, and four companies of 25th Bombay Native Infantry, the whole under command of Brigadier Stuart; my left centre facing the plain, and the village of Tehree, was guarded by No. 1 troop Bombay Horse Artillery, supported by two troops H.M.'s 14th Light Dragoons. I reinforced the picquets on the left, in the first instance, with a squadron of the 14th Light Dragoons under Lieutenant-Colonel Gall, and the 3rd Hyderabad Cavalry under Captain Abbott, and afterwards directed the troops to retire slowly before the enemy, obliquely across my front, in order to conceal my heavy guns and draw the rebel cavalry into their fire.

Disposition of force for attack on Calpee.

On the 23rd I marched long before break of day against Calpee.

Calpee captured from the rebels, 23rd May 1858. Lieutenant-Colonel Gall pursues with horse artillery and cavalry.

Once clear of the ravines, I instantly directed Lieutenant-Colonel, then Major, Gall, H.M.'s 14th Light Dragoons, to pursue the enemy as closely and as far as he could, with horse artillery and cavalry, and for this officer's very successful pursuit of the enemy I beg to mention him specially. His column took the whole of the guns with which their main body retreated from Calpee, and six caparisoned elephants. The Hyderabad Cavalry and scouts brought in more guns, which detached parties of the rebel army had abandoned in their wild flight; so that every piece of field artillery which the enemy had was taken. The pursuing cavalry made great havoc of the rebel Sepoys, the Sind Velaitees, and the mercenaries of the Nawab of Banda, till neither horse nor man could go further. The rebels, broken completely by Lieutenant-Colonel Gall's column, fled in the utmost disorder, in twos and threes across-country, throwing away their arms and accoutrements, and even their clothes, to enable them to run faster. From information furnished by Lieutenant-Colonel Gall, it was clear that the principal part of the rebels had retreated by the Jaloun road, and Sir Robert Hamilton was of opinion that they would make to the north for the Sheer Ghat, a ford across the Jumna, or another ford higher up the river. Colonel Riddell was guarding the former ford with a movable column. I detached without delay Lieutenant-Colonel Robertson with a pursuing column as follows, along the Jaloun road:—

One troop 14th Light Dragoons.
One squadron 3rd Bombay Light Cavalry.
No. 18 Light Field Battery.
160 Hyderabad Cavalry.
25th Regiment Bombay Native Infantry.

To overtake the enemy was hopeless, for European cavalry riding eighteen stone could not catch Indian cavalry riding ten or at most eleven stone. The operations of the pursuing column will be detailed in my report of the operations against Gwalior.

I am much indebted for his goodwill and assistance to Major Rickards, Political Agent for Bhopal, who was wounded when spiritedly accompanying the charge of Captain Need of H.M.'s 14th Light Dragoons at the Betwa.—I have, etc. (Signed) HUGH ROSE, *Major-General, Commanding F.D.A. and Field Force.*

From MAJOR FORBES, C.B., Commanding Rearguard, to CAPTAIN TODD, Brigade Major, 2nd Brigade, Central India Field Force.

CAMP, near DEOPORE, *16th May* 1858.

Report of Major Forbes, C.B., commanding rearguard of 2nd Brigade, in which was a troop of 14th Light Dragoons, 16th May 1858.

SIR,—I have the honour to report, for the information of the Brigadier commanding the 2nd Brigade, the arrival in camp of the rearguard, having been closely followed up from within a mile of our last encampment at Etawa[1] by 4000 to 5000 of the rebel army, of which 1000 to 1200 were cavalry. I withdrew first the guns and infantry to a position on this side of a ravine, then the troop of H.M.'s 14th Light Dragoons, and lastly, the remainder of the cavalry, at a walk, until concealed from the enemy by the nature of the ground, then at a gallop. This retrograde movement, as I expected, brought the enemy on us. For the first three miles of the remainder of our march we were almost surrounded by the rebel cavalry, and fired into by their artillery; but alternately halting and retiring, we succeeded in preventing any of the baggage from falling into their hands.—I have, etc.

(Signed) J. FORBES, *Major,*
3rd Bombay Light Cavalry, Commanding Rearguard.

From LIEUTENANT-COLONEL CAMPBELL, Commanding 2nd Brigade C. I. Field Force (Brigadier C. Steuart being on the sick-list), to the CHIEF OF THE STAFF, Central India Field Force.

CAMP DEOPORE, *18th May* 1858.

Acting-Brigadier (Lieutenant-Colonel) Campbell's Report, 18th May 1858.

SIR,—I have the honour to report, for the information of the Major-General, that a large body of upwards of 1200 of the enemy's cavalry, with three guns, moved suddenly out yesterday at 2 P.M. from the rear of a large village situated on our left flank, advancing with an evident intention of attacking our camp. The small village of Muttra on our left was then occupied by two companies of 71st Highland Light Infantry, and two guns of Bombay Light Battery No. 18, with some of the 24th Bombay Native Infantry in support. I immediately reinforced this post with the whole of the 71st Highland Light Infantry and two guns of the Bombay Light Battery No. 18, giving orders for the 14th Light Dragoons and two guns of the Royal Artillery to follow, proceeding at once to

[1] Cr Etora.

meet the enemy, taking with me the 3rd Light Cavalry and the half battery whom I met on the way to Muttra.

14th Light Dragoons, under Major Scudamore, kept up lines of communication.

The 14th Light Dragoons under Major Scudamore, and the two guns of the Royal Artillery, took up a position connecting our line with Muttra. A heavy fire commenced on both sides, but the superior fire of our artillery effectually stopped all further advance on the part of the enemy.—(*True Extract.*)

(Signed) G. E. ROSE, *A.D.C.*

From MAJOR GALL, Commanding Left Wing, 14th Light Dragoons, to the CHIEF OF THE STAFF, Central India Field Force, Calpee.

CAMP CALPEE, 25*th May* 1858.

Major Gall's report of operations of left wing, 14th Light Dragoons, dated Calpee, 25th May 1858.

SIR,—I have the honour to report, for the information of the Major-General Commanding the Central India Field Force, that on the 23rd instant, when directed by your order to pursue the enemy, supposed to be retiring from Calpee by the Gwalior or Jaloun road, with the following troops:

14th Light Dragoons, 4 troops, 153 sabres;
Six guns Horse Artillery;
3rd Regiment Hyderabad Contingent Cavalry, subsequently increased by one troop 14th Light Dragoons (48 sabres);
50 sabres 1st Regiment Hyderabad Contingent Cavalry;

Captain Need with a squadron 14th Light Dragoons detached to escort Horse Artillery.

I immediately proceeded to assemble the force placed under my command on the road indicated to me, which proved to be the high road from Calpee to Jhansi. This, however, diverges to Jaloun about three miles, as near as I can recollect, from Calpee. Here I left Captain Need, who had accompanied me thus far with his squadron, to bring up the Horse Artillery I had sent back for. Lieutenant Dowker, 1st Regiment Hyderabad Contingent, I detached a little to my right; and accompanied by Captains Abbott and Barrett at the head of their detachments, I charged through the enemy's retiring line and dispersed it. The rebels were cut up in all directions, with the loss of two guns, which they abandoned to Lieutenant Dowker on the right. The Dragoons in the centre sabred a great many of the fugitive Sepoys who, firing wildly and completely panic-stricken by the suddenness and rapidity of our advance, fell an easy prey to their pursuers, in some instances casting away their arms, in others suffering themselves to be followed into ravines where they were slain. Four elephants were soon after captured as the pursuit continued. Between 200 and 300 of the rebels had been sabred, without any casualties on our side beyond the following:—2 men wounded; 1 horse killed; 1 horse lost; 1 horse wounded.

The enemy were charged by Major Gall with a squadron of Fourteenth under Captain Barrett, and a troop of Hyderabad Cavalry under Captain Abbott, capturing 2 guns and 4 elephants, killing 200 to 300 rebels.

On my left the sound of Captain Lightfoot's guns had been heard as we approached by the Jhansi road, preceded by Captain Need, who, extending to the right and left of the road, charged the rear of an infantry column, of whom he cut down nearly 200, while Captain Lightfoot plied them with shot and shell. Captains Need and Lightfoot captured three pieces of ordnance during this advance, continued by the former to the eighth milestone on the Jhansi road.

Captain Need's squadron charges another portion of the enemy.

After watering, I joined the Horse Artillery on the Jhansi road, and proposed advancing, but the exhausted state of the Horse Artillery horses, and indeed of our men generally, would not admit of this, and I gave orders for the return of the force to camp at Calpee, which we reached after having been upwards of 13 hours in the saddle. The very weak squadron of Dragoons that I had with me was ably led by Captain Barrett, whose good conduct whilst serving under my immediate command I have already had occasion to bring to the notice of the Brigadier Commanding 1st Brigade Central India Field Force, for favourable recommendation to the Major-General. Captain Barrett's men did great execution amongst the rebels, and the sowars of the 1st Cavalry Hyderabad Contingent, led by Lieutenant Dowker, emulated them. Surgeon Stewart, 14th Light Dragoons, I have to thank for his attention to those who fell sick during the pursuit, carried on through the hottest part of the day. My thanks are also due to Assistant-Surgeons Lofthouse and Lumsdaine. Lieutenant and Adjutant Giles, left wing, 14th Light Dragoons, as on many previous occasions, distinguished himself in several personal encounters with armed Sepoys.

Major Gall's column returns to Calpee after being 13 hours in the saddle.

Officers of 14th Light Dragoons favourably mentioned.

I specially recommend Captains Abbott, Barrett, and Need, and Lieutenant Dowker, to the notice of the Major-General. Acting Regimental Sergeant-Major Clark and Private Winton, 'B' troop, 14th Light Dragoons, behaved with great gallantry. The captures were as follows:—

Large captures from the enemy at Calpee.

5 guns.	6 elephants.
1 gingall on wheels.	8 camels.
2 artillery wagons filled with ammunition.	42 bullock draughts.
1 native tumbril.	3 hackeries.
20 boxes of ammunition.	1 spring cart.

I have, etc. R. H. GALL, *Major*, *Left Wing*, 14*th Light Dragoons*.
(*True Copy.*) (Signed) G. E. ROSE, *A.D.C.*

GWALIOR DESPATCH

From MAJOR-GENERAL SIR HUGH ROSE, K.C.B., Commanding Field Forces South of the Nerbudda, to MAJOR-GENERAL SIR WILLIAM MANSFIELD, K.C.B., Chief of the Staff of the Army in India.

Extracts from Gwalior despatch, dated Poona, 13th October 1858.

POONA, BOMBAY, 13*th October* 1858.

SIR,—I have the honour to report, for the information of the Commander-in-Chief in India, the operations against Gwalior of the Central India Field

Force, and other troops placed under my command by His Excellency. After the capture of Calpee (17th to 23rd May), a short rest having enabled my European troops to recover a little, I reinforced Lieutenant-Colonel Robertson (commanding column of pursuit) with the following troops:—1 wing H.M.'s 86th Regiment; 2 squadrons H.M.'s 14th Light Dragoons.

Lieutenant-Colonel Robertson commands column of pursuit.

Lieutenant-Colonel Robertson reported to me in two expresses that the Calpee rebels had certainly taken the road to Gwalior. Not many hours after the arrival of Lieutenant-Colonel Robertson's last express, Sir Robert Hamilton received similar intelligence; when I instantly ordered off Brigadier Stuart with the following force to reinforce Lieutenant-Colonel Robertson and march after the rebels, viz.—

Brigadier Stuart ordered to march on Gwalior after the rebels.

No. 4 Light Field Battery.
Two troops Her Majesty's 14th Light Dragoons.
One wing Her Majesty's 71st Regiment.
One wing Her Majesty's 86th Regiment.
Four companies 25th Bombay Native Infantry.
Half company Bombay Sappers and Miners.
Two 18-pounders.
One 8-inch Howitzer.

An express letter received a few days later from Scindiah's agent at Gwalior removed apprehensions for his safety and that of his Government. Subsequently news came that the rebel army had attacked Scindiah at Bahadurpore, nine miles from Gwalior. His troops of all arms, with the exception of a few of his bodyguard, had treacherously gone over, the artillery in mass, to the enemy. His Highness had been obliged to fly to Agra, accompanied only by one or two attendants. The rebels had entered Gwalior, taken Scindiah's treasury and jewels—the latter said to be of fabulous value; the garrison of the fort of Gwalior had opened its gates to the rebels, and finally from 50 to 60 fine guns had fallen, as well as an arsenal, with abundance of warlike stores, into the hands of the enemy. To render this state of things still more embarrassing, Gwalior fell into rebel hands at the most unfavourable time of the year for military operations, on the eve of the great rains, and when the heat of summer was at its maximum. It was of vital importance that troops should reach Gwalior before the rains set in. I therefore, leaving by order Captain Ommaney's Royal Artillery Battery of four 9-pounders belonging to the 2nd Brigade Central India Field Force, as part of the permanent garrison of Calpee, with one troop Bombay Light Cavalry, one company Royal Engineers, a wing of the 3rd Bombay Europeans, and 400 men of the 24th Bombay Native Infantry, to garrison Calpee until relieved by Bengal troops, marched with the following force from Calpee on 6th June, following Brigadier Stuart's column by forced marches on the road to Gwalior by Jaloun, marching by night to avoid the sun, viz.:—One troop Bombay Horse Artillery; one squadron 14th Light Dragoons; one squadron 3rd Bombay Light Cavalry; Madras Sappers and Miners. One day the heat in the shade rose to

Gwalior taken by the rebels. Flight of Scindiah to Agra.

Sir Hugh Rose marches from Calpee, 6th June 1858.

130°. The officer commanding the outlying picquet of Her Majesty's 14th Light Dragoons having reported to me, on the night of the third day, that his men had fallen from their saddles from exhaustion, I had the picquet relieved by a party of Hyderabad Cavalry. My plan of attack on Gwalior was as follows: to invest it as much as its great extent would allow, and then to attack it by its weakest side, the investing troops cutting off the escape of the rebels. I directed Brigadier Smith with the Rajputana Field Force to move to a point (Kota-ke-Serai) seven miles to the east of Gwalior. I myself with Brigadier Stuart's column and the small one I had brought from Calpee marched against the Morar cantonments, which are about five miles from Gwalior on the river Morar. Once in possession of the Morar cantonments I could establish there my hospital, parks, etc., and then, joining Brigadier Smith, I intended to attack Gwalior with his force and my own. I sent Colonel Riddell to the Residency, about seven miles to the north of Gwalior, to extend his force from thence down the west side of Gwalior, and thus to invest it from that side as far as possible. On 16th June I formed my force in two lines:—First line, 1st Brigade, under Brigadier Stuart; second line, under Brigadier-General Napier, in support of the first, consisting of only a small part of the 2nd Brigade, as the rest of it was left at Calpee.

Great heat on the march, 130° in the shade.

Sir Hugh Rose's plan of attack on Gwalior.

Morar cantonments captured, 16th June 1858.

Both lines advanced, artillery in the centre, 86th Regiment on their right, 25th Bombay Native Infantry on their left, the 14th Light Dragoons on each flank. The nullahs and broken ground prevented the advance of Captain Abbott's cavalry and of the 14th Light Dragoons under Captain Thompson on the right, who reinforced my left. The success of the day was completed by the destruction of the rebels in the nullahs, and a most successful pursuit of them by Captain Thompson, with a wing of the 14th Light Dragoons. These rebels had been turned by Captain Abbott's advance from the ford of the river, across which and the bridge the main body had retreated, but Captain Thompson caught them in the plains before they could reach the hills, and made a great slaughter of them. I beg to mention specially Captain Thompson, 14th Light Dragoons, for the very good service which he did on this occasion.

Rebels pursued by Captain Thompson's wing of 14th Light Dragoons.

Brigadier Smith at Kota-ke-Serai having asked for reinforcements, I directed Lieutenant-Colonel Robertson with three troops 14th Light Dragoons, four guns, and 25th Bombay Native Infantry, to join him. The arrival of the troops from Calpee on 18th June enabled me to march from Morar that afternoon, leaving for its protection Brigadier-General Napier with the following force:—

Three troops 14th Light Dragoons reinforce Brigadier Smith at Kota-ke-Serai.

One troop Bombay Horse Artillery.	Royal Engineers.
Three troops 14th Light Dragoons.	Bombay Europeans.
Hyderabad Cavalry.	Bombay Native Infantry.
Meade's Horse.	Three guns Hyderabad Artillery.

Three troops 14th Light Dragoons left at Morar with Brigadier-General R. Napier's force, 18th June 1858.

And taking with me to Kota-ke-Serai

Two troops 14th Light Dragoons proceed with Sir Hugh Rose's force to Kota-ke-Serai for the attack of 19th June on Gwalior.

Two troops 14th Light Dragoons.	86th Regiment.
Light Field Battery.	Hyderabad Infantry.
Wing 71st Highland Light Infantry.	Two 18-pounders and one 8-inch Howitzer.
Madras Sappers and Miners.	

On the 19th June, at the attack on Gwalior, I directed Brigadier Smith with No. 3 troop Bombay Horse Artillery and a squadron of 14th Light Dragoons to be ready to attack the enemy's positions at the Phool Bagh and beyond it; and I had some time before ordered up No. 4 Light Field Battery with two troops of the 14th Light Dragoons to the heights to cover my advanced line, and to answer the enemy's batteries in position in front of Gwalior. The hilly and difficult ground prevented their arrival.

Attack on Phool Bagh by 3rd Bombay troop of Horse Artillery and a squadron 14th Light Dragoons.

The attack on Phool Bagh by the No. 3 troop of Bombay Horse Artillery and a squadron of Her Majesty's 14th Light Dragoons protected the right of the troops attacking the grand parade, and also turned the enemy's left.

Two troops of the 14th Light Dragoons were held in support on the lower slopes. Brigadier Smith speaks very highly of the steadiness with which the 14th Light Dragoons, escorting 3rd troop Bombay Horse Artillery, stood the enemy's artillery fire, shot and shell, and of the ardour with which they afterwards fell on the guns and the retreating enemy.

Scindiah returns and re-enters Gwalior escorted by a squadron 8th Hussars and a squadron of 14th Light Dragoons, 20th June 1858.

The morning after the capture of Gwalior, His Highness the Prince of Gwalior arrived there with Sir Robert Hamilton, agent to the Governor-General for Central India, and his retinue. I received Scindiah with every possible mark of respect, and, accompanied by all the superior officers of the forces and all my personal and divisional staff, had the honour of escorting His Highness to his palace in 'the Lushkar' with a squadron of Her Majesty's 8th Hussars, and another of Her Majesty's 14th Light Dragoons, most honourable representatives of my force. Our road lay through the long and handsome street which leads from the grand parade to the palace, which was lined by crowds of inhabitants who greeted Scindiah with enthusiastic acclamations.

I venture to recommend all the troops engaged in the 'Gwalior' operations, the Central India Field Force and Brigadier Smith's brigade of the Rajputana Field Force, to his Lordship's most favourable consideration.

I marched on the 6th June from Calpee for Gwalior, and on the 19th of the same month the Gwalior States were restored to the Prince.—I have, etc.

(Signed) HUGH ROSE, *Major-General,*
Commanding Field Forces of the Nerbudda.

MORAR DESPATCH

EXTRACTS from DESPATCH of BRIGADIER-GENERAL NAPIER, C.B., Commanding 2nd Brigade Central India Field Force, to the ADJUTANT-GENERAL, Central India Field Force. Morar despatch.

CAMP MORAR, *18th June* 1858.

SIR,—On the 16th inst. the 2nd Brigade, composed as follows:

1st troop Horse Artillery, No. 18 Light Field Battery;
14th Light Dragoons (10 officers, 259 men);
3rd Light Cavalry, Madras Sappers and Miners;
71st Highlanders, Hyderabad Cavalry, Infantry, and Artillery; Towana Horse;

when in sight of the cantonments of Morar, was ordered by the Major-General commanding the Central India Field Force to advance in echelon from the right, in support of the left of the 1st Brigade. Attack on Morar cantonments, 16th June 1858.

The force was disposed as follows:—No. 18 Light Field Battery on the right, supported by Johnstone's Hyderabad Horse; in the centre, Madras Sapper and Miners, and wing of Her Majesty's 71st Highland Light Infantry; while on the left was a wing of Her Majesty's 14th Light Dragoons.

On approaching the right of the cantonment the enemy opened upon us from six guns, and I directed Lieutenant Harcourt, commanding No. 18 Light Field Battery, to engage them, an order which he had barely received when he was summoned to join the 1st Brigade.

My brigade being then reduced to the wing of Her Majesty's 71st Highland Light Infantry, the right wing of Her Majesty's 14th Light Dragoons, Madras Sappers and Miners, and 100 horse of the Hyderabad Contingent, continued to advance on the enemy, who were retreating in large numbers towards their right rear. Right wing 14th Light Dragoons engaged under Major Scudamore. The ground in front was completely intersected with ravines, lined with the enemy's infantry. I therefore directed Colonel Campbell, commanding the wing of the 71st Regiment, to throw it forward in skirmishing order, supported by the 14th Light Dragoons, which was executed with great spirit by Major Rich on the right, who cleared the ravines on his front, leaving them filled with the enemy's dead, and relieving the Horse Artillery from much annoyance by their musketry. I regret to say that this service was not performed without the loss of a very promising young officer of Her Majesty's 71st Highland Light Infantry, Lieutenant Neave, who was shot whilst gallantly leading his men to the ravines. Colonel Campbell took two companies of the Seventy-first under Lieutenant Scott and cleared some ravines on his left and front, killing every man of the enemy that held them; after which he was directed to clear the top of a hill, where a party of rebels held a temple and some strong ground. This duty was thoroughly effected, and thirty of the enemy left dead on the hill. Whilst this was going on a troop of Her Majesty's 14th Light Dragoons passed round the base of the hill and cut up all the enemy who attempted to escape from it. The protection of the left of the force and the rear being placed under my especial charge by the Major-General, I moved the remainder of the 14th Light Dragoons and Johnstone's Hyderabad Horse towards the left to cover the rear, and to intercept the enemy's cavalry, who showed some disposition to move in that direction; but on observing our cavalry they rapidly disappeared through the hills to the south of Gwalior. The front being now clear of the enemy I withdrew my brigade to the shelter of the cantonments. The conduct of the whole of the troops under my command was excellent. Their perfect steadiness while under the fire of the enemy's batteries, and the gallantry with which they advanced to clear the ravines, were deserving of the Major-General's warm commendation. The Seventy-first dashed into the ravines and encountered the enemy hand to hand. I beg particularly to recommend to the Major-General's notice Colonel Campbell, commanding Her Majesty's 71st Regiment; also Major Rich, Seventy-first, and Lieutenant Scott, Seventy-first; also Major Scudamore, commanding the right wing of Her Majesty's 14th Light Dragoons, which was skilfully handled and ready for every call for its services. His skirmishers attacked and destroyed many of the enemy in the ravines. Lieutenant Gowan, with his troop, most efficiently cut off the enemy's retreat from the hill and destroyed many of them. —I have, etc. (Signed) R. Napier, *Brigadier-General, Commanding 2nd Brigade C.I.F. Force.*

Major Scudamore and Lieutenant Gowan, 14th Light Dragoons, specially mentioned.

JOWRA-ALIPORE DESPATCH

Extracts of Despatch from Brigadier-General R. Napier, C.B., commanding 2nd Brigade Central India Field Force, to the Assistant Adjutant-General, Central India Field Force.

Camp Jowra-Alipore, *21st June* 1858.

Affair at Jowra-Alipore, 22nd June 1858.

Sir,—I have to report that I received at 5.15 A.M. on the 20th June orders to pursue the enemy, with the details shown in the margin, which marched within an hour and a half after receipt of order.[1] The fort, which had been reported 'in our possession,' opened upon us as we came within range, and obliged us to make a detour to reach the Residency. We arrived late in the evening at Sumowlee, having marched about 25 miles. The enemy were reported to have 12,000 men and 22 guns, and to have marched from Sumowlee to Jowra-

[1] Lightfoot's battery Horse Artillery, Prettejohn's troop 14th Light Dragoons (60 sabres), Abbott's Hyderabad Cavalry, 3rd Light Cavalry (2 troops), Meade's Horse.

Alipore in the forenoon. We were too tired to go beyond Sumowlee, the heat of the sun having been terrific; so we rested until 4 o'clock A.M. on the 22nd, then advanced on Jowra-Alipore, where we found the enemy strongly posted, with their right resting on Alipore, guns and infantry in the centre, and cavalry on both flanks. I directed Captain Lightfoot to take up a position about 600 yards from the enemy's left flank, and enfilade their line; and to act afterwards as circumstances might dictate. Our column of march was the most convenient formation for attack: Abbott's Hyderabad Cavalry in advance; Lightfoot's troop of Horse Artillery, supported by Captain Prettejohn's troop of 14th Light Dragoons, and two troops 3rd Light Cavalry, under Lieutenant Dick, with a detachment of Meade's Horse under Lieutenant Burlton in reserve.

When the troops came into view of the enemy after turning the shoulder of the rising ground (which hid our approach), the whole were advanced at a gallop, and as soon as the artillery had reached the flank of the enemy's position, the line was formed to the left, and the guns opened on the enemy at a distance of 600 yards. After a few rounds the enemy's guns were silenced, and a rapid thinning and wavering of their ranks took place. Captain Lightfoot limbered up, and advanced at a gallop; and Captain Abbott with his Hyderabad Cavalry charged at the same moment, followed by the rest of the cavalry, who swept through the enemy's batteries and camp into the open plain, driving before them and cutting down the rebels for several miles. We advanced about six miles from our first point of attack. The enemy were dispersed in every direction, throwing away their arms. Twenty-five guns had been captured, and were lying broadcast over the plain; men and horses were exhausted, and it was necessary to retrace our steps.

Defeat and pursuit of the rebels. Twenty-five guns captured.

Besides the guns, a considerable quantity of ammunition and elephants, tents, carts, and baggage fell into our hands. Never was the route of an army more complete. I believe between 300 and 400 of the enemy were killed. The good discipline of the troops of all arms under my command has only been equalled by the courage with which they charged such a superior force. Many occasions arose when it was necessary for detached parties to act against the enemy's infantry, and they were invariably met with the promptest gallantry. Private Novell of Her Majesty's 14th Light Dragoons charged alone into the village and killed one of the enemy under a very heavy fire, for which act of gallantry I beg to recommend him for the 'Victoria Cross.' Those experienced officers, Captain Prettejohn, Her Majesty's 14th Light Dragoons, and Lieutenant Dick, 3rd Light Cavalry, were charged with the duty of supporting the guns, which they performed to my entire satisfaction.

Private Novell, 14th Light Dragoons, recommended for Victoria Cross.

To Surgeon Stewart of Her Majesty's 14th Light Dragoons, and the medical officers of the force, I am much indebted for their attention to the sick and wounded.—I have, etc.

Captain Prettejohn and Surgeon Stewart, 14th Light Dragoons, specially mentioned.

(Signed) R. NAPIER, *Brigadier-General,*
Commanding 2nd Brigade Central India Field Force.

RANODE DESPATCHES

EXTRACT from the *London Gazette*, 18th April 1859.

No. 12.

From BRIGADIER-GENERAL SIR ROBERT NAPIER, K.C.B., commanding Gwalior Division, to the CHIEF OF THE STAFF.

CAMP RANODE, *December* 21, 1858.

Extract from Ranode despatch, 21st December 1858.

SIR,—I have the honour to report, for the information of the Right Honourable the Commander-in-Chief, that I received, on the morning of the 12th instant, intimation from Captain McMahon, Her Majesty's 14th Light Dragoons, commanding a small force near the confluence of the Jumna, Chambal, and Sind rivers, that the rebels had passed into the Lohar Purgunnah of Kuchwazhur.[1] Believing that their course would be up the jungles of the Sind river, I marched from Gwalior with a force as below, on the 12th instant, intending to proceed to Dubbia,[2] on the Jhansi road, and then according to information I might receive, to intercept the enemy :—

No. 4 Light Field Battery, two guns.
14th Light Dragoons, 150 non-commissioned officers, rank and file.
2nd Gwalior Mahratta Horse, 100 sabres.
71st Highlanders, 117 non-commissioned officers, rank and file.
25th Bombay Native Infantry, 50 non-commissioned officers, rank and file.
Camel Corps, 40 camels.

At Antri, where the force rested during part of the night, I received at 2 A.M. on the 13th, when on the point of proceeding to Dubbia, an express from the political agent of Gwalior to the effect that his information led him to believe that the rebels would pass by Gohad, to the north of Gwalior. This caused me to halt until I should receive by the morning post precise information from Captain McMahon. At 10.30 A.M. the Tussildar of Antri informed me he had just ridden in from Dubbia, the very place of my destination, and had seen the smoke of the staging bungalow which the rebels were then burning, and that they were proceeding in a south-westerly direction. I immediately marched south in pursuit, and at Beettiwar,[3] where I arrived at 1.30 A.M. on the 14th, I was informed that the enemy were three kos (in this country eight or nine miles) distant. The force had been on the move for thirteen and a half hours, and required a rest. Owing to a delay in getting grass and disposing of a rear party of the enemy's infantry, in which some of the Mahratta Horse, under fire for the first time, greatly distinguished themselves, I did not get away until 10 A.M. I continued the pursuit through Nurwar, where I left behind the greater portion of the detachment of the Seventy-first and the Artillery, which could not keep up with me, and took on

[1] Or Kuchwagar. [2] Or Dutheah. [3] Or Bhitarwar.

merely the cavalry and 38 men of the 71st Highlanders on camels, and 25 of the Balandshar Horse, that were halted at Narwar on their way to Kerara,[1] owing to the news of the enemy's approach, and after a very exciting though fatiguing chase we overtook the rebels on the morning of the 17th at Ranode. Their course had been south of Nurwar and through the Amola Pass, and their direction appeared to be along the right bank of the Sind. Twice we were encamped within a few miles of them, but owing to the darkness and our ignorance of the country, and to its hilly and jungly character, we could not take advantage of it. They chose the most difficult and unfrequented paths, evidently guided by some one well acquainted with them. Once or twice we were close on their traces and cut off stragglers or took their horses, the riders throwing themselves off and darting into the dense thickets which almost close up the paths. At one moment I believed they were driven into the hands of Colonel Scudamore, Her Majesty's 14th Light Dragoons, who, according to the instructions and information I had sent him, had posted his detachments on the right bank of the Sind below Kolarus to intercept them. I expected every moment to hear his guns open fire, when I found on getting into a more open country that the enemy had turned away from the river and were going towards Ranode. They took a circuitous and difficult road through the jungles, whilst we marched by a more easy and direct one, where the country was open. I found the people of Ranode in great excitement, and was informed that the enemy was close at hand and in full march to attack them, guided by Pyroo Sing of Tehrea, a rebel Zemindar, whose fort a few miles off had been recently destroyed by Scindiah. The enemy advanced in an irregular mass, extending on a front of nearly a mile. Their numbers must have been increased since they crossed the Jumna. We had barely time to form up the 14th Light Dragoons when the enemy were within a few hundred yards. The Mahratta Horse were impeded in crossing a deep ravine by the riding camels, and were thus a little behind. Our force actually engaged consisted of—

Battle of Ranode, 17th December 1858.

133 of Her Majesty's 14th Light Dragoons, commanded by Captain Prettejohn.
60 of the Mahratta Horse, under Captain F. H. Smith.
38 of the 71st Highlanders, under Captain Smith, mounted on camels, and guided by Captain Templer, commanding Camel Corps.

133 men of 14th Light Dragoons, under Captain Prettejohn, engaged.

It was a complete surprise. The 14th Light Dragoons, excellently led by Captain Prettejohn, dashed at once into the centre of the enemy, who never attempted to stand as a body, though individuals died fighting desperately. Captain Prettejohn having received a severe wound, the command devolved on Captain Need, who, with much energy and judgment, continued the pursuit for nearly eight miles, cutting up great numbers, particularly at the end, where the fugitives were stopped by a ravine, those who could not cross it taking refuge in the jungle, impracticable for cavalry. One hundred and fifty dead bodies of the enemy have been counted immediately at Ranode, and a much larger number must have fallen during the pursuit. Captain Need estimates

14th Light Dragoons charged the enemy and repulsed them.

Captain Need pursued for eight miles.

[1] Or Karehra.

Arms, elephants, ponies, and horses captured.

these last at 300. Many of these were 12th Irregulars, the murderers of Major Holmes and his family. I fear Ferozshah has escaped for the present. Six elephants were taken, and numbers of horses, ponies, arms, etc.

It is with much pride and satisfaction that I beg permission to bring to the notice of the Right Honourable the Commander-in-Chief the admirable charge made by the 14th Light Dragoons, and the soldier-like and exemplary cheerfulness with which the troops of all arms bore the fatigue and privations inseparable from such service. With such men and officers, the General's task is easy. It was a cause of very great regret to me, and equally to themselves, that the remainder of the Seventy-first, under Major Rich, and Captain Brown's guns were necessarily left behind. I beg to recommend most particularly Captain Prettejohn, Her Majesty's 14th Light Dragoons, commanding the cavalry, for the very gallant manner in which he led his men until severely wounded; also Captain Need, 14th Light Dragoons, who succeeded to the command and most ably exercised it. Lieutenant Giles of the 14th Light Dragoons, and Lieutenant Gough of the Mahratta Horse, are very favourably mentioned by their commanding officers. I am greatly indebted to Dr. Cruickshank for his prompt and extreme attention to the wounded. Mr. Apothecary Waite, an old and excellent servant of the Government, who has been present with the 14th Light Dragoons in every engagement, was in the field and performed valuable service. I am under great obligations to Captain Todd, 14th Light Dragoons, Assistant Adjutant-General, for his invaluable aid at all hours during the pursuit and in the action. Captain Need mentions very favourably the forward conduct of Regimental Sergeant-Major Thomas Clark, 14th Light Dragoons, and Corporal George Best of 'H' troop, 14th Light Dragoons. I subjoin a list of casualties. The wounded are doing well. I have, on a former occasion, the honour to report the good service performed by two of the officers above mentioned, Captains Todd and Prettejohn, of Her Majesty's 14th Light Dragoons, for their distinguished conduct in the action with Tantia Topee at Jowra-Alipore; but I fear from some accident that my report has not reached the Right Honourable the Commander-in-Chief. I therefore beg permission to submit a copy of it, and most earnestly solicit the favour and protection of the Right Honourable the Commander-in-Chief and the Government for the officers and soldiers therein mentioned, as well as for those named in this report.—I have, etc.

Officers, etc., mentioned for gallant conduct.

(Signed) R. NAPIER, *Brigadier-General,*
Commanding Gwalior Division.

No. 13.

From CAPTAIN NEED, 14th (King's) Light Dragoons, to the ASSISTANT ADJUTANT-GENERAL, Gwalior Division.

CAMP RANODE, *December* 17, 1858.

SIR,—I have the honour to report, for the information of the General commanding, that, on Captain Prettejohn becoming disabled this morning from a severe wound, I assumed command of the squadron 14th Light Dragoons, strength 133 sabres, in pursuit of the rebels, following them for about seven miles, cutting up great numbers, and capturing several elephants, horses, etc. I beg to bring to the General's notice the good services rendered by Lieutenant Giles and all under my command, as well as Captain Lumsden, Assistant Quartermaster-General, and Lieutenant Gough of the Mahratta Horse who accompanied the squadron. The gallant conduct of Corporal Best, 14th Light Dragoons, came particularly under my notice; but when every man of the squadron behaved so well, and did such good service, it is almost unjust for me to recommend any man in particular. The pursuit was principally through low jungles, very bad ground, and full of holes, which will account for the great number of missing horses in my casualty report which accompanies this report.—I have, etc.

Captain Need's report on Ranode, 17th December 1858.

Lieut. Giles and Corporal Best, 14th Light Dragoons, mentioned.

ARTHUR NEED, *Captain*, *14th Dragoons.*

CASUALTIES TO 14TH LIGHT DRAGOONS AT RANODE, 17th December 1858.

Wounded.—Brevet-Major R. B. Prettejohn, severe sabre-cut on the outside and back of left thigh, three inches above the knee; one sergeant; one corporal, eleven men.

One officer's charger wounded; one missing; three troop-horses killed; five troop-horses wounded; thirteen troop-horses missing.

Casualties to 14th Light Dragoons at Ranode, 17th December 1858.

APPENDIX C

CASUALTIES IN SOUTH AFRICA

UP to the present time the Fourteenth have been wonderfully lucky in their casualties in action, but they have had more than their proportion of officers sick and invalided. They have also lost a large number of horses. The following extracts from the Monthly Returns rendered to the War Office supply all the information that can be obtained up to the present time (7th March 1901):—

EXTRACTS from the MONTHLY RETURNS of the 14th (King's) Hussars in South Africa, January to December 1900.[1]

	Officers.	Nos. fit for duty—W.O., N.C.O. and Men.	Nos. sick—W.O., N.C.O. and Men.	Deaths—W.O., N.C.O. and Men.	Casualties to Horses.	Remarks.
Return for January 1900. 'A' and 'C' squadrons and headquarters only.	Captain Dalton, R. A. M. Corps, severely wounded 23rd, Hussar Hill.				14 lost on voyage from England. 30 died. 70 remounts received.	

Return signed in Natal, 1st February 1900, by Lieut.-Colonel G. HAMILTON, Commanding 14th Hussars.

	Officers.	Nos. fit for duty—W.O., N.C.O. and Men.	Nos. sick—W.O., N.C.O. and Men.	Deaths—W.O., N.C.O. and Men.	Casualties to Horses.	Remarks.
Return for February 1900. 'A' and 'C' squadrons and headquarters only.	Lieut. H. Tilney joined from base, Mooi River.	280	36	4	13 died. 31 sick. 225 fit for duty. 114 remounts received.	Prevailing diseases: diarrhœa and dysentery.

Return signed at Ladysmith, 1st March 1900, by Lieut.-Colonel G. HAMILTON, Commanding 14th Hussars.

[1] These Extracts are not copies of the original Returns.

Officers.	Nos. fit for duty—W.O., N.C.O. and Men.	Nos. sick—W.O., N.C.O. and Men.	Deaths—W.O., N.C.O. and Men.	Casualties to Horses.	Remarks.	
Sec.-Lieutenant Champion with a draft of 115 men joined at Durban.	391	94	Sergt. Storer and 3 men died of enteric fever, dysentery, and abscess of the liver.	17 died and destroyed (colic and exhaustion mostly). 40 sick. 250 fit.	Prevailing diseases: enteric fev. and dysentery. 94 men transferred. 157 received by regiment.	Return for March 1900. 'A' and 'C' squadrons and headquarters only.

Return signed 1st April, Cape Colony, by Lieut.-Colonel G. HAMILTON, Commanding 14th Hussars.

Officers.	Nos. fit for duty—W.O., N.C.O. and Men.	Nos. sick—W.O., N.C.O. and Men.	Deaths—W.O., N.C.O. and Men.	Casualties to Horses.	Remarks.	
25 fit. 2 sick. Major R. M. Richardson appd. Assist. Provost-Marshal to XIth Division. Veterinary Capt. O'Donel to headquarters staff. Captain Dalton R.A.M.C., to hospital at Bloemfontein, sick from wound received in action. Capt. P. Denny, 1st Dragoon Gds., killed in action, 24th, at Roodekop. Captain D. M. Miller severely wounded at Calleberg.	1 W.O., 450 N.C.O. and men.	126	5 men died (enteric fever). Sergt. J. Cunninghame killed in action, 23rd, at Leeuwkop.	381 fit. 16 killed in action and died. 60 sick. (57 officers' chargers fit.)	'B' squadron joined 14th April, with 7 officers, 166 men, 129 horses.	Return for April 1900.

Return signed at Donkerhoek, Bloemfontein, 1st May 1900, by Capt. C. G. G. HUTCHISON, 21st Lancers, for Commdg. Officer 14th Hussars.

N.B.—A Court of Inquiry stated subsequently that Corporal G. J. Osborne (attached for duty to Roberts's Horse), reported missing, was killed in action at Sanna's Post, March 31, 1900.

Return for May 1900.

Officers.	Nos. fit for duty—W.O., N.C.O. and Men.	Nos. sick—W.O., N.C.O. and Men.	Deaths—W.O., N.C.O. and Men.	Casualties to Horses.	Remarks.
Captain Gage to base at East London. 7 sick and subsequently invalided home to England, Cpt. Brooksbank Cpt. Miller, Lieut. Henry, Lieut. Blackett, Lieut. Wright, Sec.-Lieut. Prescott-Westcar, Sec.-Lieutenant Champion. Capt. S. Robertson from Black Watch (Royal Highl[rs]) joined for duty 12th May. Capt. Hutchison to Cavalry Depôt, Bloemfontein. 19 officers fit.	449	155	1 man died at Johannesburg of wounds received in action. Sergt. Bottomley and 3 men died of enteric and dysentery.	308 fit. 94 sick.	Prevailing diseases: diarrhœa, enteric fever, and rheumatism.

Return signed at Kameel Drift, Pretoria, 1st June 1900, by Lieut.-Col. G. HAMILTON, Commanding 14th Hussars.

Return for June 1900.

Officers.	Nos. fit for duty—W.O., N.C.O. and Men.	Nos. sick—W.O., N.C.O. and Men.	Deaths—W.O., N.C.O. and Men.	Casualties to Horses.	Remarks.
Captain Stuart Robertson died of dysentery, 1st June, at Kroonstad. 6 officers sick.	576 effectives	99	Corpl. Burton and 3 men died (3 of enteric).	337 fit. 53 died, abandoned, and destroyed. 103 to sick horse depôts.	Draft received from England (3rd Hussars) of 30 men.

Return signed at Kameel Drift, Pretoria, by Lieut.-Col. G. HAMILTON, Commanding 14th Hussars, 1st July 1900.

Officers.	Nos. fit for duty—W.O. N.C O. and Men.	Nos. sick—W.O., N.C.O. and Men.	Deaths—W.O., N.C.O. and Men.	Casualties to Horses.	Remarks.	
17 fit. 3 officers sent home invalided—Major O'Brien, Captain Gage, Qr.-mr. Mugford. 3 officers joined from England—Sec.-Lt. Harvey, Sec.-Lt. Scott, Sec.-Lt. Hon. H. Robertson, 3rd Batt. Cameron Highlanders.	746 effectives	165	1 man died at Pretoria of wounds receiv'd in action.	267 fit. 64 sick. 199 at sick horse depôts. 193 remounts received.	Prevailing diseases: diarrhœa, rheumatism, veldt sores. 134 men joined.	Return for July 1900.

Return signed 1st August 1900 at Klip Pass, Pretoria, by Lieut.-Colonel G. HAMILTON, Commanding 14th Hussars.

Officers.	Nos. fit for duty—W.O., N.C.O. and Men.	Nos. sick—W.O., N.C.O. and Men.	Deaths—W.O., N.C.O. and Men.	Casualties to Horses.	Remarks.	
17 officers present (including attached); Lieut. E. J. Jameson, Acting Quarter-Master. Officers atthd.—Capt. Lockett, 1st Dragoon Guards; Capt. Arnold, 1st Madras Lancers; Capt. Hutchison, 21st Lancers; Sec.-Lieut. Hon. H. Robertson, 3rd Cameron Highlanders.			1 man died of wounds receiv'd in action. 1 man died of enteric fever.	276 fit. 49 sick. 86 at sick horse depôts. 150 received from remount depôt. 48 died and destroyed.	Prevailing diseases: diarrhœa and veldt sores. Total effectives in South Africa (including sick, etc.): 1 W.O., 719 N.C.O. and men.	Return for August 1900.

Return signed at Machadodorp, 1st September, by Lieut.-Col. G. HAMILTON, Commanding 14th Hussars.

	Officers.	Nos. fit for duty—W.O., N.C.O. and Men.	Nos. sick—W.O., N.C.O. and Men.	Deaths—W.O., N.C.O. and Men.	Casualties to Horses.	Remarks.
Return for September 1900.	2 sick. 17 fit (including attached).	554	144		173 fit. 69 destroyed, abandoned, and missing. 68 to sick horse depôts.	Health good; average daily No. of sick, 4. Prevailing diseases: rheumatism, veldt sores. 4 privates prisoners of war. 50 men sent home (time-expi'd and invalided).

Return signed, 1st October, at Machadodorp, South Africa, by Lieut.-Col. G. HAMILTON, Commanding 14th Hussars.

	Officers.	Nos. fit for duty—W.O., N.C.O. and Men.	Nos. sick—W.O., N.C.O. and Men.	Deaths—W.O., N.C.O. and Men.	Casualties to Horses.	Remarks.
Return for October 1900.	14 fit. 5 attached. 2 sick. Major R. M. Richardson from staff employment, 27.10.00.		140	1 man killed in action, Geluk, 13.10.00.	5 killed in action. 21 died. 82 to sick horse depôts. 50 destroyed and abandoned.	Health good. Total effectives in South Africa—1 warrant officer, 692 N.C.O. and men, 183 horses.

Return signed at Pretoria, 1st November, by Lieutenant-Colonel G. HAMILTON, Commanding 14th Hussars.

	Officers.	Nos. fit for duty—W.O., N.C.O. and Men.	Nos. sick—W.O., N.C.O. and Men.	Deaths—W.O., N.C.O. and Men.	Casualties to Horses.	Remarks.
Return for November 1900.	16 effectives. Lt.-Col. Hamilton commanding 4th Cav. Brigade, 21.11.00. Major Richardson to command of Cavalry Depôt, Elandsfontein. Lieuts. Wright and Walker to Maitland Camp. Capt. Arnold, 1st Madras Lancers, from staff duty. Sec.-Lieut. Lace from England, 17.11.00.		112		543 fit. 45 sick. 6 killed and destroyed. 65 to veterinary hospitals. 481 remounts.	General health good. Drafts received of 203 men. Total effectives—Men, 888. Horses, 590.

Return signed at Heidelberg, 1st December 1900, by Major E. D. BROWN, Commanding 14th Hussars.

Officers.	Total Effectives.	Deaths—W.O., N.C.O. and Men.	Casualties to Horses.	Remarks.
Officers present and doing duty with regiment, 31st Decr. 1900—Major Brown, Capt. Tottenham, Lieut. Jameson, „ Campbell, „ Wright, „ Browne, „ Hon. H. Grosvenor, „ Dawes, „ Harvey, Sec.-Lieut. Scott, „ Lace, Captain and Adj. Lawrence, D.S.O.	1 warrant officer (Regl. Sergt.-Maj., A. F. Pridgeon), 557 N.C.O. and men. Attached officers —Capt. Hutchison, 21st Lancers, Capt. Lockett, 1st Dragoon Guards.	1 man killed in action. 2 men died.	18 killed in action, destroyed, and abandoned. 25 bought as chargers. 48 died. Total effectives—530.	General health good. Prevailing diseases: veldt sores, diarrhœa, enteric fever.

Return for December 1900.

Return signed at Welgevondon, South Transvaal, 1st January 1901, by Major E. D. BROWN, Commanding 14th Hussars.

Strength of Reserve Squadron, Dec. 1900.

Reserve Squadron, 14th (King's) Hussars, at Curragh Camp, Ireland. Strength on 31st December 1900 (from the Monthly Return):—

6 officers, 1 warrant officer.
351 non-commissioned officers and men.
4 officers' chargers; 109 troop-horses.

Return signed by Captain J. MURRAY, Commanding Reserve Squadron, 14th Hussars, Curragh Camp.

EXTRACT from the *London Gazette*, 1901.

The following appeared in the *London Gazette* of Tuesday, 15th January 1901:—

REWARD FOR GALLANTRY IN SOUTH AFRICA.

Major E. D. Brown, 14th Hussars, awarded the Victoria Cross, 15th January 1901.

'The Queen has been graciously pleased to signify her intention to confer the decoration of the Victoria Cross on the undermentioned officer, whose claims have been submitted for Her Majesty's approval, for his conspicuous bravery in South Africa, as stated against his name:—

'Major E. D. Brown, 14th Hussars.—On October 13, 1900, at Geluk, when the enemy were within 400 yards, and bringing a heavy fire to bear, Major

Brown, seeing that Sergeant Hersey's horse was shot, stopped behind the last squadron as it was retiring, and helped Sergeant Hersey to mount behind him, carrying him for about three-quarters of a mile to a place of safety. He did this under a heavy fire. Major Brown afterwards enabled Lieutenant J. G. Browne, 14th Hussars, to mount, by holding his horse, which was very restive, under the heavy fire; Lieutenant Browne could not otherwise have mounted. Subsequently Major Brown carried Lance-Corporal Trumpeter Leigh out of action.'

POSTSCRIPT

Death of Second-Lieutenant Hon. H. Robertson.

Whilst these sheets were in press Second-Lieutenant the Hon. Hugh Robertson, 14th (King's) Hussars, died at Johannesburg on the 1st February 1901, from a fracture at the base of the skull, the result of an accident. He was son of Lord Robertson of Forteviot, Perthshire, N.B., and was born in 1879. He served in the 3rd Battalion (Queen's) Cameron Highlanders from March 1899 until appointed to the Fourteenth in October 1900. His premature death was much regretted by his brother officers and comrades.

From the *London Gazette*, dated War Office, Tuesday, 10th September 1901 :—

EXTRACT OF A DESPATCH *from Field-Marshal Earl Roberts, Commander-in-Chief (dated 4th September* 1901), *to the Secretary of State for War.*

'I have the honour to bring to your notice the names of the following regimental officers, non-commissioned officers and men who have rendered special and meritorious service, etc. etc. :—

'1ST (KING'S) DRAGOON GUARDS.

Names of officers, non-commissioned officers and men of 14th Hussars mentioned by Earl Roberts in his despatch for meritorious service in South Africa.

'Captain W. J. Lockett (attached to 14th Hussars).

'14TH (KING'S) HUSSARS.

'Majors E. D. J. O'Brien and E. J. Tickell, D.S.O.; Captains C. B. Tottenham, F. R. Lawrence, D.S.O., and T. E. L. Hill-Whitson; Lieutenants W. R. Campbell and the Hon. H. Grosvenor; Quartermaster (Honorary Captain) F. Mugford (now retired); Sergeant-Major (now Quartermaster and Honorary Lieutenant) A. F. Pridgeon; Squadron Sergeant-Major C. F. Perman; Sergeant W. Griffin; Corporal W. Ridgewell (Reservist); Lance-Corporal H. J. D. Leigh; Privates H. Shenton (Reservist), M. Everson (Reservist), and F. M. Grace (Reservist), etc., etc.

(Signed) 'ROBERTS, *F.-M.*'

INDEX

Printed by T. and A. CONSTABLE, (late) Printers to Her Majesty
at the Edinburgh University Press.